Lecture Notes in Computational Science and Engineering

18

Editors
M. Griebel, Bonn
D. E. Keyes, Norfolk
R. M. Nieminen, Espoo
D. Roose, Leuven
T. Schlick, New York

Springer-Verlag Berlin Heidelberg GmbH

Ursula van Rienen
Michael Günther
Dirk Hecht
Editors

Scientific Computing in Electrical Engineering

Proceedings of the 3rd International Workshop,
August 20–23, 2000, Warnemünde, Germany

With 184 Figures

Springer

Editors

Ursula van Rienen
Dirk Hecht

Fachbereich Elektrotechnik und Informationstechnik
Universität Rostock
Albert-Einstein-Straße 2
18051 Rostock, Germany
e-mail: van.rienen@e-technik1.uni-rostock.de
dirk.hecht@etechnik.uni-rostock.de

Michael Günther

Institut für Wissenschaftliches Rechnen
und Mathematische Modellbildung
IWRMM
Universität Karlsruhe
Engesserstraße 6
76128 Karlsruhe, Germany
e-mail: guenther@iwrmm.math.uni-karlsruhe.de

Cataloging-in-Publication Data applied for

Die Deutsche Bibliothek – CIP-Einheitsaufnahme

Scientific computing in electrical engineering: proceedings of the 3rd international workshop, August 20–23, 2000, Warnemünde, Germany / Ursula van Rienen ... ed.. – Berlin; Heidelberg; New York; Barcelona; Hong Kong; London; Milan; Paris; Tokyo: Springer, 2001
(Lecture notes in computational science and engineering; Vol. 18)
ISBN 978-3-540-42173-3 ISBN 978-3-642-56470-3 (eBook)
DOI 10.1007/978-3-642-56470-3

Front cover: Ute Schreiber: Two water droplets on top of epoxy resin with two embedded HV-electrodes. Potential and flux density distribution.

Mathematics Subject Classification (2000): 3406, 34B60, 35Q60, 6506, 65Fxx, 65Lxx, 65Mxx, 65Nxx, 65Z05, 78Mxx, 80A20, 94Cxx

ISSN 1439-7358
ISBN 978-3-540-42173-3

Cover Design: Friedhelm Steinen-Broo, Estudio Calamar, Spain
Cover production: *design & production* GmbH, Heidelberg
Typeset by the authors using a Springer TEX macro package

Printed on acid-free paper SPIN 10786682 46/3142/LK – 5 4 3 2 1 0

Preface

This book presents a collection of selected contributions presented at the 3rd International Workshop on Scientific Computing in Electrical Engineering, SCEE-2000, which took place in Warnemünde, Germany, from August 20 to 23, 2000. Nearly hundred scientists and engineers from thirteen countries gathered in Warnemünde to participate in the conference. Rostock University, the oldest university in Northern Europe founded in 1419, hosted the conference.

This workshop followed two earlier workshops held 1997 at the Darmstadt University of Technology and 1998 at Weierstrass Institute for Applied Analysis and Stochastics in Berlin under the auspices of the German Mathematical Society. These workshops aimed at bringing together two scientific communities: applied mathematicians and electrical engineers who do research in the field of scientific computing in electrical engineering. This, of course, is a wide field, which is why it was decided to concentrate on selected major topics. The workshop in Darmstadt, which was organized by Michael Günther from the Mathematics Department and Ursula van Rienen from the Department of Electrical Engineering and Information Technology, brought together more than hundred scientists interested in numerical methods for the simulation of circuits and electromagnetic fields. This was a great success. Voices coming from the participants suggested that it was time to bring these communities together in order to get to know each other, to discuss mutual interests and to start cooperative work. A collection of selected contributions appeared in 'Surveys on Mathematics for Industry', Vol. 8, No. 3-4 and Vol. 9, No. 2, 1999. The workshop in Berlin, organized in large part by Georg Hebermehl, Weierstrass Institute for Applied Analysis and Stochastics, had a similar scope and brought together more than eighty scientists. Both workshops showed a growing international interest. Therefore the decision was made in Berlin to continue with future biennal international workshops. The vote was to carry out the 2000 workshop at Rostock University, which has been Ursula van Rienen's place of work since the autumn of 1997. A remarkably large number of the nearly hundred SCEE-2000 participants came from industry and research laboratories and about 75% from universities. The large majority of the participants came from Europe including Eastern Europe, yet some came also from the US. We guess that the number of mathematicians and electrical engineers under the participants was about equal.

The SCEE-2000 committees decided that SCEE-2002 will be held in Eindhoven, the Netherlands. The date is already fixed: 23-28 June 2002. A regularly updated Web page with new information on the upcoming SCEE-2002 can be found under
http://www.win.tue.nl/scee2002/.

For subscription on information, send e-mail to scee2002@win.tue.nl.
Programme, logo and a photo gallery on SCEE-2000 can be found under http://www.SCEE-2000.uni-rostock.de/.
Main topics of SCEE-2000 were: Computational Electrodynamics, Circuit Design, Coupled Problems. The conference was coorganized by the Special Interest Group on Scientific Computing in Electronic Industry of the European Consortium for Mathematics in Industry. The SCEE-2000 programme committee consisted of scientists from industry and universities. Its members were Michael Günther, Karlsruhe University (TH), Germany; Ulrich Langer, Linz University, Austria; Ursula van Rienen, Rostock University, Germany; Jan ter Maten and Wil Schilders, TU Eindhoven and Philips Research Laboratories, Eindhoven, The Netherlands; Uwe Feldmann, Infineon Technologies, Munich, Germany. The programme committee invited six plenary speakers having in mind three goals: gaining leading experts, inviting mainly young scientists, which is already a kind of SCEE tradition, and gaining speakers from industry.
Another important principle of the SCEE workshops is to have only plenary sessions. Therefore also poster sessions took place which are usual in electrical engineering conferences but less usual in mathematical congresses. After reviewing the submitted abstracts the programme committee had chosen 30 contributed talks which represented the scope of contributed papers and were of broad interest. They drew attention to related posters. There was a total of 17 posters. The best posters were honoured with a poster award. The committee consisted of three invited speakers: Leszek Demkowicz, Irina Munteanu and Ronald Rietman. They chose the poster of Olaf Michelsson (Ilmenau) for the 1st prize and that of Karsten Rothemund (Rostock) for the 2nd prize. We would like to thank the participants for all the very interesting contributions on high scientific level. After careful reviewing of the full papers by two reviewers, each, we are grateful that the proceedings are published in the 'Lecture Notes in Computational Science and Engineering' by Springer (briefly Springer LNCSE), the book you have in hand now. It contains the papers of the six invited speakers and 37 contributed papers. The editors want to thank all members of the programme committee for their substantial participation in carrying out the reviewing process and Thomas Riemann for the technical help in the compilation of the book.
One "side-effort" of the local organizing committee was to find a suitable logo for SCEE-2000 and the upcoming workshops. We were fortunate to receive help from the Department for design and interior design of the Hochschule Wismar, a university for applied sciences. Students of Hanka Polkehn made several designs. The logo designed by the student Ramona Weyde-Ferch was chosen as the winner.

It is an arrangement of the four letters SCEE, some lines which might be interpreted as field lines or wave fronts and part of a bracket which stands for mathematical bracket but also symbolises the idea of this workshop as bringing together several communities, mathematicians and engineers, university and industry.
Very important in launching a scientific event such as the SCEE is substantial financial support. The Programme Committee is greatly indebted to the DFG, the central public funding organisation for academic research in Germany comparable to a Research Council or Foundation, to the Ministry of Education, Science and Culture of Mecklenburg-Western Pomerania, to Rostock University, to the town of Rostock and to several private companies which are listed in the Web: http://www.SCEE-2000.uni-rostock.de/.

Rostock, March 2001
Ursula van Rienen
SCEE-2000 Chairman

Contents

I Computational Electromagnetics

II Circuit Analysis

Part I

Computational Electromagnetics

Multi-Time PDEs for Dynamical System Analysis*

J. Roychowdhury

Bell Laboratories, Murray Hill

Abstract RF systems typically involve mixing, sampling and frequency - modulation, operations that generate signals with widely separated time scales (WSTS). SPICE-like simulation of such systems is often impossible, and even spe-cialized techniques such as harmonic balance have difficulty when there are sharp nonlinearities, e.g., switches. In this paper, we describe the use of apowerful tool for handling such systems, i.e., using multiple time scales at the differential equation level. We first describe the use of multiple time scalesfor non-oscillatory systems forced by WSTS inputs. We introduce multitime signals and the Multirate Partial Differential Equation (MPDE) andexplain how they form an efficient means of working with WSTS problems. Next, we describe how warped time scales are needed to represent WSTSsignals arising from forced oscillatory systems, like VCOs. We present the Warped MPDE (WaMPDE) for such systems.We also show how multiple time scales are useful for macromodelling RF blocks that perform frequency-translation or sampling (i.e., linear time-varying blocks).

1 Introduction

The trend towards on-chip integration is increasing the complexityof RF / analog blocks and making their design increasingly dependent on effective CAD. Unfortunately, algorithms in common CAD toolslike SPICE are often not powerful enough for useful analyses of RF systems. One important shortcoming is their inability to handle signals with widely separated time scales (WSTS), which are generated by operations like mixing, sampling and frequency-modulation in RF systems. Transient simulation of such systems is often impractically long, and even specialized techniques such as harmonic balance havedifficulty when there are sharp nonlinearities, e.g., switches.

In this paper, we review a family of recent techniques for analysingsystems with WSTS signals in a natural and efficient manner. Underlying these techniques is the concept of multiple time scales, used to reformulate WSTS problems at the differential equation level. The basic idea is to represent WSTS signals as functions of two or more artificial time variables, i.e., as multivariate functions. Using multiple time variables converts differential equations to partial differential ones, which can be solved efficiently with numerical techiques. Great improvements in computation and memory use can result from the fact that the efficiency of multitime representations is independent

* invited paper, opening talk of the workshop

of the separation of the timescales. Powerful and intuitive visualisations of WSTS signal structure are also provided by the multitime representations. For forced autonomous systems (e.g., VCOs), some of the artificial time scales must be stretched (or warped) nonuniformly in order that FM-type WSTS signals are represented efficiently. This leads toan additional unknown, the local frequency, in the multitime partial differential formulation for autonomous systems. A particularly useful feature is that the local or instantaneous frequency is solved for explicitly as it changes with time; as a result, this system-level performance metric, ordinarily difficult to determine, is obtained directly from the warped multi-time simulation. An important task in RF system design is generating macromodels of complex blocks, with the goal of replicating input-output functionality faithfully, while being simple enough for verifying entire systems. Blocks such as mixers and switched-capacitor filters which-perform frequency-translation or sampling (i.e., with essentially linear time varying (LTV) I/O properties) have traditionally been macromodelled by ad-hoc manual means. Surprisingly, applying multiple time scales leads to a useful partial differential formulation for solving this problem algorithmically. Automated macromodelling of complex LTV blocks by simple ones enables higher-level representations of communication subsystems to be coupled to detailed realizations much more tightly and quickly than previously possible. We will not provide a full discussion of the history and previous work on multi-time concepts here; the interested reader is referred to, e.g., [2,3,5,1,6,4]; [7] contains a fuller list of references. The remainder of the paper is organized as follows. We introduce artificial time scales and the Multirate Partial Differential Equation (MPDE) in Section 2, and illustrate their application on RF mixer and DC-DC power converter circuits. In Section 2, we explain warped time scales for autonomous systems, present the Warped MPDE (WaMPDE), and demonstrate its application in predicting FM in a VCO circuit. Next, in Section 3, we show how multiple time scales are useful for LTV macromodelling and present RF macromodelling examples.

2 Multiple Time Scales and PDEs

In this section, we review why it is advantageous to use two or more time scales for analyzing WSTS signals, and present the MultiratePartial Differential Equation (MPDE) for analyzing large WSTSsystems.

Consider the waveform $y(t)$ shown in Figure 1, a simple two-tone quasiperiodic signal given by

$$y(t) = \sin\left(\frac{2\pi}{T_1}t\right)\sin\left(\frac{2\pi}{T_2}t\right), \qquad T_1 = 0.02s, \qquad T_2 = 1s \tag{1}$$

The two tones are at frequencies $f_1 = \frac{1}{T_1} = 50Hz$ and $f_2 = \frac{1}{T_2} = 1Hz$, i.e., there are 50 fast-varying cycles of period $T_1 = 0.02s$ modulated by a

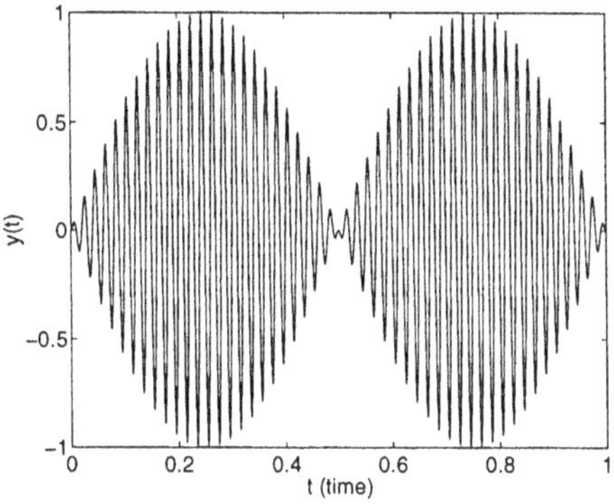

Figure1. Example 2-tone quasi-periodic signal $y(t)$

slowly-varying sinusoid of period $T_2 = 1s$. If each fast cycleis sampled at n points, the total number of time-steps needed for one period of the slow modulation is $n\frac{T_2}{T_1}$. To generate Figure 1, 15 points were used per cycle, hence the total number of samples was 750. This number can be much larger in applications where the rates are more widely separated, e.g., separation factors of 1000 or more are common in electronic circuits.
Now consider a multivariate representation of $y(t)$ using two artificial time scales, as follows: for the 'fast-varying' parts of $y(t), t$ is replaced by a new variable t_1 ; for the 'slowly-varying' parts, by t_2.

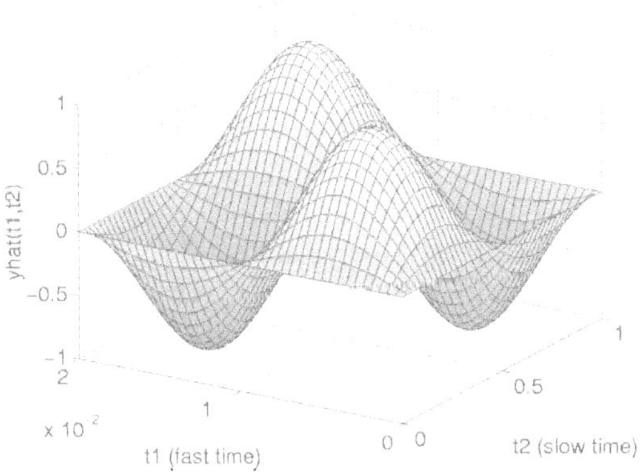

Figure2. Corresponding 2-periodic bivariate form $\hat{y}(t1, t2)$

The resulting function of two variables is denoted by

$$\hat{y}(t_1, t_2) = \sin\left(\frac{2\pi}{T_1}t_1\right)\sin\left(\frac{2\pi}{T_2}t_2\right) \tag{2}$$

The plot of $\hat{y}(t_1, t_2)$ on the rectangle $0 \le t_1 \le T_1$, $0 \le t_2 \le T_2$ is shown in Figure 2. Observe that $\hat{y}(t_1, t_2)$ does not have many undulations, unlike $y(t)$ in Figure 1. *Hence it can be represented by relatively few points, which, moreover, do not depend on the relative values of* T_1 *and* T_2 . Figure 2 was plotted with 225 samples on a uniform 15×15 grid - three times fewer than for Figure 1. This saving increases with increasing separation of the periods T_1 and T_2.

Further, note that $\hat{y}(t_1, t_2)$ is periodic with respect to both t_1 and t_2, i.e., $\hat{y}(t_1+T_1, t_2+T_2) = \hat{y}(t_1, t_2)$. This makes it easy to recover $y(t)$ from $\hat{y}(t_1, t_2)$, simply by setting $t_1 = t_2 = t$, and using the fact that $\hat{y}$ is bi-periodic. It is easy, from direct inspection of the threedimensional plot of $\hat{y}(t_1, t_2)$, to visualise what $y(t)$ looks like. As t increases from 0, the path given by $t_i = tmodT_i$ traces the sawtoothpath shown in Figure 3. By noting how $\hat{y}$ changes as this path is traced in the $t_1 - t_2$ plane, $y(t)$ can be traced. *When the time-scales are widely separated, therefore, inspection of the bivariate waveform directly provides information about the slow and fast variations of* $y(t)$ *more naturally and conveniently than* $y(t)$ *itself.*

We observe that the bivariate form can require far fewer points to represent numerically than the original quasiperiodic signal, yet it contains all the information needed to recover the original signal completely.

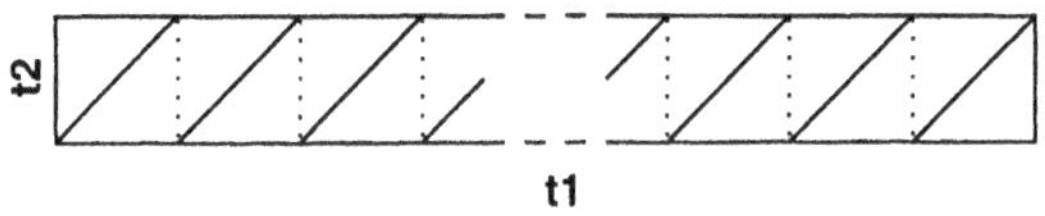

Figure3. Path in the $t_1 - t_2$ plane

This observation is the basis of the MPDE formulation, in which all the waveforms in a circuit are represented in their bivariate forms (or multivariate forms if there are more than two timescales). The key to efficiency is to solve for these waveforms directly, without involving the numerically inefficient one-dimensional forms at any point. To do this, it is necessary to first describe the circuit's equations using the multivariate functions. The traditional form of a circuit's equations, used in all simulators, is the Differential-Algebraic Equation (DAE):

$$\dot{q}(x) + f(x) = b(t) \tag{3}$$

$x(t)$ is the vector of circuit unknowns (node voltages and branch currents); q denotes the charge/flux terms and f the conductive terms; $b(t)$ is the vector of excitations to the circuit (typically from independent voltage/current sources). It can be shown that if $\hat{x}(t_1, t_2)$ and $\hat{b}(t_1, t_2)$ denote the bi-variate forms of the circuit unknowns and excitations, then the following MPDE is the correct generalization of (3) to the bi-variate case:

$$\frac{\partial q(\hat{x})}{\partial t_1} + \frac{\partial q(\hat{x})}{\partial t_2} + f(\hat{x}) = \hat{b}(t_1, t_2)$$

More precisely, if $\hat{b}$ is chosen to satisfy $b(t) = \hat{b}(t,t)$, and $\hat{x}$ satisfies (4), then it can be shown that $x(t) = \hat{x}(t,t)$ satisfies (3). Also, if (3) has a quasi-periodic solution, then (4) can be shown to have a corre-sponding bi-variate solution.

By solving the MPDE numerically in the time domain, strong non-linearities can be handled efficiently. Several numerical methods are possible, including discretization of the MPDE on a grid in the $t_1 - t_2$ plane, or using a mixed time-frequency method in which the variation along one of the time scales is expressed in a short Fourier series. Quasiperiodic and envelope solutions can both be generated, by appropriate selection of boundary conditions for the MPDE. Sparse matrix and iterative linear methods are used to keep the numerical algorithms efficient even for large systems.

Figure 4 depicts the output voltage of a switched-capacitor integrator block, obtained from a multi-time simulation based on the above concepts.

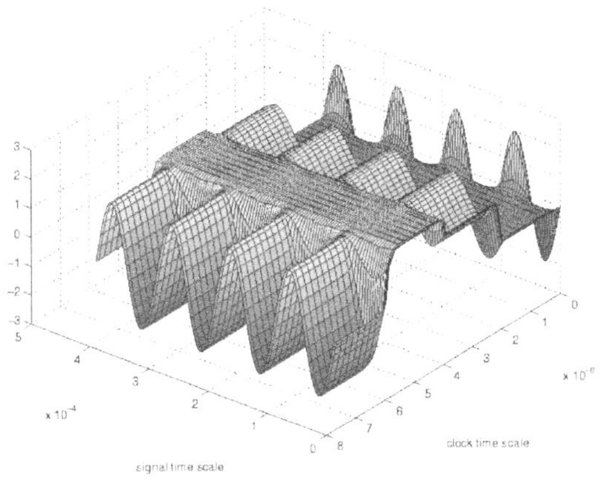

Figurc4. Multi-time output waveform of SC integrator

The cross-section parallel to the signal time scale represents the envelope of the signal riding on the switching variations. By moving these cross-sections to different points along the clock time-scale, the signal envelope at different points of the clock waveform can be seen.

A double-balanced switching mixer and filter circuit was simulated for inter-modulation distortion using a mixed time-frequency numerical method for the MPDE. The RF input was a 100kHz sinusoid with amplitude 100mV;

this sent it into a mildly nonlinear regime. The LO input was a square wave of large amplitude (1V), which switched the mixer on and off at a fast rate (900Mhz).
The third harmonic is shown in Figure 5. It contains information about the mixes $3f_1 + if_2$, i.e., the frequencies 900.3 Mhz, 1800.3Mhz, etc.. The amplitude of the 900.3 Mhz component can be seen to be about 1.1mV; hence the distortion introduced by the mixer is about 35dB below the desired signal. The multi-time simulation was about 300 times faster than traditional transient simulation techniques.
When the DAEs under consideration are oscillatory, frequency modulation (FM) can be generated. Unfortunately, FM cannot be represented compactly using multiple time scales as easily as the waveform in Figure 2. We illustrate the difficulty with an example. Consider the following prototypical FM signal

$$x(t) = \cos(2\pi f_0 t + k\cos(2\pi f_2 t)), \qquad f_0 \gg f_2 \tag{5}$$

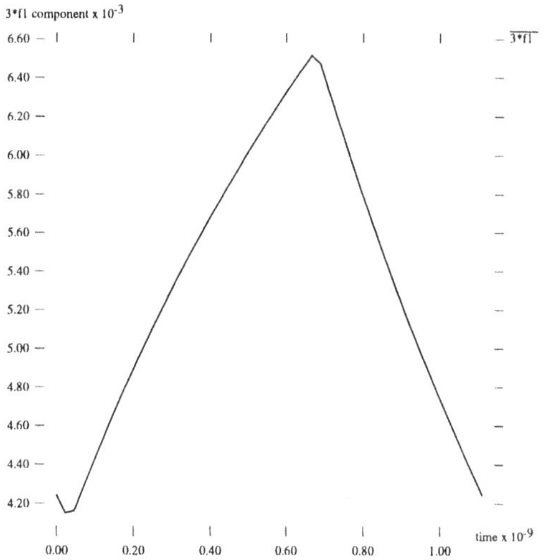

Figure5. Mixed frequency-time output: third harmonic component

with instantaneous frequency

$$f(t) = f_0 - kf_2\sin(2\pi f_2 t) \tag{6}$$

$x(t)$ is plotted in Figure 6 for f_0 = 1MHz, f_2 = 20KHz, and modulation index $k = 8\pi$. Following the same approach as for 1, a bivariate form can be defined to be

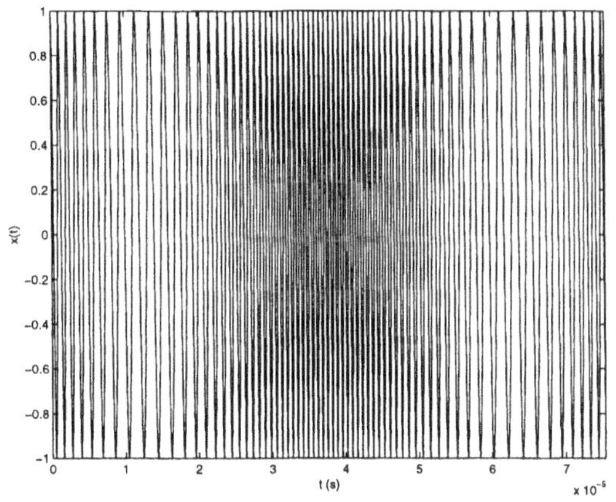

Figure6. FM signal

$$\hat{x}_1(t_1, t_2) = \cos(2\pi f_0 t_1 + k\cos(2\pi f_2 t_2)), \qquad \text{with} \quad x(t) = \hat{x}_1(t, t) \tag{7}$$

Note that $\hat{x}_1$ is periodic in t_1 and t_2, hence $x(t)$ is quasiperiodic with frequencies f_0 and f_2. Unfortunately, $\hat{x}_1(t_1, t_2)$, illustrated in Figure 7, is not a simple surface with only a few undulations like Figure 2. When $k \gg 2\pi, i.e., k \approx 2\pi m$ for some large integer m, then $\hat{x}_1(t_1, t_2)$ will undergo about m oscillations as a function of t_2 over one period T_2. In practice, k is often of the order of $\frac{f_0}{f_2} \gg 2\pi$, hence this number of undulations can be very large. Therefore it becomes difficult to represent $\hat{x}_1$ efficiently by sampling on a two-dimensional grid.

It turns out that resolving this problem requires the stretching, or warping, of one of the time scales. We illustrate this by returning to (5). Consider the following new multivariate representation

$$\hat{x}_2(\tau_1, \tau_2) = \cos(2\pi\tau_1) \tag{8}$$

together with the warping function

$$\Phi(\tau_2) = f_0\tau_2 + \frac{k}{2\pi}\cos(2\pi f_2 \tau_2) \tag{9}$$

We now retrieve our one-dimensional FM signal (i.e., (5)) as

$$x(t) = \hat{x}_2(\Phi(t), t) \tag{10}$$

Note that both $\hat{x}_2$ and Φ, given in (8) and (9), can be easily represented with relatively few samples, unlike $\hat{x}_1$ in (7). What we have achieved with (9) is simply a stretching of the time axis differently at different times, to

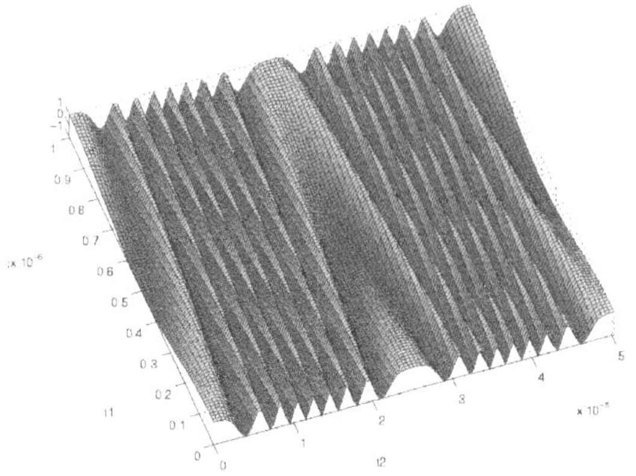

Figure7. $\hat{x}_1$: unwarped bivariate representation of FM signal

even out the period of the fast undulations in Figure 6. The extent of the stretching, or the derivative of $\Phi(\tau_2)$, at a given point is simply the local frequency $\omega(\tau_2)$, which modifies the original MPDE to result in the Warped Multirate Partial Differential Equation (WaMPDE):

$$\omega(\tau_2)\frac{\partial q(\hat{x})}{\partial \tau_1} + \frac{\partial q(\hat{x})}{\partial \tau_2} + f(\hat{x}(\tau_1, \tau_2)) = b(\tau_2) \tag{11}$$

The usefulness of 11 lies in that specifying

$$x(t) = \hat{x}(\Phi(t), t), \qquad \Phi(t) = \int_0^t \omega(\tau_2) d\tau_2 \tag{12}$$

results in $x(t)$ being a solution to (3). Furthermore, when (11) is solved numerically, the local frequency $\omega(\tau_2)$ is also obtained, which is desirable for applications such as VCOs and also difficult to obtain by any other means. As it stands, (11) is an underdetermined system, since there are more unknowns than equations. To find isolated solutions, it is augmented with an additional equation, the phase condition [4].

Figure (8) shows the changing local frequency in a VCO that was simulated with WaMPDE-based numerical techniques. The controlling input to the VCO was about 30 times slower than its nominal frequency.

Figure 9 depicts the bivariate waveform of the capacitor voltage. It is seen that the controlling voltage changes not only the local frequency, but also the amplitude and shape of the oscillator waveform.

The circuit was also simulated by traditional numerical ODE methods ("transient simulation"). The waveform from this simulation, together with the 1-dimensional waveform obtained by applying (12) to Figure 9, are shown in Figure 10. Frequency modulation can be observed in the varying density of the undulations.

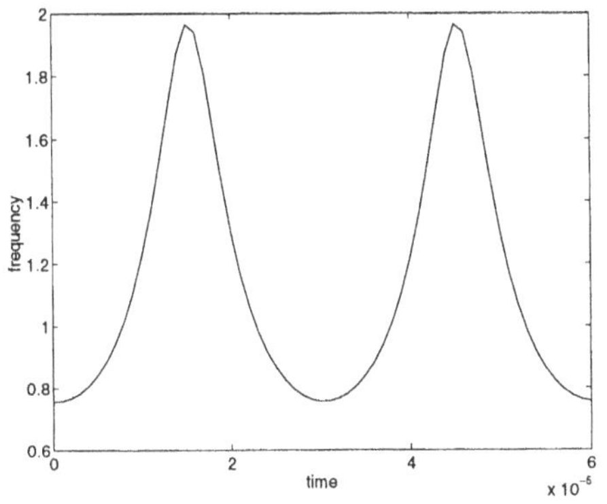

Figure8. VCO: frequency modulation frequency

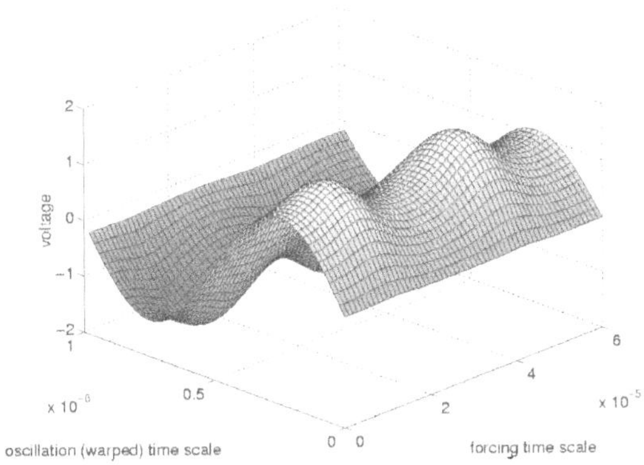

Figure9. VCO: bivariate representation of capacitor voltage

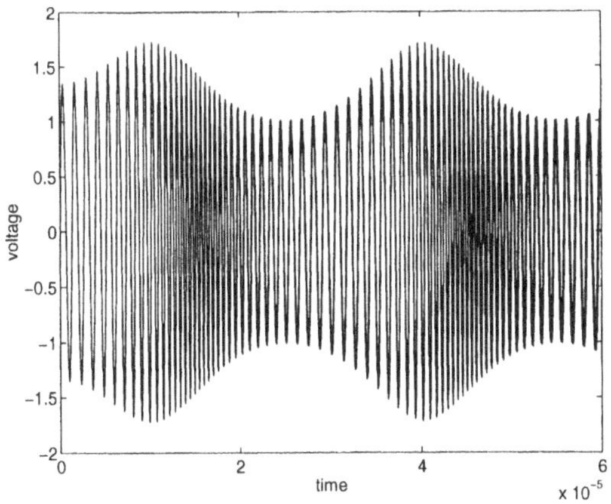

Figure10. VCO: WaMPDE vs transient simulation

3 Algorithmic macromodelling of RF systemblocks

Another useful application of multiple time scales is in macro-modelling linear time varying (LTV) systems. The main difficulty in extending LTI reduction techniques to LTV systems has been the interference of the time-variations of the system and the input. By separating the two with artificial time variables, the MPDE provides a simple and elegant solution to this problem.
The time-varying small-signal equations obtained by linearizing (3) around a steady-state solution are given by:

$$\begin{aligned} C(t)\dot{x}(t) + G(t)x(t) &= ru(t) \\ y(t) &= d^T x(t) \end{aligned} \tag{13}$$

In (13), the input to the system is the scalar $u(t)$, while the output is $y(t)$. If the above equation is Laplace-transformed (following the LTI procedure), the system time variation in $C(t)$ and $G(t)$ interferes with the I/O time variation resulting in a convolution. The LTV transfer function $H(t, s)$ is therefore hard to obtain; this is the difficulty alluded to earlier. The problem can be avoided by casting (13) as an MPDE:

$$\begin{aligned} C(t_1) \left[\frac{\partial \hat{x}}{\partial t_1}(t_1, t_2) + \frac{\partial \hat{x}}{\partial t_2}(t_1, t_2)\right] + G(t_1)\hat{x}(t_1, t_2) = ru(t_2) \\ \hat{y}(t_1, t_2) = d^T \hat{x}(t_1, t_2), \qquad y(t) = \hat{y}(t, t) \end{aligned} \tag{14}$$

Notice that the input and system time variables are now separated. By taking Laplace transforms in t_2 and eliminating $\hat{x}$, the time-varying transfer function $H(t_1, s)$ is obtained:

$$Y(t_1, s) = \underbrace{\left\{ d^T \left[C(t_1)\left\{\frac{\partial}{\partial t_1} + s\right\} + G(t_1)\right]^{-1} [r] \right\}}_{H(t_1,s)} U(s) \tag{15}$$

Observe that $H(t_1, s)$ in (15) is periodic in t_1; hence, discretizing the t_1 axis, it can also be represented as several time-invariant transfer functions $H_i(s) = H(t_1, s)$. Or, a frequency-domain discretization using harmonics of the t_1-variation can be used. Once an equivalent system of LTI transfer functions has been obtained, existing reduced order modelling techniques for LTI systems can be used to find a smaller system of equations, in the same form as (14), that have the same input-output relationship to within a given accuracy.
Details of the reduced-order modelling technique are given in [8]. We demonstrate LTV reduced-order modelling on a RFIC I-channel mixer circuit of size

about $n = 360$ nodes, excited by a local oscillator at 178Mhz. A frequency-domain discretization of the t_1 axis in (14) was employed in the model reduction process. Figure (3) shows frequency plots of $H_1(s)$, the upconversion transfer function (the first harmonic *wrt* t_1 of $H(t_1, s)$). The points marked '+' were obtained by direct computation of the full system, while the lines were computed using the reduced models of size $q = 2$ and $q = 10$, respectively. Even with $q = 2$, a size reduction of two orders of magnitude, the reduced model provides a good match up to the LO frequency. When the order of approximation is increased to 10, the reduced model is identical up to well beyond the LO frequency. The reduced models were more than three orders of magnitude faster to evaluate than the original system, hence they are useful for system-level verification.
The poles of the reduced models for $H_1(s)$, easily calculated on account of their small size, are shown in Table 1. These are useful in design because they constitute excellent approximations of the fullsystem's poles, which are difficult to determine otherwise.

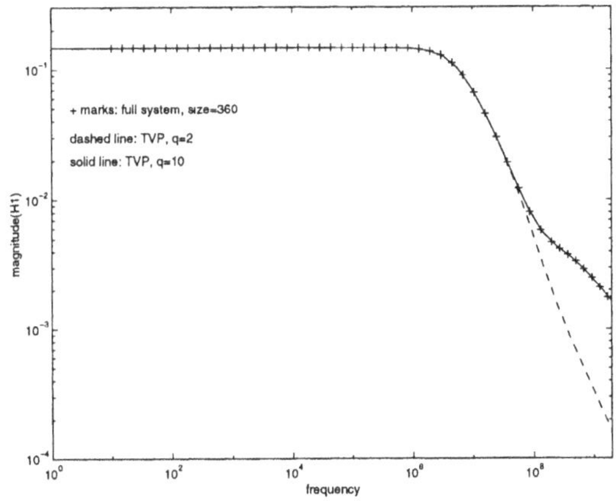

Figure11. I-channel mixer $H_1(s)$: reduced vs full system

4 Conclusion

The use of artificial time scales for analyzing RF systems addresses long-standing shortcomings of traditional CAD techniques. Using multiple times leads to compact and intuitive representations ofsignals with widely separated time scales. These are useful for analyzing a variety of RF circuits, including mixers, switched-capacitorfilters, and oscillators. Multi-time concepts are

Table1. Poles of $H_1(s)$ for the I-Channel buffer/mixer

TVP, q=2	TVP, q=10
-5.3951e+06	-5.3951e+06
-6.9196e+07 - j3.0085e+05	-9.4175e+06
	-1.5588e+07 - j2.5296e+07
	-1.5588e+07 + j2.5296e+07
	-6.2659e+08 - j1.6898e+06
	-1.0741e+09 - j2.2011e+09
	-1.0856e+09 + j2.3771e+09
	-7.5073e+07 - j1.4271e+04
	-5.0365e+07 + j1.8329e+02
	-5.2000e+07 + j7.8679e+05

also useful for automated macromodelling RF blocks in a systematic and algorithmic manner.

References

1. H.G. Brachtendorf, G. Welsch, R. Laur, and A. Bunse-Gerstner Numerical steady state analysis of electronic circuits driven by multi-tone signals. Electrical Engineering (Springer-Verlag),79:103-112, 1996.
2. J. Kevorkian and J.D. Cole. Perturbation methods in AppliedMathematics. Springer-Verlag, 1981
3. R. Mickens. Oscillations in Planar Dynamic Systems. WorldScientific, 1995.
4. O. Narayan and J. Roychowdhury. Multi-time simulation ofvoltage-controlled oscillators. In Proc. IEEE DAC, New Orleans, LA, June 1999.
5. E. Ngoya and R. Larchev'eque. Envelop transient analysis: a newmethod for the transient and steady state analysis of microwave communication circuits and systems. In Proc. IEEE MTT Symp.,1996.
6. J. Roychowdhury. Efficient methods for simulating highly non-linear multi-rate circuits. In Proc. IEEE DAC, 1997.
7. J. Roychowdhury. Analysing circuits with widely-separated time scales using numerical PDE methods. IEEE Trans. Ckts. Syst. - I: Fund. Th. Appl., March 1998. Acceptedfor publication. Preprints available at www.bell-labs.com/ jaijeet/multitime/RompdeTCAS1.pdf.
8. J. Roychowdhury. Reduced-order modelling of time-varying sys-tems. IEEE Trans. Ckts. Syst. - II: Sig. Proc., 46(10), November 1999.

Edge Finite Elements of Variable Order for Maxwell's Equations*

L. Demkowicz

Texas Institute for Computational and Applied Mathematics The University of Texas at Austin

Abstract I review the main ideas behind the construction of edge elements of variable order [16,39,18], and discuss the possibility of extending the construction to Nedelec's elements of the first kind [31]. A motivation leading to the definition of *hp*-interpolation operators is highlighted, and their impact on the *hp*-discretizations of Maxwell's equations discussed.

Introduction. The purpose of this note is twofold. First, I would like to introduce the reader to the idea of *hp*-adaptive edge finite elements for Maxwell's equations that we have been working on at Texas for the last four years, and second, I would like to respond to questions asked by Prof. Bossavit during the SCEE-2000 conference[1] on the relation between our edge elements of variable order and Nedelec's elements of the first type [31]. The note is very much an outgrowth of discussions I have had with Peter Monk and is related to the recent contributions of Ralf Hiptmair [22,23].

Why hp FEM's ? Spectral-type approximations are very effective in approximating smooth functions on coarse grids but may be hopelessly ineffective in approximating functions with low regularity (e.g. piecewise differential functions). Low order Finite Difference (FD), Finite Element (FE), or Finite Volume (FV) methods, on the other hand, do well in representing rough functions, those with jumps or piecewise smooth derivatives. In real-world applications, solutions are neither everywhere smooth or rough, but have variable regularity over the solution domain. The methodology developed to handle variations in regularity in an optimal way is called *hp* finite elements; it embodies techniques to orchestrate different orders of approximation at different regions of the domain. The result is an approach that produces exponential rates of convergence [36] for very irregular solutions, a property not attained by any traditional method. For an extensive review of basic principles behind *hp* Finite Element Methods, see [4], for the most recent approximation results, see [3].

The use of higher order elements eliminates practically the dispersion (pollution) error, see [24,25,27]. For instance, for up to (non-dimensional)

* invited paper

[1] International Workshop on Scientific Computing in Electrical Engineering (SCEE-2000), Warnemünde, Germany, August 20-23, 2000

wave number $ka = 20$, it is sufficient to model one 'cubic'(3D) wavelength with just one hexahedral element of order four. This sets a preference for using large elements of higher order (in practical 3D computations order $p = 4, 5$) as much as possible. On the other side, small elements of lower order are necessary to capture geometrical details involved when modeling cavities, thin edges or embedded antennas. Only the *hp*-discretizations allow to vary BOTH element size and element order p within the same mesh.

Time harmonic Maxwell's equations. The goal is to solve time-harmonic Maxwell's equations, in either a bounded or unbounded domain. We shall deal with the bounded domain first. The first order time-harmonic Maxwell's equations are usually reduced to the reduced wave equation expressed either in terms of electric field $\boldsymbol{E}$ or magnetic field $\boldsymbol{H}$. The choice, usually, depends upon boundary conditions. As both formulations share the same mathematical properties, we shall focus on the formulation in terms of the electric field only.

Given a *bounded* domain $\Omega \subset \mathbb{R}^3$, with boundary Γ consisting of two disjoint parts Γ_1 and Γ_2, we wish to find electric field $\boldsymbol{E}(\boldsymbol{x}), \boldsymbol{x} \in \bar{\Omega}$, that satisfies:

- the reduced wave equation in Ω,

$$\boldsymbol{\nabla} \times \left(\frac{1}{\mu} \boldsymbol{\nabla} \times \boldsymbol{E} \right) - (\omega^2 \epsilon - j\omega\sigma)\boldsymbol{E} = -j\omega \boldsymbol{J}^{imp} , \tag{1}$$

- Dirichlet (ideal conductor) boundary condition on Γ_1,

$$\boldsymbol{n} \times \boldsymbol{E} = \boldsymbol{0} , \tag{2}$$

- Neumann boundary condition on Γ_2,

$$\boldsymbol{n} \times \left(\frac{1}{\mu} \boldsymbol{\nabla} \times \boldsymbol{E} \right) = -j\omega \boldsymbol{J}_S^{imp} . \tag{3}$$

In the above, ω is an angular frequency, ϵ, μ, σ denote dielectric permittivity, magnetic permeability and conductivity of the medium, $\boldsymbol{J}^{imp}$ is a prescribed, impressed (source) current, $\boldsymbol{J}_S^{imp}$ is a prescribed, impressed surface current tangent to boundary Γ_2, $\boldsymbol{n} \cdot \boldsymbol{J}_S^{imp} = 0$, with $\boldsymbol{n}$ denoting the normal outward unit vector to Γ. Finally, j is the imaginary unit. For the sake of simplicity, we shall restrict ourselves to simply connected domains Ω only, avoiding the technical issues connected with *cohomology spaces*, see e.g. [10]. The standard variational formulation is obtained by multiplying (1) by a vector test function $\bar{\boldsymbol{F}}$, integrating over domain Ω, integrating by parts, and using the

Neumann boundary condition.

$$\left\{\begin{array}{l}\text{Find } \boldsymbol{E} \in \boldsymbol{W} \text{ such that} \\ \displaystyle\int_\Omega \frac{1}{\mu}(\nabla \times \boldsymbol{E}) \cdot (\nabla \times \bar{\boldsymbol{F}})dx - \int_\Omega (\omega^2\epsilon - j\omega\sigma)\boldsymbol{E} \cdot \bar{\boldsymbol{F}}dx = \\ \qquad\qquad -j\omega\left\{\displaystyle\int_\Omega \boldsymbol{J}^{imp} \cdot \bar{\boldsymbol{F}}dx + \int_{\Gamma_2} \boldsymbol{J}_S^{imp} \cdot \bar{\boldsymbol{F}}dS\right\} \quad \text{for all } \boldsymbol{F} \in \boldsymbol{W}\,. \end{array}\right.$$

In the above $\boldsymbol{W}$ is the *space of admissible solutions*,

$$\boldsymbol{W} := \{\boldsymbol{E} \in \boldsymbol{L}^2(\Omega) \; : \; \nabla \times \boldsymbol{E} \in \boldsymbol{L}^2(\Omega),\, \boldsymbol{n} \times \boldsymbol{E} = \boldsymbol{0} \text{ on } \Gamma_1\}\,. \tag{5}$$

The original and variational formulations are equivalent to each other. Introducing a *space of Lagrange multipliers* (scalar potentials):

$$V := \{q \in H^1(\Omega) \; : \; q = 0 \text{ on } \Gamma_1\}, \tag{6}$$

we employ a special test function $\boldsymbol{F} = \nabla q, q \in V$, to learn that solution $\boldsymbol{E}$ to (4) must *automatically* satisfy the weak form of the continuity equation,

$$-\int_\Omega (\omega^2\epsilon - j\omega\sigma)\boldsymbol{E} \cdot \nabla\bar{q}\, dx = -j\omega\left\{\int_\Omega \boldsymbol{J}^{imp} \cdot \nabla\bar{q}\, dx + \int_{\Gamma_2} \boldsymbol{J}_S^{imp} \cdot \nabla\bar{q}dS\right\}. \tag{7}$$

We also recall the Helmholtz decomposition:

$$\boldsymbol{E} = \nabla\phi + \boldsymbol{E}_0, \quad \text{where } \phi \in V \text{ and } (\boldsymbol{E}_0, \nabla q) = 0\, \forall q \in V \tag{8}$$

Here $(\;,\;)$ and $(\;,\;)_\Gamma$ will denote standard L^2 products on domain Ω and its boundary Γ.

It is well known that the standard variational equation is *not* uniformly stable with respect to wave number $k^2 = \mu(\omega^2\epsilon - j\omega\sigma)$. As $k \to 0$, we loose the control over gradients. This corresponds to the fact that, in the limiting case $k = 0$, the problem is ill-posed as the gradient component remains undetermined. A remedy to this problem is to enforce the continuity equation explicitly at the expense of introducing a Lagrange multiplier $p \in V$. The so called *regularized variational formulation* looks as follows.

$$\left\{\begin{array}{ll}\text{Find } \boldsymbol{E} \in \boldsymbol{W}, p \in V \text{ such that} & \\ \displaystyle\int_\Omega \frac{1}{\mu}(\nabla \times \boldsymbol{E})(\nabla \times \bar{\boldsymbol{F}})dx - \int_\Omega (\omega^2\epsilon - j\omega\sigma)\boldsymbol{E} \cdot \bar{\boldsymbol{F}}dx & \\ -\int_\Omega (\omega^2\epsilon - j\omega\sigma)\nabla p \cdot \bar{\boldsymbol{F}}dx = & \\ -j\omega\left\{\int_\Omega \boldsymbol{J}^{imp} \cdot \bar{\boldsymbol{F}}dx + \int_{\Gamma_2} \boldsymbol{J}_S^{imp} \cdot \bar{\boldsymbol{F}}dS\right\} & \forall \boldsymbol{F} \in \boldsymbol{W} \\ -\int_\Omega (\omega^2\epsilon - j\omega\sigma)\boldsymbol{E} \cdot \nabla\bar{q}\, dx = & \\ -j\omega\left\{\int_\Omega \boldsymbol{J}^{imp} \cdot \nabla\bar{q}\, dx + \int_{\Gamma_2} \boldsymbol{J}_S^{imp} \cdot \nabla\bar{q}dS\right\} & \forall q \in V\,. \end{array}\right. \tag{9}$$

By repeating the trick with the substitution $\boldsymbol{F} = \boldsymbol{\nabla} q$ in the first equation, we learn that the Lagrange multiplier p *identically vanishes*, and for that reason, it is frequently called the *hidden variable*. In contrary to the original formulation, the stability constant for the regularized formulation converges to one, as $k \to 0$. The regularized formulation works because gradients of the scalar-valued potentials from V form precisely the null space of the curl-curl operator.

The point about the regularized (mixed) formulation is that, whether we use it or not in the actual computations (the improved stability is one good reason to do it...), the original variational problem is *equivalent* to the mixed problem. This suggests that we cannot escape from the theory of mixed formulations when analysing the problem.

The mixed formulation and a corresponding stability analysis outlined here, led us to the idea of the variable order edge elements [16]. We start with a standard, H^1-conforming, variable order element, (triangle, quad in 2D, tetrahedron, hexahedron, or prism in 3D), identify the corresponding *space of shape functions*, and consider its image through the gradient operator. For instance, for a triangle, if the space of scalar potentials V_h consists of polynomials of order $p+1$ whose restrictions to element edges e are of *lower or equal* order p_e+1, the space of vector-valued functions $\boldsymbol{W}_h$ to approximate the $\boldsymbol{E}$-field will consist of vector-valued polynomials whose *tangential components* (and tangential only) on the element boundary will reduce to the lower order p_e. This corresponds to the simple fact that differentiation lowers the degree of polynomials.

The same assumption that $\boldsymbol{W}_h$ should include gradients of functions from V_h lies behind the construction of the parametric element [39].

Exact sequence. Spaces $H^1, \boldsymbol{H}(\text{curl})$ and the gradient operator are part of the *exact sequence*,

$$H^1 \xrightarrow{\boldsymbol{\nabla}} \boldsymbol{H}(\text{curl}) \xrightarrow{\boldsymbol{\nabla}\times} \boldsymbol{H}(\text{div}) \xrightarrow{\boldsymbol{\nabla}\circ} L^2 \,, \tag{10}$$

where the null space of each operator coincides with the range of the preeceding operator. The last operator in the sequence, div, is a surjection.

Exactly the same algebraic structure can be reproduced at the discrete level, using polynomials. We shall use, as an example, master tetrahedron T,

$$T = \{(x_1, x_2, x_3) \,:\, 0 < x_1 < 1, 0 < x_1 + x_2 < 1, 0 < x_1 + x_2 + x_3 < 1\}, \tag{11}$$

but exactly the same idea can be reproduced for hexahedra or prisms [18,35]. We introduce the following notation.

$\mathcal{P}^p_{p_e,p_f}$ space of polynomials on T of order (less of equal) p, whose traces on edges e reduce to polynomials of order p_e, and on faces f to polynomials of order p_f;

$\mathbf{P}^p_{p_e,p_f}$ space of vector-valued polynomials on T of order (less of equal) p, such that the traces of their *tangential components* to edges e reduce to (scalar valued) polynomials of order p_e, and on faces f to (vector -alued) polynomials of order p_f;

$\mathbf{P}^p_{p_f}$ space of vector-valued polynomials on T of order (less of equal) p, such that the traces of their *normal components* to faces f reduce to (scalar-valued) polynomials of order p_f;

$\mathcal{P}^p$ space of (scalar-valued) polynomials on T of order (less of equal) p.

The definitions above make sense only if $p_f \le p$, for any face f, and $p_e \le p_f$, for any face f adjacent to edge e, and any edge e. In practice, this condition is enforced by using the *minimum rule* for FE meshes. Assigning a polynomial (possibly variable) order p to each element in advance, the edge and face orders p_e, p_f are set to the minimum value of orders p for all adjacent elements. Note that for $p_e = -1$ and $p_f = -1$ we have polynomials with zero traces on the boundary.

Again we have an exact sequence:

$$\mathcal{P}^{p+1}_{p_e+1,p_f+1} \xrightarrow{\nabla} \mathbf{P}^p_{p_e,p_f} \xrightarrow{\nabla\times} \mathbf{P}^{p-1}_{p_f-1} \xrightarrow{\nabla\circ} \mathcal{P}^{p-2}\,. \tag{12}$$

Of course, spaces above make sense only, if $p \ge 2$. The only not trivial fact about the sequence is that the div operator is a surjection. This is proved simply by comparing the dimension of the range of the divergence operator,

$$\begin{aligned}
\dim\mathcal{R}(\mathrm{div}) &= \dim\mathbf{P}^{p-1}_{p_f-1} - \dim\mathcal{N}(\mathrm{div}) \\
&= \dim\mathbf{P}^{p-1}_{p_f-1} - \dim\mathcal{R}(\mathbf{curl}) \\
&= \dim\mathbf{P}^{p-1}_{p_f-1} - (\dim\mathbf{P}^{p}_{p_e,p_f} - \dim\mathcal{N}(\mathbf{curl})) \\
&= \dim\mathbf{P}^{p-1}_{p_f-1} - \dim\mathbf{P}^{p}_{p_e,p_f} + \dim\mathcal{R}(\nabla) \\
&= \dim\mathbf{P}^{p-1}_{p_f-1} - \dim\mathbf{P}^{p}_{p_e,p_f} + \dim\mathcal{P}^{p+1}_{p_e+1,p_f+1} - \dim\mathcal{N}(\nabla) \\
&= \dim\mathbf{P}^{p-1}_{p_f-1} - \dim\mathbf{P}^{p}_{p_e,p_f} + \dim\mathcal{P}^{p+1}_{p_e+1,p_f+1} - 1\,,
\end{aligned} \tag{13}$$

with the dimension of $\mathcal{P}^{p-2}$. Note that the div operator is a surjection for any p_f (including $p_f - 1 = -1$, i.e. polynomials with zero traces on the boundary), and operator **curl** maps $\mathbf{P}^p_{p_e,p_f}$ *onto* the null space of the div operator, for *any* value of p_e, including again polynomials with zero tangential traces along the edges.

Shape functions. The polynomial spaces discussed above can easily be represented as spans of shape functions associated with vertices, edges, faces and the element interior. Usually, we try to construct them in terms of *affine coordinates* λ_i [2], in order to minimize the dependence upon the denumeration of

[2] For our master element, $\lambda_1 = 1 - x_1 - x_2$, $\lambda_2 = x_1$, $\lambda_3 = x_2$.

element vertices, and the corresponding denumeration of edges and faces. We also want them to be *hierarchical*, i.e. with an increase in order p, we want to just add a few new shape functions, without altering the existing ones. The use of hierarchical shape functions is important for many different reasons, to mention a few: enforcement of global continuity requirements, a-posteriori error estimation, mesh optimization etc. The first constructions date back to the pioneering work of Szabo and his collaborators in the late seventies.

H^1-conforming triangle

$$\begin{aligned}
&\text{vertices: } \phi_i = \lambda_i, \quad i = 1,2,3 \\
&\text{edges: } \quad \phi_{i,l} = \lambda_i\lambda_{i+1}(\lambda_{i+1} - \lambda_i)^{l-1}, \quad l = 1,\dots,p_{e_i} - 1, i = 1,2,3 \\
&\text{interior: } \phi_{4,kl} = \lambda_1\lambda_2^{1+l}\lambda_3^{k-l}, \quad l = 0,\dots,k-1,\ k = 1,\dots,p-3
\end{aligned} \tag{14}$$

Index $p+1$ here is to be understood mod 3. As we can see, in the construction of the interior (bubble) shape functions, the first vertex is treated differently from the remaining two. Indeed, without modifying the spaces, it is impossible to construct shape functions that would be simulateneously hierarchical and independent of the denumeration of vertices. Also, the interior shape functions are present only for $p \geq 3$, and the edge shape functions for $p_e \geq 2$. The shape functions defined above are by no means optimal, in fact they produce horribly conditioned stiffness matrices, but they are probably the easiest choice to explain the idea of vertex, edge and interior shape functions.

A similar construction holds for the edge element [34].

$H(\text{curl})$-conforming triangle

$$\begin{aligned}
\text{edges: } \quad & \lambda^i \boldsymbol{\tau}_i^1 \\
& \lambda^{i+1} \boldsymbol{\tau}_i^2 \\
& \phi_{i,l}\boldsymbol{\tau}_i, \quad l = 1,\dots,p_{e_i} - 1, i = 1,2,3 \\
\text{interior: } \quad & \phi_{i,l}\boldsymbol{n}_i, \quad l = 1,\dots,p-1, i = 1,2,3 \\
& \phi_{4,kl}\boldsymbol{e}_i, = 0,\dots,k-1,\ k = 1,\dots,p-3, i = 1,2
\end{aligned} \tag{15}$$

Here ϕ denotes the shape functions for the scalar element, $\boldsymbol{\tau}_i$ is the unit vector tangent to i-th edge, directed from i-th to $i+1$-st vertex, $\boldsymbol{\tau}_i^2, \boldsymbol{\tau}_{i+1}^1$ denote the dual basis [3] to $\boldsymbol{\tau}_i, \boldsymbol{\tau}_{i+1}$, $\boldsymbol{n}_i$ are versors normal to element edges, and $\boldsymbol{e}_1 = (1,0), \boldsymbol{e}_2 = (0,1)$ are unit vectors of the element system of coordinates, see Fig. 1.

Again, this choice is just an illustration; in fact not even an elegant one, as the edge shape functions do not transform one into another as we rotate the vertices [4]. Also, recent study of Ainsworth and Coyle [2] show that optimal edge shape functions should be constructed enforcing different orthogonality than the one used for the H^1-conforming triangle.

[3] Vectors $\boldsymbol{a}^j$ form the dual basis to a basis $\boldsymbol{a}_i$ if $\boldsymbol{a}^j \cdot \boldsymbol{a}_i = \delta_i^j$

[4] Note that the transformations here must follow the transformation rule for gradients

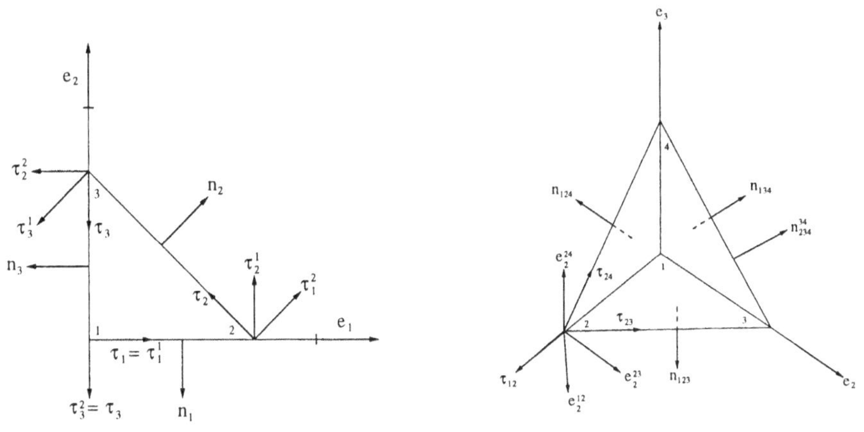

Figure1. Illustration of vectors used in the definition of $H(\mathbf{curl})$-conforming triangle and tetrahedron shape functions

Shape functions for the tetrahedral element can be constructed in a similar way.

H^1-conforming tetrahedron

$$\begin{array}{lll}
\text{vertex } i & \phi_i = \lambda_i, \quad 1 \le i \le 4 & \\
\text{edge } (ij) & \phi_{(ij),l} = \lambda_i\lambda_j(\lambda_j - \lambda_i)^{l-1}, \quad l = 1, \dots, p_{e_{ij}} - 1, & 1 \le i < j \le 4 \\
\text{face } (ijk) & \phi_{(ijk),l,m} = \lambda_i\lambda_j^l\lambda_k^m, \quad 1 \le l,m,\ 1 + l + m \le p_{f_{ijk}} & 1 \le i < j < k \le 4 \\
\text{interior} & \phi_{lmn} = \lambda_1\lambda_2^l\lambda_3^m\lambda_4^n, \quad 1 \le l,m,m,\ 1 + l + m + n \le p &
\end{array} \tag{16}$$

$H(\mathbf{curl})$-conforming tetrahedron

$$\begin{array}{ll}
\text{edge } (ij) & \lambda_i \boldsymbol{e}_i^{ij}, \\
& \lambda_j \boldsymbol{e}_j^{ij}, \\
& \phi_{(ij),l}\boldsymbol{\tau}_{ij} \quad l = 1, \dots, p_{e_{ij}} - 1, \quad 1 \le i < j \le 4 \\
\text{face } (ijk) & \phi_{(ij),l}\boldsymbol{n}_{ijk}^{ij}, \\
& \phi_{(ik),l}\boldsymbol{n}_{ijk}^{ik}, \\
& \phi_{(jk),l}\boldsymbol{n}_{ijk}^{jk}, \quad l = 1, \dots p_{f_{ijk}} - 1 \\
& \phi_{(ijk),l,m}\boldsymbol{\tau}_{ij} \\
& \phi_{(ijk),l,m}\boldsymbol{\tau}_{ik} \quad 1 \le l,m,\ 1 + l + m \le p_{f_{ijk}} \quad 1 \le i < j < k \le 4 \\
\text{interior} & \phi_{(ijk),l,m}\boldsymbol{n}_{ijk} \quad 1 \le l,m,\ 1 + l + m \le p,\ 1 \le i < j < k \le 4 \\
& \phi_{lmn}\boldsymbol{e}_i, \quad 1 \le l,m,m,\ 1 + l + m + n \le p, i = 1, \dots, 3
\end{array} \tag{17}$$

Here $\boldsymbol{\tau}_{ij}$ is the vector directed from vertex i to vertex j (tangent to edge (ij)), vectors $\boldsymbol{e}_i^{kl}$ denote the dual basis corresponding to the three tangent vectors meeting at vertex i, $\boldsymbol{n}_{ijk}$ denotes a vector normal to face (ijk), $\boldsymbol{n}_{ijk}^{ij}$ denotes the normal vector to the face adjacent to edge (ij) *other* than face (ijk), and $\boldsymbol{e}_i$ are the element system coordinates unit vectors, see Fig. 1 for illustration of selected vectors. The key to understanding the construction of the shape functions is as follows.

- Traces of (the tangential component of) shape functions associated with edge (i,j) span polynomials of order p_e. Traces of their tangential components along the remaining three edges must vanish.
- Traces of (the tangential components of) shape functions associated with face (ijk) span space $\mathbf{P}_{-1}^{p_f}$ of vector-valued polynomials with zero trace (of the tangential component) on the boundary. Traces of their tangential components along all edges and over all remaining faces must vanish.
- Interior shape functions span space $\mathbf{P}_{-1,-1}^{p}$ of vector-valued polynomials of order p with vanishing traces of their tangential components along all edges and over all faces.

$H(\mathrm{div})$-conforming tetrahedron

$$
\begin{array}{ll}
\text{face } (ijk) & \phi_i \boldsymbol{n}_i^{ijk} \\
& \phi_j \boldsymbol{n}_j^{ijk} \\
& \phi_k \boldsymbol{n}_k^{ijk} \\
& \phi_{(ij),l} \boldsymbol{n}_{ijk}^{ij}, \\
& \phi_{(ik),l} \boldsymbol{n}_{ijk}^{ik}, \\
& \phi_{(jk),l} \boldsymbol{n}_{ijk}^{jk}, \quad l = 1, \dots p_{f_{ijk}} - 1 \\
& \phi_{(ijk),l,m} \boldsymbol{n}_{ijk} \quad 1 \le l, m,\ 1 + l + m \le p_{f_{ijk}} \\
\text{interior} & \phi_{(ij),l} \boldsymbol{\tau}_{ij}, \quad l = 1, \dots p - 1,\ 1 \le i < j \le 4 \\
& \phi_{(ijk),l,m} \boldsymbol{\tau}_{ij} \\
& \phi_{(ijk),l,m} \boldsymbol{\tau}_{ik} \quad 1 \le l, m,\ 1 + l + m \le p,\ 1 \le i < j < k \le 4 \\
& \phi_{lmn} \boldsymbol{e}_i, \quad 1 \le l, m, m,\ 1 + l + m + n \le p, i = 1, \dots, 3
\end{array}
\tag{18}
$$

Here $\boldsymbol{\tau}_{ij}$ is again the vector directed from vertex i to vertex j (tangent to edge (ij)), $\boldsymbol{n}_{ijk}$ denotes a vector normal to face (ijk), vectors $\boldsymbol{e}_i^{ijk}$ denote the dual basis corresponding to the three normal vectors meeting at vertex i, $\boldsymbol{n}_{ijk}^{ij}$ denotes the normal vector to the face adjacent to edge (ij) *other* than face (ijk), and $\boldsymbol{e}_i$ are the element system coordinates unit vectors. The key to understanding the construction of the shape functions is again as follows.

- Traces of (the normal components of) shape functions associated with face (ijk) span space $\mathbf{P}^{p_f}$ of polynomials of order p_f. Traces of their normal components over all remaining faces must vanish.

– Interior shape functions span space $\mathbf{P}^p_{-1}$ of vector-valued polynomials of order p with vanishing traces of their normal components over all faces.

Nedelec's elements of the first kind. For $p_e = p_f = p$, the sequence above coincides with the sequence corresponding to Nedelec's elements of the second type [32,6]. Clearly, as it is, the construction makes sense only for $p \geq 2$. How do we generalize it to Nedelec's elements of the first kind [31]? The motivation comes from the fact that, due to the mixed formulation, approximability of *both* electric field $\boldsymbol{E}$ *and* its curl$\boldsymbol{E}$, will affect the ultimate convergence rates. This leads to the question: what is a minimal enrichment for the $\boldsymbol{E}$-fields to guarantee a p-order approximability for the magnetic field (**curl** of $\boldsymbol{E}$) as well?

It is easier to start first with the 2D case. Tetrahedron T becomes then the triangle:

$$T = \{(x_1, x_2) \; : \; 0 < x_1 < 1, \, 0 < x_1 + x_2 < 1\} \,, \tag{19}$$

and the exact sequences get shortened:

$$\begin{aligned} H^1 &\xrightarrow{\nabla} \boldsymbol{H}(\mathrm{curl}) \xrightarrow{\nabla\times} L^2 \,, \\ \mathcal{P}^{p+1}_{p_e+1} &\xrightarrow{\nabla} \mathbf{P}^p_{p_e} \xrightarrow{\nabla\times} \mathcal{P}^{p-1} \,. \end{aligned} \tag{20}$$

One obvious answer to the question raised above is simply to increase order p without changing orders p_e on the boundary,

$$\mathcal{P}^{p+2}_{p_e+1} \xrightarrow{\nabla} \mathbf{P}^{p+1}_{p_e} \xrightarrow{\nabla\times} \mathcal{P}^{p} \,. \tag{21}$$

Unfortunately, in order to maintain the exact sequence, we have also increased the space of potentials by adding extra interior shape functions of order $p+2$. A simple remedy is as follows. We can represent space $\mathcal{P}^{p+2}_{p_e+1}$ as a direct sum of $\mathcal{P}^{p+1}_{p_e}$ and an algebraic complement[5] $\tilde{\mathcal{P}}^{p+2}_{p_e+1}$,

$$\mathcal{P}^{p+2}_{p_e+1} = \mathcal{P}^{p+1}_{p_e+1} \oplus \tilde{\mathcal{P}}^{p+2}_{p_e+1} \,. \tag{22}$$

This generates a corresponding decomposition of space $\mathbf{P}^{p+1}_{p_e}$,

$$\mathbf{P}^{p+1}_{p_e} = \mathbf{P}^{p}_{p_e} \oplus \nabla(\tilde{\mathcal{P}}^{p+2}_{p_e+1}) \oplus \tilde{\mathbf{P}}^{p+1}_{p_e} \,. \tag{23}$$

Notice that tangential traces of functions from complement $\tilde{\mathbf{P}}^{p+1}_{p_e}$ must vanish on the boundary. The desired extension of the original exact sequence is then

$$\mathcal{P}^{p+1}_{p_e+1} \xrightarrow{\nabla} \mathbf{P}^{p}_{p_e} \oplus \tilde{\mathbf{P}}^{p+1}_{p_e} \xrightarrow{\nabla\times} \mathcal{P}^{p} \,. \tag{24}$$

[5] The complement may be constructed simply by taking the span of element interior shape functions of degree $p+2$.

In other words, the whole idea is to add only those polynomials of order $p+1$ from $\mathbf{P}_{p_e}^{p+1}$ that are *not* gradients. On the practical side, we would have to redefine the shape functions associated with the element interior, using explicitly gradients of scalar bubble functions, and then additional functions that are not gradients.

The construction is, of course, not unique, as there are infinitely many algebraic complements. Hiptmair [23] uses Poincare's map to map homogeneous monomials of order p,

$$\psi(\boldsymbol{x}) = x_1^k x_2^{p-k}, \quad k = 0, \dots, p, \tag{25}$$

into

$$\begin{aligned} E_1(\boldsymbol{x}) &= -x_2 \int_0^1 t\psi(t\boldsymbol{x})\, dt = x_1^k x_2^{p+1-k} \\ E_2(\boldsymbol{x}) &= x_1 \int_0^1 t\psi(t\boldsymbol{x})\, dt = -x_1^{k+1} x_2^{p-k}. \end{aligned} \tag{26}$$

The resulting homogeneous, vector-valued polynomials span Nedelec's spaces S^{p+1} (compare [31], Prop.1) . The elements of this space *do not*, however, have zero tangential traces on the element boundary. Their tangential traces vanish on the first and the third edge but not along the second one. However, the trace of the tangential component along the second edge is a polynomial of lower order p,

$$-E_1 + E_2 = -x_1^k x_2^{p+1-k} - x_1^{k+1} x_2^{p-k} = -(x_2 + x_1) x_1^k x_2^{p-k} = -x_1^k x_2^{p-k}. \tag{27}$$

This allows for an easy correction, using edge shape functions of order p corresponding to the second edge, to construct functions with vanishing tangential traces.

Nedelec's construction, however, is unique when it comes to the lowest order elements,

$$\mathcal{P}^1 \xrightarrow{\boldsymbol{\nabla}} span\{\mathbf{P}^0, (x_2, -x-1)^T\} \xrightarrow{\boldsymbol{\nabla}\times} \mathcal{P}^0. \tag{28}$$

The enriched space of constant fields ($\mathcal{R}^1$ in [31]) is the minimal one to enforce the continuity of tangential components across interelement boundaries and the resulting, global H(curl)-conformity.

I am not convinced that the Nedelec/Hiptmair construction is optimal. Ideally, we would like to deal with *orthogonal* decompositions. Requesting H_0^1 orthogonality (L^2-orthogonality of gradients) makes our decomposition of scalar potentials unique[6],

$$\mathcal{P}_{p_e+1}^{p+2} = \mathcal{P}_{p_e+1}^{p+1} \overset{\perp}{\oplus} \tilde{\mathcal{P}}_{p_e+1}^{p+2}. \tag{29}$$

[6] The condition makes sense only for $p > 2$.

Similarly, we may request for L^2 orthogonality in the second decomposition:

$$\mathbf{P}^{p+1}_{p_e} = \left(\mathbf{P}^{p}_{p_e} \oplus \nabla(\tilde{\mathcal{P}}^{p+2}_{p_e+1})\right) \overset{\perp}{\oplus} \tilde{\mathbf{P}}^{p+1}_{p_e} \,. \tag{30}$$

Construction of orthogonal shape functions is not out of question, due to the existence of a simple, closed form formula for integrating affine coordinates,

$$\int_T \lambda_1^{\alpha}\lambda_2^{\beta}\lambda_3^{\gamma} = \frac{\alpha!\beta!\gamma!}{(\alpha+\beta+\gamma+2)!} \,. \tag{31}$$

We shall return to this issue in a moment. Also, enforcing orthogonality of complement $\tilde{\mathbf{P}}^{p+1}_{p_e}$ to the gradients, for affine spaces and material properties constant within each element, would be simply equivalent to using spaces defined in (23), without the cost of introducing extra bubble shape functions for the potentials [7].

We proceed similarly in 3D. Starting with the 'enriched' diagram,

$$\mathcal{P}^{p+2}_{p_e+1,p_f+2} \xrightarrow{\nabla} \mathbf{P}^{p+1}_{p_e,p_f+1} \xrightarrow{\nabla\times} \mathbf{P}^{p}_{p_f} \xrightarrow{\nabla\circ} \mathcal{P}^{p-1} \,, \tag{32}$$

we proceed with the decomposition of potentials:

$$\mathcal{P}^{p+2}_{p_e+1,p_f+2} = \mathcal{P}^{p+1}_{p_e+1,p_f+1} \oplus \tilde{\mathcal{P}}^{p+2}_{p_e+1,p_f+2} \,, \tag{33}$$

and the corresponding decomposition of the vector-valued polynomials,

$$\mathbf{P}^{p+1}_{p_e,p_f+1} = \mathbf{P}^{p}_{p_e,p_f} \oplus \nabla(\tilde{\mathcal{P}}^{p+2}_{p_e+1,p_f+2}) \oplus \tilde{\mathbf{P}}^{p+1}_{p_e,p_f+1} \,. \tag{34}$$

The desired generalization of (24) to three dimensions looks as follows.

$$\mathcal{P}^{p+1}_{p_e+1,p_f+1} \xrightarrow{\nabla} \mathbf{P}^{p}_{p_e,p_f} \oplus \tilde{\mathbf{P}}^{p+1}_{p_e,p_f+1} \xrightarrow{\nabla\times} \mathbf{P}^{p}_{p_f} \xrightarrow{\nabla\circ} \mathcal{P}^{p-1} \,, \tag{35}$$

We have not discussed a possible modification of the $H(\mathrm{div})$-conforming space. Starting with a decomposition

$$\mathbf{P}^{p+1}_{p_f} = \mathbf{P}^{p}_{p_f} \oplus \nabla\times(\tilde{\mathbf{P}}^{p+1}_{p_e,p_f+1}) \oplus \tilde{\mathbf{P}}^{p+1}_{p_f} \,, \tag{36}$$

we can enrich space $\mathbf{P}^{p}_{p_f}$ with the additional component $\tilde{\mathbf{P}}^{p+1}_{p_f}$ and modify the exact sequence (35) to a new form:

$$\mathcal{P}^{p+1}_{p_e+1,p_f+1} \xrightarrow{\nabla} \mathbf{P}^{p}_{p_e,p_f} \oplus \tilde{\mathbf{P}}^{p+1}_{p_e,p_f+1} \xrightarrow{\nabla\times} \mathbf{P}^{p}_{p_f} \oplus \tilde{\mathbf{P}}^{p+1}_{p_f} \xrightarrow{\nabla\circ} \mathcal{P}^{p} \,. \tag{37}$$

At the end, the order of polynomials in the new sequence drops from $p+1$ to only p, as opposed to $p-2$ in the original sequence. Components $\tilde{\mathbf{P}}^{p+1}_{p_e,p_f+1}$ and

[7] Equivalently, using sequence (23) will do the same job as enriching the space for the electric field.

$\tilde{\mathcal{P}}^{p+1}_{p_f}$ can again be constructed using Poincare's map [23] and the construction coincides with Nedelec's spaces. Again, the direct sum decompositions are *not* unique, unless we request for orthogonal decompositions. It is perhaps worth to mention that for hexahedral elements and the corresponding spaces of polynomials $Q^{(p_1,p_2,p_3)}$, the order of polynomials in the original sequence drops only by one, without any modifications [18].

From the practical point of view, the modification of the approximation of the electric field consists of adding extra *both* face and interior shape functions of higher order. In order to preserve the hierarchical structure, one has to split face shape functions of order $p+1$ into those whose traces of tangential components are surface gradients of corresponding scalar-valued face shape functions of order $p+2$ and those that are not gradients, and similarly, interior shape functions of order $p+1$ have to be split into gradients of scalar bubble shape functions of order $p+2$ and those that are not gradients. Then one keeps only the non-gradient shape functions.

Definition of a Finite Element. The classical definition of Ciarlet/Raviart identifies a finite element as a triple $\{T, X, \psi_i, i = 1, \ldots, N\}$ where

- T (in our case) is the tetrahedron or triangle,
- X is a polynomial *space of element shape functions* of dimension N,
- $\psi_i : \mathcal{X} \to \mathbb{R}$, $i = 1, \ldots, N$ are linear functionals (*degrees of freedom*) defined on a *larger* space[8] $\mathcal{X}$ such that their restrictions to space X form a basis in dual space $X^\star$.

The element *shape functions* ϕ_i are then defined as a basis in X, dual to degrees of freedom ψ_j, i.e.

$$< \psi_j, \phi_i >= \delta_{ji} \,. \tag{38}$$

The corresponding *interpolation operator* $\Pi : \mathcal{X} \to X$ is then defined as follows,

$$\Pi u = \sum_{i=1}^{N} < \psi_i, u > \phi_i \,. \tag{39}$$

Alternatively, given space X, element shape functions $\phi_i \in X$, and an interpolation operator $\Pi : \mathcal{X} \to X$, there exists a *unique* set of degrees of freedom $\{\psi_i : \mathcal{X} \to \mathbb{R},\ i = 1, \ldots, N\}$ that yields the original shape functions and the interpolation operator. Since, at the end, all that matters is the interpolation, we find it more convenient to think directly in terms of the interpolation operators and spaces rather than specific degrees of freedom and shape functions [33].

[8] $\mathcal{X}$ is usually a Sobolev space including solutions to our boundary-value problem.

hp Interpolation Operators. When constructing an interpolation operator, we request the following conditions.

Locality. Values of interpolant Πu on element T should depend *only* on values of the interpolated function u on the element.

Global Continuity. When applied to a globally continuous function and a finite element mesh, element by element, the resulting global interpolant should also be globally continuous.

Optimality. The interpolant should be as close as possible to the interpolated function in an appropriate norm.

The *hp* interpolation operator Π_{hp} for H^1-conforming spaces was introduced in [33]. Conditions on locality and continuity imply that the interpolant $u^p = \Pi_{hp}u$ must match function u at the element vertices:

$$u_p(\boldsymbol{a}_i) = u(\boldsymbol{a}_i), \quad i = 1, \ldots, 4 . \tag{40}$$

By the same locality and continuity arguments, the value of u^p along an edge e must depend upon values of function u along the same edge only. With the values at the vertices fixed, the best we can do is to minimize the distance between function u and its interpolant in an appropriate norm. The approximation theory suggests norm $H_{00}^{\frac{1}{2}}$ which, however, is not local. The natural choice is then the stronger H_0^1 norm, i.e. the L^2-norm applied to the tangential derivative. We end up with the condition:

$$\int_e \left(\frac{\partial u^p}{\partial s} - \frac{\partial u}{\partial s}\right)\frac{\partial \phi}{\partial s} = 0 \quad \forall \phi \in \mathcal{P}_{-1}^{p_e}(e) , \tag{41}$$

for every edge e. Condition (41) can be viewed as the Galerkin approximation to a one-dimensional Poisson problem with Dirichlet boundary conditions. Clearly, if we agree on the choice of the norm, this is *the best* we can do! With the values of interpolant u^p along the edges already determined, the same philosophy can now be applied to faces:

$$\int_f (\boldsymbol{\nabla}_f u^p - \boldsymbol{\nabla}_f u)\boldsymbol{\nabla}_f \phi = 0 \quad \forall \phi \in \mathcal{P}_{-1}^{p_f}(f) , \tag{42}$$

and the interior of the element:

$$\int_T (\boldsymbol{\nabla} u^p - \boldsymbol{\nabla} u)\boldsymbol{\nabla} \phi = 0 \quad \forall \phi \in \mathcal{P}_{-1}^{p}(f) . \tag{43}$$

Here $\boldsymbol{\nabla}_f u = \boldsymbol{\nabla} u \cdot \boldsymbol{n}_f$ stands for the surface gradient for face f.

System of equations (40,41,42,43) defines uniquely interpolant $u^p \in \mathcal{P}^p_{p_e,p_f}$. As u^p is a linear combination of vertex, edge, face, and interior shape functions, one can solve first for the vertex d.o.f., then for the edge, face, and, finally, for the interior d.o.f. .

The choice of space $\mathcal{X}$ where the interpolation is defined is dictated by the implicitly requested regularity of the interpolated function. Condition (40) asks for the function to be continuous, condition (41) asks for the tangential derivative to be L^2-integrable[9], and so on. The issues of a sufficient regularity are by no means trivial but we shall not discuss them here.

The corresponding interpolation operators for $H(\mathbf{curl})$ and $H(\mathrm{div})$ spaces have been defined in [18]. Given a sufficiently regular vector-valued (electric) field $\boldsymbol{E}$, the corresponding hp-interpolant $\boldsymbol{E}_{hp} = \Pi_{hp}^{\mathrm{curl}} \boldsymbol{E}$ is now defined as a function $\boldsymbol{E}^p$ from the finite element space $\mathbf{P}^p_{p_e,p_f}$ that satisfies the following conditions:

- over each edge e, the trace of the tangential component of $\boldsymbol{E}^p$ is the L^2 projection of the trace of the tangential component of $\boldsymbol{E}$ onto polynomials of order p_e,

$$\int_e (\boldsymbol{E} - \boldsymbol{E}^p)_t \phi \, de = 0 \quad \forall \phi \in \mathcal{P}^{p_e}(e) \, ; \tag{44}$$

- over each face f, the tangential component of restriction $\boldsymbol{E} - \boldsymbol{E}^p$ satisfies the orthogonality conditions [10],

$$\begin{cases} \displaystyle\int_f \mathrm{curl}_f(\boldsymbol{E} - \boldsymbol{E}^p) \mathrm{curl}_f \boldsymbol{\phi} = 0 \quad \forall \boldsymbol{\phi} \in \mathbf{P}^{p_f}_{-1} \\ \displaystyle\int_f (\boldsymbol{E} - \boldsymbol{E}^p) \boldsymbol{\nabla}_f \phi = 0 \quad \forall \phi \in \mathcal{P}^{p_f+1}_{-1} \, , \end{cases} \tag{45}$$

where by $\mathbf{P}^{p_f}_{-1}$ we understand the space of vector-valued polynomials of order p_f whose tangential components vanish along the boundary ∂f, and the surface curl_f is defined as $\mathrm{curl}_f \boldsymbol{E} = \boldsymbol{n}_f \circ (\boldsymbol{\nabla} \times \boldsymbol{E})$;
- over the whole element T, $\boldsymbol{E} - \boldsymbol{E}^p$ satisfies the orthogonality conditions,

$$\begin{cases} \displaystyle\int_T \boldsymbol{\nabla} \times (\boldsymbol{E} - \boldsymbol{E}^p) \boldsymbol{\nabla} \times \boldsymbol{\phi} = 0 \quad \forall \boldsymbol{\phi} \in \mathbf{P}^p_{-1} \\ \displaystyle\int_T (\boldsymbol{E} - \boldsymbol{E}^p) \boldsymbol{\nabla} \phi = 0 \quad \forall \phi \in \mathcal{P}^{p+1}_{-1} \, , \end{cases} \tag{46}$$

where, again, by $\mathbf{P}^p_{-1}$ we understand the space of vector-valued polynomials of order p whose tangential components vanish along the whole boundary ∂T.

We refer to [18] for the remaining definition of the hp-interpolation operator Π_{hp}^{div}. In the definition above we used our original space $\mathbf{P}^p_{p_e,p_f}$, but the

[9] This assumption can be significantly weakened if the interpret the edge integral as a duality pairing

[10] The conditions are equivalent to minimizing $\|\mathrm{curl}_f(\boldsymbol{E} - \boldsymbol{E}^p)\|_{L^2(f)}$ with constraints imposed on tangential components on the boundary and the gradient component of the field.

same idea applies immediately to the extension corresponding to Nedelec's elements of the first kind.

The interpolation procedures are optimal in the sense that they are based on local (unconstrained or constrained) projections. They facilitate interpolation on variable order elements, independently of the choice of specific shape functions, and without need for special quadratures. They are used in several places in our codes, e.g. for interpolating Dirichlet data, mesh generation, unrefinements. But, first of all, the proposed interpolation procedures guarantee the commutativity of the de Rham diagram (see the following discussion).

Orthogonality of Shape Functions. One cannot orthogonalize FE basis functions to yield a diagonal stiffness matrix without loosing compact supports of the basis functions. One can, however, orthogonalize the shape functions partially, as to yield a diagonal system of equations for the interpolation procedures. Roughly speaking, for the H^1-conforming element, one would have to orthogonalize first interior shape functions, then use them to modify face shape functions to eliminate the coupling between face and interior shape functions, then orthogonalize them, and use the resulting new face shape functions to modify edge shape functions, etc. The procedure would result in new, non-hierarchical shape functions that would have to be changed after every p-refinement. It looks like that it would be more practical to enforce only partial orthogonalization, within each group of shape functions, directed at decoupling the system of equations for the interpolation. For the elliptic case, this would mean that edge shape functions corresponding to one edge would have their edge traces orthogonal in the $H_0^1(e)$ product, face shape functions would have their face traces orthogonal in the $H_0^1(f)$ product, and, finally, interior shape functions would have to be orthogonal with respect to the $H_0^1(T)$ product. A similar orthogonality could be enforced for the vector-valued shape functions. The extensions of the variable order polynomial space to Nedelec's elements of the first kind would become then uniquely defined.

There exists a huge literature on the optimal choice of higher order shape functions for spectral and hp methods, e.g. see [37,1,41,2] for recent contributions. The intention of this paragraph is merely to indicate that the orthogonality condition on the subspaces is compatible with the possibility of a partial orthogonalization of shape functions in practical computations.

De Rham diagram. The exact sequences on the continuous and discrete levels, together with the hp-interpolation operators can now be put into the De Rham diagram, either for our original sequence of polynomial spaces,

$$\begin{array}{ccccccc} H^1 & \xrightarrow{\nabla} & \boldsymbol{H}(\mathrm{curl}) & \xrightarrow{\nabla\times} & \boldsymbol{H}(\mathrm{div}) & \xrightarrow{\nabla\circ} & L^2 \\ \downarrow \Pi_{hp} & & \downarrow \Pi_{hp}^{\mathrm{curl}} & & \downarrow \Pi_{hp}^{\mathrm{div}} & & \downarrow P_{hp} \\ \mathcal{P}^{p+1}_{p_e+1,p_f+1} & \xrightarrow{\nabla} & \mathbf{P}^{p}_{p_e,p_f} & \xrightarrow{\nabla\times} & \mathbf{P}^{p-1}_{p_f-1} & \xrightarrow{\nabla\circ} & \mathcal{P}^{p-2} \end{array}, \tag{47}$$

or their modification,

$$\begin{array}{ccccccc}
H^1 & \xrightarrow{\nabla} & \boldsymbol{H}(\mathrm{curl}) & \xrightarrow{\nabla\times} & \boldsymbol{H}(\mathrm{div}) & \xrightarrow{\nabla\circ} & L^2 \\
\big\downarrow \Pi_{hp} & & \big\downarrow \Pi^{\mathrm{curl}}_{hp} & & \big\downarrow \Pi^{\mathrm{div}}_{hp} & & \big\downarrow P_{hp} \\
\mathcal{P}^{p+1}_{p_e+1,p_f+1} & \xrightarrow{\nabla} & \mathbf{P}^{p}_{p_e,p_f} \oplus \tilde{\mathbf{P}}^{p+1}_{p_e,p_f+1} & \xrightarrow{\nabla\times} & \mathbf{P}^{p}_{p_f} \oplus \tilde{\mathbf{P}}^{p+1}_{p_f} & \xrightarrow{\nabla\circ} & \mathcal{P}^{p}\,.
\end{array} \tag{48}$$

For a uniform order of approximation, the diagram reduces to the classical one [6], involving the Nedelec and Raviart-Thomas elements. The *hp* interpolation operators reduce then to those used by Nedelec and Hiptmair but the generalization is non-trivial, as for the Nedelec interpolation procedures and the variable order elements the diagram *does not* commute. The de Rham diagram generalizes to all standard elements, including curved (parametric) elements as well.

Interpolation error estimates. The de Rham diagram forms a mathematical foundation for stability and convergence analysis not only for Maxwell's equations but also for *acoustics* and various *mixed formulations.* As mentioned above, for uniform p, the interpolation operators are identical with those of Nedelec and Raviart-Thomas and the corresponding h-interpolation results are well known. One of the still open challenges (in 3D only) is to lower maximally the regularity assumptions on interpolated functions so that the interpolation error estimates could be applied to problems involving non-smooth, non-convex domains with singular solutions and h adaptive (non quasi-uniform) grids.

But the main open problem in the *hp* context is to prove interpolation error estimates for the p-method. For the H^1-conforming elements, such estimates can be derived from results of Babuška, Suri, Guo et al, but they are unknown for the remaining $H(\mathrm{curl})$ and $H(\mathrm{div})$-conforming elements. As the definitions of the interpolation procedures involve mixed problems, analysis of the p-interpolation properties can be reduced to a p-stability analysis of a mixed formulation for a Maxwell-like problem on a *single* element [20]. Numerical experiments for triangles and tetrahedra, and $p = 1, \ldots, 9$ indicate that the stability is there but the question of a general proof for arbitrary elements remains open. The case of square and hexahedral elements was succesfully analyzed by Monk [29], although with suboptimal convergence rates.

Convergence. Maxwell's eigenvalues. Stability and convergence analysis for the time-harmonic Maxwell equations leads quickly to the related (and important for itself) problem of convergence of Maxwell's eigenvalues [30]. Due to the lack of ellipticity, the problem is not covered by the standard theory for elliptic equations. Over a decade ago, Kikuchi [28] introduced the notion of *discrete compactness* that provides a sufficient condition for a convergence analysis for Maxwell's eigenvalues. In a recent contribution, Caorsi et al.

[9] demonstrate that, in a sense, the discrete compactness is not only sufficient but also necessary to assure good convergence properties. In turn, Boffi [7,8] has linked the discrete compactness with the commutativity of the de Rham diagram showing, in essence, that the two are equivalent. This strongly suggests that, eventually, we can prove the optimal *hp*-error estimates for all kinds of elements and the most general 3D problems. Our contributions to the subject include a 2D h-convergence analysis for shape regular, variable order meshes (geometrically graded meshes included) [17], and a 3D h-convergence analysis for standard Nedelec's elements of higher order on quasi-uniform meshes [30]. A general proof, including the p-convergence analysis, is linked to the p interpolation estimates discussed above, and it remains open.

The 2D discrete compactness results have been succesfully utilized by Vardapetyan [38,40] in a full wave analysis for waveguides, using the *hp* edge elements. For a given frequency ω, the problem reduces to the solution of a generalized eigenvalue problem for the constants of propagation. For *nonhomogeneous* waveguides, the corresponding operator is non self-adjoint (even for the lossless case !), and the propagation constant may be complex. Vardapetyan proved that the problem is well-posed, and derived h-convergence error estimates. The theoretical analysis has been confirmed with numerical experiments and illustrated with the solution of non-nontrivial, practical examples.

Fortran 90 implementations. In the past three years we developed four *hp* research codes, written in Fortran 90. The choice of language was motivated, among other reasons, with recycling portions of earlier Fortran 77 implementations, dynamic memory allocation, possibility of using long, self-explanatory names, and user-defined data types. The fundamental data structure for each of the implemented *hp*-adaptive discretizations [13,14,34,19,35] reduces to just two arrays of two kinds of objects: *elements* and *nodes*. The 2D code has both triangular and quadrilateral elements, the 3D code works (currently) on hexahedra only. In both 2D and 3D we have used the same strategy, starting with a *generic* version for continuous elements (2Dhp90, 3Dhp90) and then customizing it to the edge elements and Maxwell's equations (2Dhp90_EM, 3Dhp90_EM). The codes support not only the variable order approximations but the *constrained approximation* (hanging nodes) on anisotropically refined, *one-irregular* meshes. These refinements allow to produce long, needle-like elements to handle edge singularities and geometrically graded meshes for corner singularities. The source codes, along with manuals, have been placed on the Web, and are being updated regularly. Please see my web page for a further information (http://www.ticam.utexas.edu/ leszek/).

Infinite Elements. So far we have discussed interior boundary-value problems only. Most of the practical problems are formulated in unbounded domains (exterior problems, infinite waveguides). Our principal technique to discretize problems in unbounded domains has been coupled finite/infinite elements.

In [15] we proposed an infinite element for Maxwell's equations that is compatible with the *hp* discretizations in bounded domains. The main idea besides the construction is based again on including gradients in the space of test functions, so the continuity equation is implicitly satisfied. The element has been implemented in both the 2D and 3D codes [11,12].

The good news is that the computational results are promissing, and that we have been able to confirm optimal h-covergence rates. The bad news, at least on the theoretical side, is that we could not observe predicted exponential convergence [15], neither in order of approximation p, nor in number of terms N. We blame conditioning problems for it, but the issue needs a more careful analysis and further investigation.

Conclusions. Potential of the hp method. The potential of the finite element method for Maxwell's equations is best visible for problems with nonhomogeneous media. As long as the grid is aligned with material interfaces, contrary to other discretization techniques, the finite element method offers a general framework for constructing schemes of arbitrarily high order. The possibility to vary the order of approximation p and mesh size h makes the *hp* edge elements a very powerful method of discretization.

I have not discussed here any details on a-posteriori error estimation, *hp* mesh optimization, and iterative solvers for the Maxwell equations.

I believe that a succesful and competitive implementation of a fully automatic *hp* method must be based on a fully integrated approach where both the error estimation and mesh optimization are imbedded into a multigrid solver, in a similar way as this has been done for h-adaptive methods [5,21].

Acknowledgment

The work has been supported by Air Force under Contract F49620-98-1-0255.

References

1. S. Adjerid, M. Aiffa, J. E. Flaherty , "Hierarchical Finite Element Bases for Triangular and Tetrahedral Elements", *SCOREC Report 12-1999*, Rensselaer Polytechnic Institute, 1999.
2. Mark Ainsworth and Joe Coyle, "Hierarchic *hp*-Edge Element Families for Maxwell's Equations on Hybrid Quadrilateral/Triangular Meshes", Strathclyde Mathematics Research Report No. 27, September 2000.
3. I. Babuška, B. Guo, "Optimal Estimates for Lower and Upper Bounds of Approximation Errors in the *p*-version of the Finite Element Method in Two Dimensions", *Numerische Mathematik*, **85**, 219-255, 2000.
4. I. Babuška, M. Suri, "The *p* and *hp*-Versions of the Finite Element Method, Basic Principles and Properties", *SIAM Reviews*, **36**, 4, 578-632, 1994.
5. R. Beck, P. Deuflhard, R. Hiptmair, R.H.W. Hoppe, and B. Wohlmuth, "Adaptive Multilevel Methods for Edge Element Discretizations of Maxwell's Equations", *Surveys on Mathematics for Industry*, **8**, 271-312, 1999.

6. A. Bossavit, "Un noveau point de vue sur les éléments finis mixtes", *Matapli* (bulletin de la Société de Mathématiques Appliquëes et Industrielles), 23-35, 1989.
7. D. Boffi, "Discrete Compactness and Fortin Operator for Edge Elements", Preprint, March 1999.
8. D. Boffi, "A Note on the Discrete Compactness Property and the de Rham Diagram", Preprint, April 1999.
9. S. Caorsi, P. Fernandes, M. Raffetto, "On the Convergence of Galerkin Finite Element Approximations of Electromagnetic Eigenproblems", preprint 1999.
10. M. Cessenat, *Mathematical Methods in Electromagnetism, Linear Theory and Applications*, World Scientific, London 1996.
11. W. Cecot, L. Demkowicz, and W. Rachowicz, "A Two-Dimensional Infinite Element for Maxwell's Equations", *Computer Methods in Applied Mechanics and Engineering*, **188**, 625-643, 2000.
12. W. Cecot, L. Demkowicz, and W. Rachowicz, "A Three-Dimensional Infinite Element for Maxwell's Equations", *TICAM Report 00-20.*
13. L. Demkowicz, K. Gerdes, C. Schwab, A. Bajer, and T. Walsh, "HP90: A General and Flexible Fortran 90 *hp*-FE Code", *Computing and Visualization in Science*, **1**, 145-163, 1998.
14. L. Demkowicz, T. Walsh, K. Gerdes, and A. Bajer, "2D *hp*-Adaptive Finite Element Package. Fortran 90 Implementation", *TICAM Report 98-14*, July 1998.
15. L. Demkowicz and M. Pal, "An Infinite Element for Maxwell's Equations", *Computers Methods in Applied Mechanics and Engineering*, **164**, 77-94, 1998.
16. L. Demkowicz, L. Vardapetyan, "Modeling of Electromagnetic Absorption/Scattering Problems Using *hp*-Adaptive Finite Elements", *Computer Methods in Applied Mechanics and Engineering*, **152**, 1-2, 103-124, 1998.
17. L. Demkowicz, P. Monk, Ch. Schwab, and L. Vardapetyan "Maxwell Eigenvalues and Discrete Compactness in Two Dimensions" *TICAM Report 99-12*, accepted to *Mathematics and Computers with Applications.*
18. L. Demkowicz, P. Monk, L. Vardapetyan, and W. Rachowicz. "De Rham Diagram for hp Finite Element Spaces" *TICAM Report 99-06*, accepted to *Mathematics and Computers with Applications.*
19. L. Demkowicz, A. Bajer, W. Rachowicz, and K. Gerdes, "3D hp-Adaptive Finite Element Package. Fortran 90 Implementation (3Dhp90)", *TICAM Report 29.*
20. L. Demkowicz, M. Kuczma, "A Stability Analysis for the *hp* Interpolation Operators", in preparation.
21. G. Haase, M. Kuhn, U. Langer, "Parallel Multigrid 3D Maxwell Solvers", Johannes Kepler University Linz, SFB Report 99-23, 1999.
22. R. Hiptmair, "Canonical Construction of Finite Elements", *Mathematics of Computation*, **68**, 1325-1346, 1999.
23. R. Hiptmair, "Higher Order Whitney Forms", *Sonderforschungsbereich 382*, Universität Tübingen, Report 156, August 2000.
24. F. Ihlenburg and I. Babuška, "Dispersion Analysis and Error Estimation of Galerkin Finite Element Methods for the Helmholtz Equation", *Int. J. Numer. Methods Eng.*, **38**, 3745–3774, 1995.
25. F. Ihlenburg and I. Babuška, "Finite Element Solution to the Helmholtz Equation with High Wave Numbers - Part I: the h-Version of the FEM", *Computers & Mathematics with Applications*, **30**, 9, 9–37, 1995.

26. F. Ihlenburg and I. Babuška, "Finite Element Solution to the Helmholtz Equation with High Wave Numbers - Part II: the h-p-Version of the FEM", *SIAM J. Numer. Anal.*, **34**, 1, 315–358, 1997.
27. F. Ihlenburg, *Finite Element Analysis of Acoustic Scattering*, Springer-Verlag, New York, 1998
28. F. Kikuchi, "On a Discrete Compactness Property for the Nedelec Finite Elements", *J. Fac. Sci. Univ. Tokyo*, Sect. IA, Math., **36**, 479-490, 1989.
29. P. Monk, "On the $p-$ and $hp-$extension of Nédélec's Curl-Conforming Elements", *Journal of Computational and Applied Mathematics*, **53**, 117-137, 1994.
30. P. Monk, L. Demkowicz, and L. Vardapetyan, "Discrete Compactness and the Approximation of Maxwell's Equations in $doubleIR^3$" *Mathematics of Computation*, in print.
31. J.C. Nedelec, "Mixed Finite Elements in $\mathbb{R}^3$", *Numerische Mathematik*, **35**, 315-341, 1980.
32. J.C. Nedelec, "A New Family of Mixed Finite Elements in $\mathbb{R}^3$", *Numerische Mathematik*, **50**, 57-81, 1986.
33. J.T. Oden, L. Demkowicz, W. Rachowicz and T.A. Westermann, "Toward a Universal *hp* Adaptive Finite Element Strategy, Part 2. A Posteriori Error Estimation", *Computer Methods in Applied mechanics and Engineering*, **77**,113-180,1989.
34. W. Rachowicz and L. Demkowicz, "A Two-Dimensional hp-Adaptive Finite Element Package for Electromagnetics", *TICAM Report 98-15*, July 1998, *Computer Methods in Applied Mechanics and Engineering*, in print.
35. W. Rachowicz and L. Demkowicz, "A Three-Dimensional hp-Adaptive Finite Element Package for Electromagnetics", *TICAM Report 00-04*.
36. W. Rachowicz, L. Demkowicz, L. Vardapetyan, "hp-Adaptive FE Modeling for Maxwell's Equations. Evidence of Exponential Convergence" ACES' 99, Monterey, CA, March 16-20, 1999.
37. S.J. Sherwin, G.E. Karniadakis, "A New Triangular and Tetrahedral Basis for High-Order Finite Element Methods", *Int. J. Num. Meth. Eng.*, **38**, p.3775, 1995.
38. L. Vardapetyan, "*hp*-Adaptive Finite Element Method for Electromagnetics with Applications to Waveguiding Structures", *Ph.D. Dissertation*, The University of Texas at Austin, Austin, TX 78712, December 1999.
39. L. Vardapetyan and L. Demkowicz, "*hp*-Adaptive Finite Elements in Electromagnetics", *Computers Methods in Applied Mechanics and Engineering*, **169**, 331-344, 1999.
40. L. Vardapetyan and L. Demkowicz, "Full-Wave Analysis of Dielectric Waveguides at a Given Frequency", *TICAM Report 99-22*.
41. J. Webb, "Hierarchical Vector Based Function of Arbitrary Order for Triangular and Tetrahedral Finite Elements", *IEEE Antennas Propagat. Mag.*, **47** (8), 1244-1253, 1999.

A Common-Mode Skeleton Model for EMC Simulations*

R. Rietman

Philips Research Laboratories, Eindhoven, The Netherlands

Abstract Systems that consist of printed circuit boards with various digital and analog components, connecting wires and metal enclosures can show complex electromagnetic behaviour. A model is reviewed that allows for the calculation of the most important contribution to the emitted radiation from such systems. In this model the radiation is driven by *common-mode* voltage sources. The physical origin of these voltage sources is explained and the relevant electric field integral equation is derived. When planes are discretized by wire grids, the resulting equations can be analysed by means of a program like NEC. A program similar to NEC, named BERBER, is presented.

1 Introduction

In the Europe all electronic products that are to be sold have to satisfy electromagnetic compatibility (EMC) regulations. These EMC regulations are meant to reduce the risk that an apparatus disturbs its environment (emission) or that it stops functioning properly because of external electromagnetic influences (immunity). The emission norms state that the electric field at some specified distance (usually 3 or 10 metres) and frequency must not exceed some limiting value. The measurement of the radiated field must be carried out in a semi-anechoic room (SAR) with absorbing walls and ceiling and a conducting floor. This is a time-consuming procedure with the additional disadvantage that it can only be done with a more or less finished product. In practice this often means that EMC becomes an issue only in the final stages of development, when fixing the problem is expensive and time-consuming. For EMC problems to be solved (or preferably avoided) in an earlier stage of development, it is useful to have a means of predicting the radiated field from a complex structure.

Because of the vastly differing length scales involved, ranging from the 10^{-7} m level of integrated circuits to the 1 m level of the apparatus, it is clear that some simplifications have to be made for the modelling and numerical simulation to yield sensible results in a finite time. In this paper such a simplification, the *common-mode skeleton model*, is described. This model was first published in [1]. The relevant integral equation relating the sources to the excitations is derived, and it is shown how this equation can be solved numerically.

* invited paper

This paper is organized as follows: in Sect. 2 the common-mode skeleton model is presented. Two examples are given where the common-mode voltage sources can be calculated analytically. These examples are worked out in detail in the appendix. They show the physical mechanisms by which a signal on a PCB can couple to the attached cables. In Sect. 3 the model is simplified even further, by replacing the metal boards by wire grids. The electric field integral equation, suitable for the modelling of wire structures, is derived and discretized with help of the method of moments. Some choices for the basis functions are discussed. Sect. 4 discusses the EMIR suite of utilities, which was written to simplify the process of creating and visualizing the wire grid, parsing the output from the numerical solver and visualizing the results.

2 Common-Mode Skeleton

Printed circuit boards (PCBs) generally contain many small components that are connected by metal traces, the interconnect. Usually the PCB also contains one or more ground planes. A calculation of the detailed electromagnetic radiation emitted by such a PCB requires accurate knowledge of the current and charge distributions in the interconnect, the ground planes and the components. But the direct radiation from PCBs is not the most important contribution to the total radiated field.

An electrical apparatus usually contains many PCBs connected by (often multi-wire) cables. These cables can act as antennas and carry a non-zero net current, the so-called *common-mode* (CM) current. Current and charge can be also present on the metal enclosure, so that has to be taken into account as well. Even though the CM currents on the cables are usually much smaller than the currents that flow through the components and tracks on the PCBs, the CM currents cause more radiation for frequencies up to a few hundred MHz. That is because the *differential mode* (DM) currents on the PCB run in a small loop formed by a track and a return path in the ground plane or a neighbouring track, and hence are compensated locally, whereas the CM currents are not compensated by a current in a nearby return path.

When all DM currents are ignored, a much simpler configuration remains: PCBs are replaced by their ground planes and multi-wire cables become simple wires. The currents running through this simplified *skeleton* correspond to the CM currents in the original system. The original sources of radiation are replaced by CM voltage sources in the cables. The CM sources can be determined in the following way, as shown by Bergervoet in [2].

Consider a PCB with DM sources S and an attached cable. The cable carries a CM current I. First a voltage source $-V$ is inserted in the cable close to the PCB, such that the CM current in the cable becomes approximately zero. The location where the source is inserted is called a *port.* From linearity it follows that the difference with the original configuration is a PCB without sources and a voltage source V in the port. Graphically this can be represented as follows:

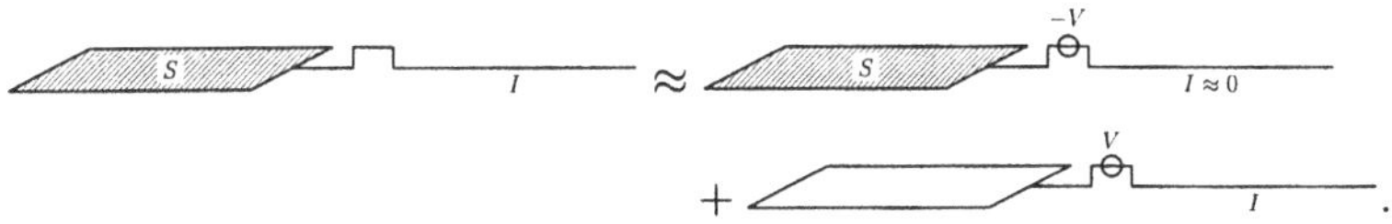

(1)

A system with zero CM current radiates negligibly:

$\approx 0,$ (2)

so as far as radiation is concerned, the CM skeleton with a voltage source is equivalent to the original configuration:

(3)

When the cable is cut from the board, the CM current on the cable becomes approximately zero, so V can be determined from the voltage in the the point at the edge of the board where the cable is attached:

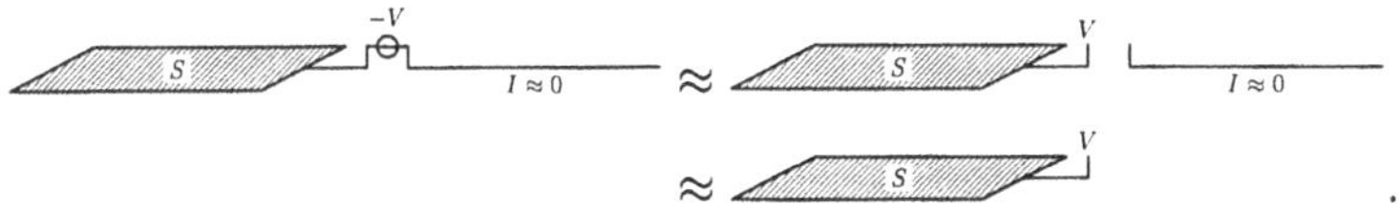

(4)

The common-mode voltage can be measured in a controlled set-up [1] with a Coupling and Decoupling Network (CDN).

The sources that are present on a PCB are often of one of the following types:

- A component on the PCB carries electric charge. The total charge on PCB and component must be zero, so the charge is compensated by an opposite charge on the PCB. The charge distribution has a non-zero electric dipole moment and due to the charges the ground plane has a non-zero potential.
- A current flows through a component and back through the ground plane, thus creating a current loop with non-zero magnetic moment. The field from this current loop induces a current in the ground plane. Due to the currents the ground plane gets a position-dependent potential.

These two kinds of source are shown in Fig. 1. For both types the common-mode potential on the ground plane can be calculated analytically for a simplified configuration with a circular ground plane and a point source. This is shown in Appendix 6, the results were also presented in [6].

3 Wire-Grid Discretization

The CM skeleton model is simplified further in the following ways:

Figure1. The figure on the left shows a component carrying a charge Q. This charge, together with the compensating charge on the PCB's ground plane, has a non-zero electric dipole moment. The figure on the right shows a differential-mode current loop, formed by the current through a track or component and the return current in the ground plane of the PCB. It has a non-zero magnetic moment.

- The boards and metal plates are replaced by a grid of cylindrical wires.
- All cables, even the flat ones, are replaced by cylindrical wires.

The resulting structure can be analysed with an antenna simulation program like NEC [3]. The publicly available version of NEC [4] can have accuracy problems for structures with short segments (much shorter than a wavelength). Such short segments are often necessary in in the modelling of complicated geometries with details much smaller than a wavelength. Therefore Jos Bergervoet of Philips Research has written a program, called BERBER, that should be more robust in this respect. BERBER is somewhat limited compared to NEC in that the current version of BERBER can only calculate the electric field, not the magnetic field. A further limitation of BERBER is that it can only deal with wire structures and not with surfaces; NEC uses a magnetic field integral equation for surfaces.

3.1 Physics of the Wire-Grid Model

The electric field in a metal wire with conductivity σ satisfies Ohm's law:

$$\boldsymbol{E}^{\text{total}}(\boldsymbol{r}) = \sigma^{-1}\boldsymbol{j}(\boldsymbol{r}). \tag{5}$$

The total electric field consists of the externally applied field $\boldsymbol{E}^{\text{ext}}$ which corresponds to the voltage sources, and the electric field $\boldsymbol{E}$ generated by the currents and charges on the wires. The external field is assumed to be non-zero only in the ports p of the system, where is has the value V_{CM}/ℓ_p with ℓ_p the length of port p. The contribution from the currents and charges on the wires can be written as in terms of a vector potential $\boldsymbol{A}$ and a scalar potential φ as (see any text on electromagnetism, e.g., [5])

$$\boldsymbol{E} = -\nabla\varphi - i\omega\boldsymbol{A} \tag{6}$$

where a time dependence $e^{i\omega t}$ is assumed and

$$\boldsymbol{A}(\boldsymbol{r}) = \frac{\mu_0}{4\pi}\int d^3r'\, G(\boldsymbol{r},\boldsymbol{r}')\boldsymbol{j}(\boldsymbol{r}') \tag{7}$$

$$\varphi(\boldsymbol{r}) = \frac{1}{4\pi\varepsilon_0}\int d^3r'\, G(\boldsymbol{r},\boldsymbol{r}')\rho(\boldsymbol{r}') \tag{8}$$

with the free space Green function G given by

$$G(\boldsymbol{r},\boldsymbol{r}') = \frac{\exp(-ik|\boldsymbol{r}-\boldsymbol{r}'|)}{|\boldsymbol{r}-\boldsymbol{r}'|} \text{ with } k = \omega/c. \tag{9}$$

The charge density ρ follows from the conservation law

$$\nabla \cdot \boldsymbol{j} + i\omega\rho = 0. \tag{10}$$

The resulting electric field integral equation (EFIE) is

$$\frac{i\mu_0\omega}{4\pi}\int d^3r'\, G(\boldsymbol{r},\boldsymbol{r}')\boldsymbol{j}(\boldsymbol{r}') + \frac{i}{4\pi\varepsilon_0\omega}\int d^3r'\, \nabla G(\boldsymbol{r},\boldsymbol{r}')\, \nabla' \cdot \boldsymbol{j}(\boldsymbol{r}') + \sigma^{-1}\boldsymbol{j}(\boldsymbol{r}) = \boldsymbol{E}^{\text{ext}}(\boldsymbol{r}). \tag{11}$$

3.2 Discretization

The EFIE (11) can be discretized with a standard method of moments (MOM) approach. Choose a set of vector-valued basis functions $\boldsymbol{\Psi}_n(\boldsymbol{r})$ and test functions $\boldsymbol{\Phi}_n(\boldsymbol{r})$ with support in the wires. The test functions may be the same as the basis functions (à la Galerkin), but they need not be the same. Expand the current density in the basis functions:

$$\boldsymbol{j}(\boldsymbol{r}) = \sum_n x_n \boldsymbol{\Psi}_n(\boldsymbol{r}). \tag{12}$$

Then take the inner product of both sides of Eq. (11) with a test function $\boldsymbol{\Phi}_m(\boldsymbol{r})$, the resulting weak form of the EFIE is the matrix equation

$$\boldsymbol{M}\boldsymbol{x} = \boldsymbol{y} \tag{13}$$

with

$$\boldsymbol{M} = \boldsymbol{L} + \boldsymbol{C} + \boldsymbol{R} \tag{14}$$

and the matrix elements are given by

$$L_{mn} = \frac{i\mu_0\omega}{4\pi}\int d^3r \int d^3r'\, G(\boldsymbol{r},\boldsymbol{r}')\, \boldsymbol{\Phi}_m(\boldsymbol{r})^* \cdot \boldsymbol{\Psi}_n(\boldsymbol{r}'), \tag{15}$$

$$\begin{aligned} C_{mn} &= \frac{i}{4\pi\varepsilon_0\omega}\int d^3r \int d^3r'\, \boldsymbol{\Phi}_m(\boldsymbol{r})^* \cdot \nabla G(\boldsymbol{r},\boldsymbol{r}')\, \nabla' \cdot \boldsymbol{\Psi}_n(\boldsymbol{r}') \\ &= -\frac{i}{4\pi\varepsilon_0\omega}\int d^3r \int d^3r'\, \nabla \cdot \boldsymbol{\Phi}_m(\boldsymbol{r})^*\, G(\boldsymbol{r},\boldsymbol{r}')\, \nabla' \cdot \boldsymbol{\Psi}_n(\boldsymbol{r}') \end{aligned} \tag{16}$$

$$R_{mn} = \sigma^{-1}\int d^3r \int d^3r'\, \boldsymbol{\Phi}_m(\boldsymbol{r})^* \cdot \boldsymbol{\Psi}_n(\boldsymbol{r}') \tag{17}$$

and

$$y_m = \int d^3r\, \boldsymbol{\Phi}_m(\boldsymbol{r})^* \cdot \boldsymbol{E}^{\text{ext}}(\boldsymbol{r}). \tag{18}$$

For numerical computations it is, of course, not feasible to use a complete basis, since infinitely many basis functions would be needed. Therefore only one or a few, suitably chosen, basis and test functions per segment are used. Regardless of the precise choice of these functions, it can be observed (see also [7]) that the matrix $\boldsymbol{M}$ is a full, complex matrix with, at least for 'low' frequencies, both 'large' eigenvalues (dominated by $\boldsymbol{C} \sim \omega^{-1}$) and 'small' eigenvalues (dominated by $\boldsymbol{R} + \boldsymbol{L}$, with corresponding eigenfunctions that lie mostly in the null space of $\boldsymbol{C}$, which corresponds to currents with zero divergence) and hence a large condition number. These properties make a straightforward implementation of a standard iterative solution method [8] ineffective and it is therefore that both NEC and BERBER currently use a direct solver, in spite of its large operation count ($O(N^3)$, where N is the dimension of $\boldsymbol{M}$).

By choosing a suitable basis that distinguishes between divergence-free and rotation-free currents and a clever pre-conditioner (a combination of a 'sparse approximate inverse' and several multi-grid steps), Menno Verbeek [7] was able to use an iterative solver that converged within a reasonable number of iterations ($O(N^{0.2})$, approximately). Thus the total operation count for this method is only slightly worse than $O(N^2)$.

A disadvantage of both the direct and Verbeek's iterative method is that the full complex matrix $\boldsymbol{M}$ has to be stored in memory (or on disk, but that is much slower), requiring $16N^2$ bytes. This limits the maximal problem size to about $N = 8000$ for a machine with 1 GB of memory. A matrix-free method that allows for the efficient calculation of a matrix-vector product $\boldsymbol{Mx}$ without storing all matrix elements of $\boldsymbol{M}$ in memory could remove this limitation, provided the pre-conditioning can be done in an efficient matrix-free way as well.

3.3 Choice of Basis and Test Functions

Each basis function is chosen to be a localized function that is non-zero on only a single segment and possibly its neighbours. Here we examine the fields coming from a single segment. The first approximation is that the current in each segment runs parallel to the cylinder axis, and that it is independent of the circumferential angle. In terms of cylindrical coordinates (ρ_n, ϕ_n, z_n) for the nth segment with endpoints at $z_n = z_{n1}$ and $z_n = z_{n2}$ and radius a_n it is written as:

$$\boldsymbol{\Psi}_n(\rho_n, \phi_n, z_n) = \hat{\boldsymbol{z}}_n \psi_n(z_n) w_n(\rho_n). \tag{19}$$

where $\psi_n(z_n) = 0$ for $z_n < z_{n1}$ and $z_n > z_{n2}$ and $w_n(\rho_n) = 0$ for $\rho_n > a_n$. The vector $\hat{\boldsymbol{z}}_n$ is a unit vector parallel to the cylinder axis of segment n. A convenient normalization is to choose $2\pi \int_0^{a_n} d\rho_n\, \rho_n w(\rho_n) = 1$, so that $\psi(z_n)$ is the total current at z_n.

The vector potential from this segment is then (suppressing the index n)

$$\boldsymbol{A}(z,\rho) = \hat{z}\frac{\mu_0}{4\pi}\int dz'\,\psi(z')\int d\rho'\,\rho' w(\rho')\int d\phi'\,\frac{1}{r'}\,e^{-ikr'}, \tag{20}$$

where $r' = \left((z-z')^2 + \rho^2 + \rho'^2 - 2\rho\rho'\cos\phi'\right)^{1/2}$. The integral over the cross-section of the segment yields a 'one-dimensional' Green function

$$G^{(1D)}(z-z',\rho) = \int_0^\infty d\rho'\,\rho' w(\rho')\int_0^{2\pi} d\phi'\,\frac{1}{r'}\,e^{-ikr'}, \tag{21}$$

so the vector potential is

$$\boldsymbol{A}(z,\rho) = \hat{z}\frac{\mu_0}{4\pi}\int dz'\,G^{(1D)}(z-z',\rho)\psi(z'). \tag{22}$$

The scalar potential is

$$\varphi(z,\rho) = \frac{i}{4\pi\varepsilon_0\omega}\int dz' G^{(1D)}(z-z',\rho)\frac{d}{dz'}\psi(z'). \tag{23}$$

The one-dimensional Green function depends on the choice of the function $w(\rho')$. Some choices for $w(\rho')$ and $\psi(z')$ will be discussed below.

Thin wire approximation The *thin wire approximation* neglects the effects of the wire thickness, all current is assumed to run on the wire axis. This leads to $\rho w(\rho) = (2\pi)^{-1}\delta(\rho)$ and

$$G^{(1D)}(z-z',\rho) = G^{\mathrm{TW}}(z-z',\rho) = \frac{1}{r}\,e^{-ikr}, \tag{24}$$

where $r = \left((z-z')^2 + \rho^2\right)^{1/2}$.

Extended thin wire approximation The *extended thin wire* (ETW) approximation assumes the current runs on the surface of the wire, so $\rho w(\rho) = (2\pi)^{-1}\delta(\rho - a)$. The wire thickness is still assumed to be negligible compared to the wavelength. This leads to

$$G^{(1D)}(z-z',\rho) = G^{\mathrm{ETW}}(z-z',\rho) = \frac{1}{2\pi}\int_0^{2\pi} d\phi'\,\frac{1}{r'}\,e^{-ikr'}$$

(where now $r' = \left((z-z')^2 + \rho^2 + a^2 - 2a\rho\cos\phi'\right)^{1/2}$)

$$\begin{aligned}
&= \frac{e^{-ikr}}{2\pi} \int_0^{2\pi} d\phi' \, \frac{1}{r'} \, e^{-ik(r'-r)} \\
&\approx \frac{e^{-ikr}}{2\pi} \int_0^{2\pi} d\phi' \, \frac{1 - ik(r'-r)}{r'} \\
&= e^{-ikr} \left(-ik + \frac{(1+ikr)}{2\pi} \int_0^{2\pi} d\phi' \, \frac{1}{r'} \right) \\
&= e^{-ikr} \left(-ik + (1+ikr) G_{\mathrm{K}}(z-z',\rho) \right) ,
\end{aligned} \tag{25}$$

where the 'elliptic kernel' G_{K} is given by

$$\begin{aligned}
G_{\mathrm{K}}(z-z',\rho) &= \frac{1}{2\pi} \int_0^{2\pi} d\phi' \, \frac{1}{r'} \\
&= \frac{2}{\pi\sqrt{(z-z')^2 + (\rho+a)^2}} \, K\left(\frac{4a\rho}{(z-z')^2 + (\rho+a)^2} \right)
\end{aligned} \tag{26}$$

and K is the complete elliptic integral of the first kind:

$$K(m) = \int_0^{\pi/2} dt \, \frac{1}{\sqrt{1 - m\sin^2 t}} . \tag{27}$$

The Rationalized Extended Thin Wire Approximation. In the Green function of the *rationalized extended thin wire* (RETW) approximation the elliptic kernel $G_{\mathrm{K}}(z-z',\rho)$ in the ETW Green function is replaced by a different kernel $G_{\mathrm{RK}}(z-z',\rho)$ that can be analytically integrated with respect to z':

$$G_{\mathrm{RK}}(z-z',\rho) = \frac{\partial}{\partial z'} F_{\mathrm{RK}}(z-z',\rho). \tag{28}$$

A good choice for F_{RK} is

$$F_{\mathrm{RK}}(z,\rho) = \ln\left(\frac{r-z}{a}\right) - \frac{z}{2\pi r} \ln\left(1 + \frac{\alpha a^2}{z^2 + (\rho-a)^2}\right) . \tag{29}$$

Then

$$\begin{aligned}
G_{\mathrm{RK}}(z,\rho) &= -\frac{\partial}{\partial z} F_{\mathrm{RK}}(z,\rho) \\
&= \frac{1}{r} - \frac{\alpha a^2 z^2}{\pi r \left(z^2 + (\rho-a)^2 + \alpha a^2\right)\left(z^2 + (\rho-a)^2\right)} \\
&\quad + \frac{\rho^2}{2\pi r^3} \ln\left(1 + \frac{\alpha a^2}{z^2 + (\rho-a)^2}\right) .
\end{aligned} \tag{30}$$

The parameter α is chosen such that G_{RK} has the same logarithmic singularity as the elliptic integral for $\rho = a$ and $z - z' \to 0$. This singularity is caused by the logarithmic divergence of the elliptic integral when its argument approaches 1: $K(1-\epsilon) \sim \ln(4/\sqrt{\epsilon})$. This gives $\alpha = 64e^{2-2\pi}$. The graphs of Fig. 2 show that the rationalized kernel is a very good approximation to the elliptic kernel for points on and outside of the cylinder.

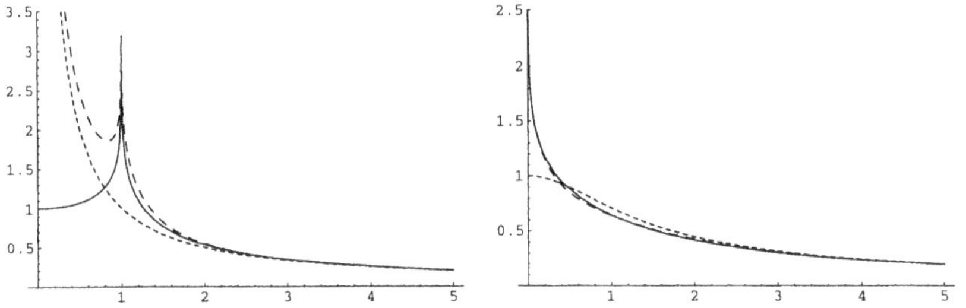

Figure2. The thin wire kernel $1/r$ (dotted line), the elliptic kernel $G_{\mathrm{K}}(z,\rho)$ (solid line) and the rationalized approximation $G_{\mathrm{RK}}(z,\rho)$ (dashed line) for cylinder radius $a=1$. On the left as a function of ρ for $z=0$, on the right as a function of z for $\rho=1$.

In the RETW approximation the integral over z' can be done analytically (albeit approximately, in a small a expansion) for a piecewise constant basis function $\psi(z) = \theta(z-z_1)\theta(z_2-z)$ (θ is Heaviside's step function). For segments that are short compared to the wavelength, a factor e^{-ikr_0}, where r_0 is the distance to a point of the segment, e.g., the midpoint, can be split off. The integrand is expanded up to second order in k, the remaining terms can then be integrated analytically:

$$
\begin{aligned}
&\int dz'\, G^{\mathrm{RETW}}(z-z',\rho)\psi(z') \\
&\quad = \int_{z1}^{z_2} dz'\, e^{-ikr}\left(-ik + (1+ikr)G_{\mathrm{RK}}(z-z',\rho)\right) \\
&\quad = e^{-ikr_0}\int_{z_1}^{z_2} dz'\, e^{-ik(r-r_0)}\left(-ik + (1+ikr)G_{\mathrm{RK}}(z-z',\rho)\right) \\
&\quad \approx e^{-ikr_0}\int_{z_1}^{z_2} dz'\left(-ik + (1+ikr_0)G(z-z',\rho)\right) \\
&\quad = e^{-ikr_0}\left[-ikz' + (1+ikr_0)F_{\mathrm{RK}}(z-z',\rho)\right]_{z_1}^{z_2}.
\end{aligned} \tag{31}
$$

In the ETW approximation the integral over z' has to be evaluated numerically. This integral is one of the terms in the electric field and there can be large cancellations when the different terms are added, leading to loss of accuracy. Therefore it is preferable to have an analytic expression for the field, even though it is only an approximation to the ETW model. The approximation is quite good, however and it must be remembered that the ETW model itself is also an approximation, even if the wires are cylindrical. Using cylindrical wires to model flat cables is a very crude approximation anyway.

Sinusoidal basis function A different, somewhat special choice of basis function is the sinusoidal function with wavenumber $\pm k$ (the same $k = \omega/c$ as in the Green function):

$$\psi(z') = e^{\pm ikz'}\theta(z - z_1)\theta(z_2 - z). \tag{32}$$

It can be useful when fairly long segments are present, because it turns out that the electric field can be calculated for arbitrary segment length. (See [9].) The z-component of the electric field

$$\begin{aligned} E_z &= -i\omega A_z - \frac{\partial}{\partial z}\varphi \\ &= -i\omega\frac{\mu_0}{4\pi}\int dz'\, G^{(1D)}(z - z', \rho)\psi(z') \\ &\quad - \frac{i}{4\pi\varepsilon_0\omega}\frac{\partial}{\partial z}\int dz'\, G^{(1D)}(z - z', \rho)\frac{d}{dz'}\psi(z') \end{aligned}$$

can be calculated for any Green function $G^{(1D)}(z - z', \rho)$. Because $G^{(1D)}$ depends on the difference $z - z'$, the derivative with respect to z can be replaced by the derivative with respect to z' (and a minus sign). Partial integration then moves this derivative to the other factor (with an additional minus sign, cancelling the previous one). Using $d^2\psi(z')/dz'^2 = -k^2\psi(z') \pm 2ike^{\pm ikz'}\left(\delta(z - z_1) - \delta(z_2 - z)\right) + e^{\pm ikz'}\left(\delta'(z - z_1) + \delta'(z_2 - z)\right)$ it is found that the first term cancels the $-i\omega A_z$ contribution, and only the boundary terms at z_1 and z_2 remain, giving

$$E_z = \frac{i}{4\pi\varepsilon_0\omega}\left[\pm ikG^{(1D)}(z - z', \rho)e^{\pm ikz'} - \frac{\partial G^{(1D)}(z - z', \rho)}{\partial z'}e^{\pm ikz'}\right]_{z_1}^{z_2}. \tag{33}$$

The ρ component of the electric field is

$$\begin{aligned} E_\rho &= -\frac{i}{4\pi\varepsilon_0\omega}\int dz'\frac{\partial}{\partial\rho}G^{(1D)}(z - z', \rho)\frac{d}{dz'}\psi(z') \\ &= \frac{i}{4\pi\varepsilon_0\omega}\left[\frac{\partial}{\partial\rho}G^{(1D)}(z - z', \rho)e^{\pm ikz'}\right]_{z_1}^{z_2} \\ &\quad \pm \frac{k}{4\pi\varepsilon_0\omega}\int_{z_1}^{z_2} dz'\,\frac{\partial}{\partial\rho}G^{(1D)}(z - z', \rho)e^{\pm ikz'}. \end{aligned} \tag{34}$$

The last term cannot be calculated analytically for an arbitrary Green function $G^{(1D)}$, but after substituting the explicit form Eq. (21) the integral over

z' is doable:

$$\begin{aligned}
&\frac{\pm k}{4\pi\varepsilon_0\omega}\int_{z_1}^{z_2} dz' \frac{\partial}{\partial\rho}\int d\rho'\,\rho' w(\rho')\int d\phi'\,\frac{e^{-ikr'}}{r'}\,e^{\pm ikz'} \\
&= \frac{\mp k}{4\pi\varepsilon_0\omega}\int d\rho'\,\rho' w(\rho')\int d\phi'\,(\rho-\rho'\cos\phi')\int_{z_1}^{z_2} dz'\,\frac{1+ikr'}{r'^3}\,e^{-ikr'\pm ikz'} \\
&= \frac{\mp k}{4\pi\varepsilon_0\omega}\int d\rho'\,\rho' w(\rho')\int d\phi'\,\frac{\rho-\rho'\cos\phi'}{s'^2}\left[\left(\frac{z'-z}{r'}\pm 1\right)e^{-ikr'\pm ikz'}\right]_{z_1}^{z_2}.
\end{aligned} \tag{35}$$

where $s' = \left(\rho^2+\rho'^2-2\rho\rho'\cos\phi'\right)^{1/2}$. For the remaining integrals over ρ' and ϕ' an expansion in the radius a can be made.

Differences between the solvers In NEC the ETW approximation is used, with both the constant basis functions, for which the z' integral is done numerically, and the sinusoidal basis functions, which can be integrated analytically. This would lead to three basis functions per segment, but this number is reduced to one basis function per segment by requiring that each basis function satisfies certain local conservation laws. These are enforced by letting the basis function corresponding to a given segment extend into the neighbouring segments, with the following requirements:

- If the segment is connected to other segments at an endpoint, then the total current into the endpoint is zero.
- If the segment is connected to other segments at an endpoint, the charge density at the junction on the jth of these segments is proportional to $[-\ln(ka_j/2)-\gamma_{\text{Euler}}]^{-1}$ where a_j is the radius of the jth segment. This relation was derived in [10] for a junction of two long wires, based on an approximate solution for a cylindrical wire with slowly varying radius.
- At the other and of each segment that is connected to the segment under consideration, both the current and its derivative (the linear charge density) are zero.
- If the segment is not connected to other segments at one end, the current at that end need not be zero, since it can flow onto the end cap. If it is required that the surface charge density near the edge of the cap be equal to the surface charge density near the end of the wire, it follows that $I = \pm(a/2)\partial I/\partial z$ for thin wires, with the + sign for the beginning of the segment and the − sign for the end point.

BERBER only uses the constant basis functions, but uses the RETW approximation to do the integral analytically. For structures with mostly short segments NEC's additional sinusoidal basis functions offer no advantage over the constant functions, and BERBER's analytic approximation is more accurate.

In both NEC and BERBER the following test functions are chosen

$$\boldsymbol{\Phi}_n(\boldsymbol{r}) = \hat{\boldsymbol{z}}_n \delta^3(\boldsymbol{r} - \boldsymbol{r}_n), \tag{36}$$

where r_n is an arbitrarily chosen point on the surface of the nth segment, e.g., a point in the middle of the segment. This function samples the parallel component of the field at $\boldsymbol{r}_n$. Another choice would be to take constant test functions, but for these the calculation of the matrix elements requires evaluating a double integral. For short segments point sampling gives a good approximation to the second integral.

4 The EMIR Suite of Utilities

Apart from the field solver BERBER, we have written a few other tools to facilitate the modelling. The whole suite of utilities is called EMIR, which might be an acronym for *Electromagnetic Interference and Radiation*, but sounds exotic anyway and fits together nicely with BERBER. The utilities are text-mode applications and work with plain text input and output files, so they should be portable to any system with decent C and FORTRAN compilers.

The user has to create two files, one containing the frequencies, the locations of the boards, wires and ports, the maximal mesh-size, various other material parameters and the points where the electric field is to be calculated, the other containing the values of the voltage sources and impedances in each of the ports as a function of frequency. The suite consists of the following programs:

emirpre reads the main input file, meshes the boards and divides the wires into segments, generates the input for either NEC or BERBER and creates files with the layout and meshed structure which can be visualized with the viewer.

viewer is a 3-d viewer for the layout and meshed structure. It is ideal for visual debugging of the input before the time consuming calculation with NEC or BERBER takes place. The viewer uses the free (LGPLed) 3-D graphics library Mesa [11], which has been ported to many platforms.

emirpost combines the results of NEC or BERBER with the voltage sources to give the electric field in the selected points and the currents through the ports.

em2plot creates plots and graphs for user-selectable quantities from emirpost's output.

Screen shots of the viewer presenting a wire-grid model are shown in Fig. 3. The wire-grid represents the PCBs, cables and metal plates in a compact all-in-one stereo tower without its metal cover. The CM sources for the complex PCBs were determined by measurement. During the post-processing step the sources can be switched on or off individually. This allows for the localization of the dominant source.

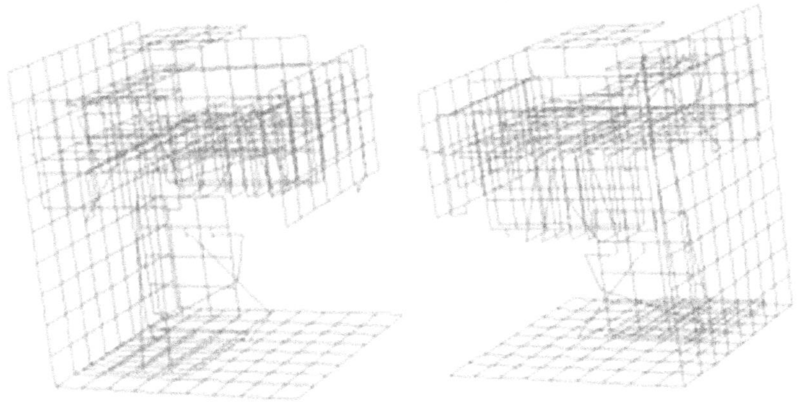

Figure3. Screen shots from the viewer, presenting a wire-grid common-mode skeleton model for a midi-set from two different angles.

EMIR can be used for predicting what will happen to the radiation if the layout is changed in a specific way. In Fig. 4 the maximum electric field at a distance of 3 metres is plotted as a function of frequency for three configurations: the complete midi-set, the midi-set without a metal cover and the midi-set without cover, but with the metal plane on the back. The limit is $100\,\mu\mathrm{V/m}$, so according to the simulations the cover for this particular stereo set helps to keep the radiation level at frequencies around 120 MHz below the limit. Interested readers are referred to [1] for more results and a comparison with experimental data.

5 Conclusion

The common-mode skeleton model is useful for modelling the EMC behaviour of electronic devices. For complex boards the common-mode voltages have to be measured, but in a few simple cases they can be estimated. Using a wire-grid discretization and the method of moments the electric field and currents can be calculated from the electric field integral equation. For structures with many short segments the basis functions can be kept simple and accuracy can be improved by using an analytic approximation to the Green-function integral instead of a numerical evaluation.

6 Calculation of the Common-Mode Voltage

The configuration we consider is the following: a PCB is modelled as a perfectly conducting circular disc with radius R. At a height $h > 0$ above the centre of the disc a point source is placed. The fields from the source induce

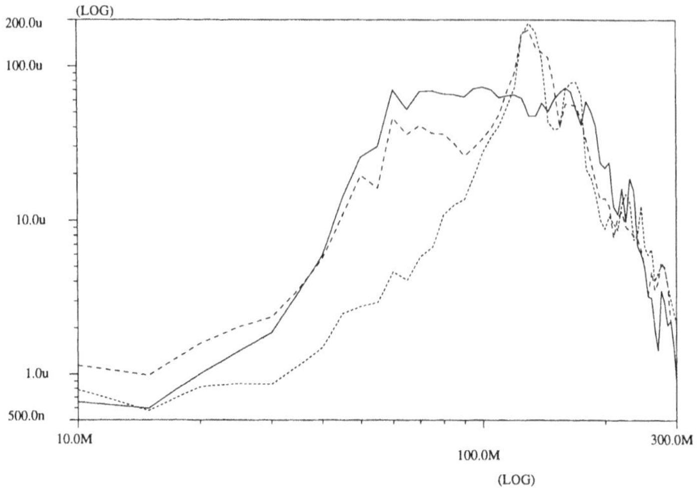

Figure4. The electric field at a distance of 3 metres as a function of frequency for three different configurations: the complete set (solid line), the set without metal cover (dashed line) and the set without cover but with the metal plane on its back (dotted line).

charge and current on the disc, such that the total parallel electric field on the disc is zero. Due to these charges and currents the disc has a non-zero potential. For simple point sources this potential can be calculated analytically in the quasi-static limit.

6.1 Electric Charge

If a point charge Q is placed at height h above the centre of the disk, the total potential in a point $\boldsymbol{r}$ on the surface of the disc is

$$V(\boldsymbol{r}) = \frac{Q}{4\pi\varepsilon_0\sqrt{h^2+r^2}} + \frac{1}{4\pi\varepsilon_0}\int_{\text{disc}} d^2r' \, \frac{\sigma(\boldsymbol{r}')}{|\boldsymbol{r}-\boldsymbol{r}'|}. \tag{37}$$

The potential on the disc must be constant, $V(\boldsymbol{r}) = V$. The integral equation (37) can be solved for the radially symmetric charge distribution $\sigma(r)$ by a method described in [12]. Central in this method is the identity, valid for $|x| < 1$ and $n \in \mathbb{N}$:

$$\int_0^{2\pi} d\phi \, \frac{\cos(n\phi)}{\sqrt{1+x^2-2x\cos\phi}} = \int_0^{2\pi} d\phi \, \frac{x^n \sin^{2n}\phi}{\sqrt{1-x^2\sin^2\phi}}. \tag{38}$$

It can be proved by expanding both sides in x. Using this identity for $n = 0$, the integral in Eq. (37) can be written as

$$\begin{aligned}
\int_{\text{disc}} d^2r' \frac{\sigma(r')}{|\boldsymbol{r} - \boldsymbol{r}'|} &= \int_0^R d\rho\, \rho\sigma(\rho) \int_0^{2\pi} d\phi \frac{1}{\sqrt{r^2 + \rho^2 - 2r\rho\cos\phi}} \\
&= \int_0^r d\rho\, \rho\sigma(\rho) \frac{1}{r} \int_0^{2\pi} d\phi \frac{1}{\sqrt{1 - (\rho\sin\phi/r)^2}} \\
&\quad + \int_r^R d\rho\, \rho\sigma(\rho) \frac{1}{\rho} \int_0^{2\pi} d\phi \frac{1}{\sqrt{1 - (r\sin\phi/\rho)^2}} \\
&= 4\int_0^r d\rho \int_0^\rho dt \frac{\rho\sigma(\rho)}{\sqrt{r^2 - t^2}\sqrt{\rho^2 - t^2}} \\
&\quad + 4\int_r^R d\rho \int_0^r dt \frac{\rho\sigma(\rho)}{\sqrt{r^2 - t^2}\sqrt{\rho^2 - t^2}} \\
&= 4\int_0^r dt \frac{1}{\sqrt{r^2 - t^2}} \int_t^R d\rho \frac{\rho\sigma(\rho)}{\sqrt{\rho^2 - t^2}},
\end{aligned} \tag{39}$$

so the integral equation for the charge density, Eq. (37) becomes

$$\frac{1}{\pi\varepsilon_0} \int_0^r dt \frac{1}{\sqrt{r^2 - t^2}} \int_t^R d\rho \frac{\rho\sigma(\rho)}{\sqrt{\rho^2 - t^2}} = V - \frac{Q}{4\pi\varepsilon_0\sqrt{h^2 + r^2}}. \tag{40}$$

This equation is equivalent to the set of equations

$$\int_0^r dt \frac{f(t)}{\sqrt{r^2 - t^2}} = \pi\varepsilon_0 V - \frac{Q}{4\sqrt{h^2 + r^2}}, \tag{41}$$

$$\int_t^R d\rho \frac{\rho\sigma(\rho)}{\sqrt{\rho^2 - t^2}} = f(t), \tag{42}$$

so it can be solved in a two-stage process: the first equation can be solved by multiplying both sides by $r/\sqrt{s^2 - r^2}$ and integrating from $r = 0$ to s. Then the integrals on the left hand side can be interchanged, resulting in

$$\int_0^s dt\, f(t) \int_t^s dr \frac{r}{\sqrt{s^2 - r^2}\sqrt{r^2 - t^2}} = \frac{\pi}{2} \int_0^s dt\, f(t).$$

Differentiating with respect to s gives $\pi f(s)/2$, so

$$\begin{aligned}
f(s) &= \frac{2}{\pi}\frac{d}{ds} \int_0^s dr \frac{r}{\sqrt{s^2 - r^2}} \left[\pi\varepsilon_0 V - \frac{Q}{4\sqrt{h^2 + r^2}}\right] \\
&= 2\varepsilon_0 V - \frac{Qh}{2\pi(h^2 + s^2)}.
\end{aligned} \tag{43}$$

The second equation can be solved in a way similar to the first: multiply both sides by $t/\sqrt{t^2 - r^2}$ and integrate from $t = r$ to R, then differentiate with respect to r. This gives $\sigma(r)$:

$$\sigma(r) = \frac{4\varepsilon_0 V}{\pi\sqrt{R^2 - r^2}} + \frac{Qh}{\pi^2 r}\frac{d}{dr}\left[\frac{1}{\sqrt{h^2+r^2}}\arctan\frac{\sqrt{R^2-r^2}}{\sqrt{h^2+r^2}}\right]. \tag{44}$$

The charge on the disc

$$Q^{\text{disc}} = 2\pi\int_0^R dr\, r\sigma(r) = 8\varepsilon_0 VR - \frac{2Q}{\pi}\arctan\frac{R}{h} \tag{45}$$

must be equal to $-Q$. This fixes the potential of the disc:

$$V = -\frac{Q}{4\pi\varepsilon_0 R}\arctan\frac{h}{R}. \tag{46}$$

For $h \ll R$ this becomes

$$V = -\frac{p}{4\pi\varepsilon_0 R^2}, \tag{47}$$

where $p = Qh$ is the dipole moment of the charge distribution.

6.2 Horizontal Magnetic Dipole

If the source is a magnetic dipole, it follows from Faraday's law $\nabla \times \boldsymbol{E} = -\partial \boldsymbol{B}/\partial t$ that (for non-zero frequencies) the component perpendicular to the disc of the total magnetic field is zero. It is assumed that frequency is so low that a quasi-static approximation is accurate for both the field from the source and the field from the induced current on the disc, i.e., $\lambda \gg \sqrt{h^2 + R^2}$. Alternatively, the quantities can be viewed as the first term in a series expansion in the frequency. Taking only the lowest order (static) magnetic field into account, the magnetic field from the source $\boldsymbol{m} = m\hat{\boldsymbol{x}}$ on the surface of the disc is simply

$$B_z^S(r,\phi) = -\frac{3\mu_0 m h r\cos\phi}{4\pi(h^2+r^2)^{5/2}}. \tag{48}$$

The current distribution $\boldsymbol{j}(r,\phi)$ on the disc has zero divergence:

$$\frac{1}{r}\frac{\partial}{\partial r}(rj_r) + \frac{1}{r}\frac{\partial}{\partial\phi}j_\phi \tag{49}$$

and its vector potential is

$$\boldsymbol{A} = \frac{\mu_0}{4\pi}\int_{\text{disc}} d^2r'\,\frac{\boldsymbol{j}(\boldsymbol{r}')}{|\boldsymbol{r}-\boldsymbol{r}'|}. \tag{50}$$

The vector potential also has zero divergence

$$\frac{1}{r}\frac{\partial}{\partial r}(rA_r) + \frac{1}{r}\frac{\partial}{\partial\phi}A_\phi = 0 \tag{51}$$

and its curl must satisfy

$$\frac{1}{r}\frac{\partial}{\partial r}(rA_\phi) - \frac{1}{r}\frac{\partial}{\partial \phi}A_r = -B_z^S. \tag{52}$$

Writing

$$A_\phi(r,\phi) = a(r)\cos\phi \tag{53}$$
$$A_r(r,\phi) = b(r)\sin\phi \tag{54}$$

the condition (51) implies that $a(r) = (rb(r))'$, where the prime denotes differentiation with respect to r. Equation (52) becomes

$$(r^3 b'(r))' = \frac{3\mu_0 m h r^3}{4\pi(h^2+r^2)^{5/2}}, \tag{55}$$

which can be integrated twice to give $b(r)$:

$$b(r) = \frac{\mu_0 m}{4\pi}\left(\frac{h}{r^2\sqrt{h^2+r^2}} - \frac{1}{r^2}\right) + b_0, \tag{56}$$

where b_0 is a yet undetermined integration constant. Assuming the current $\boldsymbol{j}$ has the same angular dependence as the vector potential $\boldsymbol{A}$, the angular and radial components of $\boldsymbol{j}$ can be written as

$$j_\phi(r,\phi) = f(r)\cos\phi \tag{57}$$
$$j_r(r,\phi) = g(r)\sin\phi \tag{58}$$

with $f(r) = (rg(r))'$, since the current on the disk must have zero divergence. Then Eq. (50) becomes

$$\begin{aligned}\frac{\mu_0}{4\pi}\int_0^{2\pi} d\phi \int_0^R d\rho\, \frac{\rho(g(\rho)+f(\rho))}{\sqrt{r^2+\rho^2-2r\rho\cos\phi}} &= b(r)+a(r)\\ &= -\frac{\mu_0 m h}{4\pi(h^2+r^2)^{3/2}} + 2b_0, \end{aligned}\tag{59}$$

$$\begin{aligned}\frac{\mu_0}{4\pi}\int_0^{2\pi} d\phi \int_0^R d\rho\, \frac{\rho(g(\rho)-f(\rho))\cos 2\phi}{\sqrt{r^2+\rho^2-2r\rho\cos\phi}} &= b(r)-a(r)\\ = \frac{\mu_0 m}{4\pi}&\left(\frac{h}{(h^2+r^2)^{3/2}} - \frac{2}{r^2} + \frac{2h}{r^2\sqrt{h^2+r^2}}\right). \end{aligned}\tag{60}$$

These two equations can be solved using the same recipe as in the previous section, using the identity Eq. (38) for $n=0$ and $n=2$ respectively, resulting in

$$\begin{aligned}g(r) = {}& \frac{4b_0}{\mu_0\pi\sqrt{R^2-r^2}} - \frac{m}{2\pi^2(h^2+R^2)\sqrt{R^2-r^2}}\\ &- \frac{m}{\pi^2}\left[\frac{\sqrt{R^2-r^2}}{(h^2+R^2)(h^2+r^2)} + \frac{1}{(h^2+r^2)^{3/2}}\arctan\frac{\sqrt{R^2-r^2}}{\sqrt{h^2+r^2}}\right]\end{aligned}\tag{61}$$

$$f(r) = \frac{4b_0}{\mu_0 \pi \sqrt{R^2 - r^2}} - \frac{m}{2\pi^2} \frac{h^2 - 3r^2}{(h^2 + R^2)(h^2 + r^2)\sqrt{R^2 - r^2}}$$
$$- \frac{m}{\pi^2} \left[\frac{(h^2 - 2r^2)\sqrt{R^2 - r^2}}{(h^2 + R^2)(h^2 + r^2)^2} + \frac{h^2 - 2r^2}{(h^2 + r^2)^{5/2}} \arctan \frac{\sqrt{R^2 - r^2}}{\sqrt{h^2 + r^2}} \right]. \quad (62)$$

The condition that the current has zero divergence leads to

$$b_0 = \frac{\mu_0 m}{8\pi(h^2 + R^2)}. \quad (63)$$

This also ensures that the current doesn't leave the disc: $g(R) = 0$.

Now that b_0 is determined, the total vector potential on the disc, including the contribution from the dipole source, is known to zeroth order in the frequency. To first order in the frequency, the total electric field is given by

$$\boldsymbol{E}^{\text{total}} = -i\omega \boldsymbol{A}^{\text{total}} - \nabla V, \quad (64)$$

where V is a potential that arises from the second order term in the expansion in frequency of the current on the disc. The potential follows from the requirement that the parallel electric field on the disc is zero. It is found that

$$V(r, \phi) = V(r) \sin \phi \quad (65)$$

with

$$V(r) = -i\omega r \left(a(r) + \frac{\mu_0 m h}{4\pi(h^2 + r^2)^{3/2}} \right) \quad (66)$$

$$= -i\omega \frac{\mu_0 m}{4\pi} r \left(\frac{1}{2(h^2 + R^2)} + \frac{1}{\sqrt{h^2 + r^2}(h + \sqrt{h^2 + r^2})} \right). \quad (67)$$

In the limit $h \to 0$ this potential is non-zero. On the edge of the disc it becomes

$$V(R, \phi) = -i\omega \frac{3\mu_0 m}{8\pi R} \sin \phi \quad (68)$$

so the CM potential is largest for points perpendicular to the magnetic dipole.

References

1. Bergervoet, J. R., G. P. J. F. M. Maas, M. J. C. M. van Doorn, *The common-mode skeleton model for assessment of electromagnetic compatibility at the system level*, Proceedings of the 12th International Zürich Symposium on EMC (1997).
2. Bergervoet, J. B., Philips Journal of Research 48 (1994), p. 63–87.
3. Burke, G., A. Poggio, *Numerical Electromagnetics Code (NEC)—Method of moments*, Lawrence Livermore National Laboratory (1981).
4. See e.g., *The unofficial Numerical Electromagnetic Code (NEC) Archives* at http://www.qsl.net/wb6tpu/swindex.html.

5. Jackson, J. D., *Classical Electrodynamics, Third Edition*, Wiley (1998).
6. Bergervoet, J. R., R. Rietman, *Combined modelling of ICs, packages and PCBs using analytical equivalent-circuit approximations*, Proceedings of the 13th international Zürich Symposium on EMC (1999).
7. Verbeek, Menno E., *Repairing near-singularity for dense EMC problems by adaptive basis techniques*, to appear in *Numerical Linear Algebra with Applications*, special issue with proceedings of the 1999 Minneapolis conference *Preconditioning Techniques for Large Sparse Matrix Problems in Industrial Applications* (2000).
8. For an overview of iterative solution methods for linear equations see e.g., Saad, Y. and H. van der Vorst, *Iterative Solution of Linear Systems in the 20-th Century*, preprint available at http://www.math.uu.nl/people/vorst/ithistory.tgz, to appear in JCAM (2000).
9. Schelkunoff, A. S. and H. T. Friis, *Antennas; theory and practice* Wiley (1952).
10. Wu, T. T. and R. P. King, *The Tapered Antenna and Its Applications to the Junction Problem for Thin Wires*, IEEE Transactions on Antennas and Propagation, Vol. AP 24 (1976).
11. The homepage of the Mesa project is http://www.mesa3d.org.
12. Kanwal, R. P., *Linear Integral Equations, Theory and Technique, Second Edition*, Birkhäuser (1997).

Computational Electromagnetism in Transformer and Switchgear Design: Current Trends

A. Blaszczyk[1], H. Ketterer[2], and A. Pedersen[3]

[1] ABB Corporate Research, Heidelberg, Germany
[2] ABB Power Trmasformers, Bad Honnef, Germany
[3] ABB Distribution, Skien, Norway

Abstract The paper is focused on efficiency of 3D electromagnetic simulation in design of power transformers and medium voltage switchgear. The dielectric and electromechanical design procedures as well as Intranet based access to simulation resources have been discussed.

1 Introduction

Computation of electromagnetic fields has a long tradition in design of high and medium voltage equipment like transformers and switchgear. In particular 2D analysis became an important part of the industrial design process. Since mid 90s a new design technology based on 3D solid modeling has been widely introduced. Designers develop virtual prototypes using efficient 3D modeling systems. These prototypes are used for creation of technical drawings and in some cases are directly applied in manufacturing. A designer experienced in 3D CAD modeling is able to create a virtual prototype of new components within few hours. Even a complex assembly representing the whole product can be accomplished within few days. Since the models are now precisely defined in a computer the designers naturally ask for simulation.

Fig. 1 shows two typical cases of design process in which simulation has been integrated. The first case is related to design of new products that are manufactured in large quantities like medium voltage switchgear. A characteristic feature is the prototype loop that represents the traditional "cut & try" design procedure: the prototype is redesigned and rebuilt until the physical tests are passed. The goal of simulation is reduction or even elimination of the prototype loop. In extreme cases the building of prototype can be skipped and the manufacturing of so called 0-series started immediately based on virtual prototype.

In Fig. 1b the order related type of design process is shown. Here the final product is developed without physical prototype. Designers follow specific rules and use simulation for evaluation of these rules. The goal of simulation is minimization of the failure probability during the final product tests and

later during operation. This type of design process is characteristic for products with very high costs per unit like large power transformers.
In both types of the design process the 3D-simulation loop is a relatively new component. It can be introduced if its duration corresponds to the whole design cycle time. That means the simulation should give the required results within few hours or days. Tasks on which an external FEM expert has to work for weeks or months cannot be integrated into these types of design process.

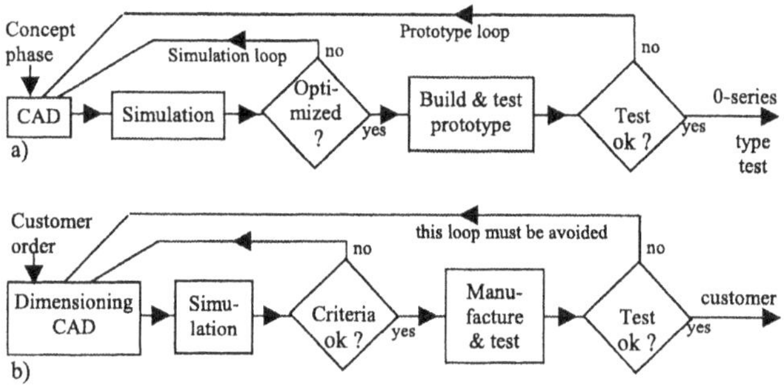

Figure1. Simulation as a component of different types of design process: a) design for quantity production, b) order related design

In the following sections we present results of selected projects aimed at introducing efficient simulation procedures into the design of power transformers and medium voltage switchgear. The main goal of these projects is improvement of efficiency and finally reduction of efforts to the level that allows a permanent use of simulation directly by designers.

2 Dielectric Design

The basic electrostatic field distribution required for dielectric design can be obtained by solving the following equation:

$$div(\epsilon \cdot grad\phi) = 0 \tag{1}$$

where Φ is the scalar electric potential and ϵ is permittivity assumed as a constant everywhere.
We solve this equation using integral approach based on the indirect boundary element method (BEM) [1]. It implies that the field in an arbitrary point can be calculated as a superposition resulting from surface charges located on the boundaries of the model. Based on boundary conditions in form of potentials on electrodes and continuity of the electric displacement on dielectric interfaces we formulate the following equation system:

$$A \cdot \sigma = b \tag{2}$$

where A is the full and unsymmetrical coefficient matrix resulting from the Fredholm integral equations of the first and second kind [1], [2]; σ is the unknown vector of charge density in discretization points on the model surfaces and $\boldsymbol{b}$ is the right hand side vector including known potential on electrodes as well as prescribed surface charge densities on dielectric interfaces (usually equal zero). Once the vector $\boldsymbol{\sigma}$ has been obtained the field can be calculated in the mesh nodes on surfaces and in arbitrary points outside the boundaries. The ABB in-house code Polopt is used for these computations [1].

The crucial advantage of the integral approach is the fact the creation of volume mesh is not required and the discretization of the model can be limited to boundary elements. According to our experience it is extremely difficult to create a volume mesh outside of solid parts for a complex CAD model, in particular for electrostatic computations that require accurate modeling of many details including rounded corners, fillets and wedges.

On the other hand the BEM formulation requires intensive computations since the matrix A in (2) is fully populated and unsymmetrical. For large problems (matrix dimension up to 30000) the sequential computation times may exceed an overnight on a standard processor. We improved the performance by introducing parallel algorithms that are well suited for integral approach [3]. The parallel efficiency on clusters up to 16 machines is larger than 80%. For example the computation time for models presented in Fig. 2 (dimension approximately 15000, 3 symmetry planes) could be reduced from 9 hours on one processor to 70 min on 8 processors (Linux, 500 MHz , 1 Gbyte RAM each). Based on this efficiency and moderate hardware costs the parallel computations on clusters of workstations have been well accepted in the industrial design environment [2,4].

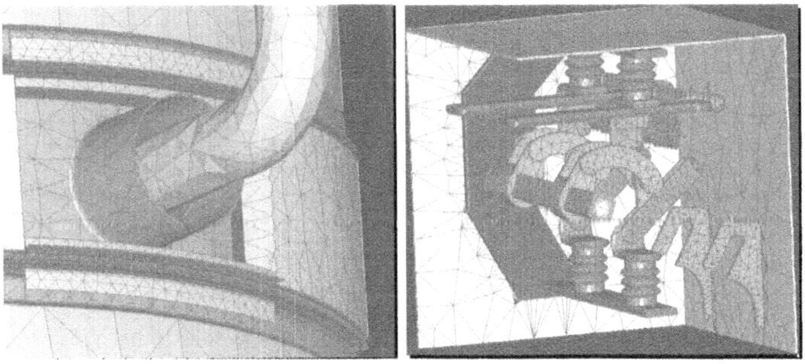

Figure2. Examples of boundary mesh for dielectric models of (a) power transformer output lead through the regulating winding for 585 kV DIL (b) 24 kV SF6-loadbreak of a ring main unit

In Fig. 2 the examples of BEM models for a power transformer and a medium voltage loadbreak are shown. Both models have been created and meshed in Pro/Engineer [5]. The integration of the mesh generator with the CAD-system is very efficient much more than the data transfer via neutral formats like STEP to other standalone meshers. In particular the following features are significant:

- An automatic detection of contact surfaces is available. Consequently, the matching mesh on interfaces between parts (necessary for BEM) can be generated automatically. The manual splitting of contact surfaces is not necessary. In some practical cases, however, the automatic meshing procedure may fail due to remaining difficulties like non-matching lines, small holes, etc. In such cases a geometry cleanup is usually required for traditional mesh generators. In an integrated system the designer looks for a workaround in the solid modeler: he changes solid features or model parameters until the meshing problems are fixed.
- The boundary conditions and user defined mesh control parameters are stored together with the original CAD-geometry. After any geometrical change the whole BEM model can be regenerated automatically. This gives an excellent oportunity for performing automatic shape optimization [6].

The goal of dielectric design computations is prediction of the withstand voltage that is specific to the discharge mechanism in the insulating media. The electrostatic solution is usually not sufficient and designers must perform evaluation of additional criteria that are based on experimental investigations of the discharge process.
For power transformers the dielectric criteria are specified for oil in the internal technical standard Electrical dimensioning of major insulation. Permissible stress" that has been used during last years in 2D analysis. Now we extend its usage to 3D models. This may require some validations and updates of the existing criteria. For example, based on 3D computations a critical volume criterion for cleats, leads and bushing terminal details has been proposed and validated.

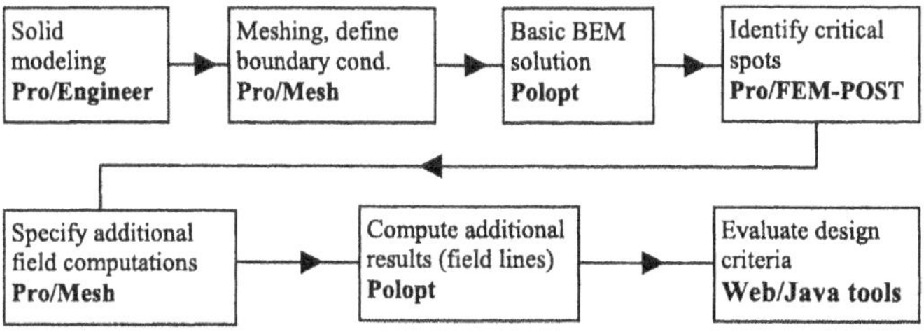

Figure3. Dielectric design simulation procedure

In case of medium voltage switchgear the design criteria have been elaborated together with the 3D-simulation procedure. The new technical standard includes rules for computation of the withstand voltage for sparkovers. It is based on the streamer inception mechanism in gases [7] and probabilistic models for of the impulse breakdown [8]. The proposed rules have been

validated for typical medium voltage switchgear designs with SF_6 and air insulation.
According to both technical standards, the main design criterion for transformers and switchgear is based on field lines. Consequently for both cases we can use the same simulation procedure that is shown in Fig. 3. Only the last step requires tools that are specific for transformers or switchgear. We implement the criteria evaluation using Java technology for easy access via Web.

3 Electromechanical design

The electromechanical design we consider in this paper is aimed at short-circuit rating of medium voltage switchgear components. It requires 2-stage procedure involving computation of electromagnetic forces and structural analysis. The last one uses the forces calculated in the first stage as mechanical loads. Since the displacements of the rigid conductors are very small the direct coupling of electromagnetic and structural solvers is not necessary. For computation of the electromagnetic forces we selected an approach that like in dielectric design avoids the mesh generation outside of solid parts but require volume mesh inside conductors. The mesh is created by the pre-postprocessing system CADfix [9]. The solver consists of the following steps that are performed by the in-house codes TWP and Polopt:

- Calculate distribution of current density $\boldsymbol{J}$ in all conductors using the Finite Element method [10]. We solve the basic equation (1) where the dielectric constant ϵ is replaced by electric conductivity σ. We assume fixed potentials on the current inlets and outlets that correspond to prescribed excitation currents and the Neumann boundary condition on all other conductor surfaces.
- Calculate the magnetic induction from the Biot-Savart law

$$\boldsymbol{B} = \frac{\mu}{4\pi} \int_V \frac{\boldsymbol{J} \times \boldsymbol{r}}{r^3} dV \tag{3}$$

  where dV is volume of an element carrying current density $\boldsymbol{J}$,$\boldsymbol{r}$ is vector directed from the element to the calculation point and μ is magnetic permeability of the outer space $\mu = \mu_0$.
- Make correction of the induction by taking into account the magnetic charge of permeable bodies. This charge can be obtained from the indirect BEM formulation based on scalar magnetic potential [11]. However, this step is required only if the influence of magnetic bodies is significant. It can be usually neglected in case of medium voltage switchgear arrangements.
- Calculate the force density as the following cross product

$$\boldsymbol{F} = \boldsymbol{J} \times \boldsymbol{B} \tag{4}$$

The above procedure does not take into account the influence of eddy currents. According to our evaluations the skin and proximity effects are not deciding for electrodynamic behavior of the medium voltage arrangements encapsulated in thin steel enclosures. The eddy current analysis based on integral approach is also available [12] but is typically associated with much larger computational effort and therefore we avoid to introduce it now into the standard design procedures.
For the mechanical computations we create a completely new model and use a standard structural solver. For our procedure we selected again CADfix as mesh generator and Abaqus [13] as a powerful tool for structural mechanics. In this way the full functionality of an advanced structural solver including plastic, dynamic and contact analyses can be offered to designers. The key point is transfer of forces from the electromagnetic to structural model. We use for this purpose the interpolation facility integrated into CADfix. However, the transfer of forces can be performed also for other preprocessors and solvers provided that the mesh topology data can be accessed. This option is interesting when application of structural solvers integrated into CAD will be considered for electromechanical analysis.
In Fig. 4a an example of electromechanical computation performed for 24 kV/62.5 kA (peak current) switchgear component is shown. The total effort required for such a computation performed according to the procedure shown in Fig. 4b is in the range of 1 man/week. Typically, for the dielectric design procedure shown in Fig. 3 the required effort varies from few hours to 2-3 man/days. In spite of simpler geometries the electromechanical design requires more effort than the dielectric. It has two following reasons: the volume mesh must be created and the mesh generation is not integrated into the CAD system.

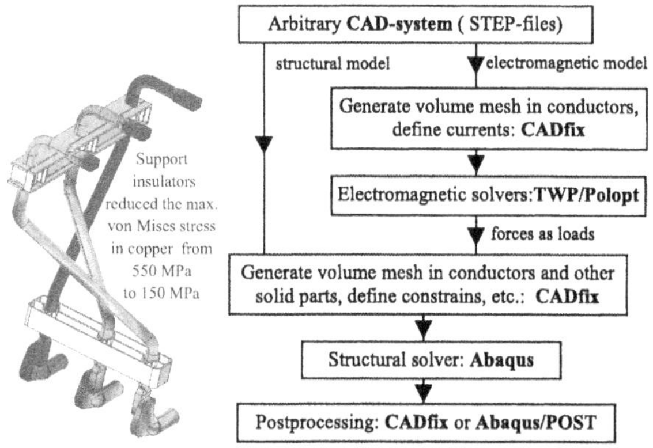

Figure4. a) Example of electromechanical results for a medium voltage switchgear component, b) electromechanical design simulation procedure

4 Intranet Based Access to Simulation Resources

The described simulation procedures require application of many specialized and expensive simulation tools running in nonstandard hardware environment. (Usually designers work with the windows-based graphical CAD-workstations while the simulation requires strong multiprocessor machines) Furthermore, the education of people and keeping the expertise up to date are also significant positions in companies' budget. On the other hand the expensive simulation tools are used in many medium size companies just few times a year. Typically a designer who is responsible for simulation cannot allocate more than 30% of his time for tasks related to computations. Due to personal fluctuations it is necessary to educate new people every few years. The creation and maintenance of the whole infrastructure associated with simulation is for a single company extremely cost inefficient. Consequently, we observe a trend to outsource the simulation resources.
Fig. 5 shows an example of the Intranet-based access to centrally installed and maintained resources that is currently being implemented for power transformer design in ABB.

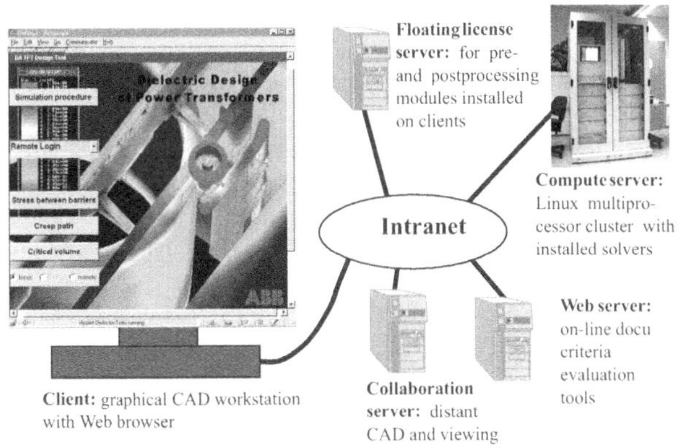

Figure5. Intranet based access to simulation resources

Included is access to: floating software licenses (Pro/Mesh and Pro/FEM-POST), compute servers with installed solvers (operated by Corporate Research Centers), Web servers with updated description of simulation procedures and technical standards for dimensioning criteria implemented as Java applets, interactive hot-line support via collaborative vieweres/modellers [14]. This kind of environment is particularly attractive for large corporations consisting of many business units distributed all over the world.

5 Conclusion

The simplified but efficient simulation procedures as well as the Intranet based access to resources have been successfully introduced in selected ABB companies. Designers can now directly apply the 3D-simulation technology to their own CAD models. It opens a new way for more advanced electromagnetic and coupled simulations and will lead to a new generation of transformer and switchgear designs within next few years.

References

1. Z. Andjelic, B. Krstajic, S. Milojkovic, A. Blaszczyk, H. Steinbigler, M. Wohlmuth: Integral Methods for the Calculation of Electric Fields, Scientific Series of the International Bureau Research Center Juelich, 1992 (ISBN 3-89336-084-0)
2. N. de Kock, M. Mendik, Z. Andjelic, A. Blaszczyk: Application of 3D boundary element method in the design of EHV GIS components, IEEE Magazine on Electrical Insulation, May/June 1998 Vol.14, No. 3, pp. 17-22.
3. A. Blaszczyk, Z. Andjelic, P. Levin, A. Ustundag: Parallel Computation of Electric Fields in a Heterogeneous Workstation Cluster, Lecture Notes on Computer Science Vol. 919, Springer Verlag Berlin Heidelberg 1995, Proc. HPCN Europe 95, pp 606-611
4. A. Blaszczyk, C. Trinitis: Experience with PVM in an Industrial Environment, Lecture Notes on Computer Science Vol. 1156, Proc. EuroPVM96 , Munich, pp. 174-179.
5. Parametric Technology Corporation: Pro/Engineer. See: www.ptc.com
6. C. Trinitis: Field optimisation of three-dimensional insulating systems in high voltage technology (in German). Fortschr.-Ber. VDI R. 21 Nr. 242. VDI Verlag 1998
7. K. Petcharaks: Applicability of the streamer breakdown criterion to inhomogeneous gas gap, Ph. D. Thesis No. 11192, ETH Zurich, 1995
8. H. Boehme: Medium voltage technology. Chapter 7. Insulation behaviour. Verlag Technik, Berlin-Muenchen 1992.
9. FEGS Ltd.: CADfix. See: www.fegs.co.uk
10. G. Sande: Computation of induced currents in three dimensions. The T-? potential formulation used in electrical power engineering", Doctorial dissertation, Norwegian Institute of Technology, 1993, ISBN 82-7119-523-9
11. B. Krstajic, Z. Andjelic, S. Milojkovic, S. Babic and S. Salon: Non-linear 3-D magnetostatic field calculation by the Integral Equation method with surface and volume magnetic charges, IEEE Trans. on Magn., Vol. 28, 1992, 1088-1091.
12. J. Shen, Z. Andjelic, B. Schaub: A hybrid single and dual simple layer boundary integral equation formulation for 3-D eddy currents, IEEE Trans. On Magnetics, Vol. 43, No. 5, 1998.
13. HKS Inc.: Abaqus. See: www.abaqus.com
14. CoCreate (a HP Company): OneSpace. FirstSpace. See: www.cocreate.com

Numerical Algorithms for the Calculation of Magneto-Quasistatic Fields Using the Finite Integration Technique

M. Clemens, S. Drobny, M. Wilke, T. Weiland

Darmstadt University of Technology, Fachbereich Elektrotechnik und Informationstechnik, Fachgebiet Theorie Elektromagnetischer Felder, Schlossgartenstrasse 8, D-64289 Darmstadt, Germany.

Abstract In this paper numerical algorithms for transient eddy current calculations based on the Finite Integration Technique are presented. The occuring differential-algebraic systems of equations, whose initial degeneration requires to deal with the problem of a suitable regularization, allows to successfully include nonlinear ferromagnetic material behavior using various linearization techniques as well as eddy current effects due to moving conductors. Techniques for error controlled adaptive time stepping and first results of a new algebraic multigrid solver for the linear systems of equations of the implicit time integration process are presented.

1 Introduction

The numerical simulation of transient magneto-quasistatic fields requires careful modelling due to the degenerate character of Maxwell's equations in magneto-quasistatic situations. These equations results in parabolic problem formulations, which are especially demanding to solve and require sophisticated algorithms from the numerical linear algebra. The following text will give a survey on the latest research efforts to tackle some main problems arising for these formulations, such as their regularization, the algorithmic treatment of additional physcial effects such as nonlinear material behavior or conductor motion and numerical schemes to improve the solution process such as optimal complexity solver for the linear systems and the usage of error control mechanisms for the implicit time stepping processes.

2 FDiTD Formulations

Based on the Finite Integration Technique (FIT), a consistent discretization scheme of Maxwell's equations in integral form onto a dual grid doublet $\{G, \widetilde{G}\}$ [15], [16], several time domain formulations for transient magneto-quasistatic problems have been presented in [6] and [5]. These so-called Finite-Difference-Implicit-Time-Domain (FDiTD) formulations typically start from the initial value problem

$$\widetilde{\mathbf{C}}\mathbf{M}_{\nu}\mathbf{C}\widehat{\mathrm{a}}(t) + \mathbf{M}_{\kappa}\frac{d}{dt}\widehat{\mathrm{a}}(t) = \widehat{\widehat{\mathbf{j}}}_{e}(t), \quad \widehat{\mathrm{a}}_0 = \widehat{\mathrm{a}}(t_0), \tag{1}$$

where $\widehat{\mathbf{a}}$ is a primitive of the electric grid voltage vector $\widehat{\mathbf{e}}$ with $\widehat{\widehat{\mathbf{b}}} = \mathbf{C}\widehat{\mathbf{a}}$ and $\widehat{\widehat{\mathbf{j}}}_e$ is the vector of excitation currents. The matrices $\mathbf{C}$ and $\widetilde{\mathbf{C}}$ represent the primal and dual curl matrices, containing only incidence information of $\{G, \widetilde{G}\}$. The material matrices $\mathbf{M}_\kappa$ and $\mathbf{M}_\nu$ correspond to the material conductivities and reluctivies of the problem, respectively. They contain all the metric and material information and thus the approximations made in the method. The differential-algebraic system of equations of Index 1 (1) is solved using unconditionally stable implicit time integration schemes [5].

The main problem of equation (1) arises from the singularity of the curlcurl-matrix, that will prohibit the uniqueness of the solutions, since in this non-gauged formulation irrotational solution parts are not considered in the non-conductive regions. One way to tackle this problem basically consists in ignoring it, while using standard preconditioned conjugate gradient methods for the solution of the linear algebraic systems with their weak gauging property, i.e., their ability to deal with consistently singular problems [5]. Another approach arises from a regularization of the full curlcurl-equation of a frequency domain formulation with an additional grad-div-operator explicitely enforcing the continuity equation. This additional operator will shift the zero eigenvalues of the curlcurl-operator into the right half of the Gaussian number plane, thus regularizing the formulation for computational domains where the boundaries of electrically conducting materials are singly connected [11]. This additional grad-div-operator reduces in the magneto-quasistatic limit to simultaneously enforcing the condition $\widetilde{\mathbf{S}}\mathbf{M}_\kappa\widehat{\mathbf{e}} = 0$ as well as a Coulomb gauge $\widetilde{\mathbf{S}}\mathbf{D}_{\tilde{A}}\mathbf{D}_{\tilde{S}}^{-1}\widehat{\mathbf{e}} = 0$ in the non-conductive parts. Adopting this approach to (1) results in the regular equation

$$\left[\widetilde{\mathbf{C}}\mathbf{M}_\nu\mathbf{C} + \mathbf{M}_1\widetilde{\mathbf{S}}^T\mathbf{M}_2\mathbf{S}\mathbf{M}_1\right]\widehat{\mathbf{a}}(t) + \mathbf{M}_\kappa\frac{d}{dt}\widehat{\mathbf{a}}(t) = \widehat{\widehat{\mathbf{j}}}_e(t), \quad \widehat{\mathbf{a}}_0 = \widehat{\mathbf{a}}(t_0), \quad (2)$$

where the matrices in the discrete grad-div operator are defined as

$$[\mathbf{M}_1]_{ii} = \begin{cases} [\mathbf{M}_\kappa]_{ii} & : \ [\mathbf{M}_\kappa]_{ii} \neq 0, \\ \left[\mathbf{D}_{\tilde{A}}\mathbf{D}_{\tilde{S}}^{-1}\right]_{ii} & : \ [\mathbf{M}_\kappa]_{ii} = 0, \end{cases} \quad (3)$$

$$\mathbf{M}_2 = \mathbf{D}_{\tilde{V}}^{-1}\mathbf{D}_{<\nu>}\mathbf{D}^2_{<\kappa,1>}, \quad \text{with} \quad (4)$$

$$[\mathbf{D}_{<\kappa,1>}]_{ii} = \begin{cases} [\mathbf{D}_{<\kappa>}]_{ii} & : \ \text{Node with index i in conductor,} \\ 1 & : \ \text{Node with index i in non-conductor,} \end{cases} \quad (5)$$

Note, that the matrices $\mathbf{D}_{<\nu>}$ and $\mathbf{D}_{<\kappa>}$ contain averaged material information of the dual cells $\tilde{V} \in \tilde{G}$.

Another method of regularization of (1) is the introduction of a small artificial conductivity in the non-conductive regions, such that (1) forms a system of ordinary differential equations. With this approach the zero eigenvalues in the nullspace of the curlcurl matrix will be shifted onto the positive real axis. Unfortunately, this system will become very stiff in the presence of metallic conductors: The effective condition number of the time discrete system matrix, i.e., the quotient of the maximum and the minimal non-zero eigenvalue, is typically increased and hence the convergence rate of CG-type solvers will deteriorate. To improve this situation, that also may naturally arise from geophysical modelling, again a grad-div-augmentation was presented [2], which forms a special case of the grad-div operator augmentation in (2).

Table 2 shows typical effects of these regularization for a small eddy current test problem (11.475 dof), for which 180 time steps where performed and a transient simulation of the TEAM 7 problem (139.482 dof) with 30 time steps using a ramped sinusoidal current excitation at 50 Hz [6]: The non-gauged configuration exhibits the fastest convergence, whereas the formulation (2) requires more time due to the more expensive operator-type matrix-vector-multiplications (cf. [6]). An artificial regularization with $\kappa'_{air'} = 1$ S/m significantly increases the effective condition number and the PCG-solver needs more time to reach the prescribed accuracy. Beneficial spectral effects of the grad-div-augmentation as predicted in [2] are largely compensated with the the more expensive MxV-operations.

Table1. Effects of Grad-Div-regularization on the PCG-convergence

	Equation (1) non-gauged	Equation (2) grad-div augmentation
Small copper plate, 180 timesteps		
$\kappa_{air} = 0$ S/m	136 s	161 s
$\kappa'_{air'} = 1$ S/m (artificial)	306 s	294 s
Team 7 problem (30 timesteps)		
$\kappa_{air} = 0$ S/m	2.572 s	2.987 s
$\kappa'_{air'} = 1$ S/m (artificial)	14.602 s	15.898 s

2.1 Error-Controlled Adaptive Time Stepping

Other than the explicit leapfrog time integration scheme, typically referred to as the FDTD method, where the time step length typically is derived from the Courant-Friedrich-Levy stability criterion, the implicit time integration schemes used within the FDiTD formulations are unconditionally stable [5]. Here, the truncation errors of the time integration methods provide the determining factors to the time step lengths. Remarkably enough, only few recent publications so far seem to have tackled this issue for transient eddy currents (see e.g. [3]), whereas in mathematical literature adaptive time stepping schemes for implicit systems are widely covered (cf. [10], [12], [13]). For the error-controlled adaptive stepsize integration of (1) a stiffly accurate, L-stable, third order Rosenbrock-3(2)-method (ROW3(2)) with an embedded time integration scheme of second order [13] is used. For the prediction of the time steps so-called I-controller and more sophisticated PI-controller schemes are used [10]. Results of an eddy current test example with a ramped sinusoidal current excitation for $f = 50\,\mathrm{Hz}$ depicted in Table 2.1 show a significant influence of the user defined absolute tolerance parameter on the overall number of accepted and rejected steps for a transient eddy current test problem configuration excited by a ramped sinusoidal current of 50 Hz. for three periods. Although the number of rejected time steps of the I-controller scheme shown in Table 2.1 is larger than that of the PI-controller, the total number of executed time steps is smaller. These first results show that additional developments for an automatic *atol* parameter detection is required for robust, yet econonic time integration schemes.

Table2. Results of PI- and I-controller strategies for an eddy current (number of accepted/rejected (total) steps):

*rtol**atol*	1e-4	1e-5	1e-6
1e-2	I : 16/6 (22) PI : 10/1 (11)	I : 73/106 (179) PI : 153/38 (191)	I : 672/1143 (1815) PI : 1623/406 (2029)
1e-3	I : 64/59 (123) PI : 113/33 (146)	I : 814/1450 (2265) PI : 1965/531 (2265)	I : $> 10^4 / > 10^4$ (–) PI : $> 10^4 / > 10^4$ (–)

2.2 Algorithms for Nonlinear Magnetic Materials

Industrial magneto-quasistatic eddy current problems typically feature ferromagnetic materials involving a nonlinear constitutive material relation. This renders (1) a nonlinear formulation

$$\widetilde{\mathbf{C}}\mathbf{M}_{\nu}(\widehat{\mathbf{a}})\mathbf{C}\widehat{\mathbf{a}} + \mathbf{M}_{\kappa}\frac{d}{dt}\widehat{\mathbf{a}} = \widehat{\widehat{\mathbf{j}}}_{e}, \tag{6}$$

which typically requires the repeated solution of a nonlinear system at each time step within the time discrete formulations.

In case of ferromagnetic materials with a nonlinear magnetization curve the reluctivity matrix $\mathbf{D}_{\nu}$ depends on the unknown vector potential $\widehat{\mathbf{a}}$ resulting in the nonlinearity of the formulation (1). The relation between material values and field intensities is normally given by discrete characteristics, and thus cannot be incorporated into system (1) by simple operators.

A stable but in some cases slow method is the successive approximation technique [4], which solves the nonlinear algebraic equation by splitting it up in a sequence of linear problems for which the material distribution of the previous iteration step is used. In order to prevent convergence problems in saturated regions and to reduce the number of nonlinear cycles per time step a dynamic relaxation process is introduced which takes into account the nonlinear material behaviour. Convergence problems can also be avoided by considering the form of the magnetization curve and choosing an appropriate interpolation scheme for the magnetization curve [8]. For this the choice of the employed scheme for computing the new reluctivity has an authoritative influence on the convergence of the iterative algorithm due to the form of the magnetization curve of the material.

The Newton–Raphson method typically offers a faster convergence behaviour than the Successive Approximation technique by solving the nonlinear equations arising from the formulation (1) using a Taylor–expansion. The iteration consists of an incremental update $\widehat{\mathbf{a}}^{(i+1)} = \widehat{\mathbf{a}}^{(i)} + \alpha \Delta \widehat{\mathbf{a}}^{(i)}$. The solution increment $\Delta \widehat{\mathbf{a}}^{(i)}$ results from solving the linear system of equations

$$\left[\mathbf{F} + \frac{\partial \mathbf{F}}{\partial \widehat{\mathbf{a}}} \widehat{\mathbf{a}}\right]^{(i)} \Delta \widehat{\mathbf{a}}^{(i)} = \widehat{\widehat{\mathbf{j}}} - \mathbf{F}^{(i)} \widehat{\mathbf{a}}^{(i)} \tag{7}$$

with the relaxation factor $\alpha \in (0, 1]$. The Jacobian matrix can be written as

$$\left[\mathbf{F} + \frac{\partial \mathbf{F}}{\partial \widehat{\mathbf{a}}} \widehat{\mathbf{a}}\right] = \widetilde{\mathbf{C}} \left[\mathbf{M}_{\nu}(\widehat{\mathbf{a}}) + \mathbf{M}_{\Delta\nu}(\widehat{\mathbf{a}})\right] \mathbf{C} = \widetilde{\mathbf{C}} \mathbf{M}_{\nu_d}(\widehat{\mathbf{a}}) \mathbf{C}, \tag{8}$$

with the differential reluctivity ν_d given by $\Delta \nu = 2B^2 \frac{\partial \nu}{\partial B^2} = \nu_d - \nu$.

Thus the modelling of nonlinearity of the material is restricted to the material matrices. Note, that again the FI Technique allows to separate the mesh topology from the material relations in the nonlinear case. The calculation of the numerical derivatives required for the Jacobian matrix is performed using a $\nu(B^2)$–curve for which a Lagrange interpolation is applied.

An underrelaxation factor is introduced in order to either achieve convergence or to improve the speed of convergence.This factor is computed for each nonlinear iteration by trying values $\alpha = 2^{-n}$ until the residual is reduced when compared with the step before. The nonlinear iteration is stopped when the condition $\frac{||\Delta \widehat{\mathbf{a}}||}{||\widehat{\mathbf{a}}||} \leq \varepsilon$ holds.

As an industrial example a nonlinear transient calculation of a thin film read write magnetic head benchmark problem of the Storage Research Consortium (SRC) of Japan is presented. A full description of the problem, FEM solutions and measurements are given in [9]. Fig. 1(a) shows the computer model of the magnetic head design, which was fully discretized in 3D amounting to 750.000 degrees of freedom. Fig. 1(b) shows a very good agreement between the measured values for the vertical magnetic field $B_z(t)$ at $(x, y, z) = (0.0, -0.07, 0.2)\,\mu m$, here calculated with the successive approximation scheme.

2.3 Solution of the Linear Systems of Equations

With the number of degrees of freedom in FDiTD simulations ranging between $\mathcal{O}(10^3) - \mathcal{O}(10^7)$, typically the most time in transient eddy current simulations is spent with the repeated solution of the linear algbraic systems of equations. In [6] the SSOR-CG method was proposed as solution

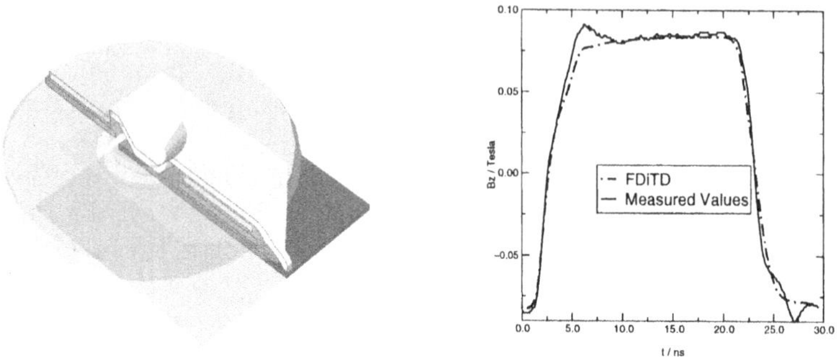

(a) (b)

Figure1. Fig. (a) shows an industrial design of an read write magnetic head [6]. Fig. (b) depicts the results of a nonlinear transient simulation taken for the vertical component of the magnetic flux density B_z in fromt of the tip of the magnetic head.

method, but it suffers from its non-optimal asymptotical complexity for large-scale problems and it does not very well benefit from good start solutions. Asymptotically optimal multilevel methods, however, require careful consideration of the nullspace of the curlcurl-operator. Recently the algebraic multigrid (AMG) package PEBBLES [14] was enhanced for $H(curl)-$conformal magneto-quasistatic curlcurl formulations. It uses an Falk-Arnold-Winter-smoother [1] for the solution parts in the nullspace of the curlcurl operator. First test results depicted in Fig. 2 for one solution step within a transient simulation of a c-shaped magnet configuration shows the asymptotical complexity behavior of $(O(n^j), j \approx 1)$ of the PEBBLES AMG in comparison with the complexities for SSOR-CG ($j \approx 1.3$) and a sparse direct solver ($j \approx 2.8$).

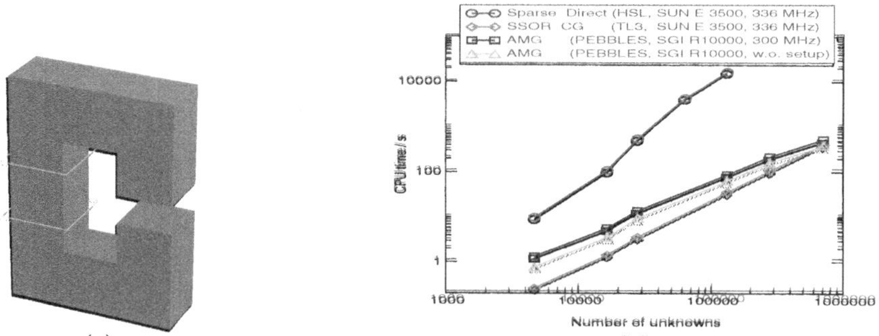

(a) (b)

Figure2. Fig. (a) shows the shape of the c-magnet, which was discretized with different grid resolutions and corresponding cpu-times shown in Fig. (b)

2.4 Modelling of Conductor Motion

The calculation of eddy currents introduced by the unidirectional motion of conducting media inside a magnetic field (as e.g. in the design of linear eddy current railway brakes) requires an extension of the FDiTD formulation (1) using a moving-coordinate system formulation

$$\frac{d}{dt}\widehat{\mathbf{a}}(t^{n+1}) \approx \frac{\widehat{\mathbf{a}}(t^{n+1}) - \mathbf{O}_{IP}\widehat{\mathbf{a}}(t^n)}{\Delta t}, \tag{9}$$

where the operator $\mathbf{O}_{IP}$ projects the path integrals of the modified vector potential inside the moving conductor to an upwind position onto the grid, which has moved during the timestep Δt by a distance $\boldsymbol{v} \cdot \Delta t$. The moving-coordinate formulation is suited for stable simulations even with high conductor velocities $\boldsymbol{v}$ [7].

The application of a Galilean transformation ($t \to t', \boldsymbol{r} \to \boldsymbol{r}'$) for non-relativistic velocities $\boldsymbol{v}$ yields a fixed-coordinate system formulation, where the reference frame is tied to the non-moving parts of the system corresponding to an Eulerian coordinate description. The resulting formulation using the matrix notation of the Finite Integration method is given by

$$\widetilde{\mathbf{C}}\mathbf{M}_\nu\mathbf{C}\widehat{\mathbf{a}}' + \mathbf{M}_\sigma\frac{d}{dt'}\widehat{\mathbf{a}}' = \widehat{\widehat{\mathbf{j}}}'_s + \mathbf{M}_\sigma\mathbf{O}_{\boldsymbol{v}\times B}\mathbf{C}\widehat{\mathbf{a}}' = \widehat{\widehat{\mathbf{j}}}'_s + \widehat{\widehat{\mathbf{j}}}_{\mathrm{mov}} \tag{10}$$

with a discretization of $\boldsymbol{v} \times \boldsymbol{B}$ in the operator $\mathbf{O}_{\mathbf{v}\times B} := \mathbf{D}_s\mathbf{X}(\boldsymbol{v})\mathbf{O}_{\mathbf{AE}}$. Here $\mathbf{O}_{AE}$ is a grid operator interpolating facet-flux vector components $\widehat{\widehat{\mathbf{b}}}'_i = (\mathbf{C}\widehat{\mathbf{a}}')_i$ to averaged magnetic flux densities $\mathbf{b}_i$. The matrix $\mathbf{X}(\boldsymbol{v})$ evaluates the expressions $\mathbf{e}_i := (\boldsymbol{v} \times \mathbf{b}_i)$ and thus yields averaged electric field intensities. The matrix $\mathbf{X}(\boldsymbol{v})$ depends only on $\boldsymbol{v}$ and the grid topology of G, is skew-symmetric and has purely imaginary conjugate complex pairs of eigenvalues alone. Left application of the diagonal matrix $\mathbf{D}_s$ containing the lengths of the cell edges of G results in a vector of grid voltages $\widehat{\mathbf{e}}$, which is then multiplied with the material matrix of conductivities $\mathbf{M}_\sigma$ to obtain the grid vector of the motion induced currents $\widehat{\widehat{\mathbf{j}}}_{\mathrm{mov}}$.

Within the implicit time integration schemes BDF1 or BDF2 [12], the convection current term $\mathbf{M}_\sigma\mathbf{O}_{\boldsymbol{v}\times B}\mathbf{C}$ adds a non-symmetric part to the system matrix systems. This convection term scales with the velocity v and is known to cause numerical instabilities in fixed-coordinate formulations if it becomes dominant with respect to the symmetric part of the system matrix. Typically this case arises for Peclet numbers $P = \mu\sigma v l/2$ larger than one, where l is the maximum grid edge length in the direction of motion.

3 Conclusion

The paper presented formulations and algorithms for the numerical simulation of transient magneto-quasistatic fields using the FI technique with extensions to also cover nonlinear material behavior as well as eddy current effects arising from moving conductors. The presented results cover efforts to overcome the degenerate character of the formulation by regularization techniques and its implicit nature by proposing the usage of optimal complexity methods for the solution of the algebraic systems of equations as well as error controlled variable stepsize time integration schemes. For the latter methods further research for the suitable choice of parameters is required to achieve robust, yet economic schemes.

References

1. Arnold, D., Falk, R., Winther, R.: Multigrid in $h(div)$ and in $h(curl)$ (1999). Numer. Math., to appear.
2. Bossavit, A.: 'stiff' problems in eddy-current theory and the regularization of Maxell's equations. In Conference Records of the CEFC 2000, Milwaukee (1997) 497. Full paper submitted to IEEE Transactions on Magnetics.
3. Cameron, F., Piché, R., Forsman, K.: Variable step size time integration methods for transient eddy current problems. IEEE Transactions on Magnetics **34** (1998) 3319–3323
4. Clemens, M., Drobny, S., Weiland, T.: Time integration of slowly-varying electromagnetic field problems using the finite integration technique. In Proceedings of the ENUMATH 97, Heidelberg (1999) 246–253
5. Clemens, M., Weiland, T.: Numerical algorithms for the FDiTD and FDFD simulation of slowly-varying electromagnetic fields. Int. J. Num. Mod. **12** (1999) 3–22
6. Transient eddy current calculation with the FI-method. IEEE Transactions on Magnetics **35** (1999) 1163–1166
7. Clemens, M., Weiland, T., Wilke, M.: Transient eddy current formulation including moving conductors using the finite integration method. In Honma, T., (ed.) Proceedings of the CEFC'99, Sapporo, Japan volume 2 (1999) 592–593
8. Drobny, S., Weiland, T.: Iterative algorithms for nonlinear transient electromagnetic field calculation (1999). Proceedings of the ISEM 99, Pavia, Italy.
9. Fujiwara, K., et al.: Thin film write head field analysis using a benchmark problem. In Proc. Compumag99 Conference, Sapporo (1999) 744–745
10. Gustafsson, K.: Control-theoretic techniques for stepsize selection in implicit runge-kutta methods. ACM Transactions on Mathematical Software **20** (1994) 496–517
11. Hahne, P.: Zur Numerischen Berechnung Zeitharmonischer Elektromagnetischer Felder. Ph.D. thesis Technische Hochschule Darmstadt (1992)
12. Hairer, E., Wanner, G.: Solving Ordinary Differential Equations II, Stiff and Differential-Algebraic Problems. Springer-Verlag, Wien, New York (1996)
13. Lang, J.: Two-dimensional fully adaptive solutions of reaction diffusion equations. Applied Numerical Mathematics **18** (1995) 223–240
14. Reitzinger, S., Schöberl, J.: Algebraic multigrid for edge elements (2000). Preprint
15. Weiland, T.: A discretization method for the solution of Maxwell's equations for six-component fields. Electronics and Communications AEÜ **31** (1977) 116–120
16. Time domain electromagnetic field computation with finite difference methods. Int. J. Num. Mod. **9** (1996) 259–319

Parallel Maxwell Solvers *

G. Haase, M. Kuhn, U. Langer, S. Reitzinger, J. Schöberl

SFB F013 "Numerical and Symbolic Scientific Computing", Johannes Kepler University Linz, A-4040 Linz, Austria

Abstract The numerical simulation of technical 3D magnetic field problems requires fast solvers for the resulting systems of equations. Geometric or algebraic multigrid methods can be used depending on whether a mesh hierarchy is available or not. The paper presents the main ideas of the sequential methods and describes the necessary adaptations for the parallelization. Performance results for both, sequential and parallel solvers underline the high efficiency of the methods.

1 Introduction

The computer simulation of 3D technical magnetic field problems leads easily to systems of equations with several hundreds of thousands or even millions of unknowns. Using finite element (FE) discretizations, the most efficient solvers are certainly multigrid or multilevel methods. Typically, the multigrid convergence rate is independent of the mesh size parameter, and the arithmetical complexity grows linearly with the number of unknowns. However, the standard multigrid algorithms fail for the Maxwell finite element equations in the sense that the convergence rate deteriorates as the mesh size decreases. To overcome this drawback, R. Hiptmair proposed to modify the smoothing iteration by adding a smoothing step in the discrete potential space [1]. Similarly, D. Arnold, R. Falk and R. Winther suggested a special block smoother that has the same effect [2]. However, the efficiency of geometric multigrid methods relies essentially on explicit hierarchy information, i.e., on a sequence of hierarchical grids. The basic idea is to reduce high and low frequency components of the error by smoothing and coarse grid correction, respectively. If such a sequence of grids is not available, geometric multigrid methods (GMG) cannot be applied anymore. Instead, algebraic multigrid methods (AMG) [3] can be used. These methods generate appropriate coarse grid problems from single-grid information only. In simple cases, the AMG algorithm may reproduce the GMG algorithm, i.e., it is possible to choose the components of AMG such that the efficiency of GMG is preserved. Then the parallelization of these, or, more precisely, of appropriately modified multigrid solvers [4] is certainly the only feasible way to enhance the efficiency of these algorithms.

In this paper we are interested in the fast solution of the magneto-static problem $\mathrm{curl}(H) = J, \quad H = \nu B, \quad \mathrm{div}(B) = 0$, where H and B denote

* This research has been supported by the Austrian Science Fund FWF within the SFB "Numerical and Symbolic Computing" under the grant SFB F013.

the magnetic field intensity and the magnetic flux density, respectively. The permeability μ ($\nu = 1/\mu$) and the impressed current density J ($\operatorname{div}(J) = 0$) are given. Furthermore, we assume a bounded domain Ω such that $J = 0$ outside Ω. We impose boundary conditions $B \cdot n = 0$ on $\partial\Omega$, where n is the normal vector on $\partial\Omega$. Introducing the vector potential $B = \operatorname{curl}(u)$ we obtain the variational formulation for the magneto-static problem in Ω and the boundary conditions on $\partial\Omega$: Find $u \in H_0(\operatorname{curl}, \Omega)$ such that

$$\int_\Omega \nu \operatorname{curl}(u) \cdot \operatorname{curl}(v)\, dx + \int_\Omega \sigma\, u \cdot v\, dx = \int_\Omega J \cdot v\, dx \quad \forall v \in \mathbb{V}, \tag{1}$$

where $\mathbb{V} := H_0(\operatorname{curl}, \Omega) := \{v \in L^2(\Omega)^3 |\ \operatorname{curl}(v) \in L^2(\Omega)^3,\ v \times n = 0 \text{ on } \partial\Omega\}$. The small positive artificial conductivity $\sigma \ll 1$ has been inserted in (1) in order to ensure uniqueness of u. In the case of eddy current problems it is required in non-conductive regions only. Alternatively, we can derive a regularized mixed variational formulation based on the exact Coulomb gauge. Eliminating the corresponding Lagrange multiplier, we again arrive at some primal variational formulation that is similar to (1), see [5]. Now, let $\mathbb{V}_h \subset \mathbb{V}$ be the lowest order edge element space (see [6]) on a shape-regular triangulation of Ω with mesh width h. Then the finite element approximation to the variational problem (1) in $\mathbb{V}_h$ is equivalent to the system of finite element equations

$$\mathsf{K}_h \underline{\mathsf{u}}_h = \underline{\mathsf{f}}_h. \tag{2}$$

In Section 2 we discuss the general ideas of GMG and AMG for solving (2) and emphasize the peculiarities of magnetic field problems. The main ideas of the parallelization of the algorithms are presented in Section 3. In Section 4, we discuss numerical results. Conclusions are drawn in Section 5.

2 Geometric and Algebraic Multigrid Solvers

If geometric multigrid solvers are to be applied, a hierarchy of meshes is required. Starting with a coarse mesh, a sequence of meshes is usually obtained by uniform or adaptive refinement. In this paper we restrict the considerations to the case of uniform refinement. Fig. 1 shows the coarsest mesh modelling the geometry of a transformer and a mesh resulting after 2 uniform refinement steps. For adaptive methods and results we refer to [7,5]. Based on a hierarchy, a sequence of problems (matrices) together with intergrid-transfer operators can be constructed in a canonical way. The main idea is to reduce high frequency components of the error on the fine grid by smoothing and reduce the remaining components by a coarse grid correction. This idea can be applied recursively resulting in Algorithm 1. Here K_h, K_H denote the fine and coarse grid system matrix, respectively. The intergrid transfer operators are called prolongation ($\mathfrak{P}$) and restriction ($\mathfrak{P}^T$). As mentioned before, a special smoother has to be chosen for the considered problem class. On the

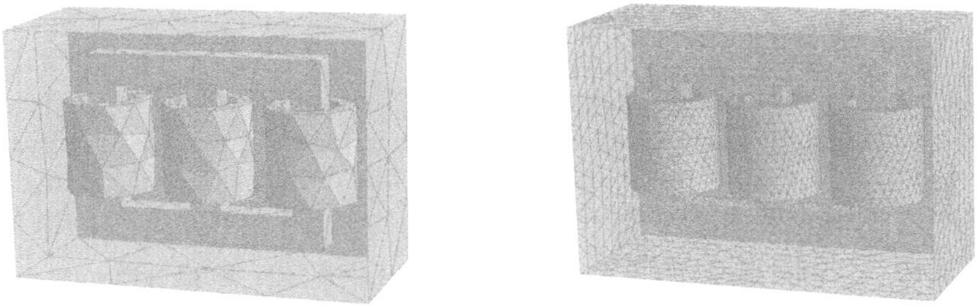

Figure1. Transformer: coarse mesh and refined mesh adapted to the geometry.

Algorithm 1 (Parallel) Multigrid $\mathrm{MG}(\mathsf{K}_h, \underline{u}, \underline{f})$

if CoarseGrid **then**	
$\quad \underline{u} \Leftarrow \mathrm{DirectSolve}(\mathsf{K} \cdot \underline{u} = \underline{f})$	coarsest grid, direct solver
else	
$\quad \widetilde{\underline{u}} \Leftarrow \mathrm{Smooth}(\mathsf{K}_h, \underline{u}, \underline{f})$	pre-smoothing
$\quad \underline{d} \leftarrow \underline{f} - \mathsf{K}_h \cdot \widetilde{\underline{u}}$	compute the defect
$\quad \underline{d}^H \leftarrow \mathfrak{P}^T \cdot \underline{d}$	restriction to the coarser grid
$\quad \underline{\mathfrak{w}}^H \leftarrow 0$	initial guess for the coarse grid correction
$\quad \underline{\mathfrak{w}}^H \Leftarrow \mathrm{MG}(\mathsf{K}_H, \underline{\mathfrak{w}}^H, \underline{d}^H)$	compute the coarse grid correction
$\quad \underline{\mathfrak{w}} \leftarrow \mathfrak{P} \cdot \underline{\mathfrak{w}}^H$	prolongation to the finer grid
$\quad \widehat{\underline{u}} \leftarrow \widetilde{\underline{u}} + \underline{\mathfrak{w}}$	update the fine grid solution
$\quad \underline{u} \Leftarrow \mathrm{Smooth}^T(\mathsf{K}_h, \widehat{\underline{u}}, \underline{f})$	post-smoothing
end if	

coarsest grid, a direct solver is usually applied. This may limit the efficiency, if the coarse grid has to many unknowns. Alternatively, AMG can be used as coarse grid solver. The main idea is to design the required AMG-components

1. Coarsening, i.e., reducing the dimension of the finite element space,
2. Transfer operators: prolongation $\mathfrak{P} : \mathbb{V}_H \mapsto \mathbb{V}_h$ and restriction $\mathfrak{P}^T$,
3. Definition of the coarse matrix $\mathsf{K}_H := \mathfrak{P}^T \mathsf{K}_h \mathfrak{P}$,

by using single-grid information only. It turnes out that these components have to be adapted to the problem class under consideration in order to obtain robust algorithms. AMG for M-matrices is well known [3]. However, in our case the system matrix is far from being an M-matrix. In [8] an AMG method for edge-element discretizations has been presented. The main idea is to consider the associated nodal space for describing the kernel of the curl-operator. The coarsening is driven by a agglomeration in the nodal space. The coarse edge-space is defined by edges between those agglomerates which are connected by edges of the finer space. The prolongation is chosen such that the application to a curl-free coarse grid function results in a curl-free fine grid function [8].

3 Parallelization

The aim of parallelization is to split both data and operations to the P processors available. The consistency of the algorithms is preserved by message passing. In our case, the parallelization is based on a non-overlapping domain decomposition, i.e., we decompose $\overline{\Omega}$ into P subdomains $\overline{\Omega}_s$ such that $\overline{\Omega} = \bigcup_{s=1}^{P} \overline{\Omega}_s$ with $\Omega_s \cap \Omega_q = \emptyset \; \forall q \neq s,\; s,q = \overline{1,P}$ holds. Each subdomain Ω_s is discretized by a mesh $\tau_{h,s}$ such that the whole triangulation $\tau_h = \bigcup_{s=1}^{P} \tau_{h,s}$ of Ω forms a conforming mesh. A global finite element space $\mathbb{V}_h$ is defined with respect to τ_h and the local spaces $\mathbb{V}_{h,s}$ are projections of $\mathbb{V}_h$ onto $\tau_{h,s}$. The index set of edges in Ω and Ω_s is denoted by ω_h and ω_s, respectively. Let $N_h := |\omega_h|$ and $N_s := |\omega_s|$. Then the mapping of a vector $\underline{u}_h \in \mathbb{R}^{N_h}$ in global numbering onto a local vector $\underline{u}_s \in \mathbb{R}^{N_s}$ in subdomain $\overline{\Omega}_s$ $(s = \overline{1,P})$ is represented symbolically by subdomain connectivity matrices $\mathcal{A}_s : \mathbb{R}^{N_h} \mapsto \mathbb{R}^{N_s}$ with entries

$$\mathcal{A}_s^{[i,j]} := \begin{cases} 1 & \text{if} \quad j = \text{global number of } i \\ 0 & \text{else} \end{cases} \qquad \forall i \in \omega_s\,,\; \forall j \in \omega_h\,. \tag{3}$$

The transpose $\mathcal{A}_s^T$ of these binary matrices $\mathcal{A}_s$ maps a local vector back onto the global one. The index set of all those subdomains to which an unknown $u^{[j]}$, $j \in \omega_h$ belongs, is denoted by

$$\sigma^{[j]} := \{s \,|\, u_h^{[j]} \in \overline{\Omega}_s\} = \{s \,|\, \exists i \in \omega_s : \mathcal{A}_s^{[i,j]} \neq 0\}\,. \tag{4}$$

We store the data related to the ith edge in the subdomain Ω_s if $s \in \sigma^{[i]}$. This approach results in local data denoted by index s of two types [9]: accumulated data (vector $\underline{\mathfrak{u}}$, matrix $\mathfrak{M}$) represented by

$$\underline{\mathfrak{u}}_s := \mathcal{A}_s \cdot \underline{\mathfrak{u}}\,, \qquad \mathfrak{M}_s := \mathcal{A}_s \cdot \mathfrak{M} \cdot \mathcal{A}_s^T \tag{5}$$

and distributed data (vector $\underline{\mathsf{v}}$, matrix K) represented by

$$\underline{\mathsf{r}} = \sum_{s=1}^{P} \mathcal{A}_s^T \cdot \underline{\mathsf{r}}_s, \qquad \mathsf{K}_h := \sum_{s=1}^{P} \mathcal{A}_s^T \cdot \mathsf{K}_s \cdot \mathcal{A}_s\,. \tag{6}$$

It turns out, that functionals are represented as distributed data whereas functions are represented as accumulated data. Moreover, the local FE-accumulation with respect to $\mathbb{V}_{h,s}$ produces automatically distributed right hand sides $\underline{\mathsf{f}}_s$ and distributed matrices K_s. It can be shown that the multiplication of a distributed matrix with an accumulated vector

$$\mathsf{K}_h \cdot \underline{\mathfrak{w}} = \sum_{s=1}^{P} \mathcal{A}_s^T \mathsf{K}_s \mathcal{A}_s \cdot \underline{\mathfrak{w}} = \sum_{s=1}^{P} \mathcal{A}_s^T (\mathsf{K}_s \cdot \underline{\mathfrak{w}}_s) = \sum_{s=1}^{P} \mathcal{A}_s^T \underline{\mathsf{v}}_s = \underline{\mathsf{v}} \tag{7}$$

results in a distributed vector. The realization requires no communication at all because we only have to compute $\underline{\mathsf{v}}_s = \mathsf{K}_s \cdot \underline{\mathfrak{w}}_s$ locally. If an accumulated matrix $\mathfrak{M}$ fulfills the condition

$$\forall i,j \in \omega_h : \qquad \sigma^{[i]} \not\subseteq \sigma^{[j]} \implies \mathfrak{M}^{[i,j]} = 0, \tag{8}$$

then the operations $\underline{\mathfrak{w}} = \mathfrak{M} \cdot \underline{\mathfrak{u}}$ and $\underline{\mathsf{d}} = \mathfrak{M}^T \cdot \underline{\mathfrak{r}}$ can be performed locally without any communication [9].

Now we observe that Algorithm 1 is also the appropriate parallel formulation, where double-line arrows "$\Leftarrow$" indicate that communication is required for the corresponding operation. The coarse grid system is accumulated globally once in the setup phase. During the iteration only a vector has to be assembled for computing the coarse grid solution. Furthermore, the smoother requires communication and has to be adapted appropriately. In the simplest case Jacobi steps are used instead of Gauss-Seidel steps. Then, the complete GMG algorithm is independent of the partitioning, i.e., independent of the number of processors. The smoother involves a vector conversion from distributed to accumulated type, i.e., one next neighbour communication across interfaces is required per smoothing step. Again, the only special treatment required by parallel Maxwell solvers is the correct choice of the smoother. We apply the smoother proposed by Hiptmair [1] since it is well suited for parallel implementation. It consists of two subsequent classical smoothing steps, one in the edge-space and one in the corresponding nodal space. The matrix representation of the mapping (lifting) between the two spaces fulfills again a pattern condition, i.e. is free of communication [4]. In the case of GMG, the prolongation $\mathfrak{P}$ is of overlapping type and satisfies naturally the pattern condition (8). Hence, prolongation and restriction are free of communication.

In AMG, coarsening and prolongation operators are components which can be chosen. The main idea in the design of parallel AMG is to choose these components such that the resulting prolongation operators are of overlapping type satisfying the pattern condition. For this purpose, the coarsening starts at interfaces involving more then 2 processors, and continues with faces between two processors and finally the coarsening of inner edges is realized. Moreover, the coarsening has to be synchronized such that the coarse grid problem is conforming across interfaces between processors. This synchronization requires next neighbour communication. Once the setup phase for AMG is completed, the solver follows exactly the same principles as in parallel GMG. However, since the partitioning of the mesh influences the coarsening of AMG, the performance of AMG depends slightly on the number of processors.

4 Numerical Results

In this section we will present 2 examples, each of them being solved by GMG and AMG. Since GMG makes use of the exact grid hierarchy, the components, in particular the prolongation, can be chosen in an optimal way. On the contrary, AMG uses only the fine grid information to set up the coarse grid problems. As described in [8], the AMG prolongation is chosen in a simple way preserving only constant fields. That is why GMG gives always better convergence results. Hence, GMG is the method of choice if a grid hierarchy is available. The AMG algorithm has been shown to be superior to classical preconditioners as ILU or SSOR in many applications [10].

In the examples we use GMG and AMG as preconditioners in a preconditioned conjugate gradient solver. Always a V22-cycle (2 pre- and post-smoothing steps) is used which improves the performance in terms of the required CPU-time of AMG compared to a V11-cycle, whereas the performance of GMG is similar in both cases, V11- and V22-cycle for the examples being considered. The stopping criterion is a relative accuracy of 10^{-6}. The artifical conductivity is $\sigma := 10^{-6}$ in all cases. The experiments have been carried out using the software tools NETGEN (3D mesh generator), FEPP (finite element simulation package), PEBBLES (AMG toolbox) and DDComm (communication library), (see [11] for further information). In the parallel case MPI is used as message passing library.

Sequential Algorithms. First we consider again the model of the transformer, see Fig. 1. We have $\nu = 10^{-3}$ for the iron core and $\nu = 1$ elsewhere. The field is generated by currents in the coils. Block-smoothers as proposed in [2] are used. The coarsening of AMG is matrix-dependent. Table 1 shows the

Table1. Number of unknowns, CPU time in seconds for the generation of the system matrix, the setup of the preconditioner, number of iterations ($\epsilon = 10^{-6}$), time for the solver.

		Geometric MG			Algebraic MG		
dof	SysMat	Setup	It.	Solver	Setup	It.	Solver
3466	0.8	4.5	2	0.2	0.7	19	1.2
26907	6.3	5.4	8	7.4	3.0	32	20.0
212597	51.1	11.5	9	70.0	21.1	52	273.6

results for an SGI OCTANE 250 MHz, the CPU time is given in seconds. For GMG and AMG, the setup involves the LU-decomposition of the coarse grid problem and the initialization of the block-smoother. The setup involves additionally the coarsening in the case of AMG.

Parallel Algorithms. Fig. 2 shows the geometry and the mesh of a C-magnet being considered now. The problem has 180426 unknowns. The decomposition into subdomains has been obtained by the recursive spectral bisection [12].

We have $\nu \equiv 1$. The field is generated by a current in the coil. The hybrid smoother proposed in [1] is used, see also [4]. In order to analyze the

Figure2. C-Magnet: geometry and mesh with 153408 elements and 206549 nodes.

parallel performance we consider the following components of the algorithm: generation of the system matrix, setup and solver, see Table 2. The time has been measured for an SGI ORIGIN 2000, 300 MHz using MPI_WTime. As expected, the generation of the system matrix shows optimal speedup since this component is free of communication. In the case of GMG, the setup phase involves the setup of the smoother and the LU-decomposition of the coarse-grid system matrix with 2907 unknowns. Since the setup is dominated by the sequential LU-decomposition, only low speedups can be observed. In the case of AMG, the setup involves the coarsening itself and the LU-decomposition of the coarse-grid system which has less than 500 unknowns. Hence, the setup shows reasonable speedups since it is dominated by the coarsening of the inner nodes. In both cases, AMG and GMG, the solver shows a similar speedup behavior. Although most of the gain is due to the additional CPU capacity, the additional cache which comes with each processor also contributes to the acceleration. Further results can be found in [4,13].

5 Conclusions

We have presented efficient parallel solvers for 3D magnetic field problems. The methods being discussed apply directly to the case of eddy current problems. Moreover, advanced problems such as nonlinear or/and transient problems, as well as coupled magneto-mechanical field problems and optimal design problems require the solution of linearized problems several times (see, e.g. [14]). Then the algorithms presented here can be used immediately.

Table2. Number of processors (P), wall clock time (T) in seconds for generating the system matrix, the setup phase and the solver. Corresponding speedup (S), number of iterations (It.) and speedup with respect to 1 iteration (S(1)).

			Geometric MG					Algebraic MG					
	SysMat		Setup		Solver			Setup		Solver			
P	T	S	T	S	It.	T	S	T	S	It.	T	S	S(1)
1	40.3	1.0	13.1	1.0	13	99.4	1.0	44.4	1.0	60	247.3	1.0	1.0
2	19.6	2.0	10.8	1.2	13	69.2	1.4	24.4	1.8	52	127.1	1.9	1.7
4	9.7	4.1	9.1	1.4	13	31.8	3.1	12.6	3.5	52	48.9	5.0	4.3
8	4.7	8.5	8.8	1.5	13	16.2	6.1	7.6	5.8	58	28.2	8.8	8.4

References

1. R. Hiptmair, Multigrid methods for Maxwell's equations, SIAM J. Numer. Anal. 1999 Vol. 36 pages:204-225
2. D. Arnold and R. Falk and R. Winther, Multigrid in H(div) and H(curl), Numer. Math. 2000 Vol. 85 pages: 197-218
3. J. W. Ruge and K. Stüben, Multigrid Methods, Algebraic Multigrid (AMG), SIAM 1986 Vol.5 pages:73-130
4. Gundolf Haase and Michael Kuhn and U. Langer, Parallel Multigrid 3D Maxwell Solvers, University Linz, SFB Report 1999 No. 99-23
5. Michael Kuhn and Ulrich Langer and Joachim Schöberl, Scientific Computing Tools for 3D Magnetic Field Problems, The Mathematics of Finite Elements and Applications X 2000 pages: 239-258
6. J. Nédélec, A new family of mixed finite elements in R^3, Numer. Math. 1986 Vol.50 pages:57-81
7. R. Beck and P. Deuflhard and R. Hiptmair and R.H.W. Hoppe and B. Wohlmuth, daptive multilevel methods for edge element discretizations of Maxwell's equations, Surv. Mat. Ind. 1999 Vol.8 pages: 271-312
8. S. Reitzinger and J. Schöberl, Algebraic Multigrid for Edge Elements, University Linz 2000 No.00-15
9. Gundolf Haase, Parallelisierung numerischer Algorithmen für partielle Differentialgleichungen, Teubner 1999 ISBN 3-519-02970-7
10. M. Kaltenbacher and S. Reitzinger, Nonlinear 3D Magnetic Field Computations using Lagrange FE-functions and Algebraic Multigrid, 2000
11. SFB F013, Software tools, SFB "Numerical and Symbolic Scientific Computing" http://www.sfb013.uni-linz.ac.at 1999
12. G. Haase and M. Kuhn, Preprocessing for 2D FE-BE Domain Decomposition Methods, Comp. Vis. Sci. 1999 Vol.2 pages 25-35
13. G. Haase and M. Kuhn and S. Reitzinger, Parallel AMG on Distributed Memory Computers, University Linz SFB Report No.00-16 2000
14. M. Schinnerl and J. Schöberl and M. Kaltenbacher and U. Langer and R. Lerch, Multigrid Methods for the fast Numerical Simulation of Coupled Magnetomechanical Systems, ZAMM 2000 Vol. 80 pages: 117-120

Simulation of Electric Field Strength and Force Density on Contaminated H–V Insulators

U. Schreiber*[1], S. Keim[2], and U. van Rienen[1]

[1] Institute of General Electrical Engineering, Rostock University, Albert–Einstein–Str. 2, D–18051 Rostock, Germany
[2] High Voltage Laboratory, Darmstadt University of Technology, Landgraf–Georg–Str. 4, D-64283 Darmstadt, Germany

Abstract This paper reports on the simulation of the behavior of single water droplets on a horizontally arranged polymeric surface under the influence of an applied electric a.c. field. The goal of this investigation are predictions for the aging performance of contaminated insulators in high-voltage power problems.

1 Introduction

High–voltage insulators are stressed by the applied electric field as well as by other environmental factors. As a result of this stress, the surface of the insulating material gets aged and the dielectric material looses it's hydrophobic and insulating characteristics. The contamination of the object with water droplets accelerates the aging process. Experimental investigations have shown that with increase of applied voltage, droplets vibrate first, they are then extended to the direction of the applied electric field and finally flashover bridging water droplets occurs (see [6]). To improve the understanding of aging phenomena it is advisable to observe single droplets on an insulating surface. The shape of the droplets supplies more information about the status of the insulating material.
This paper presents first the calculation of electric field strength and force density around the droplets. We model our problem as an electro–quasistatic 3D-problem (see [9]). For discretisization we use the Finite Integration Theory (FIT, see [10]). FIT was especially developed to solve Maxwell's equations. It guarantees a consistent conversion of Maxwell's equations on a staggered grid, the so–called Maxwell–Grid–Equations. Our insulator problem leads to an almost singular complex symmetric system of linear equations. We investigate various solution methods for our complex symmetric system in order to compare these methods and find sufficiently robust ones.
Second, we simulate our water droplet as a rotational ellipsoid with geometric values from experiments. We are looking for a relation between force density at the droplets and the measured shape of the ellipsoid. The goal is to predict the droplet behavior without experiments.

* supported by Deutsche Forschungsgemeinschaft

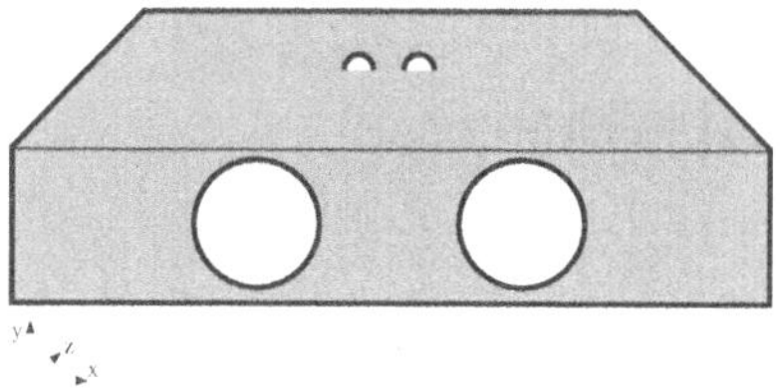

Figure1. The unaged solid epoxy resin sample with horizontally embedded electrodes and two water droplets on the top of the test object.

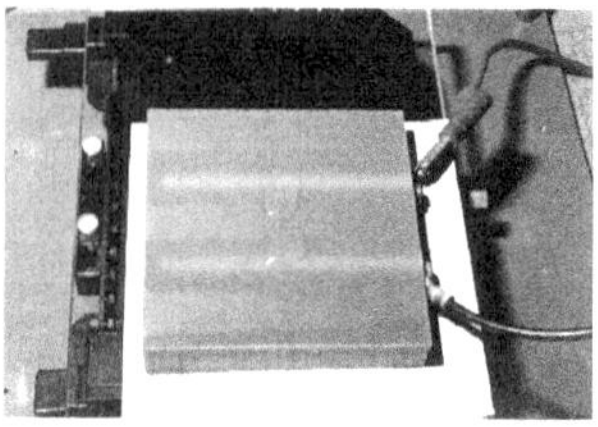

Figure2. Two water droplets after 20 min application of high voltage (8 kV). The droplets are deformed during the experiment .

2 Modelling for Electro–Quasistatics

The considered devices are driven with 50 Hz a.c. high–voltage, i.e. the electromagnetic field is slowly varying. The displacement current plays a crucial role. It's density $\partial \boldsymbol{D}/\partial t$ is linked up with a time dependent magnetic field, which again is linked to the electric field $\boldsymbol{E}$ via curl $\boldsymbol{E} = -\partial \boldsymbol{B}/\partial t$. The time variation of the magnetic flux density, and thus the eddy current, is neglectable. So we can formulate our problem as an electro–quasistatic one. As first approximation the rather slow droplet vibration is neglected here.

2.1 Problem Formulation

The low frequency field with significant displacement current leads to the assumption

$$\frac{\partial \boldsymbol{B}}{\partial t} = 0, \qquad \frac{\partial \boldsymbol{D}}{\partial t} \neq 0.$$

For a time harmonic field $\boldsymbol{E}(\boldsymbol{r},t) = \boldsymbol{E}(\boldsymbol{r}) \cos(\omega t + \Phi)$ we use the representation $\boldsymbol{E}(\boldsymbol{r},t) = \mathrm{Re}(\underline{\boldsymbol{E}}(\boldsymbol{r}) \mathrm{e}^{i\omega t})$ with the complex amplitude $\underline{\boldsymbol{E}}(\boldsymbol{r}) = \boldsymbol{E}(\boldsymbol{r}) \mathrm{e}^{i\Phi}$. Under these conditions Maxwell's equations for the time harmonic field reduce to

$$\mathrm{curl}\, \underline{\boldsymbol{E}} = 0, \tag{1}$$

$$\mathrm{curl}\, \underline{\boldsymbol{H}} = i\omega\, \underline{\boldsymbol{D}} + \sigma\, \underline{\boldsymbol{E}} + \underline{\boldsymbol{J}}_0, \tag{2}$$

$$\mathrm{div}\, \underline{\boldsymbol{D}} = \rho, \tag{3}$$

$$\mathrm{div}\, \underline{\boldsymbol{B}} = 0. \tag{4}$$

From (1) – (4) we get

$$\mathrm{div}\, \big((i\omega\varepsilon + \sigma)\, \underline{\boldsymbol{E}} \big) = -\, \mathrm{div}\, \big(\underline{\boldsymbol{J}}_0 \big).$$

Equation (1) says that the electric field $\underline{\boldsymbol{E}}$ may be described as the gradient of a scalar complex potential. Note that this is a complex potential: $\underline{\boldsymbol{E}} = -\, \mathrm{grad}\, \underline{\varphi}$. The final relation for the electric scalar potential results to

$$\mathrm{div}\, \big((i\omega\varepsilon + \sigma)\ \mathrm{grad}\, \underline{\varphi} \big) = \mathrm{div}\, \big(\underline{\boldsymbol{J}}_0 \big). \tag{5}$$

The force density $\boldsymbol{f}$ is closely related to the energy density and may be calculated from Maxwell's stress tensor as

$$\boldsymbol{f} = \frac{\mathrm{d}\boldsymbol{F}}{\mathrm{d}A} = \varepsilon \boldsymbol{E}(\boldsymbol{E} \cdot \boldsymbol{n}) - \frac{1}{2}\varepsilon \boldsymbol{E}^2 \boldsymbol{n}.$$

A discrete evaluation scheme for the force density and their components can be found in [2].

2.2 Discretization

The Finite Integration Theory (FIT, see [10]) is a proven consistent scheme which converts Maxwell's equations in integral form to systems of linear equations, the so–called Maxwell–Grid–Equations. In our electro–quasistatic insulator problem FIT transforms the continuous equation (5) into the discretized one

$$\widetilde{\mathbf{S}}(i\omega\,\mathbf{M}_\varepsilon + \mathbf{M}_\sigma)\widetilde{\mathbf{S}}^{\mathrm{T}}\,\underline{\boldsymbol{\Phi}}_E = \widetilde{\mathbf{S}}\underline{\widehat{\widehat{\mathbf{j}}}}_0.$$

The notation

$$\mathbf{A}_\sigma := \widetilde{\mathbf{S}}\mathbf{M}_\sigma\widetilde{\mathbf{S}}^{\mathrm{T}}, \quad \mathbf{A}_\varepsilon := \widetilde{\mathbf{S}}\mathbf{M}_\varepsilon\widetilde{\mathbf{S}}^{\mathrm{T}}, \quad \underline{\mathbf{p}}_0 := \widetilde{\mathbf{S}}\underline{\widehat{\widehat{\mathbf{j}}}}_0$$

leads to the resulting complex linear system for our field calculations

$$(\mathbf{A}_\sigma + i\omega\,\mathbf{A}_\varepsilon)\underline{\boldsymbol{\Phi}}_E = \underline{\mathbf{p}}_0. \tag{6}$$

The boundary conditions and the applied potentials are explicitly included during the matrix–setup.
Notice that the real part of the complex symmetric matrix $\mathbf{A} = \mathbf{A}_\sigma + i\omega\,\mathbf{A}_\varepsilon$ is just the matrix $\mathbf{A}_\sigma$ for stationary currents and the imaginary part is the matrix $\mathbf{A}_\varepsilon$ of electrostatics, scaled with the frequency ω.
The software package MAFIA (see [7]) is based on FIT. It covers the complete range of electrodynamics from statics to high-frequency. All calculations presented below were performed with MAFIA.

2.3 Iterative Methods

For the calculation of a three-dimensional structure, the rank of the system matrix in (6) is usually in the order of 10^3 – 10^7. The matrix is a band matrix with band width seven and a large condition number. We are going to investigate several iterative methods for the solution of the discretized complex symmetric potential problem.
One important class of iterative methods for solving the insulator problem are the Krylov–subspace methods. For the problem described above, the convergence speed of these methods is only acceptable in combination with minimal residual smoothing techniques. Numerical studies showed that the Jacobi–preconditioned Krylov–subspace methods can be used. Other preconditioners are too expensive by negligible advantage. An explicit description

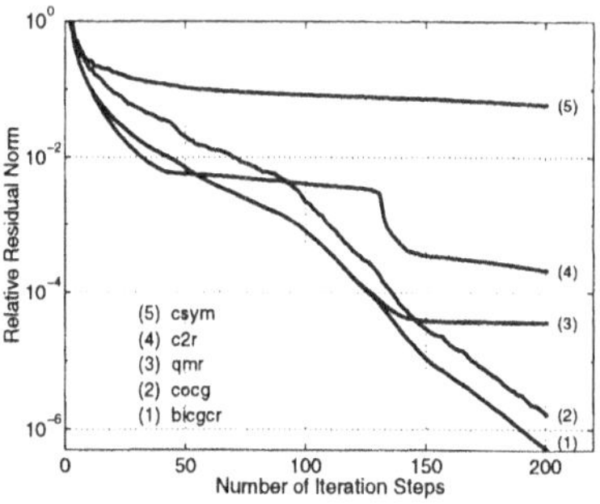

Figure3. Comparison of preconditioned iterative algorithms for solving complex symmetric systems. Method (1)–(3) and (5) are with minimal residual smoothing.

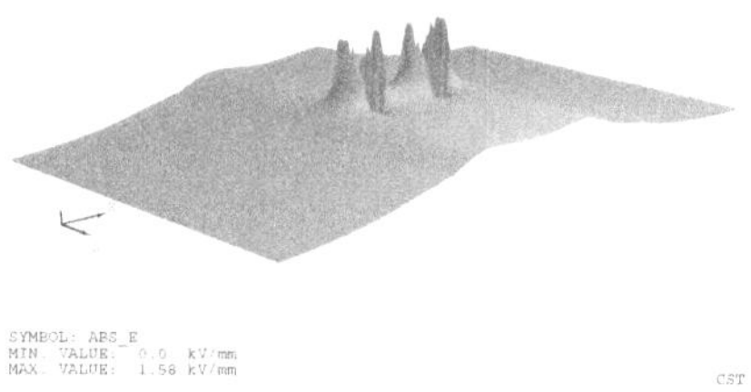

Figure4. Absolute field strength of the electric field on the surface of the epoxy resin with two water droplets.

of Krylov–subspace methods, references and a detailed investigation for complex symmetric systems with similar results may be found in [4,8].

A Krylov–subspace method which has already proven it's applicability is the Conjugate Orthogonal Conjugate Gradient Algorithm (COCG) together with the Jacobi–preconditioner. COCG is a symmetric variant of the BiCG algorithm and requires the least arithmetical work of all investigated methods. The algorithm achieves a relatively high level of accuracy but the relative residual norm still suffers somewhat from oscillations even after minimal residual smoothing.

The second method is the Bi-Orthogonal Conjugate Gradient Conjugate Residual Method (BiCGCR, see [4,8]) which is also a symmetric variant of the BiCG algorithm. The algorithm is tested with the Jacobi–preconditioner and requires more work and more memory than COCG. On the other hand BiCGCR frequently attains the best accuracy of all investigated methods. Through the minimal residual smoothing the relative residual norm is scarcely oscillating and the curves are smoother than COCG.

The third investigated method is the Quasi Minimal Residual Method (QMR). The two-term version of this method using the Jacobi–preconditioner with minimal residual smoothing is more stable than the classical three–term version. QMR usually produces very smooth convergence curves but sometimes doesn't reach the accuracy as BiCGCR and COCG.

An iterative method called CSYM (see [3]) which also exploits the complex symmetric matrix structure of the system is investigated. This method is based on unitary equivalence transformations of the system matrix to a tridiagonal form and works similar to the Lanczos method. The preconditioned CSYM shows a good convergence behavior for small problems. But for system matrices with large rank as in our case the Jacobi–preconditioner scarcely influences the iteration speed. Especially CSYM shows the necessity of choosing

an adapted preconditioner. In each calculated example the convergence curve is very smooth but the accuracy is not sufficient enough.
An investigated method by Axelsson solves the problem as a real one (see [1]). In this procedure the condition of the system matrix is usually squared and the real algorithm is not recommendable. Axelsson generalized the method by introducing a real parameter and minimized the corresponding spectral condition number near two by optimizing this parameter. With Axelsson's method our calculations show strongly varying convergence behavior. In some cases smooth curves of convergence with sudden residual reduction may be observed. In our calculations, the convergence speed of Axelsson's method is not better than the convergence speed of COCG, BiCGCR and QMR.

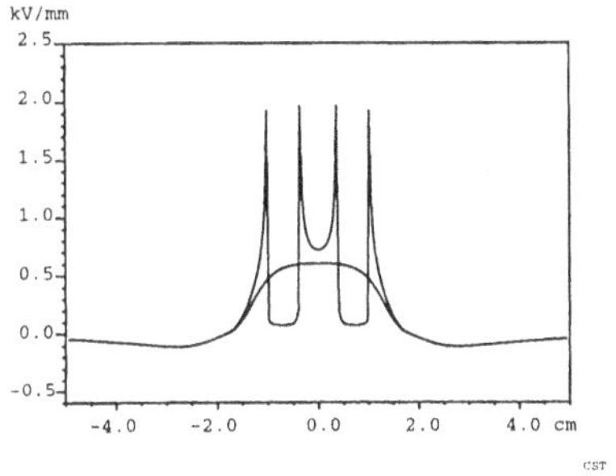

Figure5. Field strength component in x–direction on the surface of the epoxy resin with and without water droplets.

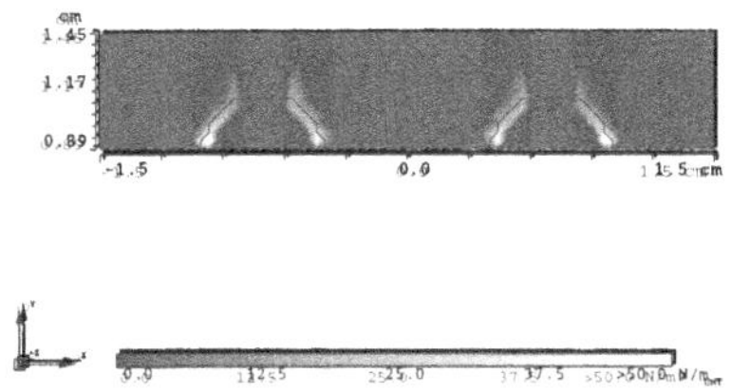

Figure6. Force density around the water droplets during application of 15 kV.

We tested the introduced algorithms for the following example. The sample is a block of epoxy resin with length and width of 100 mm and a height of 20 mm. The test object has horizontally embedded electrodes with a center distance of 35 mm and a radius of 7.5 mm. We put only two droplets on it with a diameter of 3 mm (hemispheres) and a center distance of 10 mm. The discretization yields a system of 124608 complex unknowns. The epoxy resin has a relative permittivity of $\varepsilon_r = 4$ and a conductivity of $\sigma = 10^{-12}$ S/m. The water drops have a relative permittivity of $\varepsilon_r = 81$ and a conductivity of $\sigma = 10^{-6}$ S/m. The permittivity of the air surrounding the structure is $\varepsilon_r = 1.000576$. A frequency f of 50 Hz and a voltage of 15 kV are assumed. The convergence behavior, which is characteristic for the methods, is shown in Fig. 3. Finally we can state that an appropriate choice of the preconditioner is crucial for *all* solvers in order to speed up the solution. For this reason we will focus our attention on this task while the setup of new algorithms seems not to be too profitable.

3 Forces on Droplets

We know that rotating droplets of oil in water have the shape of an ellipsoid. Also the contact angle measuring instrument (see [5]) approximates the droplet shape as an elliptic shape. Therefore we express the droplets as a rotational ellipsoid. Elliptic shapes are easy to calculate and bring acceptable results for our calculations. We will study more difficult shape forms in future after more experiments.

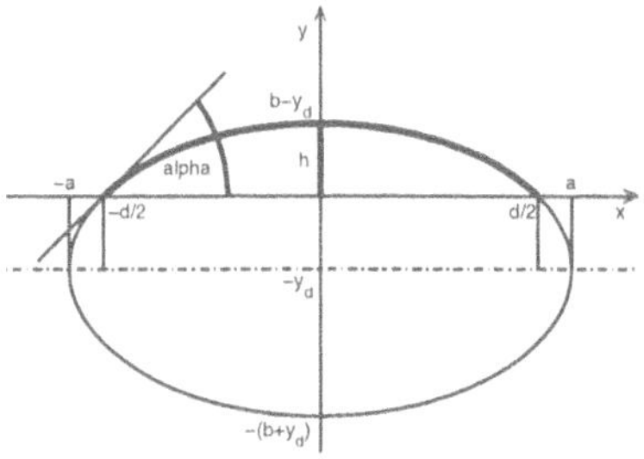

Figure7. Shape of droplet as an ellipsoid with contact angle alpha.

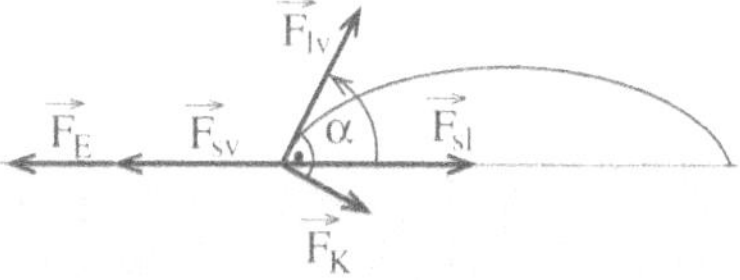

Figure8. Forces acting at the triple point.

From experiments we get the diameter d, the height h and the contact angle α of the water droplets. From this we can exactly calculate the diameters a and b of the ellipse as

$$a = M\left(2M - \left(\frac{d}{2}\right)^2\right)^{-1/2}, \qquad M := \frac{d}{2}\left(\frac{d}{2} - \frac{h}{\tan\alpha}\right),$$

$$b = a\tan\alpha\left(\frac{d}{2}\right)^{(-1)}\left(a^2 - \left(\frac{d}{2}\right)^2\right)^{1/2}$$

and the position $(0, -y_d)$ of the center from $y_d = b - h$. The resulting requirement $\tan\alpha < 4h/d$ is acceptable.

If we have an interface between two different media a surface tension is acting. It tries to minimize the surface of a droplet in air like the skin of a ballon. The microscopic reason are the intermolecular forces. A triple point is the contact point between three different media, in our case air, water and epoxy resin. At this triple point we have three surface forces caused by the surface tension (see Fig. 8). They are in equilibrium $|F_{\mathrm{sv}}| = |F_{\mathrm{sl}}| + |F_{\mathrm{lv}}|\cos\alpha$ and result in a specific contact angle α. An additional pressure of the curvature p_{K} acts perpendicular to the water surface. The pressure of the curvature p_{K} depends on the surface tension σ ($\sigma_{\mathrm{H_2O}} = 72.8$ mN/m) and the radii of curvature R_1 and R_2 as

$$p_{\mathrm{K}} = \sigma\left(\frac{1}{R_1} + \frac{1}{R_2}\right).$$

The radii R_1 and R_2 can very easily be calculated for points in xy–cut as

$$R_1 = a^2 b^2 \left(\frac{x^2}{a^4} + \frac{y^2}{b^4} \right)^{3/2}, \qquad R_2 = b^2 \left(\frac{x^2}{a^4} + \frac{y^2}{b^4} \right)^{1/2}.$$

The electric force changes the shape of the droplet and rises the pressure of the curvature. We are now looking for the relation of the electric force density calculated with FIT and the pressure of curvature calculated with measured height h, diameter d and contact angle α.

4 Example

Two experiments with one single droplet were carried out – one with a centered droplet and one with a droplet near the high voltage electrode (see [5]). The decrease in height and the increase in diameter of the droplet by a rising voltage is to be seen in Fig. 9. For these two experiments we calculated the force density at the triple points (see Fig. 10).

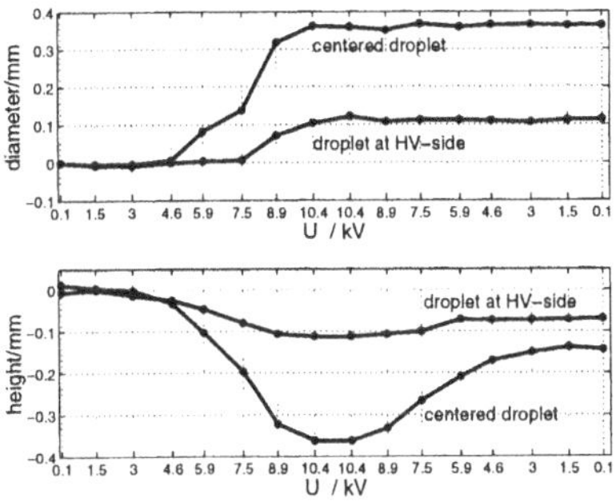

Figure9. Droplets during increase of the voltage, upper picture: changes of diameter, lower picture: changes of height.

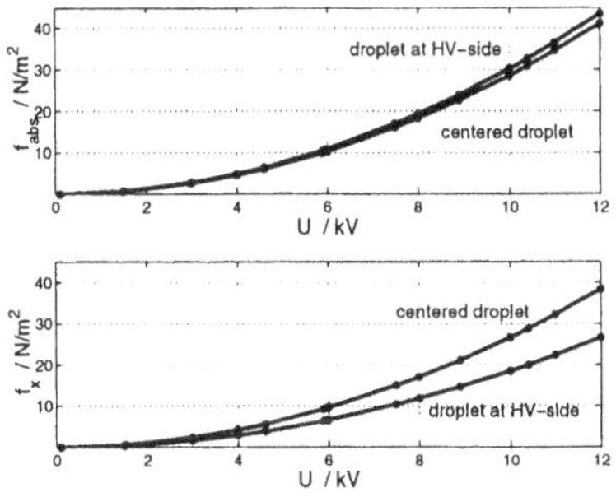

Figure10. Force density at triple point, upper picture: absolute value of force density, lower picture: force density in direction of the applied electric field.

The absolute force density at the triple points is nearly the same for both droplet arrangements. But the force density in direction of the applied electric field is greater for the centered droplet. This confirms the experimental differences seen between the centered droplet and the droplet at the HV–side.

Additionally, for two voltages we know the contact angle for the centered droplet. From these values we calculated the pressure of curvature from the relations above:

	height/[mm]	diameter/[mm]	contact angle	$p_K/[N/m^2]$
0.1 kV	1.49	4.05	70.14	59.04
10.4 kV	1.11	4.41	60.50	92.74

It is conspicuous, that the difference $p_K(10.4\text{ kV})$ - $p_K(0\text{ kV})$ in the pressure of curvature is nearly the same as the force density $\boldsymbol{f}(10.4\text{kV})$.

5 Conclusion and Outlook

The simulation of the behavior of water droplets on polymeric surfaces under the influence of an applied high–voltage field is a demanding task. The first step, the numerical calculation of electric field strength and force density, is presented here. The result of the droplet measurements is our assumption, that the difference $p_K(\text{x kV})$ - $p_K(0\text{ kV})$ of pressure of curvature approximates well the force density during increasing voltage. We will validate this hypothesis by more measurements and simulations.

References

1. Owe Axelsson, *Real Valued Iterative Methods for Solving Complex Linear Systems*, University Nijmegen, January 1999.
2. Michael Bartsch and Thomas Weiland, *2D and 3D Calculation of Forces*, IEEE Transactions on Magnetics, vol. 30, No. 5, September 1994.
3. Angelika Bunse–Gerstner and Ronald Stoever, *On a Conjugate Gradient–Type Method for Solving Complex Symmetric Linear Systems*, Linear Algebra and its Applications, 1992.
4. Markus Clemens, Rolf Schuhmann, Ursula van Rienen and Thomas Weiland, *Modern Krylov Subspace Methods in Electromagnetic Field Computation Using the Finite Integration Theorie*, Applied Computational Electromagnetics Society Journal, vol. 11, no. 1, pp. 70 – 84, 1996.
5. Karsten Herrbach, *Study of Behavior of droplets on Polymeric Surfaces under Electrical Stress Conditions with the Aid of a Contact Angle Measuring System* Studienarbeit, High Voltage Laboratory, University of Technology, Darmstadt.
6. Sabine Keim and Dieter König, *Study of the Behavior of Droplets on Polymeric Surfaces under the Influence of an Applied Electrical Field*, IEEE Conference on Electrical Insulation and Dielectric Phenomena (CEIDP), Austin, USA, October 17 – 20, 1999, pp. 707 – 710.
7. MAFIA Version 4.020, CST GmbH, Büdinger Str. 2a, D–64289 Darmstadt.
8. Ursula van Rienen, *Numerical Methods in Computational Electrodynamics – Linear Systems in Practical Applications*, Springer, Lecture Notes in Computational Science and Engineering, Vol. 12, 2000.
9. Ursula van Rienen, Markus Clemens and Thomas Weiland, *Computation of Low–Frequency Electromagnetic Fields*, ZAMM, Applied Mathematics and Mechanics, vol. 76, pp. 567 – 568, 1995.
10. Thomas Weiland, *A discretization method for the solution of Maxwell's equation for six–component fields*, Electron. Commun. AEÜ, **31**(3), pp. 116 – 120, 1977.

Commercial EM Codes Suitable for Modeling of Microwave Heating - a Comparative Review

V.V. Yakovlev

Worcester Polytechnic Institute, 100 Institute Road, Worcester, MA 01609

Abstract Electromagnetic modeling software available in the market is studied in accordance with specific criteria to compose contemporary database of codes applicable to designing industrial systems of microwave thermal processing. Operating systems, kernel computational methods, interfaces, optimization options, possibilities of solution of associated thermal problem and financial information are among the basic characteristics addressed. It is shown that sixteen selected codes are very differently adjusted to practical needs of the field. Vendors are rated with respect to their dedication to the microwave power engineering. Issues related to further evaluation of the codes (solution of benchmark problems) are discussed.

1 Introduction

The recent remarkable increase in the quality of modeling and the certain decrease in the costs of software and hardware have caused a substantial growth of the use of advanced computer simulations in designing wireless telecommunication equipment, computer systems, networking and other products.

Although engineers dealing with microwave non-communication applications have also become more interested in modeling, for most, this arena remains new and unexplored. The review of commercial EM codes [1] written for users in the telecommunications field does not include any indication as to what extent these codes are convenient for modeling of microwave heating. The only paper [2] outlining the concepts of numerical methods suitable for microwave thermal processing was published 10 years ago. Many cases of successful modeling in R&D projects have been reported, however, they have not specified the use of simulations in design of industrial systems. The microwave power community currently seems to be lacking not just specific technical data, but very general information on modern computational opportunities [3].

The present paper looks at the software market and considers the capabilities of particular codes with respect to the needs of microwave power engineering. The results are represented in the form of a software database with discussion of the issues associated with the further analysis of the codes.

2 Rationales

Despite all the progress in numerical mathematics and computational technologies, microwave thermal processing still cannot be modeled in its entirety. No modeling packages can provide complete information about processes, which necessarily involve thermal conduction. Neither can one take into account such phenomena as water transport, evaporation, mass transfer or chemical transformations and reactions. Corresponding research does exist, but there is no computer code implementing numerical solutions of all these problems at once.

Today, valuable results can be obtained from simulation of electromagnetic and thermal processes. All applicable software can be divided into two groups in accordance with this feature. Another classification identifies academic and commercial codes. Judging the latter appears to be difficult: published information is far from complete. The academic codes can be often working tools in the hands of no one but the authors. It would be perhaps unrealistic to suppose that much university software can be used in industry.

As for codes for analysis of thermal processes, they might be useful in practical situations. However, in the present review, we assume that the electromagnetic phenomenon plays the primary role in efficient microwave heating and is essentially responsible for the entire process.

Therefore, we focus here on EM commercial codes. Initial purpose was to discover what software available in the market can be efficiently used to help analyze and design systems of microwave processing. The subsequent objective is to study the most promising codes and work out certain generalizations and recommendations to the microwave power community.

3 Selection Criteria

All codes can be divided into two groups as implementing algorithms for low frequency and high frequency (HF) electromagnetics. Since at HF mathematics is more complicated, the number of packages applicable to this field is smaller (some 30%). Among codes in this smaller group, there are packages dealing with 2D approaches and oriented on planar structures, as well as codes addressing open problems and beneficial for electromagnetic compatibility, antennas, etc. We ignore these tools as irrelevant to the present study and reduce the size of the market to a very limited number of codes.

The remaining packages were exposed to the criteria on the software capabilities. The characteristics determined by numerical methods can be listed in order of increasing complexity. The part of the list relevant to problems of microwave heating may be presented as follows:

- Eigenfields in presence of lossy materials; power of the eigenmodes;
- Fields excited by the given source;
- Dissipated power of the excited fields;
- Specific absorption ratio (SAR) patterns;
- Level of coupling.

It seems for practitioners in the microwave power industry it would be feasible to have software with the capability of calculating the dissipated power of the excited fields *as a minimum*. Finding eigenfields may not be enough for the purposes of industrial design: these data require certain interpretation; it might make sense in R&D projects though. So only codes able to determine at least patterns of dissipated power are analyzed below.

Commercial Electromagnetic Software of Actual/Potential Use in Microwave Power Engineering (August 2000)

Vendor	Code	Operating System & License/Maintenance per year	Status, Major Technique, Performances, Requirements, etc.
Agilent Eesof EDA[1] www.tm.agilent. com/tmo/hpeesof	*Agilent HFSS Designer 5.5*	UNIX, Windows 95/NT4 *call vendor*	Actual use. FEM.DP.OSA[2] optimization, PC-to-UNIX simulations
Ansoft, Corp. www.ansoft.com	*Ansoff HFSS 7.0*	UNIX, Windows 95/98/NT4 $42K/12%	Actual use. FEM. DP, SAR (for plane wave). Eigenmode solver for anisotropic materials.
ANSYS, Inc.	*ANSYS/ EMAG*	UNIX,Windows 95/98/NT4 $25-30K/ *call vendor*	Actual use. FEM. DP. Compatibility with ANSYS CAD codews, advanced animation.
CRC Research Institute, Inc. www.crc.co.jp	*MAGNA/ TDM*	UNIX, Windows NT4 *call vendor*	Potential use. FDTD.DP. 2 years on Japanese market; no English manual
CST GmbH www.cst.de	*MAFIA 4*	UNIX, Windows 95/98/NT4 $9-50K/10%	Actual use. FIM.DP,SAR. PBA: up to 20 mil.cells. Optional temperature analysis
	Microwave Studio 2	Windows 95/98/ NT4 $10-50K/10%	Actual use. FIM. DP, SAR PBA, AutoCAD export/import Full CAD design
ElectroMagnetic Applications, Inc. www.csn.net/ ~emaden	*EMA3D 3.0*	UNIX, Windows NT4 $14-27K/ $2.5-4K	Potential use. FDFD & FDTD. EMA-FAM[3] interface
IMST GmbH www.imst.de	*EMPIRE 2.2*	Linux, UNIX, Windows 95/98/ NT4 $12-20K/ from $1.5K	Actual use. FDTD. DP, SAR. Auto CAD import (limited to 3D boxes.) 300MB hard-disk space
Infolytica, Corp. www.infolytica.com	*FullWave*	Windows 95/98/ NT4 *call vendor*	Potential use. FEM. DP. Advanced interface
The Japan Research Institute	*JMAG-*	UNIX,Windows NT4	Potential use. FEM & FDTD.

Commercial Electromagnetic Software of Actual/Potential Use in Microwave Power Engineering (August 2000)

www.jri.co.jp	*Works*	$35K/*call vendor*	DP. Thermal solver; 2GB hard-drive.
Matra Systems & Information emc2000. matrasi-tls.fr	*EMC2000-VF*	Windows 95/98/ NT4 $15-20K (w/interface $30-35K)/15%	Potential use. TDFV.DP,SAR. 4GB hard-disk space. *HyperMesh*[4] interface
Remcom, Inc. www.remcom.com	*XFDTD 5.1*	UNIX, Windows 95/98/NT4 $15K/$3K	Actual use. FDTD. DP, SAR. iSIGHT optimization[5]. Multiprocessor for FDTD.
QWED www.qwed.com.pl	*QuickWave 1.9*	Windows 95/98/ NT4 $15K/15%	Actual use. Conformal FDTD, DP, SAR. 15MB hard-disk space ACIS export/import, AutoCAD import, optimization.
Tech Univ. of Hamburg-Harburg www.tu-harburg.de/ ~tebr	*CONCEPT II 8.0*	Linux, UNIX $26K/ 2 years free	Potential use. MoM. DP, SAR. 100MB hard-drive space
Weidingler Assoc., Inc. www.wai.com	*EMFlex*	UNIX $40K/ *call vendor*	Potential use. FETD. DP. Optional thermal solver
Zeland Software, Inc. www.zeland.com	*FIDELITY 2.0*	Windows 95/98/ NT4 $20K/ *call vendor*	Potential use. FDTA. DP, SAR. 1GB hard-dsk space

NOTE: DP = Dissipated Power. SAR = Specific Absorption Ratio. FEM = Finite Element Method. FETD = FE in Time Domain. FDFD, FDTD = Finite-Difference in Frequency Domain and Time Domain. MoM = Method of Moments. FIM = Finite Integration Method. TLM = Transmission Line Method. TDFV = Time Domain Finite Volume. PBA = Perfect Boundary Approximation.

[1]Formerly Hewlett Packard Eesof. [2]Optimization Systems Associates, Inc.; purchased by Hewlett Packard in 1997. [3]FEGs, Ltd., www.fegs.co.uk. [4]Altair Engineering, Inc., www.altair.com. [5]Engineous Software, www.engineous.com.

4 Analysis

Table 1 includes the names and main characteristics of the 16 full-wave 3D codes, which fulfill the selection criteria. Each code exists in either one, or several versions for various operating systems. Today, 14 out of 16 packages work under Windows NT. A few years ago, UNIX associated with workstations or supercomputers was the only option for efficient EM codes. Now this big disadvantage for the industry is over.

The software status in the microwave power engineering is specified by two terms. "Actual" means that it has been used at least once in some R&D or industrial project whereas "Potential" indicates that all capabilities outlined above are available. Currently, the Actual/Potential ratio is equal to 8/8.

Among the kernel computational methods, FEM and FDTD prevail, but the method traditionally considered less powerful and flexible (Method of Moments) is also available. Dissipated power and SAR are identified as the key parameters, which can be calculated and visualized. SAR is not now an option in all codes; however, vendors say that since calculating one on the basis of the other is simple, this would be added if any customers request it.

The interfaces appear to be very different. Agilent and Ansoft have developed for their own advanced interfaces. *Microwave Studio, Empire*, and *QuickWave-3D* offer advanced functions for export/import data to work with some CAD software. *EMC2000-VF* merely works with another piece of commercial software, *HyperMesh*, serving as an interface.

Another important feature is an optimization option. It is present in *Agilent HFSS 7.0*, *Ansoft HFSS 5.5*, *QuickWave-3D 1.9*, and *XFDTD 5.1*. The option implements a certain procedure of computer optimization, i.e., a subsequent solution for various scenarios with the choice of parameters supposed to be the best in accordance with some specified criteria. Running this option may take much time, but it might be useful.

Two vendors have implemented practical solutions of how to overcome the major disadvantage of the classical FDTD: the stair-like approximation of the mesh cells. The incorporation of the so-called Perfect Boundary Approximation in *Microwave Studio* and the Conformal FDTD in *QuickWave-3D* allows an accurate modeling of curvilinear regions.

Interested users can work now with at least 3 vendors exploring an opportunity of solving the EM problems in association with the thermal ones. CST and Weidingler Associates promote thermal packages completely compatible with their *MAFIA, Microwave Studio* and *EMFlex. JMAG Works* exists as the set of compatible modules including the thermal code.

The interactive profile of this study has allowed us to determine the vendors' rating regarding their dedication to the MW heating applications (Table 2). The companies in the first column exhibit very little interest in the area. The 2-star vendors are aware of microwave heating applications and admit that they may have certain involvement in corresponding technical support and further development of their products. QWED heads the list,

Table2. Rating of Vendors' Dedication to MW Heating Applications

*	**	* * *
Agilent, CRC RI, EMA, IMST, Infolytica, JRI, Remcom, TU H-H, WAI, Zeland	Ansoft, ANSYS, CST, Matra S&I	QWED

demonstrating a real desire to work for the field. Some features specifically appropriate for the microwave heating (specified modal excitation, field envelope, etc.) are already available in their code. Also, the company has recently announced implementation of special regimes for running *QuickWave-3D* in applications with thermal changes of media parameters.

Each code has a variety of options and configurations, and the license costs depend dramatically on that; the average is $26,500. The average cost of the maintenance (technical support and upgrade) is $3,000.

5 Examples: Benchmark Problems

To make specific evaluation of the selected codes in the context of particular needs of the microwave power engineering and determine recommendations, further study of the database is needed. It is supposed that solutions of a series of benchmark problems (BMPs) should give data for comparison of the efficiency of the codes and their computational characteristics. The problems have to be associated with typical systems of the field a possibility of independent (e.g., experimental) verification of the results.

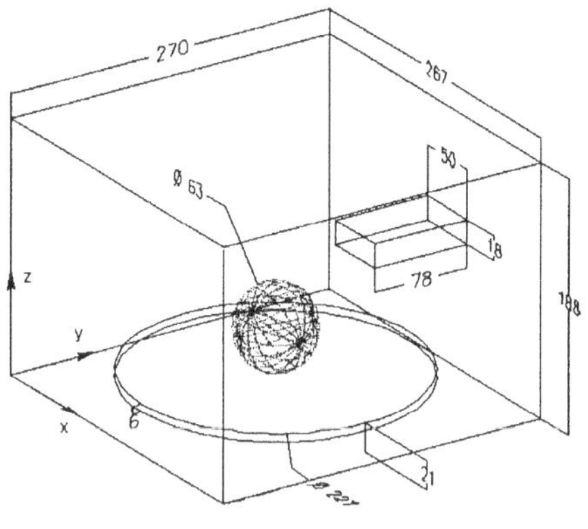

Figure1. The configuration of the BMP models with the potato.

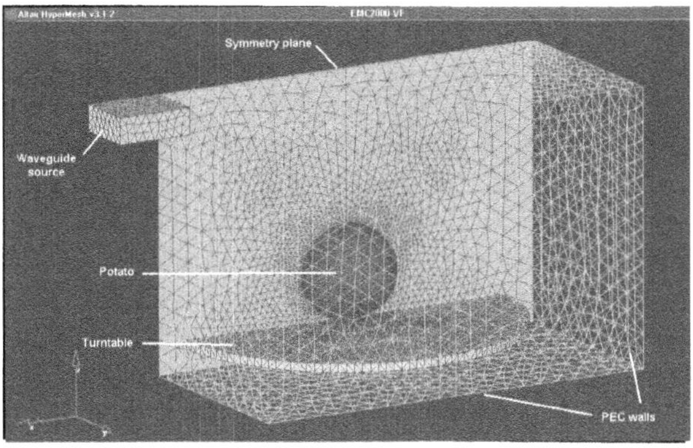

Figure2. Interface image of the BMP (a) created by *EMC2000-VF/HyperMesh*. *Courtesy of EADS - Matra Systemes & Information, www.emc2000.org, 2000.*

The structure chosen for the first BMP is based on the microwave oven designed by Per Risman [4]. The cavity is excited by one magnetron (frequency 2.45 GHz, average power 1 kW) through a waveguide feeder; the walls are of perfect electric conductivity. The problem deals with two different processed materials: (a) uniform potato centered on the shelf; (b) two-layered pizza centered on a microwave-transparent platform located on the shelf. The sketch of the oven is shown in Fig. 1.

The potato and pizza are approximated by spherical and cylindrical two-layer models respectively. The potato is supposed to be uniform and characterized by relative permittivity $\epsilon = 65 - i20$ and density $\rho = 1.0\ g/cm^3$. In the pizza, the lower compact layer is of $\epsilon = 6 - i3$ and $\rho = 0.25g/cm^3$; the upper soft layer possesses $\epsilon = 45 - i20$ and $\rho = 1.1g/cm^3$.

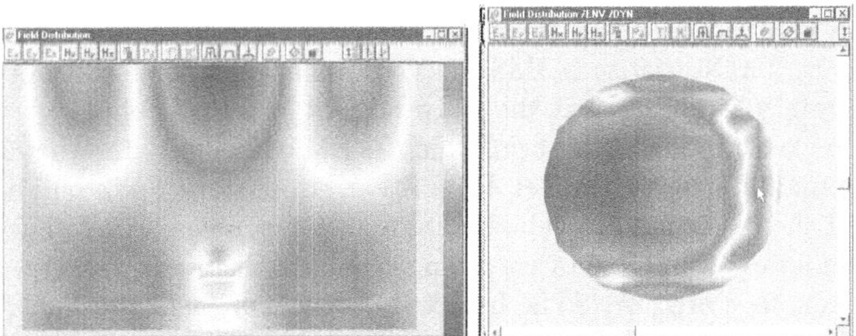

Figure3. Examples of *QuickWave-3D* simulation: the electric field (E_z) in the vertical cut (yz-plane)(left) and the SAR in the horizontal cut (right); the central planes of the potato. *Courtesy of QWED, www.qwed.com.pl, 2000.*

The expected output includes: (a) patterns of the electric field, dissipated power and SAR in the vertical and horizontal cuts; (b) coupling, i.e., the reflection factor for the frequency range and specifically at the operating frequency and percentage of the magnetron power dissipated in the materials.

The full-scale solutions of the BMPs by all 16 codes is probably unfeasible. The luck of the some vendors' interest in the microwave heating applications may mean that at the present time their products could not really provide comparable/competitive data. Initial results are currently available from the two- and three-star vendors. Fig. 2 reproduces the model of the BMP (a) as it is shown in the *EMC2000-VF's* interface. The patterns of the electric field and SAR calculated by *QuickWave-3D* are shown in Fig. 3. These examples do not pretend to be either genuine benchmarking results, or meaningful simulation data; they are rather simple illustrations of what commercial codes can provide for the microwave power industry.

6 Conclusion

All 16 full-wave 3D EM commercial codes applicable to microwave heating modeling were originally developed for communication applications, so they adapt to MW heating problems quite differently and thus should be carefully considered in association with the current application. Expenses associated with the software should be analyzed in the context of possibility to reduce the traditional expensive and inefficient cut-and-try phase in industrial design of the systems.

This review has suggested software database for non-communication microwave engineering. Presently, with the set of general characteristics and features, the paper might be *food for thought* for engineers and practitioners. Anticipated results of benchmarking should shed light on efficiency of the codes and show to what extent some of them could be convenient in the use in practical industrial projects.

Acknowledgments. The study was supported by the Center for Industrial Mathematics & Statistics (CIMS) at WPI. The web site of Electromagnetic Compatibility Laboratory at the University of Missouri-Rolla (www.emclab.umr.edu/csoft.html) and the EM community Internet database EMLIB maintained by the Jet Propulsion Laboratory (emlib.jpl.nasa.gov) provided the directories to valuable information sources. The following software vendors delivered data regarding characteristics and features of their codes: Ansoft, Corp.; ANSYS, Inc.; CRC Research Institute; CST, GmbH; EMA, Inc.; IMST, GmbH; The Japan Research Institute; Matra Systemes & Information; Remcom, Inc.; QWED; Technical University Hamburg- Harburg; Weidingler Associates, Inc.; Zeland Software, Inc. Yang Cai, David Dibben, Erik Esveld, Vyacheslav Komarov, and Per Risman informed the author about particular issues relevant to the software considered.

References

1. Mitrotzik, M.S., Prather, D: How to choose EM software. IEEE Spectrum **34** (Dec 1997) 53–58
2. Lorenson, C.: The why's and how's of math modeling for microwave heating. Microwave World **11** (March 1990) 14–23
3. Palombizio, A., Yakovlev, V.V.: Parallel worlds of microwave modeling and industry: a time to cross? Microwave World **20** (Sept 1999) 14–19
4. Risman, P.: A microwave oven model. Examples of microwave heating computations. Microwave World **19** (Summer 1998) 20-21

On the Representation of Differential Forms by Potentials in Dimension 3

A. Bossavit[1]

Électricité de France, 1 Av. Gal de Gaulle, 92141 Clamart, France

Abstract The notions of scalar and vector potential of a layer of charge or currents are presented in differential geometric language, using a notation which aims at clarifying the connection with standard vector algebraic formalism.

1 Introduction

The reinterpretation of Maxwell's equations in terms of differential forms which has occurred these last years in computational electromagnetism has proved beneficial to the interested community. We now understand edge elements, for instance, as a natural brand of finite elements for differential forms of degree one (electric and magnetic field). Well-established methods of "generalized finite differences" flavor, such as those implemented in the MAFIA codes [6], can now be understood as a natural discretization of the differential-geometric version of Maxwell's equations, thanks to the Finite Integration Technique and to the view of network constitutive laws as a discretized version of the Hodge operator. Since the marriage between finite elements and boundary-integral methods is also a natural thing to do in electromagnetism, one feels the need to reinterpret the latter in geometric language.

A few steps in this direction were taken in an earlier paper [1] where I studied generalizations to differential forms of the "Dirichlet-to-Neumann map", that connects the trace of a harmonic function with the values of its normal derivative at the boundary. A classification of all integral methods liable to be used in hybrid B.I.M.–F.E.M. methods was thus achieved. The paper—to which I have to refer for this—was only descriptive, and relied on a small number of basic statements of the potential theory of differential forms. The aim of the present note is to prove these. Basic notions, for which there are precious few sources (among these, [3]) are given in Section 2. In view of applications (briefly suggested in Section 4), I have taken care to set bridges between the notations of differential geometry and of standard vector calculus. The core of the paper, potential theory, is Section 3.

2 Differential Forms

In what follows, E denotes three-dimensional oriented Euclidean space, V is the associated (real) vector space, with dot product "$\cdot$" and cross product

"×". The distance of two points x and y, i.e., the norm $(v \cdot v)^{1/2}$ of the vector $v = y - x$, is denoted $|x - y|$. At times, it will be insightful to pretend we work in generic dimension d, but the case we do have in mind is $d = 3$.

A smooth bounded surface S, equipped with a continuous field of unit normals n, is embedded in E. In many cases of interest, S is the image of a sphere, enclosing some bounded domain, and hence ∂S is empty. For simplicity, we shall assume this is so, although this restriction could easily be lifted.

By "$f = x \to \mathrm{Expr}(x)$" one means "$f$ is the function that assigns to x the value of the expression Expr". We'll keep in mind the distinction between "vector" (an element v of V) and "vector at a point" (a pair $\{x, v\}$, with v considered as "anchored at x"), without stressing it too heavily. It applies of course to all kinds of geometric objects, including those coming now.

2.1 Algebra of Forms

A *p-covector* ω is a multilinear alternating map of type $V \times \ldots \times V \to \mathbb{R}$, sending p vectors $\{v_1, \ldots, v_p\}$ to a real number that we shall as a rule denote by $\langle \omega; v_1, \ldots, v_p \rangle$. (One may see this as the "oriented p-volume" of the parallelotope built on the v_is.) If u is a given fixed vector, the maps $v \to u \cdot v$ and $\{v, w\} \to u \cdot (v \times w)$ are p-covectors with $p = 1$ and $p = 2$ respectively, which we denote by ${}^1u$ and ${}^2u$. If φ is a given real number, the map $\{u, v, w\} \to \varphi\, u \cdot (v \times w)$ is a 3-covector, similarly denoted ${}^3\varphi$. To φ also corresponds the 0-covector ${}^0\varphi$, mapping an empty list of vector arguments to the real number φ. A vector *field* u, now, and a scalar field φ, generate fields of covectors, the obvious way (The covector ${}^pu(x)$, i.e., the value at x of the form ${}^pu$, operates on vectors anchored at x.) We still denote them by ${}^1u$, ${}^2u$, and ${}^3\varphi$, ${}^0\varphi$. A smooth field of p-covectors is called a *differential form* (DF) *of degree p*, or "p-form". A 0-form, thus, is just a function. Obviously, the foregoing examples, where φ or u act as "proxies" for the forms ${}^0\varphi$, ${}^1u$, etc., exhaust all possibilities when $d = 3$.

Covectors thus being "p-slot machines", some of the slots can be filled up with fixed, given vectors, hence covectors of lower degree. In particular, $i_v\omega$ will denote the $(p-1)$-covector $\{v_2, \ldots, v_p\} \to \langle \omega; v, v_2, \ldots, v_p \rangle$. If v is a vector field, the corresponding form $i_v\omega$ is called *inner product* of v and ω.

We shall also consider covectors, and forms, "living on S", in the sense that vectors on which they act are always tangent to S. Curiously perhaps, the situation is more involved here than in 3D, for in addition to the 0-covectors ${}^0\varphi$, the 1-covectors ${}^1u = v \to u \cdot v$ and the 2-covectors ${}^2\varphi = \{v, w\} \to \varphi\, n \cdot (v \times w)$, we also have the so-called "twisted" 1-covectors ${}^{\tilde{1}}u = v \to u \cdot (n \times v)$, and the corresponding forms. (Twisted forms exist in all dimensions, but involve considerations on orientation that we chose to ignore here, except for the previous example.)

Notice that φ and u, in what precedes, could well have been the traces on S of 3D fields (in the case of u, the "tangential" trace, i.e., the orthogonal projection to the tangent plane, which we whall denote by u_S), which motivates what follows. A smooth form living in E has a *tangential part* or *trace* on S, which one might define as its pullback by the injection of S in E, but we'll prefer the following direct description: the trace $t\omega$ is the field, living on S, of covectors $t\omega(x) = \{v_1, \ldots, v_p\} \to \langle \omega(x); v_1, \ldots, v_p \rangle$, where all the v_is are tangent to S at x. Remark that $t\,{}^1u = {}^1u_S$, but $t\,{}^2u = {}^2(n \cdot u)$. The *normal part* on S of a p-form ω, with $p > 0$, is defined as $n\omega = t\, i_n\omega$. Thus, $n\,{}^1u = {}^0(n \cdot u)$ and $n\,{}^2u = {}^{\bar{1}}u_S$, while $n\,{}^3\varphi = {}^2\varphi$.

It may happen that ω, though smooth outside S, is discontinuous across S. Then traces $t_+\omega$ and $t_-\omega$ can be defined on the "upstream" and "downstream" sides of S (taking n as the direction of the imagined flow). The *jump* of $t\omega$ [resp. $n\omega$] is then $[t\omega] = t_+\omega - t_-\omega$ [resp. $[n\omega] = n_+\omega - n_-\omega$]. Be aware that $[t\,{}^2u] = {}^2[n \cdot u]$, and $[n\,{}^2u] = {}^{\bar{1}}[u_S]$.

The *exterior derivative*, in standard differential geometry, is an operator d from p-forms to $(p+1)$-forms, so defined as to make valid the Stokes theorem, $\int_{\partial\Omega} \omega = \int_\Omega \mathrm{d}\omega$, where Ω is some $(p+1)$-manifold. A form ω is *closed* if $\mathrm{d}\omega = 0$. It's easy to see that

$$\mathrm{d}\,{}^0\varphi = {}^1(\mathrm{grad}\varphi),\ \mathrm{d}\,{}^1u = {}^2(\mathrm{rot}u),\ \mathrm{d}\,{}^2u = {}^3(\mathrm{div}u). \tag{1}$$

On S, similar appeal to the Stokes theorem turns this into

$$\mathrm{d}\,{}^0\varphi_S = {}^1(\mathrm{grad}_S\varphi),\quad \mathrm{d}\,{}^1u_S = {}^2(\mathrm{rot}_S u_S), \tag{2}$$

where grad_S and rot_S are the obvious tangential operators ($\mathrm{rot}_S u_S = n\cdot\mathrm{rot}u$). All this can thus be subsumed:

Lemma 1. $\mathrm{d}t = t\,\mathrm{d}$.

Proof. (The symbol d is overloaded here, to be understood as acting on spatial forms on the right-hand side and on surface forms on the left-hand side.) Take a p-form ω and a smooth $(p+1)$-manifold Ω embedded in S. Then, by Stokes and the very definition of t, one has $\int_\Omega \mathrm{d}t\omega = \int_{\partial\Omega} t\omega = \int_{\partial\Omega} \omega = \int_\Omega \mathrm{d}\omega = \int_\Omega t\mathrm{d}\omega$, for any Ω, hence the result. (Note that $\mathrm{d}\,[t\omega] = [t\,\mathrm{d}\omega]$.) $\diamond$

The *Hodge operator* $*$ maps a p-covector ω to the $(d-p)$-covector $*\omega$ such that $\langle *\omega; v_{p+1}, \ldots, v_d\rangle = \langle \omega; v_1, \ldots, v_p\rangle$ when the v_is form a direct orthonormal basis (which is enough of a definition, by linearity). Reiterating this yields $** = (-1)^{p(d-p)} = (-1)^{p(d-1)}$. One then sees that $*\,{}^0\varphi = {}^3\varphi$ and $*\,{}^1u = {}^2u$ and vice versa ($** = 1$ when $d = 3$). On S, $*\,{}^0\varphi = {}^2\varphi$ and $*\,{}^2\varphi = {}^0\varphi$, but $*\,{}^1u = -{}^{\bar{1}}u$ and $*\,{}^{\bar{1}}u = {}^1u$.

Lemma 2. $*n = t*$.

Proof. (Same kind of overloading, $*$ being $*_E$ on the right and $*_S$ on the left.) At a point x in S, take the orthonormal basis $\{n, v_2, \ldots, v_d\}$, where all the v_is are tangent to S. Then $\langle *n\omega; v_{p+1}, \ldots, v_d\rangle = \langle n\omega; v_2, \ldots, v_p\rangle = \langle \omega; n, v_2, \ldots, v_p\rangle = \langle *\omega; v_{p+1}, \ldots, v_d\rangle = \langle t * \omega; v_{p+1}, \ldots, v_d\rangle$, according to the definitions of n and of the two $*$'s. $\diamond$

As a corollary, $*t = (-1)^p n*$, where p is the degree of the form onto which this acts (as will be our convention in all cases). This comes from $*_E*_E = (-1)^{(d-1)p}$ and $*_S*_S = (-1)^{dp}$.

The *wedge product* $\omega \wedge \eta$ of a p-covector by a q-covector is a $(p+q)$-covector, namely the map

$$\{v_1, \ldots, v_p, v_{p+1}, \ldots, v_{p+q}\} \to \sum_{\sigma \in \mathcal{C}_p^{p+q}} \pm \langle \omega; v_{\sigma(1)}, \ldots, v_{\sigma(p)} \rangle \langle \eta; v_{\varsigma(1)}, \ldots, v_{\varsigma(q)} \rangle.$$

Here, $\mathcal{C}_p^{p+q}$ is the set of increasing injections from $[1, p]$ into $[1, p+q]$, ς is the complementary injection, and the sign is the parity of the number of swaps necessary to reorder the v_is. One may check that $\eta \wedge \omega = (-1)^{pq} \omega \wedge \eta$. Again, the definition via proxies makes all this easier: ${}^0\varphi \wedge \omega = \varphi\omega$, ${}^1u \wedge {}^1u' = {}^2(u \times u')$, and ${}^1u \wedge {}^2u' = {}^3(u \cdot u')$. On S, ${}^{\bar{1}}u \wedge {}^1u' = {}^2(u \cdot u')$. One also verifies, using (1) and (2), that

$$\mathrm{d}(\omega \wedge \eta) = \mathrm{d}\omega \wedge \eta + (-1)^p \omega \wedge \mathrm{d}\eta, \tag{3}$$

where p is the degree of ω. It's also easy to check this, to be used later:

Lemma 3. $\langle \omega \wedge *\omega; v_1, \ldots, v_d \rangle = \sum_{\sigma \in \mathcal{C}_p^d} \langle \omega; v_{\sigma(1)}, \ldots, v_{\sigma(p)} \rangle^2 \geq 0.$

The *scalar product* of two covectors ω and ω' is $\langle \omega \wedge *\omega'; v_1, \ldots, v_d \rangle$ where the v_is form a direct orthonormal basis. It is equal to the scalar product of the proxies, in all cases. We'll take the liberty to denote it by $\omega \cdot \omega'$.

2.2 Analysis of Forms

Differential forms of degree p can be integrated on p-manifolds, by the standard limit process on Riemann sums, but a definition in terms of proxies helps bypassing what would be a tedious development: $\int_c {}^1u = \int_c \tau \cdot u$ and $\int_\Sigma {}^2u = \int_\Sigma n \cdot u$, where τ and n are unit fields, tangent to the curve c and normal to the surface Σ, respectively. For a volume Ω, one has $\int_\Omega {}^3\varphi = \int_\Omega \varphi$, and the "integral" of ${}^0\varphi$ on the 0-manifold that a point x makes is just $\varphi(x)$.

We can now define the prehilbertian scalar product $(\omega, \omega') = \int_E \omega \wedge *\omega'$, on the space $\mathcal{F}^p(E)$ of smooth compactly supported p-forms. This makes sense, because $\omega \wedge *\omega' \equiv \omega' \wedge *\omega$ is a d-form, meant to be integrated on a d-manifold, and $(\omega, \omega) \geq 0$ by Lemma 3. Completion of $\mathcal{F}^p(E)$ with respect to the corresponding norm gives a Hilbert space which we shall call $F^p(E)$. Similarly, but with a deliberately distinct notation, $\langle \omega, \omega' \rangle = \int_S \omega \wedge *\omega'$, hence $F^p(S)$ by completion. In terms of proxies, one has $({}^0\varphi, {}^0\varphi') = ({}^3\varphi, {}^3\varphi') = \int_E \varphi\varphi'$ and $({}^1u, {}^1u') = ({}^2u, {}^2u') = \int_E u \cdot u'$. On S, $\langle {}^0\varphi, {}^0\varphi' \rangle = \langle {}^2\varphi, {}^2\varphi' \rangle = \int_S \varphi\varphi'$ and $\langle {}^1u, {}^1u' \rangle = \langle {}^{\bar{1}}u, {}^{\bar{1}}u' \rangle = \int_S u \cdot u'$. Sobolev spaces of forms are the next obvious step: $F^p_{\mathrm{d}}(E)$ will be the space $\{\omega \in F^p(E) : \mathrm{d}\omega \in F^{p+1}(E)\}$, equipped with its natural scalar product, $(\omega, \omega') + (\mathrm{d}\omega, \mathrm{d}\omega')$. Same thing with $F^p_{\mathrm{d}}(S)$, with $\langle \, , \rangle$ substituted for $(\, ,)$. "Trace theorems" hold as in the scalar theory [4]. In particular, $\omega \in F^p_{\mathrm{d}}(E)$ implies that $t\omega$ is well defined and the same on both sides of S, hence $[t\,\omega] = 0$.

The *codifferential* is the operator $\delta = (-1)^{d(p-1)+1} * \mathrm{d}*$, or more simply said, such that $*\delta = (-1)^p \mathrm{d}*$. (Equivalently, $\delta * = (-1)^{d-p+1} * \mathrm{d}$.) The sign is there to make $(\mathrm{d}\omega, \eta)$ equal, thanks to (3) and the Stokes theorem, to $(\omega, \delta\eta)$, up to boundary terms, thus casting δ in the role of *adjoint* of d. A form ω

is *coclosed* if $\delta\omega = 0$. In terms of proxies, $\delta^3\varphi = -{}^2(\mathrm{grad}\varphi)$, $\delta^2 u = {}^1(\mathrm{rot}u)$, and $\delta^1 u = -{}^0(\mathrm{div}u)$. Again, if we call $F^p_\delta(E)$ the space $\{\omega \in F^p(E) : \delta\omega \in F^{p-1}(E)\}$, equipped with its natural scalar product, $\omega \in F^p_\delta(E)$ implies that the normal part $n\omega$ is well defined and $[n\,\omega] = 0$.

On S, $\delta^1 u = -{}^0(\mathrm{div}_S u)$, where $\mathrm{div}_S u$ is defined as $\mathrm{rot}_S(n \times u)$, and $\delta^2\psi = {}^1(\mathrm{rot}_S\psi)$, where rot_S is reused (let's accept the slight risk of confusion that goes with such overloading) to denote the surface operator $\psi \to -n \times \mathrm{grad}_S\psi$. Remark that $\delta^{\bar{1}} u = -{}^0(\mathrm{rot}_S u)$.

Lemma 4. $\delta n = (-1)^d n\delta$ (i.e., $\delta n = -n\delta$ in 3D).

Proof. Pre- and post-multiply $\mathrm{d}\,t = t\,\mathrm{d}$ by $*$, and use $t* = *n$. The only difficulty is to keep track of the minus signs. ◇

Duly accounting for boundary terms, now, yields integration-by-parts formulas. For instance, let $\omega \in F^p_{\mathrm{d}}(E)$ and $\eta \in F^{p+1}_\delta(E - S)$. The jump $[t\omega]$ across S must then vanish, but $[n\eta]$ needs not. By Stokes, $\int_{E-S} \mathrm{d}(\omega \wedge *\eta) = \int_S [t(\omega \wedge *\eta)] = \int_S t\omega \wedge [t * \eta] = \int_S t\omega \wedge [*n\eta]$ after Lemma 2. Hence

$$(\mathrm{d}\omega, \eta) - (\omega, \delta\eta) = \langle t\omega, [n\eta]\rangle. \tag{4}$$

The two interesting 3D interpretations, $\int_{E-S}(u\cdot\mathrm{grad}\varphi - \varphi\,\mathrm{div}u) = \int_S \varphi\,[n\cdot u]$ and $\int_{E-S}(u\cdot\mathrm{rot}v - v\cdot\mathrm{rot}u) = \int_S u\cdot[n \times v]$, are well known. Symmetrically, if $\omega \in F^p_{\mathrm{d}}(E - S)$ and $\eta \in F^{p+1}_\delta(E)$,

$$(\mathrm{d}\omega, \eta) - (\omega, \delta\eta) = \langle [t\omega], n\eta\rangle. \tag{4'}$$

On S, the formula is $\langle \mathrm{d}\omega, \eta\rangle = \langle \omega, \delta\eta\rangle$ (no boundary terms under our simplifying assumptions), which translates as $\int_S \varphi\,\mathrm{div}_S u = -\int_S u\cdot\mathrm{grad}_S\varphi$ and $\int_S \varphi\,\mathrm{rot}_S u = \int_S u\cdot\mathrm{rot}_S\varphi$.

Proposition 1 (Green's formulas). *If both* ω and ω' *are in both* $F^p_{\mathrm{d}}(E)$ *and* $F^p_\delta(E)$ *and their* d's *in* $\in F^{p+1}_\delta(E - S)$, *then*

$$(\delta\mathrm{d}\omega, \omega') - (\omega, \delta\mathrm{d}\omega') = \langle [n\mathrm{d}\omega], t\omega'\rangle - \langle t\omega, [n\mathrm{d}\omega']\rangle. \tag{5}$$

Under symmetrical assumptions,

$$(\mathrm{d}\delta\omega, \omega') - (\omega, \mathrm{d}\delta\omega') = -\langle [t\delta\omega], n\omega'\rangle - \langle n\omega, [t\delta\omega']\rangle. \tag{5'}$$

Proof. Set $\eta = \mathrm{d}\omega'$ in (4) and use the symmetry of $(\mathrm{d}\omega, \mathrm{d}\omega')$. Treat (4') the same way, with $\omega = \delta\eta'$, and swap symbols. ◇

3 Potentials

The *Green operator* $\mathcal{G}$, in dimension 3, sends $\alpha \in \mathcal{F}^p(E)$ to the form $\mathcal{G}\alpha$ such that, for any p vectors $v_1, \ldots, v_p$ anchored at x,

$$\langle \mathcal{G}\alpha(x); v_1, \ldots, v_p\rangle = \frac{1}{4\pi}\int_E \frac{\langle \alpha(y); v_1, \ldots, v_p\rangle}{|x - y|}\,dy \tag{6}$$

where dy refers to the Lebesgue measure. (Notice how the vectors need be transported to point y, which makes the theory hard to generalize beyond the Euclidean context.) This is convolution by the kernel $1/(4\pi|x|)$, so derivations such as d and δ "slip through", as usual, by under-the-summation-sign differentiation: $\mathrm{d}\mathcal{G} = \mathcal{G}\mathrm{d}$, $\delta\mathcal{G} = \mathcal{G}\delta$, and

$$(\mathrm{d}\delta + \delta\mathrm{d})\mathcal{G}\alpha = \alpha, \tag{7}$$

showing that $\mathcal{G}\alpha$ is *harmonic* outside the support of α. (Indeed, $-\Delta = \mathrm{d}\delta + \delta\mathrm{d}$ is the Laplacian on forms, as the formulas $(\mathrm{d}\delta + \delta\mathrm{d})^0\varphi = -^0(\mathrm{div}\,\mathrm{grad}\varphi)$ and $(\mathrm{d}\delta + \delta\mathrm{d})^1u = {}^1(\mathrm{rot}\,\mathrm{rot}u - \mathrm{grad}\,\mathrm{div}u)$, and the similar ones obtained when $p = 2$ and $p = 3$, testify.) Also note that $*\mathcal{G} = \mathcal{G}*$.

It's tempting to call $\mathcal{G}\alpha$ the "potential" of α, but we'll refrain to do so: physics suggests that the proper definition for the *potential* of the *charge* q is rather $\mathcal{G} * q$, with prior hodging, a p-form if q was a $(d-p)$-form. (Notice how $\mathcal{G}*$ behaves like a kind of "non-local Hodge".) Anyway, we need to consider potentials of charges living on S or other zero-measure sets, for which (6) is not an adequate formula. This prompts the following generalization to so-called "currents".

Currents, in de Rham's sense [5], generalize differential forms just as ordinary distributions generalize functions. The idea is strictly the same: on $\mathcal{F}^p(E)$, take linear functionals C such that $\lim_k C(\omega'_k) = C(\omega')$ when ω'_k tends to ω' in the same sense as test functions in Schwartz's theory (uniform convergence of all derivatives on all compact sets). Clearly, the map $\omega' \to (\omega, \omega')$ satisfies these requirements, which allows one to inject smooth forms into the space of currents. The latter is much larger than $\mathcal{F}^p(E)$: For instance, a p-manifold M generates a p-current, by $\omega' \to \int_M \omega'$. Also, and of particular interest to us here, the map $\omega' \to \langle\alpha, t\omega'\rangle$, where α is a p-form living on S, is a current, which we shall denote by $t^*\alpha$, like the adjoint it is (for $\langle\alpha, t\omega'\rangle = (t^*\alpha, \omega')$). Operators d and δ extend to currents (no need to create new symbols) via $\mathrm{d}\omega = \omega' \to (\omega, \delta\omega')$, and the like. Same trick with the star, and also with the Green operator, defined as $\mathcal{G}\omega = \omega' \to (\omega, \mathcal{G}\omega')$. It's now natural to let it act on the current $t^*\alpha$, hence a new operator $G = \mathcal{G}t^*$:

Definition. *$G\alpha$ is the current $\omega' \to \langle\alpha, t\,\mathcal{G}\omega'\rangle$.*

The *potential* of the surface charge q is then defined as $G * q$.

Such a definition will help keep proofs short, but lacks intuitive appeal. In search of an alternative, let's first prove the following Lemma. At a point y of S, let's denote by π the orthogonal projection to the tangent plane, and by $\pi^*\alpha$ the covector $\{v_1, \ldots, v_p\} \to \langle\alpha; \pi v_1, \ldots, \pi v_p\rangle$. Let $t_y\omega(x)$ be the p-covector at y, acting on vectors tangent to S at this point. Then,

Lemma 5. $(\pi^*\alpha)(y) \cdot \omega(x) = \alpha(y) \cdot t_y\omega(x)$.

Proof. Use Lemma 3, with special orthonormal bases which relate as follows: on the right, $d-1$ vectors tangent to S at y. On the left, the basis obtained by placing $n(y)$ ahead of the previous list. All terms containing n vanish, hence the result. ◇

Proposition 2. *An alternate expression of $G\alpha$ is*

$$\langle G\alpha(x); v_1, \ldots, v_p\rangle = \frac{1}{4\pi}\int_S \frac{\langle\alpha(y); \pi v_1, \ldots, \pi v_p\rangle}{|x-y|}\,dy, \tag{8}$$

where dy is the surface measure. (This corrects (8) in [1], where the π's were missing.)

Proof. More compactly, (8) is $G\alpha(x) = (4\pi)^{-1} \int_S dy\,(\pi^*\alpha)(y)/|x-y|$. Take a smooth p-form ω'. Then $4\pi\,(G\alpha, \omega') = \int_E dx \int_S dy\,((\pi^*\alpha)(y)\cdot\omega'(x))/|x-y|$. On the other hand, $4\pi\,\langle\alpha, t\,\mathcal{G}\omega'\rangle = \int_S dy \int_E dx\,(\alpha(y)\cdot(t_y\omega')(x))/|x-y|$, and the two integrals are equal, thanks to Fubini and Lemma 5. ◊

Properties of G differ from those of $\mathcal{G}$: Some commutativity relations have been lost in the process. Now, $\mathrm{d}G \neq G\mathrm{d}$ and $*G \neq G*$, as a rule (the two d's and the two stars refer to different dimensions). It remains true that $[tG\alpha] = 0$, however (it can be seen from (8), assuming $\alpha \in F^p(S)$), and that

Proposition 3. $G\delta = \delta G$.

Proof. On the one hand, $\delta G\alpha = \omega' \to (G\alpha, \mathrm{d}\omega')$, on the other hand $G\delta\alpha = \omega' \to \langle\delta\alpha, tG\omega'\rangle$. But $(G\alpha, \mathrm{d}\omega') = \langle\alpha, tG\mathrm{d}\omega'\rangle = \langle\alpha, \mathrm{d}tG\omega'\rangle = \langle\delta\alpha, tG\omega'\rangle$. ◊

Corollary. *If* $\delta\alpha \in F^{p-1}(S)$, *then* $[nG\alpha] = 0$ *and* $[t\delta G\alpha] = 0$.

Proof. By Prop. 3, $[nG\alpha]$ has a (locally, at least) square summable δ, hence the same normal part $n\delta G\alpha$ on both sides. Next, $[t\delta G\alpha] = [tG\delta\alpha]$. ◊

As for (7), it must now be replaced by $(\mathrm{d}\delta + \delta\mathrm{d})G\alpha = t^*\alpha$, a current "supported on S", hence

Proposition 4. *Outside* S, $\mathrm{d}\delta G = \mathrm{d}G\delta = -\delta\mathrm{d}G$.

The status of all possible jumps has thus been clarified, except for the most important one, relative to what can be called the "normal derivative", $n\mathrm{d}G\alpha$, of $G\alpha$:

Proposition 5. $[n\mathrm{d}G\alpha] = \alpha$.

Proof. Let $\omega' = \mathcal{G}\alpha'$, where α', unlike α, is a smooth form with compact support in E. Then $-\Delta\omega' = \alpha'$, by (7), $[n\mathrm{d}\omega'] = 0$ and $[t\delta\omega'] = 0$ owing to the smoothness of ω'. On the other hand, $\Delta\omega = 0$ in $E - S$ by Prop. 4, and $[t\delta\omega = 0]$ (corollary to Prop. 3). Adding (5) and (5'), we find that $(\omega, \alpha') = \langle[n\mathrm{d}\omega], t\mathcal{G}\alpha'\rangle$. Since on the other hand $(G\alpha, \alpha') = \langle\alpha, t\mathcal{G}\alpha'\rangle$, the result follows. ◊

4 3D Illustrations

The three possibilities in dimension 3 are

$$^0\varphi \equiv G * {}^2q = {}^0(x \to \frac{1}{4\pi}\int_S \frac{q(y)}{|x-y|}\,dy), \tag{9}$$

$$^1a \equiv G * {}^{\tilde{1}}j = {}^1(x \to \frac{1}{4\pi}\int_S \frac{j(y)}{|x-y|}\,dy), \tag{10}$$

$$^2u \equiv G * {}^0\psi = {}^2(x \to \frac{1}{4\pi}\int_S \frac{\psi(y)\,n(y)}{|x-y|}\,dy), \tag{11}$$

respectively corresponding to a single layer of charge, a single layer of currents, and to what is sometimes called the "curl potential" of a scalar distribution over S. Let's focus on a in (10), in the case where the charge $^{\tilde{1}}j$ is a closed form, i.e., $\text{div}_S j = 0$. Then $\text{div}a = 0$ by Prop. 3, and Prop. 5 says that if one sets $h = \text{rot}a$, then $[n \times h] = -j$, in conformity with Ampère's theorem. (We'll ignore such minus signs, which are an unavoidable nuisance in this theory, for the rest of the discussion.) Now, if $\text{div}_S j = 0$, there is a stream-function ψ from which j derives, $j = \text{rot}_S \psi$ (note how considerations of topology would intrude at this point if S was not an embedded sphere as assumed), i.e., ${}^1j = \text{d}^0\psi$. By Prop. 3, then, one has ${}^1a = \delta^2 u$, where u is the curl-potential of ψ, as given by (11), i.e., $a = \text{div}u$. But now let's invoke Prop. 4, as applied to ${}^2\psi$: It says that $\text{d}\,{}^1a = \delta \text{d}\,{}^2u$, that is, in more understandable language, that the proxy field $h = \text{rot}a$ has another representation outside S, as $h = \text{grad}\phi$, where $\phi = \text{div}u$. The latter potential, ϕ, is *not* continuous across S, and its jump is equal to ψ: it's what is called a "double-layer" potential (with a kernel in $|x - y|^{-3}$ instead of $1/|x - y|$).

What we have observed here is general: A single layer of charge on S (a closed $(d-1-p)$-form $q = \text{d}k$) generates a harmonic p-form ω, its potential. Then $\text{d}\omega$ happens to be, outside S of course, the δ of another kind of potential, namely $\text{d}G * k$, the double-layer potential of k. When $d = 3$ and $p = 1$, we thus find a dual representation of a harmonic vector field, $h = \text{rot}\,a = \text{grad}\,\phi$, for instance, in terms of one or the other potential. Since the same vector field can represent a 2-form as well, we also have a dual representation of h of the same appearance, $h = \text{grad}\varphi = \text{rot a}$, but where now φ is the scalar potential of a single layer of magnetic charge, and a (not the same as the above a) the (vector) double layer-potential of a surface magnetic current.

Graphic illustrations of these dual representations can be found in [1], with an application to the construction of "Dirichlet-to-Neumann" maps.

Acknowledgments

Thanks to R. Hiptmair for his critical reading of [1].

References

1. Bossavit, A: The 'scalar' Poincaré-Steklov operator and the 'vector' one (...). *in* Domain Decomposition Methods for Partial Differential Equations (R. Glowinski et al., eds.), SIAM, Philadelphia (1991) 19-26.
2. Costabel, M.: Boundary integral operators on Lipschitz domains. SIAM J. Math. Anal. **19** (1988) 613-26.
3. Morrey, C.B.: Multiple integrals in the calculus of variations. Springer-Verlag, New York (1966).
4. Paquet, L.: Problèmes mixtes pour le problème de Maxwell. Annales Fac. Sc. Toulouse **4** (1982) 103-41.
5. de Rham, G.: Variétés différentiables. Hermann, Paris (1960).
6. Weiland, T.: Time domain electromagnetic field computation with finite difference methods. Int. J. Numer. Modelling: Electronic Networks, Devices and Fields **9** (1996) 295-319.

Hysteresis Models for Transient Simulation*

L.R. Dupre[1], R. Van Keer[2], J.A.A. Melkebeek[1],
Y.I. Moroz[3], S.E. Zirka*[3]

[1] Ghent Univ., Dept. Elect. Power Eng., St. Pietersnieuwstraat 41, B-9000 Gent, Belgium
[2] Ghent Univ., Dept. Mathemat. Analysis, Galglaan 2, B-9000 Gent, Belgium
[3] Dnepropetrovsk State Univ., Naukova 13, Dnepropetrovsk 49050, Ukraine

Abstract A simple alternative to the history-dependent (HD) hysteresis models is studied. It is a history-independent (HI) model, where any trajectory in the *B*-*H* plane is constructed by means of interpolation between the first-order reversal curves obtained by experiments. It is shown that such a model, where the path of any transition curve is completely determined by the current reversal point, may be used under certain conditions for modeling the transients in conducting ferromagnetic media. These conditions are the large swing of the average induction, B_a, during the cycle and the smoothness of the dependence $B_a(t)$. It was found that in the regime where the derivative dB_a/dt changes its sign no more than twice during a period, the difference between the results obtained with the accurate HD- and the simplified HI-models is almost negligible.

1 Introduction

The reliability of modeling and quality of design of a wide range of devices containing magnetic cores is determined by the proper accounting for hysteresis. At the same time there is an open question concerning the specific hysteresis model used. The most advanced HD-hysteresis models [1] reproduce all main features of magnetization, including the 'return-point-memory' and the 'wiping-out' property. The wages of the model accuracy is a significant increase of the input (material) data required for the model identification and the complication of the procedure of constructing the transition curves. For this reason we have studied a possible alternative to these models, namely a simple hysteresis model which does not follow the properties above. In such a HI-model, any trajectory in the *B*-*H* plane is completely determined by the current reversal point. In order to investigate the possibilities of the HI-model, we compare the transients and the steady-state regimes in a pulse generator, as well as the losses in a stator tooth of an asynchronous machine (AM). While numerically studying the transients in a conducting ferromagnetic sheet, we explain the reason of the distinction between the processes, calculated with the HD- and HI-models, and state the conditions for the applicability of the history-independent hysteresis model.

* The work was supported financially by the GOA-project 99-200/4 and by the research project No 3604209B of FWO-Vlaanderen. Luc Dupre is a postdoctoral researcher of FWO-Vlaanderen

2 History-Dependent and History-Independent Hysteresis Models

The hysteresis model problem is the problem how to construct (predict) the transition curves which correspond to any changes of the magnetic field H. Inspite the variety of characteristics among different magnetic materials, some general features are observed in their magnetization processes. These features have been described already in 1905, [2], and are known as Madelung's rules (see e.g. [3]). Considering the hysteresis curves in Fig. 1, these experimentally established rules can be stated as follows:

1. The path of any transition (reversal) curve is uniquely determined by the coordinates of the reversal point, from which this curve emanates.
2. If any point 7 of the curve 3-7-1 becomes a new reversal point, then the curve 7-8-3 originating at point 7 returns to the initial point 3 ('return-point memory').
3. If the point 8 of the curve 7-8-3 becomes the newest reversal point and if the transition curve 8-7 extends beyond the point 7, it will pass along the part 7-1 of curve 3-7-1, as if the previous closed loop 7-8-7 did not exist at all ('wiping-out property').

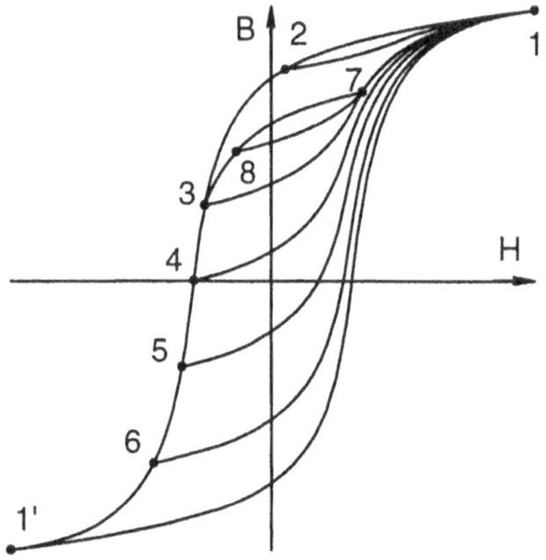

Figure1. Transition curves illustrating Madelung's rules

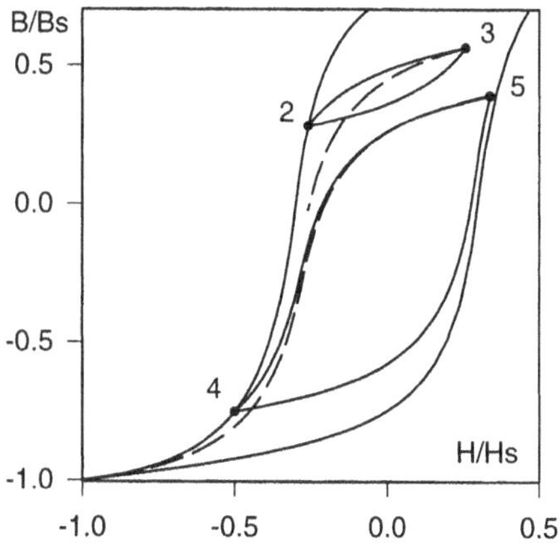

Figure2. Magnetic trajectories in the HD (solid lines) and HI-models (dashed lines)

The choice between rule 1 and rule 2 above determines the choice between the history-dependent and history-independent hysteresis models. If, irrespective of the reversal point location, we follow strictly the first rule, then we ignore the history of the magnetization. Since only the last reversal point is kept by such a *history-independent* model, the trajectory originating from that reversal point must lead immediately to the major loop tip. In Fig. 2, such trajectories originating from points 3 and 5 are shown by dashed lines. They lead directly to a negative saturation, but not to the points 2 and 4, respectively, as in the case of the HD-model.

When constructing a *history-dependent* model satisfying the second (return-point-memory) and the third (wiping-out) rules, the first rule may be limited to the first-order reversal curves, which are the curves that originate from the points lying on the major loop branches (points 2 till 6 in Fig. 1).

3 Implementation of the Hysteresis Models

For an accurate material model, the use of a major hysteresis loop and of a family of first order curves is advisable. Any transition curve of an HI-model is constructed by means of interpolation between the first-order reversal curves obtained by experiments. This leads to a simple history-independent hysteresis model. When the first-order curves are unavailable, they can be generated as outlined in [4] (this procedure is carried out only once for a given material).

Our attention to the first-order reversal curves is also caused by their key role in constructing the history-dependent hysteresis models. In the history-dependent model [4] used in this paper, the reversal curve of any order (its overall dimensions, ΔH and ΔB, are always known in the model satisfying rule 2 above) is constructed by copying the segment of the first-order curve which has the same overall dimensions and is the nearest to the modeled curve in the B-direction. The realization of the HI-model reduces to a simple interpolation. It is interesting to compare the results obtained with these models, and to determine the conditions of applicability for the simple HI-model.

4 The Study of Transients with the HD- and HI-Hysteresis Models

An example of a device for which the transients and steady-state regime can not be evaluated without hysteresis modeling, is a line-type generator designed for continuous operation at the pulse rate f=250 pps ($T = 1/f = 4000\mu$s) and the pulse width t_p=50 μs. During the time t_p, the pulse voltage U_2=250 kV should be generated across the unidirectionally conducting load, R_L=120 Ohm. Pulse waveform requirements for this generator are: the maximum rise and fall times of the voltage are 2 and 4 μs, respectively; the pulse-top nonflatness (the difference between the maximum and minimum values of load voltage over the pulse top duration with respect to the nominal magnitude of the pulse) is less than 3.0%; the maximum backswing is 10%.

We examine a device on the base of a 12-section pulse forming network (PFN) discharging in the load through a step-up pulse transformer (PT). The PFN impedance, $\rho = \sqrt{L_1/C_1}$, is 12 Ohm; the voltage U_1 on the PFN capacitors, C_1, is 55 kV. Therefore the PT turn ratio, $n=w_2/w_1$, is about 5.

An equivalent circuit for studying the transients in the generator is shown in Fig. 3, where the distributed parameter magnetic circuit of the PT is symbolically shown by the inductance L_μ. The switch S represents a power tetrode used as a controlled commutator. The networks $R_k L_k$ and $R_v V$ are used for suppressing the pulse overshoot and backswing, respectively. The current source I_r is used to supply a reset field, H_r, to allow a maximum swing ΔB_a during the pulse. During the time $t_c \cong 0.95T$, the PFN is charged to

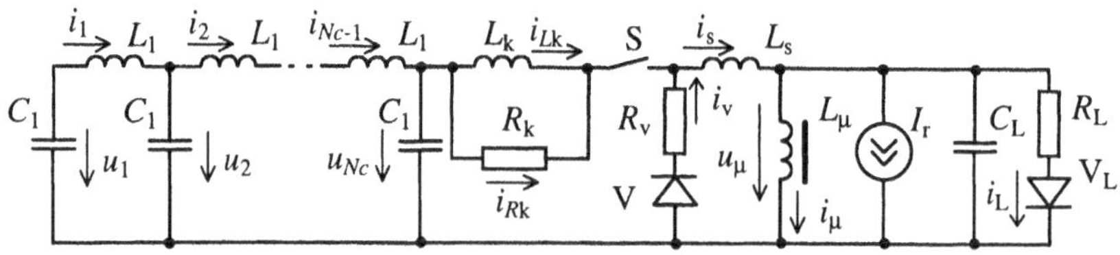

Figure3. Equivalent circuit of the pulse generator

the voltage U_1. When the switch S is closed, the PFN is partially discharged into the resistive-capacitive load R_L-C_L. The analysis that follows assumes that the voltage on the PFN is recovered and that the currents in the $R_k L_k$-network die out completely during the interpulse period, as will normally be true. Using the state-variables denoted in Fig. 3 and employing the finite-difference scheme in the notations from [5] for the evaluation of transients in the conducting ferromagnetic sheets of the core, lead to the following simultaneous equations for the discharging circuit of the generator concerned:

$$\frac{du_1}{dt} = \frac{-i_1}{C_1}, \quad \frac{du_k}{dt} = \frac{i_{k-1} - i_k}{C_1}, \quad (k = 2, ..., N_c - 1), \tag{1}$$

$$\frac{du_{Nc}}{dt} = \frac{i_{Nc-1} - i_{Lk} - i_{Rk}}{C_1}, \quad \frac{di_k}{dt} = \frac{u_k - u_{k+1}}{L_1}, \quad (k = 1, ..., N_c - 1), \tag{2}$$

$$i_v = \frac{i_s R_k - u_{Nc} - i_{Lk} R_k}{R_k + R_v}, \text{ if } i_s R_k > u_{Nc} + i_{Lk} R_k, \text{ otherwise } i_v = 0, \tag{3}$$

$$\frac{di_s}{dt} = \frac{u_{Nc} - i_{Rk} R_k - u_\mu}{L_s}, \quad \frac{di_{Lk}}{dt} = \frac{i_{Rk} R_k}{L_k}, \quad i_{Rk} = i_s - i_{Lk} - i_v, \tag{4}$$

$$\frac{du_\mu}{dt} = \frac{i_s - i_L - i_\mu l_a H_1/w_1 - I_r}{C_L}; \; i_L = \frac{u_\mu}{R_L}, \text{ if } u_\mu > 0, \text{otherwise } i_L = 0, \tag{5}$$

$$\frac{dH_1}{dt} = \frac{8H_2 - 7H_1 - H_3}{2\mu_0 \mu_d(H_1)\sigma h^2} + \frac{3(N-1)u_\mu}{\mu_0 \mu_d(H_1) w_1 S}, \quad \frac{dH_N}{dt} = \frac{2(H_{N-1} - H_N)}{\mu_0 \mu_d(H_N)\sigma h^2}, \tag{6}$$

$$\frac{dH_i}{dt} = \frac{H_{i-1} - 2H_i + H_{i+1}}{\mu_0 \mu_{\rm d}(H_i)\sigma h^2}, \ (i = 2, ..., N-1) \ , \tag{7}$$

Fifty microseconds later, the switch tube is commanded to cut off, and the pulse trailing edge and backswing are formed. When the switch is open, we only solve (5)-(7) together with the equation

$$\frac{di_{\rm s}}{dt} = \frac{-i_{\rm s}R_{\rm v} - u_\mu}{L_{\rm s}}, \text{ if } i_{\rm s} \geq 0, \text{ otherwise } i_{\rm s} = 0. \tag{8}$$

The solution of (1)-(8), obtained for N=13 and N_c=12 (the number of the PFN capacitors), using the HD-model [4], is illustrated by a dynamic magnetization curve shown in Fig. 4 and by a transient voltage across the load (Fig. 5), where $U_2 = u_\mu n$. These dynamic curves correspond to the first 5 periods of the generator work, when the initial state of the magnetic material is characterized by point 1 in Fig. 4 (the reset field $H_{\rm r}$=−300 A/m is assumed). The static hysteresis loop of the material (non-oriented steel with thickness 0.15 mm and resistivity $0.38 \cdot 10^{-10}$ Ohm·m) is shown in Fig. 4 by a dashed line. The dashed line in Fig. 5 is the first pulse in the absence of the network $R_{\rm k}L_{\rm k}$. As seen from Figs. 4 and 5, two periods of the excitation are sufficient to establish a steady-state regime in the magnetic core and, consequently, in the device as a whole.

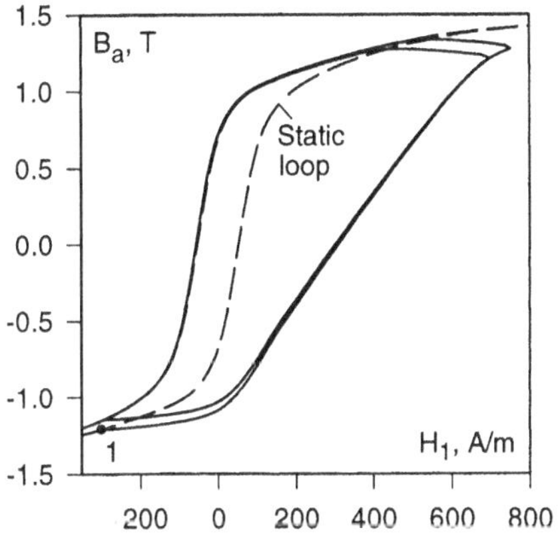

Figure4. Start-up transient in the PT magnetic circuit for $\Delta B_{\rm a}$=2.5 T

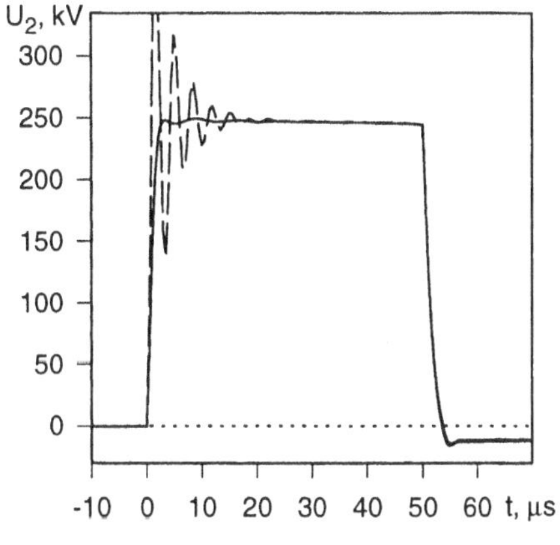

Figure5. Output voltage waveforms with (solid line) and without (dashed line) $R_{\rm k}L_{\rm k}$ network

Due to the appropriate reset field, the swing $\Delta B_{\rm a}$=2.5 T is allowed. It is remarkable that the results obtained for such a $\Delta B_{\rm a}$ with the HI-model, nearly coincide with the above results obtained with the HD-model. For this reason, they are not represented here.

The difference between the HD- and HI-models becomes significant when the swing $\Delta B_{\rm a}$ is limited to 0.5 T, which is only acceptable in the absence of the reset field. The start-up transient in a laminated core, made from 0.35 mm thick sheets, and the corresponding succession of the first nine transformed pulses calculated with the HD-model, are shown in Figs. 6 and 7 respectively. Note that it is impossible to predict the location and shape of the steady-state dynamic loop, and hence the voltage waveform in the load, without using an adequate hysteresis model.

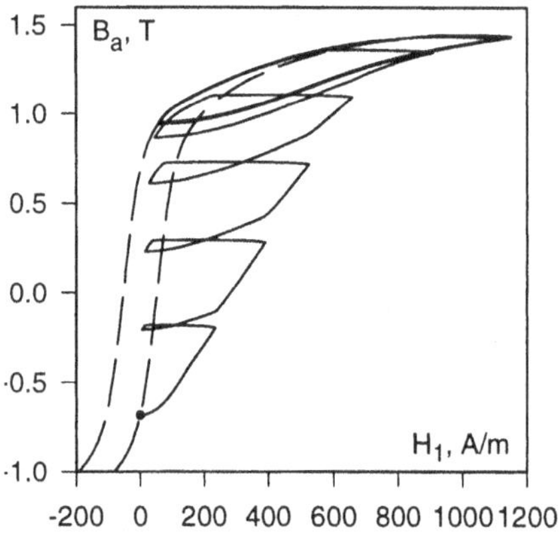

Figure6. Start-up transient in the PT magnetic circuit for ΔB_a=0.5 T

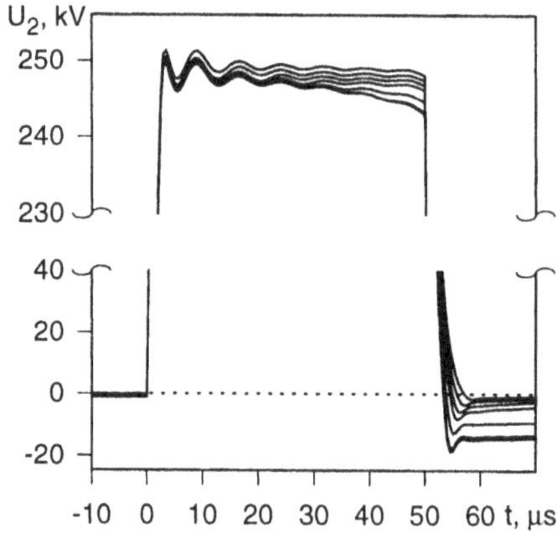

Figure7. The voltage across the generator load for ΔB_a=0.5 T

The difference between the results obtained with the HD- and HI-models, is seen from Figs. 8 and 9, where steady-state hysteresis loops and corresponding pulses in the load are superposed. The area of the steady-state dynamic loop (specific losses per cycle) corresponding to the HD-model is by 6% larger than that for the HI-model. Total nonflatness of the pulse-top is 2.7% for the HD-model, and 2.3% for the HI-model. If the requirements to the pulse shape are less severe, the use of the HI-model would be admissible (the 6% difference in the losses is not crucial in the case when the minor loop area is small). As can be seen from Fig. 2, with the increase of minor loop area, the difference in the losses computed with the HD- and HI- models, becomes smaller. For this reason, the above profiles calculated with the HD- and HI-models for ΔB_a=2.5 T are almost the same.

This is no longer the case (even for a large ΔB_a) when the derivative dB_a/dt changes its sign more than twice during a period. As an example, the magnetization processes in a stator tooth of an asynchronous machine (AM) can be examined. Fig. 10 shows the steady-state loops for the HD- and HI-models when the average induction in the tooth changes, as shown in Fig. 11 by curve 1 (one positive half-cycle is depicted only).

These results have been obtained with the system (6)-(7), where

$$u_\mu = U_m \sin(\omega t) + kU_m \sin(18\omega t) \ . \tag{9}$$

The difference in specific losses per cycle for the HD- and HI-models, is 26% for maximum slot ripples (k=2 in (9)). For smaller slot ripples, shown in

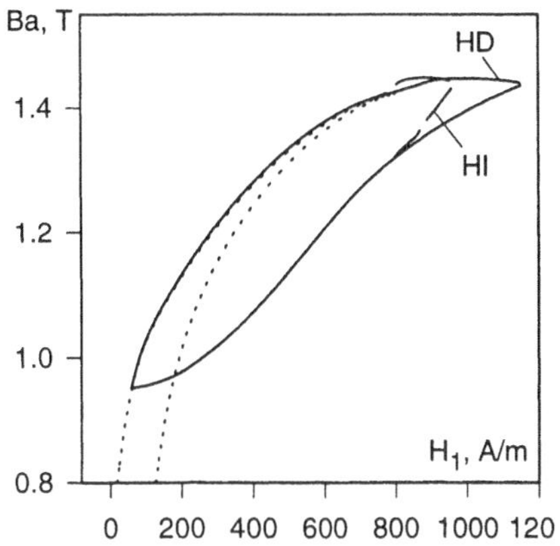

Figure8. Steady-state dynamic loops calculated with HD and HI models

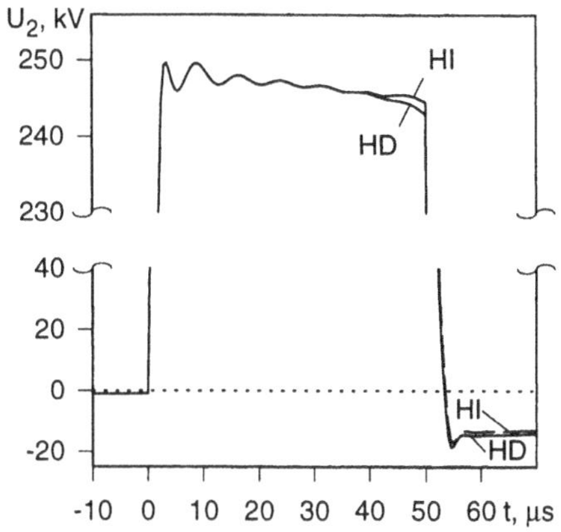

Figure9. Voltage waveforms for HD and HI models (ΔB_a=0.5 T)

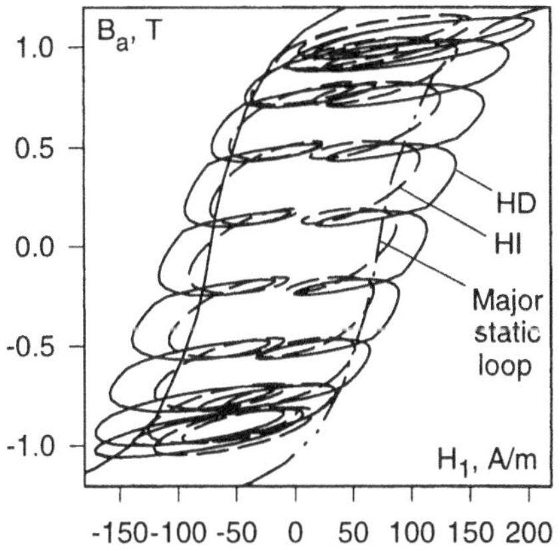

Figure10. Steady-state dynamic loops in the stator tooth of AM

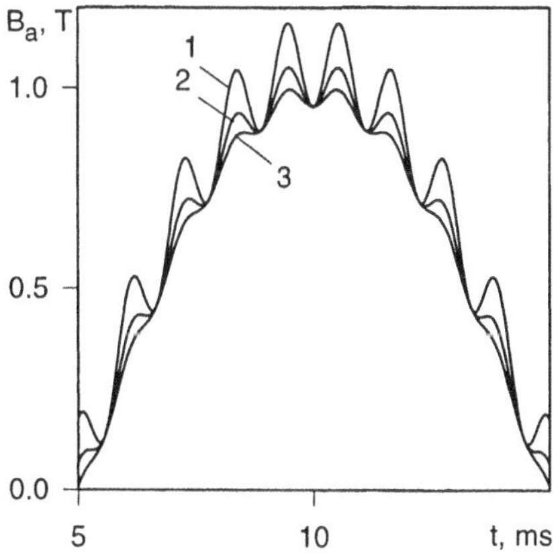

Figure11. The average inductions in the stator tooth

Fig. 11 by curves 2 (k=1) and 3 (k=0.5), this difference is 16% and 1%, respectively. Thus, the smaller is the ripple, the smaller is the difference between the results, obtained with the HD- and HI-models.

5 Conclusions

We have proposed a simple HI-hysteresis model, where any trajectory in the B-H plane is constructed by means of interpolation between the first-order reversal curves obtained experimentally. It was shown that such a model is sufficiently accurate when the amplitude of the average induction, B_a, in the magnetic core is large and the function $B_a(t)$ is smooth. The accuracy of the HI-model decreases with decreasing minor loop area. However, this mode is typical for the low-power devices where the power losses are not crucial. Only in the special case, when the contribution of higher harmonics is substantial (in this regime the derivative dB_a/dt changes its sign more than twice during a period), the HD-hysteresis model should be used. This may suggest to start the design of a new device, where the temporal behavior of B_a is not evident, by means of the simple HI-model. The necessity of using more accurate (but much more involved and time consuming) HD-models becomes clear after analyzing the time variation of the induction B_a.

References

1. Mayergoyz, I.D.: Mathematical models of hysteresis. New York: Springer-Verlag, 1991
2. Madelung, E.: On the magnetization by a fast current and an operation principle of the magnetodetectors of Rutherford-Marconi, (in German). Ann. Phys. **17 (5)** (1905) 861-890
3. Brokate, M., Sprekels, J.: Hysteresis and phase transitions. New York: Springer-Verlag, 1996
4. Zirka, S.E., Moroz, Y.I.: Hysteresis modeling based on similarity. IEEE Trans.Magn. **35 (4)** (1999) 2090-2096
5. Van Keer, R., Dupre, L.R., Melkebeek, J.A.A., Moroz, Y.I., Zirka, S.E.: On the Evaluation of Transients in Conducting Ferromagnetic Cores. These proceedings

Calculation of Frequency Domain Parameters in Lossfree and Lossy Structures Using the FI-Technique and a Modal Approach

R. Schuhmann and T. Weiland

Fachgebiet Theorie Elektromagnetischer Felder
Darmstadt University of Technology, Germany
schuhmann@temf.tu-darmstadt.de

Abstract In this paper we present a new approach for the calculation of scattering parameters of microwave devices based on the discretization with the Finite Integration (FI)-Technique. The macroscopic properties of such devices are extracted from field solutions, which are calculated from a superposition of the eigenvectors of the associated system matrix. Thus the simulation time for highly resonant structures can be considerably reduced compared to conventional time domain analysis.

1 Introduction

The usage of modal approaches is a well-known and very general technique, which can be applied to the solution of both continuous and discrete field problems. The underlying discretization method we use in this paper is the Finite Integration (FI)-Technique [1,2,3], which could be substituted by a wide range of similar methods leading to large algebraic matrix equations characterizing the electromagnetic behavior of the structure to be simulated. As our special interest is in the calculation of scattering parameters, however, some special properties of the FI-Technique are utilized, resulting in a highly efficient and accurate extraction of waveguide mode related data from the field solutions.

Basing on a dual grid doublet $\{G, \widetilde{G}\}$, the FI-Technique supplies a consistent transformation of Maxwell's Equations into a set of matrix equations:

$$\operatorname{curl} \boldsymbol{E} = -\frac{\partial \boldsymbol{B}}{\partial t} \quad \leftrightarrow \quad \mathbf{C}\,\widehat{\mathrm{e}} = -\frac{d}{dt}\widehat{\widehat{\mathrm{b}}} \qquad \operatorname{div} \boldsymbol{D} = \rho \quad \leftrightarrow \quad \widetilde{\mathbf{S}}\,\widehat{\widehat{\mathrm{d}}} = \mathrm{q} \qquad (1)$$

$$\operatorname{curl} \boldsymbol{H} = \frac{\partial \boldsymbol{D}}{\partial t} + \boldsymbol{J} \quad \leftrightarrow \quad \widetilde{\mathbf{C}}\,\widehat{\mathrm{h}} = \frac{d}{dt}\widehat{\widehat{\mathrm{d}}} + \widehat{\widehat{\mathrm{j}}} \qquad \operatorname{div} \boldsymbol{B} = 0 \quad \leftrightarrow \quad \mathbf{S}\,\widehat{\widehat{\mathrm{b}}} = 0 \qquad (2)$$

$$\begin{aligned} \boldsymbol{D} &= \varepsilon\,\boldsymbol{E} & \leftrightarrow \quad \widehat{\widehat{\mathrm{d}}} &= \mathbf{M}_\varepsilon\,\widehat{\mathrm{e}} \\ \boldsymbol{B} &= \mu\,\boldsymbol{H} & \leftrightarrow \quad \widehat{\widehat{\mathrm{b}}} &= \mathbf{M}_\mu\,\widehat{\mathrm{h}} \qquad (3) \\ \boldsymbol{J} &= \boldsymbol{J}_S + \kappa\,\boldsymbol{E} & \leftrightarrow \quad \widehat{\widehat{\mathrm{j}}} &= \widehat{\widehat{\mathrm{j}}}_S + \mathbf{M}_\kappa\,\widehat{\mathrm{e}} \end{aligned}$$

The discrete vectors $\widehat{\mathrm{e}}$ and $\widehat{\widehat{\mathrm{b}}}$ are the electric grid voltages and magnetic grid fluxes, defined on the edges and facets of the primary grid G, respectively, and the vectors $\widehat{\widehat{\mathbf{d}}}$, $\widehat{\mathbf{h}}$, and $\widehat{\widehat{\mathbf{j}}}$ are the electric grid fluxes, the magnetic grid voltages and the electric grid currents, defined on facets and edges of the dual grid $\widetilde{G}$. The matrices $\mathbf{C}$ and $\widetilde{\mathbf{C}}$ are the discrete curl-operators for the primary and the dual grid (with $\mathbf{C} = \widetilde{\mathbf{C}}^T$); $\mathbf{S}$ and $\widetilde{\mathbf{S}}$ are the discrete div-operators ('source'). The discrete vectors are related to each other by the material matrices in (3), which can be established in diagonal form for so-called 'dual orthogonal grid' (DOG)-systems. For more details on the method cf. [2].

An important issue in the analysis and design of microwave components is the computation of macroscopic quantities, such as the amplitudes of incoming and reflected waves at the ports and the associated scattering parameters. Fig. 1 shows the notation of these quantities for a two-port system ($n = 2$), where one *port* is defined as one propagating mode in one of the waveguide lines connected to the structure. The desired scattering matrix $\underline{\mathbf{S}}(\omega)$ is then defined by

$$\begin{pmatrix} \underline{b}_1 \\ \underline{b}_2 \end{pmatrix} = \underline{\mathbf{S}} \begin{pmatrix} \underline{a}_1 \\ \underline{a}_2 \end{pmatrix}, \qquad \underline{\mathbf{S}} = \begin{pmatrix} \underline{S}_{11} & \underline{S}_{12} \\ \underline{S}_{21} & \underline{S}_{22} \end{pmatrix}, \tag{4}$$

where $\underline{a}_i$ and $\underline{b}_i$ are the amplitudes of incoming and outgoing waves, respectively (the underline denoting complex quantities).

To deal with such quantities within a full-wave simulation technique like the FI-Technique, the knowledge of the field patterns of the waveguide modes in the port planes is mandatory. This requires the solution of a *two-dimensional* (related to the number of spatial dimensions) algebraic eigenvalue problem,

$$\mathbf{A}_{2\mathrm{D}}(\omega)\, \widehat{\mathbf{e}}_{i,t} = k_i^2\, \widehat{\mathbf{e}}_{i,t}, \tag{5}$$

to determine the transversal electric fields $\widehat{\mathbf{e}}_{i,t}$ and the propagation constants k_i of at least the fundamental mode in each of the waveguides. The system matrix $\mathbf{A}_{2\mathrm{D}}(\omega)$ of this equation is real and non-symmetric for lossfree ports. Further, for homogeneous ports the eigenvectors $\widehat{\mathbf{e}}_{i,t}$ in (5) do not depend on the frequency ω [1].

An important property of the FI-Technique is, that the *discrete* solutions of this eigenvalue problem — if proprerly normalized — fulfill the orthogonality properties of continuous waveguide modes [4,5]

$$\langle \mathrm{mode}_i, \mathrm{mode}_j \rangle = \widehat{\mathbf{e}}_{i,t}^T \mathbf{K}\, \widehat{\mathbf{h}}_{j,t} = \delta_{ij} \quad \leftrightarrow \quad \int_A (\boldsymbol{E}_i \times \boldsymbol{H}_j) \cdot d\boldsymbol{A} = \delta_{ij} \tag{6}$$

where $\mathbf{K}$ is a simple matrix operator modeling the cross-product of the transversal mode fields.

This enables an *exact* (up to the level of numerical noise) extraction of the wave amplitudes refering to each 2D-mode from any field solution in the discrete space. This can be utilized for several highly accurate techniques for the computation of scattering parameters.

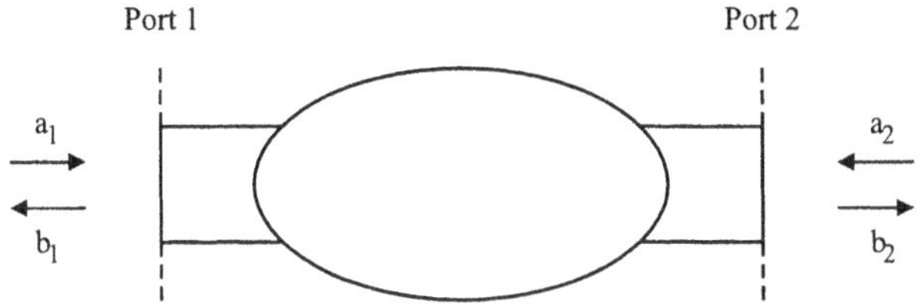

Figure1. Microwave structure with two waveguide ports. For each port (each waveguide mode) the wave amplitudes of incoming (a_1, a_2) and out-going (b_1, b_2) waves are to be calculated.

2 Calculation of S-Parameters in Time and Frequency Domain

Applying the FI-Technique in *time domain* on a cartesian grid using the leap-frog time iteration scheme, we obtain the same update equations for the electric and magnetic field vectors as in the well-known FDTD algorithm [2,6]. The orthogonality property of the waveguide modes mentioned above can then be applied to establish a special absorbing boundary condition for waveguides [4]: The fields are decomposed at the boundary planes of the computational grid using (6), and each mode can be treated separately using its previously calculated propagating constant k_i. This procedure not only minimizes spurious reflections at the boundary planes, but also leads (after $m = 1 \ldots M_T$ time steps) to a time series of wave amplitudes $a_i(m \cdot \Delta t)$ and $b_i(m \cdot \Delta t)$ at all ports ($i = 1 \ldots n$), which can easily be transformed in the frequency domain using a Discrete Fourier Transform (DFT). If one (and only one) of the ports is excited by a given wave amplitude $a_1(m \cdot \Delta t)$, one column of the $n \times n$–S-parameter matrix can be calculated from each simulation run.

In many cases this highly efficient time domain approach can be considered as the current 'state of the art'-technique, as no algebraic solver has to be applied, and broadband results can be obtained by a single simulation run. However, in some cases it suffers from the high number of time steps to be performed, especially when the structure includes sharp resonance peaks (leading to long settling times, cf. Fig. 2), or if fine geometric details have to be resolved by the mesh (leading to a small time-step width Δt due to the Courant-criterion).

Alternatively, the same boundary operator can be included in a *frequency domain* approach. The governing equation there is the inhomogeneous discrete curl-curl-equation [7]

$$\left(\mathbf{C}^T\mathbf{M}_\mu^{-1}\mathbf{C} + i\omega\mathbf{M}_\kappa - \omega^2\mathbf{M}_\varepsilon\right)\widehat{\mathbf{e}} = -i\omega\,\widehat{\widehat{\mathbf{j}}}_S$$

$$\leftrightarrow \quad \left(\text{curl}\;\mu^{-1}\text{curl}\; + i\omega\kappa - \omega^2\varepsilon\right)\boldsymbol{E} = -i\omega\,\boldsymbol{J}_S. \quad (7)$$

If the absorbing boundary condition based on the expansion in waveguide modes is implemented in this formulation, the system matrix becomes non-symmetric and complex and has complex eigenvalues due to the energy flow through the port planes [8]. This makes the solution of the system much more expensive, and there are only some rare cases, where this approach can compete with time domain calculations, even if modern solvers and preconditioning techniques are applied.

To stick to real-valued systems at least for lossfree structures, also closed (PEC or PMC) boundary conditions can be applied in the frequency domain [9]. After calculating a field solution, again the wave amplitudes can be extracted, using the orthogonality of the waveguide modes. Due to the reflection of the waves at the closed boundaries, now now all incoming and outgoing wave amplitudes $\underline{a}_i(\omega)$ and $\underline{b}_i(\omega)$ are unequal to zero. To calculate the $n \times n$–S-matrix (with $n = 2$ for the two-port system in Fig. 1) therefore n linear independent field solutions from different excitation terms in (7) are required.

These excitations are somehow arbitrary; the only limitation is, that they have to be located outside the reference planes of the ports. A good choice is to calculate the equivalent surface currents at the port planes from the modal fields (the solutions of the two-dimensional eigenvalue problem) at these ports. If the calculation domain contains materials with dielectric or finite-conductivity losses, again the system matrix becomes complex, which considerably increases the computation time.

The main disadvantage of both of the frequency domain approaches is, that the scattering parameters at only one frequency point are obtained from one computation run. To obtain S-parameter curves for a given frequency range, therefore a series of linear systems of equations have to be solved.

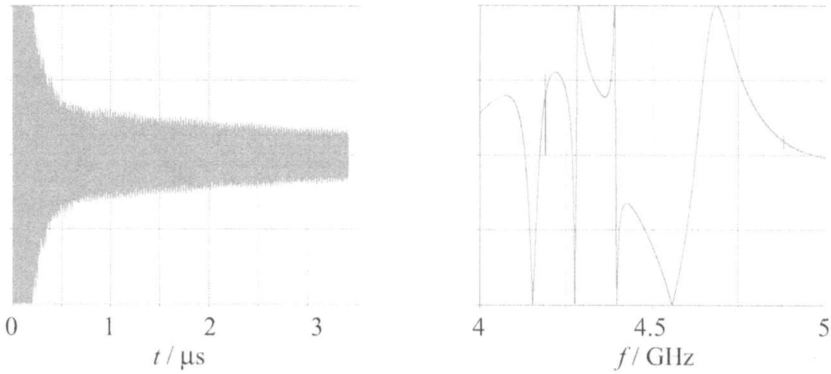

Figure2. Typical time signal (wave amplitude at one of the ports and DFT-spectrum) for a highly resonant structure. After the excitation at one of the ports has vanished, the energy is leaving the structure only very slowly.

3 Modal Approach

The modal approach introduced in the following chapter does not have this disadvantage, but allows the efficient calculation of scattering parameters over a certain frequency range. A similar approach for a Finite Element discretization has already proposed in [10]. An alternative formulation basing on the impedance matrix of multi-port systems was presented in [11].

The use of a superposition of eigenmodes for the solution of inhomogeneous problems is a very general approach, which can also be adopted to the discrete curl-curl-equation (7)

$$\mathbf{A}_{3\mathrm{D}}\,\widehat{\mathbf{e}} - \omega^2 \mathbf{M}_\varepsilon\,\widehat{\mathbf{e}} = -i\omega\,\widehat{\widehat{\mathbf{j}}}_S. \tag{8}$$

The system matrix (for lossfree structures) $\mathbf{A}_{3\mathrm{D}} = \mathbf{C}^T\mathbf{M}_\mu^{-1}\mathbf{C}$ is also the system matrix of the (three-dimensional) eigenvalue problem for cavities:

$$\mathbf{C}^T\mathbf{M}_\mu^{-1}\mathbf{C}\,\widehat{\mathbf{e}}_j = \omega_j^2\,\mathbf{M}_\varepsilon\,\widehat{\mathbf{e}}_j. \tag{9}$$

The solutions of (9) are the eigenfrequencies ω_j and the associated eigenmodes $\widehat{\mathbf{e}}_j$. Within the FIT approach, the eigenvalue problem (9) is symmetric positive semidefinite, and thus the (properly normalized) modes fulfill the orthogonality relation

$$\widehat{\mathbf{e}}_i^T\mathbf{M}_\varepsilon\,\widehat{\mathbf{e}}_j = \delta_{ij}. \tag{10}$$

If we insert the superposition of a number p of these eigenmodes

$$\widehat{\mathbf{e}} = \sum_{j=1}^{p} \underline{\alpha}_j\,\widehat{\mathbf{e}}_j \tag{11}$$

in the inhomogeneous problem (8), we get

$$\sum_{j=1}^{p} \underline{\alpha}_j(\omega_j^2 - \omega^2)\mathbf{M}_\varepsilon\,\widehat{\mathbf{e}}_j = -i\omega\,\widehat{\widehat{\mathbf{j}}}_S. \tag{12}$$

where the unknowns are now the scalar, in general complex coefficients $\underline{\alpha}_j$. Multiplying this equation by $\widehat{\mathbf{e}}_k^T$ finally leads to an explicit formula for these coefficients:

$$\underline{\alpha}_k = \frac{-i\omega}{\omega_k^2 - \omega^2}\,\widehat{\mathbf{e}}_k^T\,\widehat{\widehat{\mathbf{j}}}_S. \tag{13}$$

Once the eigensolutions of (9) have been calculated, this formula can be easily evaluated for arbitrary frequencies ω. From the frequency dependence in (13) in can be expected, that the main influence on the field solution is due to eigensolutions with $\omega_k \approx \omega$. In practice, usually some 10 to 40 modes with eigenfrequencies in or near the desired frequency range are sufficient for accurate solutions.

The main computational cost of this approach is required by the calculation of p solutions of the three-dimensional eigenvalue problem. For lossfree structures, (9) can easily be transformed in a symmetric, real-valued eigenproblem $\mathbf{A}\mathbf{x}_j = \omega_j^2 \mathbf{x}_j$, which can be efficiently solved by modern algebraic algorithms. In our implementation we use a simultaneous subspace method [12], which is part of the EM-software package MAFIA [3].

If conductivity losses have to be considered — introducing a nonzero matrix $\mathbf{M}_\kappa$ in (7) — the same procedure leads to a linear system of dimension $p \times p$ for the coefficients $\underline{\alpha}_k$:

$$\frac{\omega_k^2 - \omega^2}{-i\omega} \underline{\alpha}_k - \sum_{j=1}^{p} \left((\widehat{\mathbf{e}}_k^T \mathbf{M}_\kappa \, \widehat{\mathbf{e}}_j) \underline{\alpha}_j \right) = \widehat{\mathbf{e}}_k^T \, \widehat{\widehat{\mathbf{j}}}_S, \tag{14}$$

which has to be solved at each frequency point. As the order p of the field expansion in (11) is usually a low number, the efficiency of the method is only slightly decreased.

However, this formula with a finite number p of eigenmodes only leads to accurate results, if the fields in the structure are mainly determined by those eigenvectors, which can exist in both the lossy and the lossfree model. In other words, having $\mathbf{M}_\kappa \neq \mathbf{0}$, the frequency dependence expressed in (14) does not provide a clear criterion, which eigensolutions of the lossfree problem (9) should be considered in the expansion, as the system's off-diagonal entries (the sum-terms) do not depend on the eigenfrequencies ω_k. Thus this *power-loss*-type approach is only applicable to systems with small losses.

As an alternative, especially for structures with higher losses, the eigenmodes of the *complex* eigenvalue problem

$$\mathbf{C}^T \mathbf{M}_\mu^{-1} \mathbf{C} \, \underline{\widehat{\mathbf{e}}}_j = \underline{\omega}_j^2 \, \underline{\mathbf{M}}_\varepsilon \, \underline{\widehat{\mathbf{e}}}_j \qquad \text{with} \quad \underline{\mathbf{M}}_\varepsilon = \mathbf{M}_\varepsilon + \frac{1}{i\omega_0} \mathbf{M}_\kappa \tag{15}$$

can be applied. To stick to a linear formulation, the complex permittivity matrix $\underline{\mathbf{M}}_\varepsilon$ is approximated by an evaluation at one frequency ω_0 in the middle of the desired frequency range. Note, that for DOG-systems this material matrix is diagonal and can be easily inverted. The eigenvalue problem can then be transformed into a standard problem with a symmetric, complex (but not Hermitian) system matrix

$$\underline{\mathbf{A}}_{3D} = \underline{\mathbf{M}}_\varepsilon^{-1/2} \widetilde{\mathbf{C}} \mathbf{M}_\mu^{-1} \mathbf{C} \underline{\mathbf{M}}_\varepsilon^{-1/2}. \tag{16}$$

This approach again leads to the explicit formula (13), but now the eigenvectors $\underline{\widehat{\mathbf{e}}}_j$ as well as the eigenfrequencies $\underline{\omega}_j$ have to be calculated from the complex formulation in (15) and (16).

4 Example

As an example, the scattering parameters of the so-called 'combline'-filter in Fig. 3 are calculated. To get a validation of the modal approach above, the result from a time domain calculation using exactly the same grid model serves as a reference for the reflection and the transmission coefficients.

The frequency curves in Fig. 3 show a small passband with a middle frequency of about 1.67 GHz and steep slopes (note the small frequency range), yielding long settling times in the time domain calculation.

The modal approach already leads to qualitatively good results for an expansion with only $p = 10$ modes with eigenfrequencies between 1.6 GHz and 8.1 GHz. Using $p = 30$ eigenmodes (eigenfrequencies up to 17 GHz) the results agree very well with the reference, and the calculation with $p = 200$ modes has nearly perfectly converged. The calculation times, including mesh generation and matrix calculation, behave like 1 : 5.9% : 7.3% : 55% (time domain, modal approach with 10, 30, 200 modes).

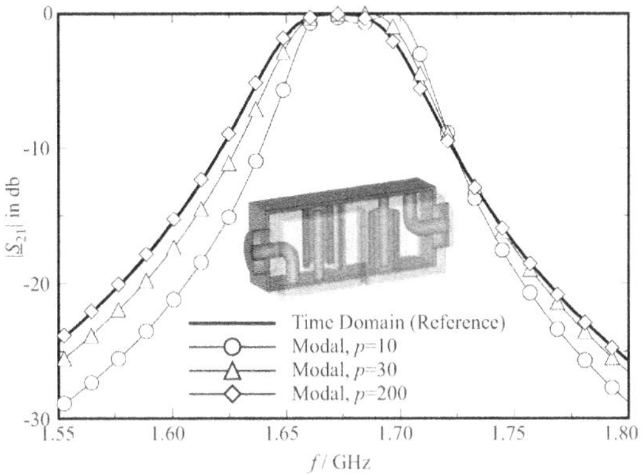

Figure3. Combline filter driven by two coaxial waveguide ports and transmission coefficient $|\underline{S}_{21}|$ in db. Results from a time domain calculation (reference) and from the modal approach using $p = 10, 30, 200$ eigenmodes, respectively.

5 Conclusion

The modal approach is an efficient method for the calculation of scattering parameters. The main numerical cost is due to the solution of the three-dimensional eigenproblem of the closed structure, whereas for the final extraction of the S-parameters only an explicit formula or a small linear system has to be evaluated at each frequency point. Further investigations will focus on validation and error estimation techniques, especially in the field of structures with small losses, where two variants of the basic approach can be applied.

References

1. Weiland, T.: A Numerical Method for the Solution of the Eigenvalue Problem of Longitudinally Homogeneous Waveguides. Electronics and Communication (AEÜ) **31** (1977) 308
2. Weiland, T.: Time Domain Electromagnetic Field Computation with Finite Difference Methods. International Journal of Numerical Modelling: Electronic Networks, Devices and Fields **9** (1996) 295–319
3. MAFIA 4, available from: CST GmbH, Büdingerstr. 2a, 64289 Darmstadt, Germany (www.cst.de)
4. Dohlus, M.: Ein Beitrag zur numerischen Berechnung elektromagnetischer Felder im Zeitbereich. Ph.D.-Thesis, Darmstadt University of Technology (1992)
5. Schuhmann, R.: Die Nichtorthogonale Finite-Integrations-Methode zur Simulation elektromagnetischer Felder. Ph.D.-Thesis, Darmstadt University of Technology (1999)
6. Yee, K.S.: Numerical Solution of Initial Boundary Value Problems Involving Maxwell's Equations in Isotropic Media. IEEE Transactions on Antennas and Propagation **17** (1966) 585–589
7. Hahne, P., Weiland, T.: 3D Eddy Current Computation in the Frequency Domain Regarding the Displacement Current. IEEE Transactions on Magnetics **28,2** (1992) 1801–1804
8. Schuhmann, R., Clemens, M., Thoma, P., Weiland, T.: Frequency and Time Domain Computations of S-Parameters Using the Finite Integration Technique. Proceedings of the 12th ACES Conference, Monterey (1996) 1295–1302
9. Christ, A., Hartnagel, H.: Three-Dimensional Finite-Difference Method for the Analysis of Microwave-Device Embedding. IEEE Transactions on Microwave Theory and Technique **35,8** (1987) 688–696
10. Brauer, J.R., Lizalek, G.C.: Microwave Filter Analysis Using a New 3-D Finite-Element Modal Frequency Method. IEEE Transactions on Microwave Theory and Techniques **45,5** (1997) 810–818
11. Dohlus, M., Schuhmann, R., Weiland, T.: Calculation of Frequency Domain Parameters Using 3D Eigensolutions. Special Issue of International Journal of Numerical Modelling: Electronic Networks, Devices and Fields **12** (1999) 41–68
12. Timm, M., Schmitt, D., Schuhmann, R., Weiland, T.: Simultaneous Computation of Higher Eigenmodes in Lossfree Waveguides and Cavities. IEEE Transactions on Magnetics **34,5** (1998) 3240–3243

Statistical Simulation Approach to Inverse Geoelectric Problem

I. Belinskaya-Abnizova

Institute of Computational Mathematics and Mathematical Geophysics Siberian Branch of Russian Academy Sciences, Novosibirsk, Russia

Abstract In this paper we consider the inverse problem arising in parameter identification for the Maxwell equations. Namely, we want to identify the conductivity of soil from surface measurements of the electric and magnetic fields forced by a ground penetrating electric current. Starting from the 3D setting the problem is finally reduced to the 1D Volterra integral equation of the second kind.

Introduction

The inverse problems for Maxwell's equations are central to many areas of science and technologies: remote sensing, near-surface radar-location, medical imaging, geophysical exploration etc. The main idea of inverse methods is to find some details about the medium considered from the measurements of electromagnetic fields. From the mathematical point of view we consider the determination of the coefficients for the PDE system with the help of additional information about the solution at some time-like subspace. This kind of problems are known to be ill-posed in the Adamart sense, and require special regularization methods to apply. There are a lot of usual numerical methods to solve these problems: inversion of finite difference scheme, linearization, optimization, etc. They have different advantages and disadvantages, namely they are expensive. We decided to apply weight Monte Carlo method at the last step of the algorithm because this method is known to have some advantages towards other numerical methods, namely: small dependence on the dimension of the problem; it is not memory hungry, it is fast; it corresponds to the peculiarity of inverse problems where it is required to find the functional over solution not knowing the whole solution.

The determination of the soil conductivity

Assume that 3D space is divided with the plane $x_3 = 0$ into two half-spaces "air": $R^3_- = \{x \in R^4 | x_3 < 0\}$ and "earth": $R^3_+ = \{x \in R^4 | x_3 > 0\}$. Consider the system of Maxwell's equations

$$curl\ H = \varepsilon \frac{\partial E}{\partial t} + \sigma E + j, \qquad curl\ E = -\mu \frac{\partial H}{\partial t} \tag{1}$$

Here $H = (H_1, H_2, H_3), E = (E_1, E_2, E_3)$ and $j(x,t)$ are functions depending on spatial coordinate $x \in R^3$ and time. The coefficients ε, μ, σ depend on $x \in R^3$ These coefficients are constants in R^3_- and smooth in R^3_+.

The inverse problem is to determine $\varepsilon, \mu, \sigma, \quad x \in R^3_+$ out of relations (1) and some additional measurements of the electromagnetic field.
Mentioned above type of inverse problems arise frequently in geophysical exploration. In our model we assume the surface $x_3 = 0$ of Earth to be almost flat, the function $j(x,t)$ stands for an exterior current generating an electromagnetic field. The functions E, H are treated as the electric and magnetic strength vectors of the electromagnetic field described by equations (1). The coefficients ε, μ, σ are permittivity, permeability and conductivity of a medium, respectively. They are of the main interest in geophysical exploration. To determine them, an exterior current is generated and the strength components of the electromagnetic field are measured on the surface. Let us consider the simple case with the source of an exterior current in the form $j(x,t) = (0,1,0)'g(x_1)\delta(x_3)\theta(t)$, which corresponds to the case when the current is switched on instantaneously, directed along the x_2 axis, concentrated on the earth surface $x_3 = 0$ and distributed over the x_1 with the density $g(x_1)$. Here $\theta(t)$ is Heaviside theta-function, $\delta(x_3)$ is Dirac delta-function.

Assume that all coefficients of Maxwell system do not depend on x_2, x_1. Let us determine $\sigma \in R_+$. It is known [8-10] that with some assumptions the system (1) is reduced to the following system:

$$\sigma \frac{\partial}{\partial t} E = \frac{\partial}{\partial x}\left(\frac{1}{\mu}\frac{\partial}{\partial x} E\right) - \frac{\lambda^2}{\mu} E$$

$$E\mid_{t<0} \equiv 0, \quad E\mid_{x=0} = f(t), \quad \left(\frac{\partial E}{\partial x} - |\lambda| E\right)\mid_{x=0} = g(\lambda)\delta(t).$$

Here we consider the system of Fourier images of the corresponding functions with respect to x_1 with fixed Fourier parameter λ.

As known [11], the solution of the above problem is connected with the solution to mixed problem for wave equation (Fourier parameter λ with respect to horizontal variable is assumed to be zero):

$$v_{tt} = \frac{1}{\mu\sigma} v_{xx}, \quad x, t \in R^+ \tag{2}$$

$$v|_{t<0} \equiv 0 \tag{3}$$

$$v_x|_{x=+0} = g(0)\sigma(t) \tag{4}$$

Apply dynamical variant of Gelfand-Levitan method to solve problem (2)-(4). This methods allows to determine the sufficient and necessary conditions of existence of the solution in the large. The basic idea of the method is reducing of the original differential problem to the one-parameter system of linear integral equations of the second kind. Let us introduce new variable and new functions, assuming μ to be positive constant

$$z = \int_0^x \sqrt{\mu\sigma(\xi)}, s(z) = \sqrt{\frac{\sigma(+0)}{\sigma(x)}}, -q(z) = \frac{s''(z)}{s(z)} + 2\frac{s'}{s}.$$

If we use even continuation with respect to z into R_-, then $u(z,t)$ is solution of the system

$$u_{tt} = u_{zz} - q(z)u, \quad z \in R, \ t \in R_+ \tag{5}$$

$$u|_{t=0} = 0, \quad u_t|_{t=0} = \gamma\delta(z) \tag{6}$$

$$u|_{z=0} = f(t), \ \gamma = \frac{-2g(0)}{\sqrt{\mu\sigma(+0)}} \tag{7}$$

We obtain finally the integral equation

$$\gamma w(z,t) + \int_{-z}^{z} f'(t-s)w(z,s)ds = -\frac{1}{2}[f'(t-z) + f'(t+z)], \tag{8}$$

here $z \in R_+$, $t \in (-z, z)$. The solution of (8) is connected with (5)-(7) problem solution as follows: $q(z) = 4\frac{\partial}{\partial z}w(z, z-0)$, $z \in R_+$. The equation (8) is the classical object in the inverse problem theory, it is called Gelfand-Levitan-Krein-Marchenko equation. Here $f'(t)$ is derivative of the odd continuation of the function $f(t)$ in the points of its continuity.

With known $q(z)$ it is possible to determine $s(z)$, and respectively, the sought coefficient $\sigma(x)$.

According to the theory of weight Monte Carlo methods [7] consider Monte Carlo approach to solve the equation (8). Rewrite it in more convenient form:

$$w(z,t) - \frac{1}{\gamma}\int_{-z}^{z} f'(t-s)w(z,s)ds = -\frac{1}{2\gamma}[f'(t-z) + f'(t+z)], \qquad (9).$$

We have to check the conditions of Monte Carlo applicability, i.e., the convergence of the Neumann series to the solution of this equation. Here we have the required conditions because of Volterra type of the integral equation (9).

Let us seek the solution of (9) in the space L_2 with appropriate norms. As $f'(t-s)$ is an even function then its Fourier series can be written down with the cosines $f'(t) = \frac{a_0}{2} + \sum\limits_{k=1}^{\infty} a_k \cos(\frac{k\pi t}{2z})$, $t \in (-z, z)$. We take N_1 first members of Fourier transform with the accuracy required. To find the local solution to (9) let us present this equation in the form $\varphi = (k_x, \varphi) + 1$, where $k_x(t,s) = \frac{1}{2}f'(t-s)$ is the kernel of (9).

We construct the unbiased collision estimator ξ such that

$$M\xi = w(z) = M\sum_{n=0}^{N} Q_n f'(2x_n) + \frac{1}{2\gamma}f'(2z),$$

This means that we construct the Markov chain $x_0, x_1, ..., x_N$, with random weights $Q_0, ..., Q_N$, $x_0 = z$, given by formulas

$$Q_0 = \frac{1}{\pi(x_0)}, \qquad Q_n = Q_{n-1}\frac{k(x_{n-1}, x_n)}{p(x_{n-1}, x_n)(1 - g(x_{n-1}))}.$$

Here N is the random number of the final statement before breaking the chain, $k(x_{n-1}, x_n)$ is the kernel of (9), $g(x_n)$ is the probability of breaking the Markov chain in the statement x_n, $\pi(x_0)$ is the initial distribution density of the first statement $x_0 = -z$ of the Markov chain, $p(x_{n-1}, x_n)$ is the density of transition between any two neighbouring statements:

$$p(x_{n-1}, x_n) = \sum_{i=1}^{N_1} |b_i(x_0)||\cos(\frac{k\pi(x_{n-1}, x_n)}{2T})|, \quad or$$

$$p(x_{n-1}, x_n) = \sum_{i=1}^{N_1} |b_i|\theta(\cos(\frac{k\pi(x_{n-1}, x_n)}{2T}),$$

where b_i are chosen according to $\int\limits_{-x_0}^{x_0} p(t,s)ds = 1$.

Due to the Volterra type of integral equation (9) we have fulfilled the condition $||K_1|| < 1$ which is sufficient for the convergence of Monte Carlo algorithm. Here K_1 is the integral operator with the kernel $k_1(t,s) = |k(t,s)|$.

Note that with these Markov chain characteristics we obtain the unbiased estimation [4] as well as finiteness of the mean number of the statements $M(N)$. This means our Markov chain will break after the finite number of transitions with probability 1.

We apply the well-known superposition method to construct ξ from the formula $\int_{-x_0}^{\xi} p(t,s)ds = rand$. Here $rand \in U_{(0,1)}$ is the uniformly distributed in the interval $(0,1)$ random value. It is known, that if Neuman series converges to the variance equation

$$\chi(x) = \int_{-x}^{x} \frac{K^2(t,s)\chi(s)}{p(t,s)} ds + \frac{1}{m^2\pi(x)} \equiv K_p\chi(x) + \frac{1}{m^2\pi(x)},$$

the variance of the main estimation is finite. We also have this condition fulfilled because of the Volterra type of the above equation.

The numerical comparison was made with the help of the finite element method and the inversion of the difference scheme method. The difference scheme inversion method is the usual tool to solve inverse problems, it was proposed and proved in [1]. The original difference problem is reduced to the appropriate Gursat problem with data at the characteristics. For this problem, a finite difference scheme of the nonlinear algebraic equations of order $O(h)$ is constructed. Its solution is considered to be an approximate solution to the original problem. This scheme has weak stability. We have studied several model problems by the three above mentioned methods. In each case, a single test was used to make a comparison. Dividing the interval $(-T, T)$ into 2000 points, we obtain satisfactory (about 5-10 percent) accuracy for a relatively small partial variable x by using finite element method. As x further increases, satisfactory accuracy is not attained. There is also a considerable growth of storage with the sampling. In order to obtain the accuracy 0.01-10 percent by Monte Carlo, the averaging over 200000 trajectories was used. The algorithm variance increases linearly as x increases, the average number of steps per trajectory increasing similarly.

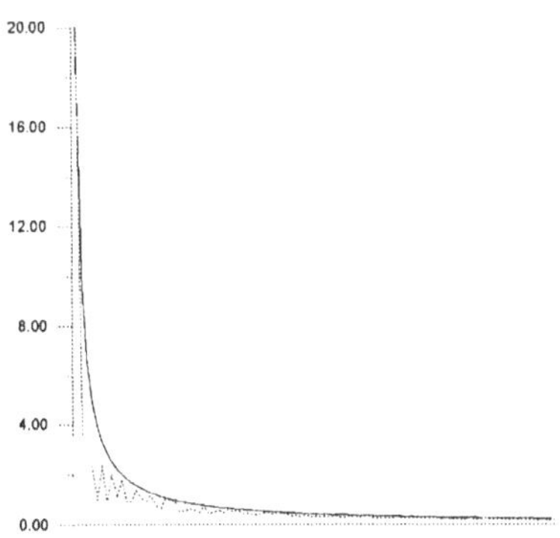

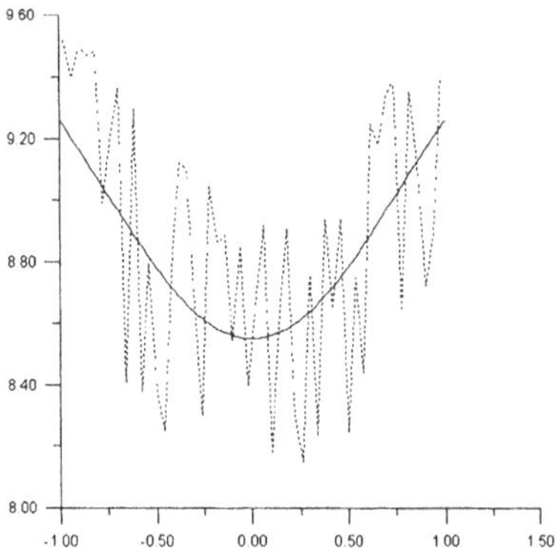

On the plots above there are two examples of Monte Carlo estimation of inverse problems solution with additional information in the form of $f(x) = a + cos(x^2)$, $f(x) = 1/((ax^2) + b)$, $a, b = const$. With x increasing, the estimate accuracy decreases slowly and remains satisfactory for relatively large (100-150) x. The storage size for the Monte Carlo algorithm is comparatively small. To attain the appropriate accuracy by the difference scheme inversion method a grid of 200×200 size is needed. In this case, the solution is estimated accurately only for small x from the interval $(0, 1)$. There is a comparatively slow growth of variance, and the average number $M(N)$ of Markov chain statements in the Monte Carlo algorithm. In addition Monte Carlo method turns to be the most speedy among the three methods. The illustration of the comparative numerical results are in the Plot. The comparative tests for the function, which describes the the additional information: $f(t) = b + sin(ct)$, $b, c = const$, are presented. Here we can see that with

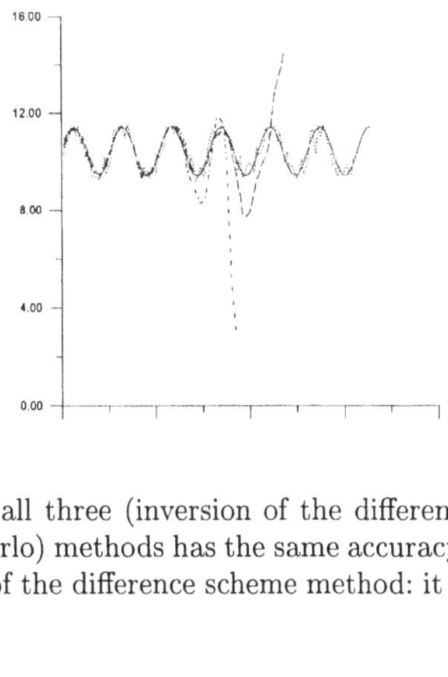

small spatial values all three (inversion of the difference scheme, finite elements and Monte Carlo) methods has the same accuracy. The first method to diverge is inversion of the difference scheme method: it diverges with z equal

15. The next to diverge is finite elements methods, it keeps the accuracy up to $z = 45$. And only Monte Carlo obtains satisfactory accuracy up to the comparatively big for inverse methods spatial values $z = 150$.

The Cauchy problem solution

The corresponding direct problem is usually solved by difference methods, but we suggest to solve it with probabilistic methods too. Consider the Cauchy problem in the following form:

$$U_{tt} = \Delta U - q(x)U \tag{10}$$

$$U|_{t=0} = 0, \;\; U_t|_{t=0} = \varphi(x), \;\; x = (x_1, x_2, x_3) \in R^3, \quad t \in R^+ \tag{11}$$

Assume that functions $q(x), \varphi(x)$ satisfy the conditions of uniqueness and existence of the classical Cauchy problems solution.

Using the formulas for the classical Cauchy problems solution [3] we reduce our problem to the equivalent integral equation

$$U(x,t) = U_0(x,t) - \frac{1}{4\pi}\int_0^t \rho[\int\int_\omega q(x+\alpha\rho)U(x+\alpha\rho, t-\rho)d\omega]d\rho, \tag{12}$$

here

$$U_0(x,t) = \frac{t}{4\pi}\int\int_\omega \varphi(x+\alpha t)d\omega \tag{13}$$

ω – the surface of the unique radius sphere, $\alpha = (\alpha_1, \alpha_2, \alpha_3) \in R^3$ – the unique vector. Assume $U_k(x,t) = \frac{1}{4\pi}\int_0^t \rho[\int\int_\omega q(x+\alpha\rho)U_{k-1}(x+\alpha\rho, t-\rho)d\omega]d\rho,$, $U_0(x,t)$ is defined in (13). The sum of infinite series $g(x,t) = U_0(x,t) - U_1(x,t) + U_2(x,t) - \ldots$ gives the solution of the integral equation (12). It follows from the [5] that

$$|U_k(x,t)| \leq \frac{M^{k+1}t^{2k+1}}{(2k+1)!} \tag{14}$$

Because the series $g(x,t)$ can be twice differentiated termwise, then the function $g(x,t)$ is the solution of the Cauchy problem under consideration. We can prove directly the fact of satisfying of the $U(x,t)$ function to the initial conditions.

Therefore we have the integral Volterra equation of the second kind (12). We require the convergence of the Neuman series for the solution of this equation to use Monte Carlo method. As follows from the estimation (14) the series $g(x,t)$ at every fixed x converges uniformly at every finite interval of varying t.

Write down (12) in the operator form $U = KU + U_0$, K – the integral operator of the equation (12) and consider it as adjoint to it's adjoint equation:

$$U = KU + U_0, \quad U^* = K^*U^* + \delta(x - x_0).$$

Here K^* – operator adjoint to K.

We can obtain the local estimator of a functional in the point x_0 by Monte Carlo method: $I_\delta = (U, \delta(x - x_0)) = (U_0, U^*) = U(x_0, t_0)$. The estimator has the form:

$$\xi = U_0(x_0, t_0) + \sum_{n=1}^{N} Q_n U_0(x_n, t_n), \quad M\xi = U(x_0, t_0),$$

here Q_n – some random weights, we will define them later; U_0 is defined in (13). To solve the equation (12) we construct the algorithm of random walk. Let us define the Markov's chain. It is sufficiently to define a) the initial distribution density. In our case it is δ-function $\delta(x - x_0)$; b) the transitional density from the statement (x, t) into (x', t'); c) the probability of the path break.

The kernel of the integral equation (12) is $k(x, t) = -\frac{1}{4\pi} t q(x + \alpha t)$. Taking into accordance the form of the kernel we may model the time transition from t into t' by formula $t' = t(1 - \sqrt{rand})$.

The transitional density with respect to time on the interval $[0, t_0]$ is $r_1(t) = \frac{2t}{t_0^2}$, here t_0 is fixed. With fixed t' we model the spatial transition from x into x' by formula: $x' = x + w(t - t')$.Here w – unique isotropic vector. The conditional transitional density in space with fixed t is equal to: $r_2(x|t) = \frac{1}{4\pi}$, e.g. we choose x uniformly on the sphere of the radius t.

Therefore, the transitional density from the statement (x, t) into (x', t') is defined with formula $r(x, t) = r_1(t) r_2(x|t)$. The Markov chain is broken when the current time point is occur to be in the ε-neighborhood of zero.

Define the weights Q_n by formula

$$Q_0 = 1, ..., N, \;\; Q_n = Q_{n-1} \frac{k(x_{n-1}, t_{n-1})}{r(x_{n-1}, t_{n-1})}.$$

Consider the estimator's ξ mean value as the approximation of the equation (12) solution in the fixed point (x_0, t_0). This algorithm allows to make parallel calculations of the solution in several points. The disadvantage of the method proposed is comparatively fast growth of algorithm variance.

Conclusion

As the conclusion I would like to note that Monte Carlo application happened to be effective to solve inverse geoelectric problem even in 1D case. It is obvious that Monte Carlo algorithms are of special interest in multidimensional problems, where their small dependence on the dimension is the main advantage. Though some numerical results about the solution of 2D acoustic inverse problems were obtained by author [2], the work on this topic is in progress.

References

1. 1. Alexeev A.S., Inverse Dynamic Problems of Seismology. Methods and algorithms for the Interpretation of geophysical Data. "Nauka", Moscow, 1967, pp.9-84 (in Russian)
2. 2. Belinskaya I.I, The solution to some inverse problems by Monte Carlo method, Russ.J.Num.Anal.Math.Modelling, 1999, v.14, N2, pp.109-123.
3. 3. Courant, R., Hilbert, D., The methods of mathematical physics, M.-L, *GTTI*, v.2, 1965. (in Russian)
4. 4. Ermakov, S.M.,G.A.Mikhajlov, G.A., The statistical simulation, Moscow *"Nauka"*, 1982.(in Russian)
5. 5. B.M.Levitan, B.M., About the asymptotical behaviour of the spectral function and decomposition by the eigenfunctions of the equation $\Delta u+\lambda - q(x)u = 0$, *Works of Moscow Math. Soc.*, v.4, 1955. (in Russian)
6. 6. Marchuk G.I., Agoshkov V.I., Introduction to Projection Grid Methods, Nauka, Moscow, 1984 (in Russian)
7. 7. Mikhailov, G.A., The Optimization of Weight Monte Carlo Methods, Novosibirsk, *"Nauka"*, 1987.(in Russian)
8. 8. Oristaglio, M.L., Hohmann, G.W., Diffusion of electromagnetic fields into a 2D earth: a finite-difference approach, Geophysics 49, N 7, 870-894, 1984
9. 9. Romanov, V.G., The Inverse Problems of Mathematical Physics, Novosibirsk, *"Nauka"*, 1984. (in Russian)
10. 10. Romanov, V.G., Kabanikhin S.I., Pukhnacheva T.P., Inverse problems of Electrodynamics, Computer Center, SB of USSR Academy Sci., Novosibirsk (in Russian) 1984
11. 11. Svetov B.S., Barsukov P.O., Transformation of quasistationary geolectric transient processes into equivalent wave ones, Fizika Zemli 8, 29-37, 1984

A Scalar BEM for Time Harmonic Eddy Current Problems with Impedance Boundary Conditions

O. Sterz[1] and C. Schwab[2]

[1] Interdisziplinäres Zentrum für Wissenschaftliches Rechnen, Universität Heidelberg, Im Neuenheimer Feld 368, 69120 Heidelberg, Germany
[2] Seminar für Angewandte Mathematik, ETH Zürich, HG G 58.1, 8092 Zürich, Switzerland

Abstract The time harmonic eddy current model together with an impedance or Leontovich[6] boundary condition is used to describe the electromagnetic field exterior to a lossy, highly conducting and possibly magnetic body at low frequencies. The so called penetration depth is assumed to be small. The problem will be reduced to a scalar, hypersingular boundary integral equation (BIE) on the surface Γ of the conductor. Convergence results and numerical examples are given. Some aspects on the computation are discussed in detail.

1 Introduction

An important class of problems in power engineering can be described by the eddy current approximation which is obtained by neglecting the displacement current in the Maxwell-equations. Usually this results in a transmission problem in $\Omega^+ \cup \Omega^-$, where by $\Omega^- \subset \mathbb{R}^3$, we denote a bounded, contractible domain (i.e. simply connected with connected boundary) with smooth boundary Γ occupied by a lossy, highly conductive medium. The complement $\Omega^+ = \mathbb{R}^3 \setminus \overline{\Omega^-}$ is assumed to be nonconductive. The material parameters κ and $\mu = \mu_0 \mu_r$ are positive and real valued L^∞-integrable scalar fields on $\mathbb{R}^3$. Here κ is the conductivity and μ the permeability, $\mu_0 = 4\pi \, 10^{-7} \, \frac{\mathrm{Vs}}{\mathrm{Am}}$ is the permeability of free space, μ_r is dimensionless. In the following we focus on the frequency domain with unknown complex amplitudes $\boldsymbol{E}$ (electric field) and $\boldsymbol{H}$ (magnetic field) with a fixed angular frequency $\omega \in \mathbb{R}^+$.

When dealing with a highly conductive body where the so-called *penetration depth* $\delta = \sqrt{\frac{2}{\omega\mu\kappa}}$ is small, one can approximately describe the behaviour of the fields at the conductor's boundary by impedance boundary conditions (IBCs),

$$\boldsymbol{n} \times \boldsymbol{E} = \eta \, \boldsymbol{n} \times (\boldsymbol{n} \times \boldsymbol{H}) \ \text{ on } \ \Gamma , \tag{1}$$

that allow one to study an exterior problem instead of a transmission problem (e.g. [1,11]). Here $\eta = (1+i) \sqrt{\frac{\omega\mu}{2\kappa}}$ is the so-called *surface impedance* and $\boldsymbol{n}$ is the unit normal vector field on Γ pointing into Ω^+. For simplicity we assume μ and κ in the conductor to be constant, but our proofs can be easily extended to the more general case.

We chose the magnetic field $\boldsymbol{H}$ as the unknown in our model. The advantage of modelling in $\boldsymbol{H}$ is that in the nonconductive region the eddy current problem can be reduced to a scalar one (see, e.g., [2]) - and thus, by using impedance boundary conditions, the whole problem becomes scalar. This idea has been introduced by Mayergoyz and Bredosian [8] in a finite element context. In a detailed paper [13] we analysed this reformulation, presented a hypersingular boundary integral formulation and proved existence and uniqueness. Here we resume the main results and discuss some details of a Galerkin boundary element method, focussing on the implementation. For practical needs we also sketch how to generalize the presented method to the setting of multiply connected domains. A numerical example is given in the last section.

2 Impedance Boundary Value Problem

2.1 Elimination of the Electric Field

The electric field can be eliminated from the eddy current model. Using (see, e.g., [3])

$$\boldsymbol{n} \cdot \mathbf{curl}\, \boldsymbol{E} = \mathrm{div}_\Gamma(\boldsymbol{E} \times \boldsymbol{n}), \tag{2}$$

and inserting (1) into the Farady law $\mathbf{curl}\, \boldsymbol{E} = -i\,\omega\mu\boldsymbol{H}$, we get a boundary condition equivalent to (1). In vector notation it reads

$$\mathrm{div}_\Gamma\, \boldsymbol{H}_t + \alpha\, H_n = 0 \ \text{ on } \ \Gamma, \tag{3}$$

where $\boldsymbol{H}_t = -\boldsymbol{n} \times (\boldsymbol{n} \times \boldsymbol{H})$ is the projection of $\boldsymbol{H}$ onto the tangent plane at Γ, $H_n = \boldsymbol{H} \cdot \boldsymbol{n}$ is its normal component, div_Γ is the surface divergence (see, e.g., [3]), and $\alpha := i\omega\mu_0/\eta$. Note that because of $\omega, \mu, \kappa > 0$, we have that $\alpha =: (1+i)\,\beta$ with $0 < \beta \in \mathbb{R}$.

2.2 The H-Boundary Value Problem

Assuming that in $\overline{\Omega^+}$ the field $\boldsymbol{H}$ can be split into a (known) exciting field $\boldsymbol{H}_0$ that can be computed by the Biot-Savart law and an unknown secondary field $\boldsymbol{H}_s$, we find for the eddy current model with an IBC the formulation

$$\mathrm{div}\, \boldsymbol{H}_s = 0 \ \text{ in } \ \Omega^+, \tag{4a}$$

$$\mathbf{curl}\, \boldsymbol{H}_s = 0 \ \text{ in } \ \Omega^+, \tag{4b}$$

$$\mathrm{div}_\Gamma\, \boldsymbol{H}_{st} + \alpha H_{sn} = g \ \text{ on } \ \Gamma \tag{4c}$$

and the condition at infinity

$$\boldsymbol{H}_s(\boldsymbol{x}) = \mathcal{O}\big(|\boldsymbol{x}|^{-2}\big) \quad \text{as } |\boldsymbol{x}| \to \infty, \tag{4d}$$

where

$$g := -\,\mathrm{div}_\Gamma\, \boldsymbol{H}_{0t} - \alpha\, \boldsymbol{H}_{0n}\,. \tag{5}$$

Condition (4d) is a consequence of $\boldsymbol{H}_s$ being a harmonic vector field with zero value at infinity (see [7]).

2.3 Scalar Potential Formulation

By (4b) and the assumption of Ω^- being contractible, we have in Ω^+ that

$$\boldsymbol{H}_s = -\mathbf{grad}\,\phi, \quad \boldsymbol{H} = \boldsymbol{H}_0 - \mathbf{grad}\,\phi\,. \tag{6}$$

Inserting into (4), we get the BVP:

$$\Delta\phi = 0 \text{ in } \Omega^+, \tag{7a}$$

$$-\Delta_\Gamma\phi - \alpha\frac{\partial\phi}{\partial n} = g \text{ on } \Gamma, \tag{7b}$$

$$\phi = \mathcal{O}(|x|^{-1}) \text{ as } |x| \to \infty \tag{7c}$$

where $\Delta_\Gamma = \mathrm{div}_\Gamma\,\mathbf{grad}_\Gamma$ denotes the surface Laplacian.

Problem (7) is a nonstandard problem, since the principal part of the PDE lives on the boundary Γ. In particular, the boundary operator is not subordinate to the Laplacian in Ω^+.

Proposition 1. *Any solution ϕ of* (7) *gives, via* (6), *a solution of* (4).

A variational formulation of problem (7) as well as existence and uniqueness results can be found in [13].

2.4 Extension to Non-Contractible Domains

Up to now we have assumed that the conductive domain Ω^- is contractible, i.e. having neither holes nor loops, — a very restrictive assumption in view of conductor geometries in real applications. Since it is assumed that δ is small and therefore an inner boundary has no influence, there are no problems with holes (e.g. a conductive hollow sphere). But there are severe problems with loops:

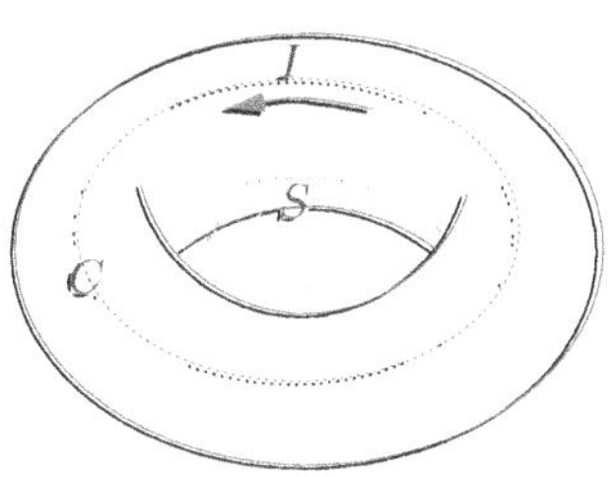

Figure1. A conductive torus and cut S.

Let us consider a conductive torus. First problem: Due to nonzero loop currents the exterior field cannot be the gradient of a single valued scalar potential. A well known possible solution to this problem is the introduction of a cut with a constant jump in the potential (along surface S in Fig. 1). That makes the nonconductive domain simply connected. Another possibility — dual to the former – is the introduction of the vector potential due to an (unknown) filament current along a closed loop in Ω^- (e.g. along the path C in Fig. 1).

Second problem: For the multiply connected case problem (4) fails to possess a unique solution. But if we add the condition (see [5])

$$i\omega\,\mu_0 \int_S \boldsymbol{n}\cdot\boldsymbol{H} + \int_{\partial S} \boldsymbol{E}\cdot\boldsymbol{\tau} = 0\,, \tag{8}$$

where $\boldsymbol{\tau}$ is the unit vector along the curve ∂S, the solution is unique. Equation (8) is Faraday's law in the exterior domain integrated along the cut S.

Having sketched the way to extend the method for the more general case, we further restrict our attention to the contractible case for the sake of simplicity.

3 Boundary Formulation

We reduce problem (7) to an integrodifferential equation on Γ. We use the single layer potential

$$\phi(\boldsymbol{x}) = (\mathcal{S}\,\sigma)(\boldsymbol{x}) := \frac{1}{4\pi} \int\limits_{\boldsymbol{y}\in\Gamma} \frac{1}{|\boldsymbol{x}-\boldsymbol{y}|}\,\sigma(\boldsymbol{y})\,ds_{\boldsymbol{y}}\,, \quad \boldsymbol{x}\notin\Gamma \tag{9}$$

with unknown density σ on Γ. Then equations (7a) and (7c) are automatically satisfied. Inserting (9) into (7b), we get

$$-\Delta_\Gamma\, V\sigma - \alpha\,\frac{\partial}{\partial n}\,(V\sigma) = g\,, \tag{10}$$

where $\mathrm{V}\sigma := (\mathcal{S}\,\sigma)|_\Gamma$ is the single layer operator on the boundary Γ. This is an integrodifferential equation on Γ for the density σ. The jump relations for $(\mathcal{S}\,\sigma)(\boldsymbol{x})$ imply

$$-\Delta_\Gamma\, V\sigma + \alpha\Big(\frac{1}{2}\,I - K'\Big)\sigma = g \ \text{ on } \Gamma \tag{11}$$

where

$$(K'\sigma)(\boldsymbol{x}) := \int\limits_{\boldsymbol{y}\in\Gamma} \frac{\partial}{\partial n_{\boldsymbol{x}}}\Big(\frac{1}{4\pi\,|\boldsymbol{x}-\boldsymbol{y}|}\Big)\,\sigma(\boldsymbol{y})\,ds_{\boldsymbol{y}} \tag{12}$$

is the adjoint operator to the classical double layer potential (modulo a factor $\frac{1}{2}$).

A variational formulation of (10) reads:
Find $\sigma \in H^{\frac{1}{2}}(\Gamma)$ such that

$$a(\sigma, \widetilde{\sigma}) := \left\langle \mathbf{grad}_\Gamma V\sigma, \overline{\mathbf{grad}_\Gamma \widetilde{\sigma}} \right\rangle_\Gamma + \alpha \left\langle \left(\frac{1}{2} I - K'\right) \sigma, \overline{\widetilde{\sigma}} \right\rangle_\Gamma = \left\langle g, \overline{\widetilde{\sigma}} \right\rangle_\Gamma \quad (13)$$

for all $\widetilde{\sigma} \in H^{\frac{1}{2}}(\Gamma)$.

Theorem 2. *For every $g \in H^{-\frac{1}{2}}(\Gamma)$, the variational problem* (13) *admits a unique solution $\sigma \in H^{\frac{1}{2}}(\Gamma)$ and its potential $\phi(\boldsymbol{x}) = (\mathcal{S}\,\sigma)(\boldsymbol{x})$ is the unique solution of the problem* (7).

All proofs can be found in [13].

4 Galerkin Discretization

Let $\{\mathcal{T}_h\}_h$ be a quasi-uniform, shape-regular family of triangulations on Γ consisting of triangular curvilinear boundary elements.
By S_h^p, we denote the space of continuous, piecewise polynomial functions in local coordinates on Γ of total (separate) degree p, $p \geq 1$. Then, for any $\varepsilon > 0$,

$$S_h^p \subset H^{\frac{3}{2}-\varepsilon}(\Gamma), \ \dim S_h^p = \mathcal{O}(p^2 |\mathcal{T}_h|)$$

where $|\mathcal{T}_h|$ = # of elements in $\mathcal{T}_h$. The space S_h^p has the **approximation property**

$$\min_{v \in S_h^p} ||\sigma - v||_{H^s(\Gamma)} \leq C\, h^{\min(p+1,k)-s} ||\sigma||_{H^k(\Gamma)} \quad (14)$$

for $\sigma \in H^k(\Gamma)$, where C is independent of h, p and σ, and $0 \leq s \leq 1$, $k \geq s$.

The Galerkin-BEM for the problem (13) reads:
Find $\sigma_N \in S_h^p$ such that

$$a(\sigma_N, \widetilde{\sigma}) = \langle g, \overline{\widetilde{\sigma}} \rangle_\Gamma \quad \forall \widetilde{\sigma} \in S_h^p . \quad (15)$$

Proposition 3. *For h sufficiently small,* (15) *admits a unique solution $\sigma_N \in S_h^p$ and*

$$||\sigma - \sigma_N||_{H^{\frac{1}{2}}(\Gamma)} \leq C \inf_{v \in S_h^p} ||\sigma - v||_{H^{\frac{1}{2}}(\Gamma)} .$$

If, in particular, $g \in H^{-\frac{1}{2}+s}(\Gamma)$, $s \geq 0$, as $h \to 0$:

$$||\sigma - \sigma_N||_{H^{\frac{1}{2}}(\Gamma)} \leq C\, h^{\min(p+\frac{1}{2}, s)} ||g||_{H^{-\frac{1}{2}+s}(\Gamma)} . \quad (16)$$

For the calculation of losses, we shall be also interested in $H^{-\frac{1}{2}}(\Gamma)$-error estimates. We have

Proposition 4. *If $g \in H^{-\frac{1}{2}+s}(\Gamma)$, $s \geq 1$, then*

$$||\sigma - \sigma_N||_{H^{-\frac{1}{2}}(\Gamma)} \leq C\, h^{\min(p+\frac{3}{2}, s+1)} ||g||_{H^{-\frac{1}{2}+s}(\Gamma)} . \quad (17)$$

Proposition 4 gives immediately a result on the convergence of the Ohmic losses, i.e. the quantities

$$P = \frac{\mathrm{Re}\,\eta}{2} \int\limits_{\Gamma} |\boldsymbol{H}_t|^2 \, ds = \frac{\mathrm{Re}\,\eta}{2} \int\limits_{\Gamma} |\boldsymbol{H}_{0t} - \mathbf{grad}_{\Gamma} \, \mathrm{V}\, \sigma|^2 \, ds \,. \tag{18}$$

Let P_N be as in (18) with σ replaced by σ_N. Then we have

Proposition 5. *If $g \in H^{-\frac{1}{2}+s}(\Gamma)$, the Ohmic losses converge, as $h \to 0$, like*

$$|P - P_N| \leq C \, h^{\min(p+\frac{3}{2}, s+1)} \|g\|_{H^{-\frac{1}{2}+s}(\Gamma)} \,.$$

In particular, if $p = 1$ and g is smooth, the Ohmic losses converge as $\mathcal{O}(h^{\frac{5}{2}})$.

5 Remarks on the Implementation

For a computational realisation of the presented method we apply triangular boundary elements with piecewise quadratic geometry representation and a piecewise linear approach for the unknown density and the test functions. That is we use $\sigma_N, \overline{\sigma} \in S_h^1$, $\sigma_N(\boldsymbol{x}) = \sum_{j=1}^{N} \varphi_j(\boldsymbol{x}) \, \sigma_{Nj}$, with φ_j being the usual hat functions. After rewriting the sesquilinear form (13) and doing an integration by parts on the RHS, the discrete Galerkin form reads

$$\sum_{j=1}^{N} a_{Nij} \, \sigma_{Nj} = g_{Ni} \,, \qquad i = 1..N \,, \tag{19}$$

where

$$a_{Nij} := \int\limits_{\mathrm{supp}\,\varphi_i} \Big(\big(\mathbf{grad}_{\Gamma} \, \varphi_i - \alpha \, \varphi_i \, \boldsymbol{n}\big) \cdot \boldsymbol{K} \, \varphi_j + \frac{\alpha}{2} \, \varphi_i \, \varphi_j \Big) \, ds \,, \tag{20}$$

$$g_{Ni} := \int\limits_{\mathrm{supp}\,\varphi_i} \big(\mathbf{grad}_{\Gamma} \, \varphi_i - \alpha \, \varphi_i \, \boldsymbol{n}\big) \cdot \boldsymbol{H}_0 \, ds \,. \tag{21}$$

Here the operator $\boldsymbol{K}$ is defined as

$$(\boldsymbol{K} \, \sigma)(\boldsymbol{x}) := -p.v. \int\limits_{\Gamma} \sigma(\boldsymbol{y}) \, \frac{\boldsymbol{x} - \boldsymbol{y}}{4\pi \, |\boldsymbol{x} - \boldsymbol{y}|^3} \, ds_{\boldsymbol{y}} \,.$$

For the computation of the matrix coefficients (20) we have to pay special attention to the *outer integration*, which is naturally done element by element. The inner integrals considered as functions of the source point $\boldsymbol{x}$ are not smooth. This is contrary to common belief. In our case we have a $\log(\mathrm{dist}(\boldsymbol{x}, \partial\pi_{\boldsymbol{x}}))$ behaviour of the inner integrals, with $\pi_{\boldsymbol{x}}$ denoting an element. Thus, standard cubature techniques for the outer integration will fail to give good accuracy.

To overcome this problem we adopt the cubature scheme presented in [4] and [10]. There the double surface integrals over pairs of elements of the Galerkin form are considered as integrals over four dimensional domains and the singularities are cancelled by domain transformations introducing relative coordinates. We have to distinguish between 4 cases for an element pair $\{\pi_{\boldsymbol{x}}, \pi_{\boldsymbol{y}}\}$:

1. identical elements
2. the elements share exactly one edge
3. the elements share exactly one vertice
4. the elements are disjoint

E.g. for the identical case the double integral is transformed into a sum of 6 integrals over the domain $(0,1) \times T_3$ where T_3 is the 3-dimensional unit simplex.

The linear equation system (19) is solved by GMRES [12] with simple scaling as a preconditioner.

6 Numerical Results

Sphere in Constant Field
As a test problem we consider an approximated sphere of radius 1 m in a constant exciting field $\boldsymbol{H}_0 = 1\,\mathrm{A/m}\,\boldsymbol{e}_z$ and $\omega = 2\,\pi\,50\,\mathrm{s}^{-1}$ with material parameters typical for steel, i.e. $\kappa = 0.666 \cdot 10^7\,\mathrm{S/m}$ and $\mu_r = 200$.

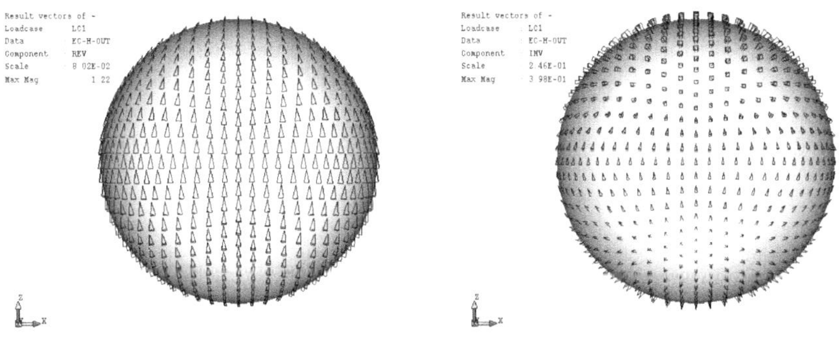

Figure2. Real (left) and imaginary (right) part of the magnetic field at the surface

With the resulting penetration depth $\delta = 1.95$ mm, the eddy current model with an IBC reasonably resembles physical behaviour. For this case the constant in the sesquilinear form yields $\beta = 2.56\,\mathrm{m}^{-1}$. Both terms of the form contribute numerically. The magnetic field at the surface is shown in Fig. 2, numerical values are given in A/m.

Acknowledgments The authors wish to acknowledge K.V. Boos, B. Schaub and Z. Andjelic at ABB Corporate Research Center Heidelberg, making this work possible. Special thanks to A. Bossavit and R. Hiptmair for helpful discussions during the conference.

References

1. H. Ammari, C. Latiri-Grouz and J.C. Nédélec: The Leontovich boundary value problem for the time-harmonic Maxwell equations. Asymp. Anal. **18** (1998), no. 1&2, 33-48.
2. A. Bossavit: *Computational electromagnetism.* Academic Press, San Diego (1998).
3. M. Cessenat: *Mathematical methods in electromagnetism.* World Scientific, Singapore (1996).
4. S. Erichsen and S. Sauter: Efficient automatic quadrature in 3-d Galerkin BEM. Computer methods in applied mechanics and engineering, **157** (1998), 215-224.
5. P.R. Kotiuga: Topological considerations in coupling magnetic scalar potentials to stream functions describing surface currents. IEEE Trans. Mag.**25**, no 4, (1989).
6. L.M. Leontovich: On approximate boundary conditions for electromagnetic fields on the surface of highly conducting bodies (in russian), *Research in the propagation of radio waves* (Academy of Sciences of the USSR, Moscow 1948).
7. E. Martensen: *Potentialtheorie* (Teubner, Stuttgart 1968).
8. Mayergoyz and G. Bredosian: On finite element implementation of impedance boundary conditions. J. Appl. Phys. **75**, 10 (1994), 6027-6029.
9. K.M. Mitzner: An integral equation approach to scattering from a body of finite conductivity. Radio Science **2** (1967), 1459-1470.
10. T. v. Petersdorff and C. Schwab: Fully discrete multiscale Galerkin BEM, in Vol. 6 of *Wavelet analysis and applications* (Dahmen, Kurdila and Oswald, Academic Press 1997), 287 ff.
11. S.M. Rytov: Calcul du skin-effet par la méthode des perturbations. Journal de Physique USSR, **2** (1940), 233-242.
12. Y. Saad and M.H. Schultz: GMRES: A generalized minimal residual algorithm for solving nonsymmetric linear systems. SIAM J. Sci. Stat. Comput., **7** (1986), 856-869.
13. O. Sterz and C. Schwab: A scalar boundary integrodifferential equation for eddy current problems using an impedance boundary condition. Submitted to Computing and Visualization in Science (2000).

Appraisal of Asymptotics in Electromagnetic Field Calculations

M. Kedde, P.-P. Borsboom, and C.R. Traas

[1] Thomson CSF Signaal, P.O. Box 42, 7550 GD Hengelo, The Netherlands
[2] University of Twente, P.O. Box 217, 7500 AE Enschede, The Netherlands

Abstract An expression for the electric field of a dipole near a perfectly conducting wedge has been derived to appraise the Uniform Theory of Diffraction (UTD) in regions where it is expected to deviate from the exact solution, e.g. the near field. This expression, obtained using the theory of Green's functions, has been evaluated using numerical integration. First the solution is verified for the case of a dipole near a ground plane, which can also be calculated using image theory. Next the numerical solution for the near field is compared to results obtained with the UTD-method for different parameter configurations. The most striking difference occurs when the angle of the wedge is sharp and both the source and observer are located within a few wavelengths from the edge.

1 Introduction

The electromagnetic radiation of antennas in a realistic environment (e.g. fixed to a mast or surrounded by buildings) is influenced by reflections of surrounding structures and by diffraction at the edges of these structures. This edge diffraction can be calculated with asymptotic methods (like Uniform Theory of Diffraction (UTD) or Geometrical Theory of Diffraction (GTD)). One of the assumptions on which they are based is that the observer is located many wavelengths away from the edge. When the observer and the source are situated within a few wavelengths of the edge, the accuracy of the asymptotics is unclear.

The first aim of this paper is to get a quantitative idea of this accuracy by studying the simplified problem of a dipole near a perfectly conducting wedge for which an expression is available in closed-form. The closed-form expression is quite involved. Hence, the second aim is to find ways for successfully handling the complicated closed-form expression numerically, which is far from trivial.

In the remainder of this paper we shall give the expression for the electric field using Green's functions. As only a limited time was available, we selected one representative case of an axial dipole. This presents all the numerical difficulties of the general case, so that the results here can be used as a stepping stone to the general case. Finally, after evaluating the accuracy of this expression against an analytic example we conclude with a comparison with the UTD-method.

2 Dyadic Green's Functions for a Perfectly Conducting Wedge

2.1 Configuration

The configuration that is considered in this paper is a two-dimensional perfectly conducting wedge in vacuum, with a dipole used as a source, see Fig. 1. The positions of the source $\bar{R}' = (r', \phi', z')$ and observation point $\bar{R} = (r, \phi, z)$ are given in cylindrical coordinates, because of the configuration of the wedge.

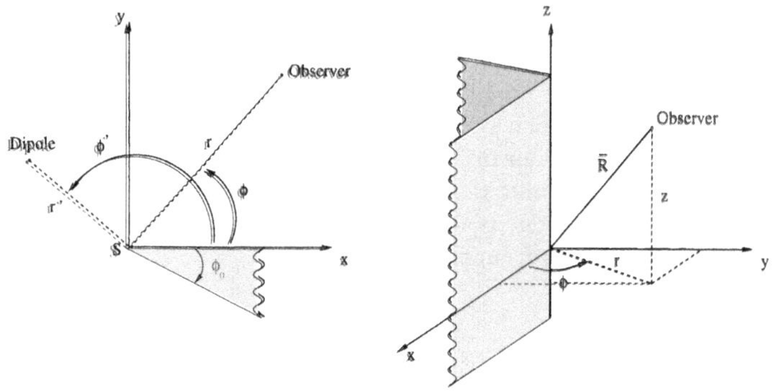

Figure1. Geometry of the wedge

2.2 The Helmholtz Equation

From the Maxwell equations with time-dependence $e^{-i\omega t}$, see [6], the Helmholtz equation,

$$\nabla \times \nabla \times \bar{E} - k_0^2 \bar{E} = i\omega\mu_0 \bar{J}, \tag{1}$$

can be derived. In (1) $\bar{E}$ is the electric field, $\bar{J}$ the electric current density, ω the angular frequency, $k_0 = \omega\sqrt{\mu_0\epsilon_0}$ the free-space wavenumber and ϵ_0 and μ_0 the electric and magnetic constant, respectively.
To ensure uniqueness the boundary condition on the faces of a perfectly conducting wedge is given by

$$\hat{n} \times \bar{E} = \bar{0}. \tag{2}$$

This condition characterizes $\bar{E}$ on the faces of the wedge except the edge, where the field has a singularity in the r-derivative, see [2]. Another boundary condition is the radiation condition, which states that the electrical field approaches a spherical wave as the observer goes to infinity.

2.3 The Electric Field of a Dipole

To solve (1), we introduce dyadic Green function $\bar{\bar{G}}$ which satisfies,

$$\nabla \times \nabla \times \bar{\bar{G}} - k_0^2 \bar{\bar{G}} = \bar{\bar{I}} \delta(\bar{R} - \bar{R}'), \tag{3}$$

where $\bar{\bar{I}}$ the dyadic identity and δ the delta-function. Using the principle of superposition, the electric field for an arbitrary source can be represented by:

$$\bar{E}(\bar{R}) = i\omega\mu_0 \iiint_V \bar{J}(\bar{R}') \cdot \bar{\bar{G}}(\bar{R}, \bar{R}')\, dV. \tag{4}$$

This means that the dyadic Green function $\bar{\bar{G}}$ describes the field of an arbitrarily directed dipole. This Green function is derived in [3][7] and is given by:

$$\begin{aligned} \bar{\bar{G}}(\bar{R}, \bar{R}') &= -\frac{1}{k_0^2} \hat{r}\hat{r}\delta(\bar{R} - \bar{R}') \\ &+ \frac{i}{4(2\pi - \phi_0)} \int_{-\infty}^{\infty} dh \sum_{\nu} \frac{2 - \delta_0}{\lambda^2} \\ &\cdot \begin{cases} \bar{M}^{(1)}_{e\nu\lambda}(h)\bar{M}'_{e\nu\lambda}(-h) + \bar{N}^{(1)}_{o\nu\lambda}(h)\bar{N}'_{o\nu\lambda}(-h), & r > r' \\ \bar{M}_{e\nu\lambda}(h)\bar{M}'^{(1)}_{e\nu\lambda}(-h) + \bar{N}_{o\nu\lambda}(h)\bar{N}'^{(1)}_{o\nu\lambda}(-h), & r < r' \end{cases} \end{aligned} \tag{5}$$

where the vector wave functions are

$$\bar{M}^{(1)}_{e\nu\lambda}(h) = \nabla \times [H^{(1)}_{\nu}(\lambda r)\, \cos(\nu\,\phi)\, e^{ihz} \hat{z}], \tag{6}$$

$$\bar{M}_{e\nu\lambda}(h) = \nabla \times [J_{\nu}(\lambda r)\, \cos(\nu\,\phi)\, e^{ihz} \hat{z}], \tag{7}$$

$$\bar{N}^{(1)}_{o\nu\lambda}(h) = \frac{1}{k_0} \nabla \times \nabla \times [H^{(1)}_{\nu}(\lambda r)\, \sin(\nu\,\phi)\, e^{ihz} \hat{z}], \tag{8}$$

$$\bar{N}_{o\nu\lambda}(h) = \frac{1}{k_0} \nabla \times \nabla \times [J_{\nu}(\lambda r)\, \sin(\nu\,\phi)\, e^{ihz} \hat{z}], \tag{9}$$

and

$$\delta_0 = \begin{cases} 1 & \text{for} \quad \nu \quad \text{or} \quad \lambda = 0, \\ 0 & \text{for} \quad \nu \quad \text{and} \quad \lambda \neq 0. \end{cases} \tag{10}$$

In these equations λ is the wavelength, $J_\nu(\lambda r)$ and $H^{(1)}_\nu(\lambda r)$ are the Bessel function and Hankel function of the first kind, respectively, and $\nu = \frac{n}{2 - \phi_0/\pi}$ with $n = 1, 2, 3, \ldots$.The $\hat{r}$, $\hat{\phi}$ and $\hat{z}$ are the unit vectors in cylindrical coordinates. The integrals in eq. (5) are understood in the Principal Value sense and the convergence of the series in eq. (5) is pointwise.

3 Axially Placed Dipole near a Perfectly Conducting Wedge

One representative case of an axially placed dipole has been selected. From the similarity of the expressions, it can be seen that the more general case can be solved straightforwardly from the special case. This special case is obtained by taking the equations in the $\hat{r}\hat{z}$, $\hat{\phi}\hat{z}$ and $\hat{z}\hat{z}$- direction of (5). This gives,

$$\bar{E}(r,\phi,z) = \frac{-IL}{2\omega\epsilon_0(2\pi-\phi_0)} \cdot \int_{-\infty}^{\infty}\sum_{\nu}\begin{bmatrix} ihJ_\nu(\lambda r')\frac{\partial H_\nu^{(1)}(\lambda r)}{\partial r}\sin(\nu\phi)\sin(\nu\phi')e^{ih(z-z')}dh\,\hat{r} \\ \frac{ih\nu}{r}J_\nu(\lambda r')H_\nu^{(1)}(\lambda r)\cos(\nu\phi)\sin(\nu\phi')e^{ih(z-z')}dh\,\hat{\phi} \\ \lambda^2 J_\nu(\lambda r')H_\nu^{(1)}(\lambda r)\sin(\nu\phi)\sin(\nu\phi')e^{ih(z-z')}dh\,\hat{z} \end{bmatrix}, \quad (11)$$

where $r > r'$, L is the length of the dipole and r and r' are interchanged for $r < r'$. To solve (11) for the near field, the order of integration and summation is reversed. This gives

$$E(r,\phi,z) = \frac{-IL}{2\omega\epsilon_0(2\pi-\phi_0)}\sum_{\nu}\begin{bmatrix} iI_\nu^r\sin(\nu\phi)\sin(\nu\phi')\hat{r} \\ \frac{i\nu}{r}I_\nu^\phi\cos(\nu\phi)\sin(\nu\phi')\hat{\phi} \\ I_\nu^z\sin(\nu\phi)\sin(\nu\phi')\hat{z} \end{bmatrix}, \quad (12)$$

where the integrals I_ν^r, I_ν^ϕ and I_ν^z are given by

$$I_\nu^r = \int_{-\infty}^{\infty} hJ_\nu(\lambda r')\frac{\partial H_\nu^{(1)}(\lambda r)}{\partial r}e^{ih(z-z')}dh, \quad (13)$$

$$I_\nu^\phi = \int_{-\infty}^{\infty} hJ_\nu(\lambda r')H_\nu^{(1)}(\lambda r)e^{ih(z-z')}dh, \quad (14)$$

$$I_\nu^z = \int_{-\infty}^{\infty} \lambda^2 J_\nu(\lambda r')H_\nu^{(1)}(\lambda r)e^{ih(z-z')}dh. \quad (15)$$

The numerical evaluation of these integrals is discussed in the next section. Because the argument of the Bessel function $J_\nu(\lambda r')$ is smaller than the argument of the Hankel function $H_\nu^{(1)}(\lambda r)$, the summation converges.

4 Numerical Evaluation of the Green's Function

The integration interval can be reduced using symmetry, giving

$$I_\nu^r = 2i\int_0^{\infty} hJ_\nu(\lambda r')\frac{\partial H_\nu^{(1)}(\lambda r)}{\partial r}\sin(h(z-z'))dh, \quad (16)$$

$$I_\nu^\phi = 2i\int_0^{\infty} hJ_\nu(\lambda r')H_\nu^{(1)}(\lambda r)\sin(h(z-z'))dh, \quad (17)$$

$$I_\nu^z = 2\int_0^{\infty} \lambda^2 J_\nu(\lambda r')H_\nu^{(1)}(\lambda r)\cos(h(z-z'))dh. \quad (18)$$

Because the integrand vanishes rapidly for $h > k_0$, these integrals can be computed numerically using an adaptive recursive Newton-Cotes 8 panel rule, which standard is available in Matlab,

$$\int_a^b f(x)dx \doteq \frac{4h}{14175}[989f(x_0) + 5888f(x_1) - 928f(x_2) + 10496f(x_3) - 4540f(x_4) + 10496f(x_5) - 928f(x_6) + 5888f(x_7) + 989f(x_8)], \quad (19)$$

with h the stepsize.
These integrations must be handled very carefully, because of the strongly oscillating nature of the integrands. The technique described above has been applied to several configurations. In this paper only two configurations are discussed. The first configuration is used to check the numerical solution and the second configuration is used to appraise the UTD-method.

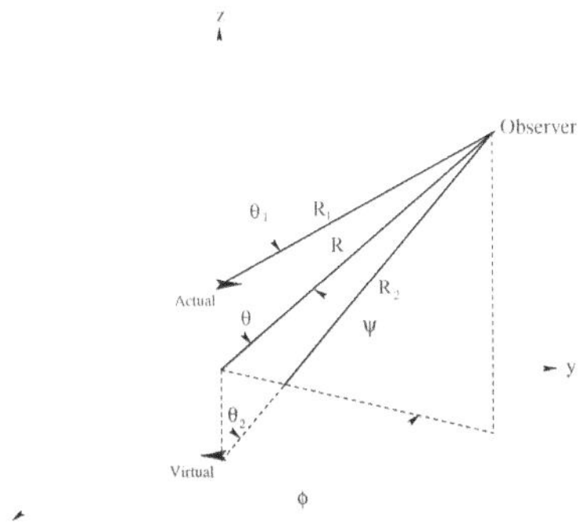

Figure2. Horizontal dipole above a plane in three dimensions

4.1 Validation of the Numerical Solution

For the validation of the numerical solution the case of a dipole above a plane is used, because an exact solution for this case can be found using the *image theory*, see [1]. The image theory is based on the fact that the field of a source above a ground plane equals the superposition of the original source and its mirror image below the plane, see Fig. 2.
The electric near field in spherical coordinates is in [1] given by:

$$E_R = \eta \frac{IL\cos\theta}{2\pi R^2}\left[1 - \frac{i}{k_0 R}\right] e^{ik_0 R}, \quad (20)$$

$$E_\phi = 0, \quad (21)$$

$$E_\theta = -i\eta \frac{k_0 IL \sin\theta}{4\pi R}\left[1 - \frac{i}{k_0 R} - \frac{1}{k_0^2 R^2}\right] e^{ik_0 R}. \quad (22)$$

with η the free space impedance.
The numerical evaluation of (12) is compared to the solution of the image theory (20-22) for the following parameter configuration: $\phi_0 = 180°$, $\bar{R}' = (\frac{1}{3}\lambda, 90°, 0)$ and $\bar{R} = (r, \phi, \frac{10}{3}\lambda)$, where ϕ varies from 0 to 180°. Several values for the distance r are taken. For radar applications the wavelength λ varies between 0.3 and 120 cm. The results of the simulations are depicted in table 1. Note that the expressions (20-22) are given in spherical coordinates. A transformation has been used to compare both solutions. The relative error

Table1. Relative error between the image theory and numerical integration for several values of r

r	Mean error	r	Mean error
$\frac{2}{3}\lambda$	0.1965	$\frac{8}{3}\lambda$	5.7170e-8
$\frac{4}{3}\lambda$	5.8354e-5	$\frac{10}{3}\lambda$	3.8516e-9
2λ	1.5763e-5	$\frac{20}{3}\lambda$	1.1975e-8

increases if r approaches r'. This is the consequence of the chosen solution method. The Green function solution (11) is the combination of two solutions: one solution for $r < r'$ and one solution for $r > r'$ joined together at $r = r'$. Hence, the solution should become singular at $r = r'$.
If $r \geq \frac{4}{3}\lambda$ the solution of the numerical integration is in excellent agreement with the image theory. This good agreement improves confidence in the numerical evaluation of eq. (12). This justifies its use as an appraisal of the UTD-method for an axial dipole in the next subsection.

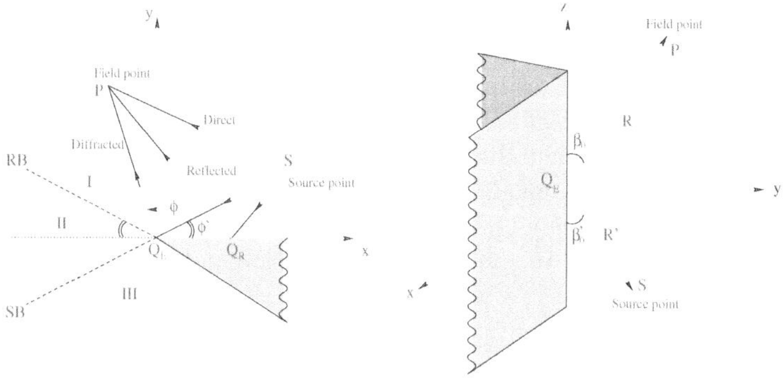

Figure3. Wedge geometry for edge diffraction using UTD.

5 Comparison with UTD-method

In this subsection the Green function method with numerical integration is compared with the Uniform Theory of Diffraction (UTD), see [4]. The parameter configuration considered in this case is: $\phi_0 = 30°$, $\bar{R}' = (\frac{\lambda}{3}, 7.2°, 0)$ and $\bar{R} = (r, \phi, \frac{10}{3}\lambda)$, with ϕ from 0 to 330°.
The UTD-method splits the region into three parts (see Fig. 3). The boundaries shown are the shadow boundary (SB) and the reflection boundary (RB). In region I the electric field is split into a direct, a reflected and a diffracted part. In region II the direct and diffracted field remain and in region III only the diffracted field exists. Expressions for the direct, reflected and diffracted fields are given in [5].The absolute relative error for several values of r are given in table 2. The errors are quite large if they are compared to the errors

Table2. Relative error between the UTD and numerical integration

r	Mean error	r	Mean error
		$\frac{8}{3}\lambda$	0.0871
$\frac{4}{3}\lambda$	0.2895	$\frac{10}{3}\lambda$	0.0657
2λ	0.1426	$\frac{20}{3}\lambda$	0.0433

in table 1. Because of the confidence obtained in the numerical integration, we may say that the UTD-method is considerably in error for small distances in the considered configuration. More work is needed, however, on the systematic comparison between UTD and numerical integration to check whether this applies for different wegde/dipole configurations as well.

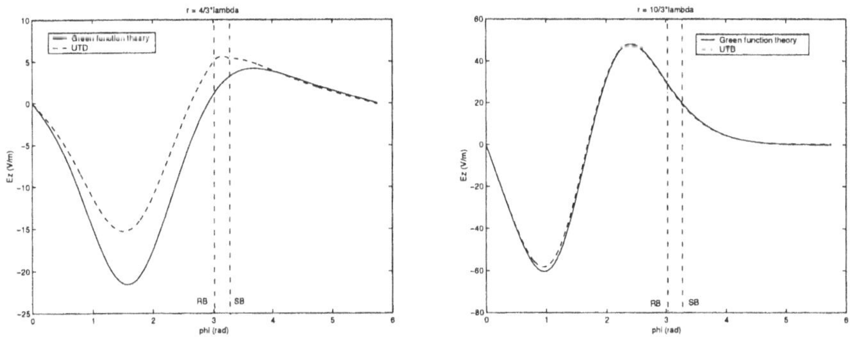

Figure4. The real part of the z-component of the electric field for $r = \frac{4}{3}\lambda$ (left) and $r = \frac{10}{3}\lambda$ (right).

In Fig. 4 the solution for $r = \frac{4}{3}\lambda$ and $r = \frac{10}{3}\lambda$ is depicted for both methods. The wavelength λ is chosen to be 30 cm. In this figure we see that the most striking difference occurs in the regions I and II. In the region III, where only the diffracted field exist, the difference is quite small.

6 Conclusion

In this paper a comparison between the numerical solution, obtained by the Green function theory, and the UTD method for a dipole near a perfectly conducting wedge is made. The most striking difference occurs when both the dipole and the observer are located within two wavelengths from the edge. In the regions of the direct and reflected fields the largest errors are found. It has been shown that the complicated closed-form expression for the electric field can be handled quite well by choosing suitable numerical methods.

References

1. Balanis, C.A.: Antenna Theory. Harper and Row New York (1982)
2. Büyükdura, O.M.: Radiation from Sources and Scatters near the Edge of a Perfectly Conducting Wedge. Ph.D. dissertation, The Ohio State University, Columbus, Ohio (1984)
3. Kedde, M.: The electric field of a dipole near a perfectly conducting wedge. Master's thesis, University of Twente (1999)
4. Kouyoumjian, R.G., Pathak, P.H.: A Uniform Geometrical Theory of Diffraction for an Edge in a Perfectly Conducting Surface. Proc. IEEE **62** 1448–1461 (1974)
5. Pathak, P.H., Kouyoumjian, R.G.: The Dyadic Diffraction Coefficient for a Perfectly Conducting Wedge. ElectroScience Lab, Dept. of Electrical Engineering, The Ohio State University, Columbus, Ohio (1970)
6. Stratton, J.A.: Electromagnetic Theory. McGraw-Hill, New York (1941)
7. Tai, C.T.: Dyadic Green's Functions in Electromagnetic Theory, second edition. IEEE Press Piscataway (1994)

Shape Optimization of Homogeneous Electromagnets *

D. Lukáš

SFB F013, University of Linz, Austria
Dept. of Applied Mathematics, VŠB-TU Ostrava, Czech Republic

Abstract Magneto-optical effects are investigated among others for their application in storage media. Measurements of Kerr effect require magnetic field as homogeneous as possible. This is generated by so-called homogeneous electromagnets. The optimization aims at the optimal shape of the pole heads. The governing linear magnetostatic problem is approximated by the Finite Element Method (FEM) where quadratic triangular elements or edge elements are used in the 2-dimensional (2D) or 3-dimensional (3D) case, respectively. The solver is either a direct or multigrid Preconditioned Conjugate Gradient method (PCG), depending on the problem size. The Sequentional Quadratic Programming (SQP) method with the BFGS update of Hessian matrix was used for the optimization. We computed an optimized 2D coarse design which was produced afterwards. The measurements show significant improvements of the homogeneity. We also computed an optimized 2D fine design by a hierarchical strategy, which is an iterative process where a coarse optimized shape is used as an initial design for the optimization on a finer grid. This approach seems to suit our class of problems very well. Finally, a coarse approximation of the 3D optimal shape was calculated.

1 Introduction

A number of applications needs homogeneous magnetic fields. They can be produced by electromagnets which impress such a field among the pole heads. These are called homogeneous electromagnets. In Fig. 1 there are two examples of them. In our case the homogeneous electromagnets are used for measurements of Kerr effect. The research of magneto-optical effects is interesting among others for the applications in magneto-optical storage media. This is investigated by the research team around prof. J. Pištora at Department of Physics at VŠB - Technical University of Ostrava, see [1], with whom we closely cooperate.

This paper presents mathematical modeling and shape optimization of an electromagnet of a so-called Maltese Cross geometry. First, let's describe the physical problem. A sample of magneto-optical material is placed to the center of the Maltese Cross, see Fig. 2 (left). Rays coming to the center are polarized in the polarization plane by the magnetic field and then they

* This research has been supported by the Austrian Science Fund FWF within the SFB "Numerical and Symbolic Computing" under the grant SFB F013

Figure1. Homogeneous electromagnets - geometries "O-ring" (left) and "Maltese Cross" (right)

reflect on the sample. The components of the reflected rays are measured. The field is to be as homogeneous as possible. By switching the sense of two currents we can polarize the rays in the orthogonal polarization plane as well. The device, see Fig. 2 (left), consists of a ferromagnetic yoke, 4 poles, and 4 windings. The shape of the pole heads influences the magnetic field in the plane significantly.

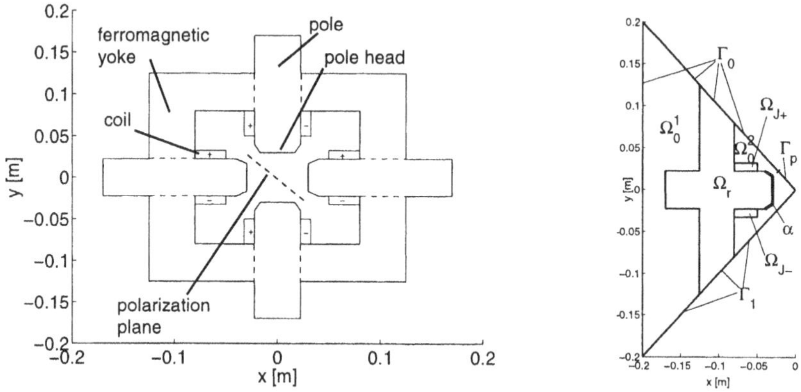

Figure2. Cross section of the Maltese Cross (left), computational domain Ω_{2D} (right)

2 Mathematical Model of the Maltese Cross

2.1 Linear Magnetostatic Problem

First we introduce Maxwell equations for 3D linear magnetostatics. Let $\overline{\Omega} := \cup_{i=1}^{k} \overline{\Omega_i} \subset \mathbf{R}^3$ be a material-wise decomposition where Ω is bounded and

Ω_i are Lipschitz domains such that $\mu(x,y,z) = \mu_i$ denote domain-wise constant permeabilities of the appropriate materials. Further let $\mathbf{B}$ denotes the magnetic induction, $\mathbf{H}$ is the magnetic strength density, and $\mathbf{J}$ is the current density. Then the Maxwell equations formally read

$$\left.\begin{aligned} \mathbf{rot}(\mathbf{H}) &= \mathbf{J} \\ \mathbf{B} &= \mu\mathbf{H} \\ \mathbf{div}(\mathbf{B}) &= 0 \end{aligned}\right\} \text{ in } \Omega \ . \tag{1}$$

The corresponding mixed boundary conditions are

$$\begin{aligned} \mathbf{B}\cdot\mathbf{n} &= 0 \text{ on } \Gamma_{\mathrm{D}} \ , \\ \mathbf{H}\times\mathbf{n} &= \mathbf{0} \text{ on } \Gamma_{\mathrm{N}} \end{aligned} \tag{2}$$

where $\partial\Omega = \overline{\Gamma_{\mathrm{D}}} \cup \overline{\Gamma_{\mathrm{N}}}$, $\Gamma_{\mathrm{D}} \cap \Gamma_{\mathrm{N}} = \emptyset$ and $\mathbf{n}$ is the unit normal vector of $\partial\Omega$. Finally the interface conditions

$$\begin{aligned} [\mathbf{B}\cdot\mathbf{n}]_{i,j} &= 0 \ , \\ [\mathbf{H}\times\mathbf{n}]_{i,j} &= \mathbf{0} \end{aligned} \tag{3}$$

hold where $[\,]_{i,j}$ denotes the jump on the interface $\Gamma_{i,j} := \overline{\Omega_i} \cap \overline{\Omega_j}$ and $\mathbf{n}$ is the corresponding unit normal vector of $\Gamma_{i,j}$.

Now we introduce the magnetic vector potential $\mathbf{A}$ as follows

$$\mathbf{rot}(\mathbf{A}) = \mathbf{B} \ , \ \operatorname{div}(\mathbf{A}) = 0 \ . \tag{4}$$

We obtain the reduced 2-dimensional problem by looking at a typical cross section $\Omega_{2\mathrm{D}}$. There hold $\mathbf{J} = (0, 0, \mathrm{J}(x,y))$, $\mathbf{B} = (\mathrm{B}_x(x,y), \mathrm{B}_y(x,y), 0)$, and the magnetic potential is uniquely given by

$$\mathbf{A} = (0, 0, u(x,y)) \ , \ \mathbf{B} = \left(\frac{\partial u}{\partial y}, -\frac{\partial u}{\partial x}, 0\right)$$

where u stands for the scalar potential.

We apply the 2D linear magnetostatics to the Maltese Cross problem. Since the magnetic field should be orthogonal to the polarization plane and, after switching the currents, to the orthogonal polarization plane as well, we prescribe the symmetric shapes of the pole heads. Therefore, the magnetic field is symmetric with respect to both of the polarization planes and we will deal only with a quarter of the original domain, see Fig. 2 (right). By using the scalar potential, the Maxwell setting of the 2D Maltese Cross problem formally reads as follows :

$$\begin{aligned} -\mathbf{div}\left(\tfrac{1}{\mu}\nabla u\right) &= J \text{ in } \Omega_{2\mathrm{D}} \ , \\ u &= 0 \text{ on } \Gamma_0 \cup \Gamma_{\mathrm{p}} \ , \\ \tfrac{\partial u}{\partial n} &= 0 \text{ on } \Gamma_1 \end{aligned} \tag{5}$$

with the interface conditions

$$\begin{aligned} [\nabla u \times \mathbf{n}]_{i,j} &= 0 \ , \\ \left[\left(\tfrac{1}{\mu}\nabla u\cdot\mathbf{n}\right)\right]_{i,j} &= 0 \ . \end{aligned} \tag{6}$$

The parameters are $\mu_0 = 4\pi.10^{-7}\ \mathrm{H.m^{-1}}$, $\mu_\mathrm{r} = 5100$,
$\mu = \begin{cases} \mu_0 & \text{in } \overline{\Omega} \setminus \overline{\Omega_\mathrm{r}} \\ \mu_0.\mu_\mathrm{r} & \text{in } \overline{\Omega_\mathrm{r}} \end{cases}$, and

$$\mathrm{J} = \begin{cases} 12.5\ \mathrm{MA.m^{-2}} & \text{in } \Omega_{\mathrm{J}_+} \\ 0\ \mathrm{MA.m^{-2}} & \text{in } \Omega \setminus (\Omega_{\mathrm{J}_+} \cup \Omega_{\mathrm{J}_-}) \\ -12.5\ \mathrm{MA.m^{-2}} & \text{in } \Omega_{\mathrm{J}_-} \end{cases} .$$

Note that the ferromagnetic material is a kind of steel and J is given by the current I = 6.3 A and the wire diameter d = 0.8 mm.

Finally, we set the weak formulation. Since the domains are Lipschitz, we can form the Hilbert ansatz space

$$V = \{v \in H^1(\Omega) \mid v = 0 \text{ on } \Gamma_0 \cup \Gamma_\mathrm{p}\} . \tag{7}$$

Since $J \in L^2(\Omega_{2\mathrm{D}})$ and $1/\mu \in L^\infty(\Omega_{2\mathrm{D}})$ hold, the weak formulation of the 2D linear magnetostatic problem is well-defined and reads as follows :

$$\text{Find } u \in V \ : \ A(u, v) = b(v) \text{ for all } v \in V \tag{8}$$

where

$$A(u, v) := \int_{\Omega_{2\mathrm{D}}} \frac{1}{\mu}.\nabla u.\nabla v \, dx \ , \ b(v) := \int_{\Omega_{2\mathrm{D}}} \mathrm{J}.v \, dx \ . \tag{9}$$

Note, that in the 3D case the boundary conditions are very similar and the weak formulation is built in $H(\mathbf{rot})$, see [2].

2.2 Shape Optimization Problem

Let Γ_p be the polarization line in Fig. 2 (right), in which inhomogeneities of the magnetic field are to be minimized. Let $\alpha \subset \partial\Omega_\mathrm{r}$ be the graph of the shape of the pole head. Independently of the geometry of an electromagnet, we consider the following optimization problem :

$$\min_{\alpha \in \mathcal{F}} \varphi(\alpha) \tag{10}$$

under the constraints

$$\mathrm{B}_{\mathrm{min}}^{\mathrm{avg}} \le \|\mathbf{B}_\alpha^{\mathrm{avg}}\| , \tag{11}$$

$$\alpha_\mathrm{l} \le \alpha \le \alpha_\mathrm{u} \tag{12}$$

where

$$\varphi(\alpha) := \frac{1}{\mathrm{meas}(\Gamma_\mathrm{p}).\, \|\mathbf{B}_\alpha^{\mathrm{avg}}\|^2} \cdot \int_{\Gamma_\mathrm{p}} \|\mathbf{B}_\alpha(\mathbf{r}) - \mathbf{B}_\alpha^{\mathrm{avg}}\|^2 \, ds , \tag{13}$$

$$\mathbf{B}_\alpha^{\mathrm{avg}} := \frac{1}{\mathrm{meas}(\Gamma_\mathrm{p})} \cdot \int_{\Gamma_\mathrm{p}} \mathbf{B}_\alpha(\mathbf{r}) \, ds , \tag{14}$$

$$\mathcal{F} := \{\alpha \subset \partial\Omega_\mathrm{r} \mid \alpha \text{ is Lipschitz continuous and symmetric}\} , \tag{15}$$

and where $\alpha_\mathrm{l} = -0.05$ m, $\alpha_\mathrm{u} = -0.03$ m, and the typical average induction $\mathrm{B}_{\mathrm{min}}^{\mathrm{avg}} = 0.25$ T. This 2D formulation can be easily extended to 3D where α is a 2D function and Γ_p is the polarization plane.

3 Components of the Solver

The solver of the shape optimization problem uses the software of SFB F013. There are, namely, the finite element package FEPP [2], the mesh generator NETGEN [3], and the algebraic multigrid package PEBBLES [4] involved. This framework is mainly used to solve the state problem by the FEM method. In the 2D case we use quadratic triangular elements rather than linear to reach a better accuracy of the B-field. In the 3D case the edge elements [2] are used. The arising linear system is basically solved by the Conjugate Gradients method with a multigrid preconditioner. In the case when the number of unknowns is not large, a direct solver is applied.

The optimization method is based on the SQP method with the BFGS update of Hessian matrix. The optimization problem described in Sect. 2.2 is discretized and the design variables are $x-$coordinates of the nodes along the shape α which is interpolated linearly. The integrals in (13) and (14) are replaced by a quadrature formula of the 0th order. Derivatives are calculated by finite differences of the 3rd extrapolation order.

In shape optimization, once the shape has changed, the mesh is deformed. We use a simple strategy where nodes are displaced in the horizontal bar around the pole head only. The horizontal displacements are linearly dependent on the shape displacements as long as the node is far from the boundary α. For big displacements overlapping elements can appear. This happens within the line-search procedure and we simply exclude such designs α for which an element has flipped. Since only the feasible shapes are involved, the number of design variables is half the number of nodes along α such that the design is symmetric with respect to the $x-$axis. The shape α should also be Lipschitz continuous and, moreover, smooth enough because, otherwise, the linear magnetostatics is not valid and the shape is hardly producable. That's why, an additional regularity constraint on the maximal curvature of α has been implemented. Just for the purposes of an evaluating this constraint we interpolate the design nodes by cubic splines $\{\alpha_i\}$ and the curvature is calculated at each of the nodes along α. The constraint reads

$$-\frac{1}{\rho_{\min}} \leq \frac{\alpha_i''(x_i)}{\sqrt{[1+(\alpha_i'(x_i))^2]^3}} \leq \frac{1}{\rho_{\min}} \quad \text{for all the design nodes } i \qquad (16)$$

where $\rho_{\min} > 0$ is a minimal curvature radius. Note that the discretized shape itself remains piece-wise linear as we consider linear elements only.

In order to get fine enough results efficiently, a hierarchical optimization strategy has been implemented. We say that the classical approach is applied if only one discretized optimization problem is solved, i.e. the topology of the mesh doesn't change. By the hierarchical strategy we mean that several discretized optimization problems are solved sequentially such that they approximate the problem finer at higher levels and the optimized design is used as the initial one at the next level. There are a refinement and a prolongation strategy involved. The refinement strategy deals with the refinement of the grid in order to get a better approximation of the cost functional and of the constraints. In our case we refine the discretization of Γ_p and use NETGEN for generation of the finer mesh. The prolongation strategy prolonges the optimized design to the finer level. It's done by using the interpolation of the shape by cubic splines such that the middle points of the splines are used at the finer level. A stopping criterion of the hierarchical strategy can be a convergence criterion of the cost function or a maximal number of levels. The hierarchical strategy is successful as far as the coarse optimized design approximates the fine one well.

4 Numerical Results

4.1 Comparisons of the 2D Coarse Optimized Design to Measurements

First, we present the 2D optimized design which we computed by the classical approach on a uniform mesh of the step size 2.5 mm and 25600 unknowns. The computational domain was the whole square. The initial design was the one in Fig. 2 (left). There were 5 design variables and the optimized design is shown in Fig. 3 (left). This was solved by MATLAB's Optimization Toolbox [5] and we reached the result in 5 iterations. Both of the constraints (11) and (12) were involved where $B^{avg}_{min} = 0.18\,T$. The derivatives were computed by the adjoint method of analytical sensitivity analysis.

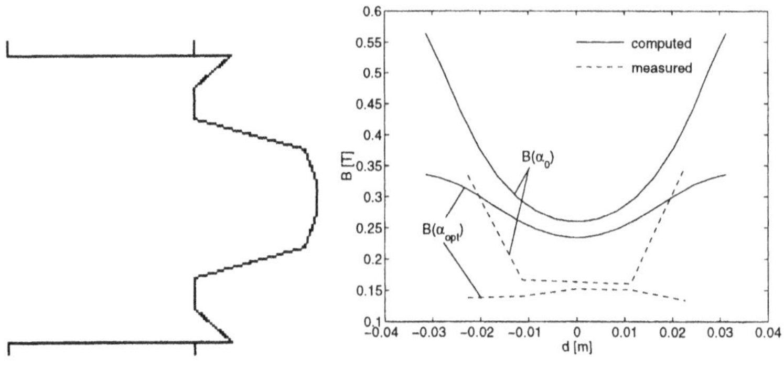

Figure3. 2D optimized pole head (left) and magnetic field comparisons (right)

This optimized design was produced and the magnetic field was measured at Department of Physics at VŠB-TU Ostrava. The measurements show significant improvements of the homogeneity. In Fig. 3 (right) there is drawn the magnetic induction along the polarization line Γ_p. The homogeneity of the optimized computed field has improved by the factor 10, and even much more in the case of the measured field. Considering the state problem, we have observed that the nonlinear model would approximate the measurements well.

4.2 Hierarchical Optimization Strategy

We computed the 2D fine optimized design by the hierarchical optimization strategy. The constraint parameters were $B^{avg}_{min} = 0.25\,\mathrm{T}$ and the minimal curvature radius $\rho_{\min} = 1\,\mathrm{cm}$. The computation proceeded at 3 levels with 3, 6, and 11 design variables, respectively. Properties of the computation are presented in Fig. 4. We can see that there is no need of a remeshing. The last column shows improvements of the cost function at each of the levels. Even if the discretized optimization problem at different levels is different, we can observe some convergence of the cost function $\varphi_l(\alpha_l)$. Finally, note that the classical approach takes 10 mesh regenerations when applied for the 11 design variables. Thus, the hierarchical approach is more efficient for our optimization problems.

level l	optimized designs α_l	# of des. variables	# of unknowns	# of remeshs.	$\varphi_l(\alpha_{l-1}) \downarrow \varphi_l(\alpha_l)$
0 (init.)		3	482		0.1068
1	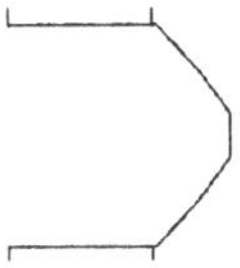	3	482	0	0.1068 ↓ 0.0162
2		6	1469	0	0.0166 ↓ 0.0140
3	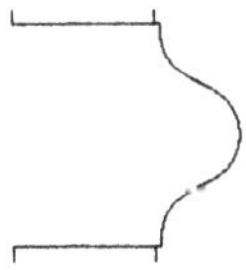	11	3331	0	0.0145 ↓ 0.0135

Figure4. Hierarchical optimization strategy

4.3 Optimized 3D Shape

At the end we present a result computed in 3D. As far as the state problem is considered, we use edge elements and the number of unknowns is 2516. Considering the optimization, there are 4 design variables, $B^{avg}_{min} = 0.2\,\mathrm{T}$. We used the classical approach with 2 mesh regenerations. The optimized design is in Fig. 5. From the first glance it seems that the 2D optimized design is a good approximation of the 3D one.

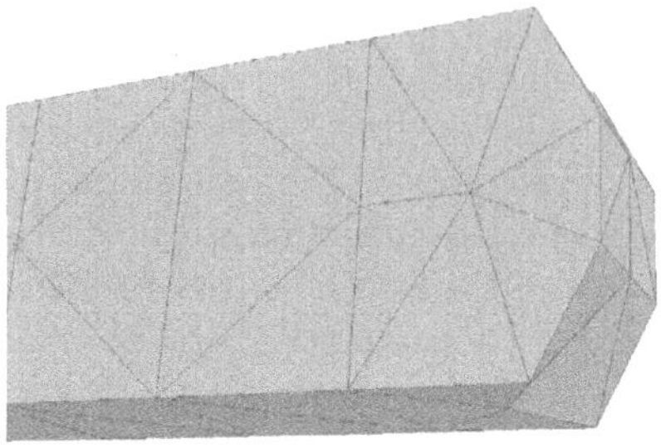

Figure5. Optimized 3D shape

5 Conclusions

We formulated a setting of the Maltese Cross problem in 2D. The 2D coarse optimized pole head was produced and the measurements have shown significant improvements of the homogeneity. We have found the hierarchical optimization strategy efficient for our problems. At last, the optimized coarse 3D result is quite similar to the 2D one.

One of the most important tasks for further work is an implementation of the direct and the adjoint technique of analytical sensitivity analysis. We will also investigate the hierarchical strategy more and apply it to the 3D case. We expect that with those properties the optimization solver will suit the problems of optimal shaping of homogeneous electromagnets very well. Finally, we will compute nonlinear problems and test the method for other geometries of homogeneous electromagnets.

References

1. J. Pištora and K. Postava and R. Šebesta, Optical guided modes in sandwiches with ultrathin metallic films, Journal of Magnetism and Magnetic Materials 1999 Vol. 198-199 pages:239-259
2. M. Kuhn and U. Langer and J. Schöberl, Scientific Computing Tools for 3d Magnetic Field Problems, The Mathematics of Finite Elements and Applications 2000 pages:239-259
3. J. Schöberl, NETGEN - An advancing front 2D/3D-mesh generator based on abstract rules, Comput.Visual.Sci 1997 pages:41-52
4. S. Reitzinger, PEBBLES – User's Guide, SFB "Numerical and Symbolic Scientific Computing", http://www.sfb013.uni-linz.ac.at.
5. MATLAB Optimization Toolbox User Manual, MATLAB Optimization, The MathWorks, Inc. 1993

Field Calculations of High Accuracy by BEM Using Extrapolation

G. Martínez[1] and R. Becker[2]

[1] on leave from Dept. Fsica Aplicada III, Fac. de Fsica,UCM, Madrid, Spain
[2] Institut für Angewandte Physik der Universität Frankfurt/M, Germany

Abstract For ray tracing with high accuracy and for the possible application of higher order ray tracing algorithms, field calculations are needed with a precision of better than 10^{-10}. Since field calculations using the boundary element method (BEM) can be easily performed for a given number of boundary elements, we have explored the feasibility and the improvement of accuracy by extrapolation to arbitrary fine discretisation. As examples for electrostatic fields we investigated a spherical condenser and specially shaped cans, where the exact fields are known. By doubling the number of elements, BEM calculations usually improve by more than a factor of 4 in accuracy, as shown by our examples. However, improvements by orders of magnitude become possible by extrapolating a set of two or more calculations with increasing number of elements. For a spherical condensor and a magnetic lens we can demonstrate, how even artifacts, originating from insufficient numerical analysis, are suppressed by the extrapolation technique.

1 Introduction

Numerical field calculations can be performed by applying the methods of finite differences (FDM), finite elements (FEM), or boundary elements (BEM). In any case, the solution is obtained by discretisation of space, transforming differential or integral equations into a set of algebraic equations. Inherently, accuracy of calculation is lost and artefacts may be created. For problems which are well-behaved, one may expect an improvement of the accuracy with finer discretisation. The gain of accuracy in this way, however, is low as compared to the numerical work involved.

It has been established by L.F. Richardson [1] as early as in 1910, that the systematic improvement of numerical solutions by decreasing the scale of discretisation can be used to extrapolate a set of calculations (two and more) to an accuracy corresponding to arbitrary fine discretisation. Depending on the problem, the gain in accuracy can reach orders of magnitude and provide results, which otherwise are obtained with much higher computational effort. Figure 1 shows for a point inside a cylindrical metallic can the dependencies of electrostatic field and potential errors on the number of boundary elements used. Except for very few elements, the variation looks very systematically like a power law with an exponent between 2 and 3 and worth of extrapolation.

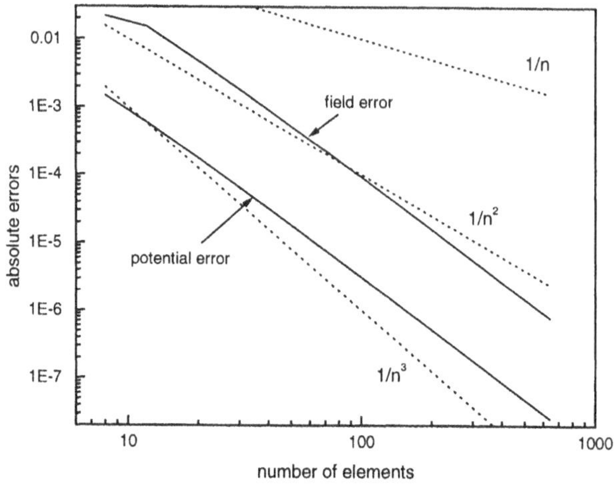

Figure1. Absolute error of the BEM solutions for electrostatic potential and field as functions of the number of elements n used in generating the boundary. These values correspond to a point situated at the centre of the shaded area of Fig. 2a; other points investigated showed a similar behaviour. Dotted lines indicate the trend by successive powers of $1/n$

Ray tracing of charged particles with high accuracy is possible using advanced high order methods [2]. This requires fields - to be obtained by numerical simulations - of correspondingly high precision. FDM and FEM need interpolation in meshes, which is spoiling the improvement of extrapolation. In contrast to this, BEM offers the significant advantage of direct field and potential calculation everywhere, making it suitable for efficient extrapolation.

In this paper we compare different extrapolation schemes with respect to the possible improvement of numerical solutions for electrostatic and magnetostatic problems, calculated by BEM-type programs.

2 Extrapolation Techniques

As demonstrated in Fig. 1, an increase of discretisation elements will systematically improve the accuracy of a value y to be computed, as long as there are not too few elements. This "convergence" behaviour, to be observed towards an infinite number of elements n, may be subjected to standard mathematical treatment by using the inverse of the number of elements $x = 1/n$ as the argument for a fitting function. The Richardson method uses the interpolation polynomial [1,3]:

$$y = \sum_{i=0}^{k-1} a_i x^{pi} \tag{1}$$

with $p = 1, 2$, depending on the problem and k being the number of required calculations with different values of n, giving a polynomial of degree $(k-1)*p$. The extrapolated value then is obtained at $x = 0$, namely $y = a_0$.

The convergence of the value y for n going to infinity may be also expressed with another very general function:

$$y = y_0 + \frac{a}{(n - n_0)^b} \tag{2}$$

where it is suggested that $n_0 = 0$, because an infinite error could be expected only as n approaches 0. Inverting the number of elements to variable x then gives a simple power law:

$$y = y_0 + ax^b \tag{3}$$

This power law requires 3 calculations to determine the coefficients of the fitting function, being comparable in computing effort to the case $k = 3$ of the Richardson method. For the set of points $(x_1, y_1), (x_2, y_2), (x_3, y_3)$ fulfilling the condition

$$\frac{x_1}{x_2} = \frac{x_2}{x_3} \tag{4}$$

it is particularly simple to find the extrapolated value from eq. (3)

$$y_0 = \frac{y_1 y_3 - y_2^2}{y_1 + y_3 - 2y_2} \tag{5}$$

leading to a straightforward scheme, not needing the evaluation of a and b.

We have applied these techniques to several problems and tested that all of them (except Richardson with $p = 1$) produce an acceleration of the convergence of the approximate solutions, but their efficiency depends on the characteristics of each case as we shall see below.

3 Analysis of Computational Errors

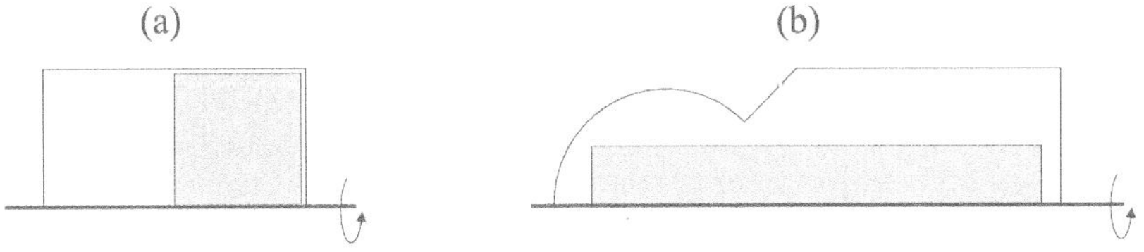

Figure2. Geometries of the axis-symmetric systems used in the study. (a) Cylindrical can. (b) Can with a keyhole cross section. The set of points at which calculations where performed are grid points inside the shaded areas.

Two examples have been selected for which the exact solution is known *a priori*. Figures 2(a)-(b) show the studied geometries. They consist of hollow conductors having rotational symmetry. The simplest one is a closed cylindrical can. The other is a modification of this, including shapes in the form of a

Table1. Average error and gain of accuracy of extrapolated solutions

a) Average over 10201 values computed inside the shaded area of the cylindrical can (Fig. 2a)

	Method	Rich. $k = 2$	Rich. $k = 3$	Rich. $k = 4$	Power law
error	potentials	3.18E-8	2.30E-8	2.18E-8	5.57E-9
	fields	1.17E-5	1.19E-5	1.26E-5	7.40E-6
gain	potentials	1.57	2.16	2.28	8.94
	fields	1.86	1.83	1.73	2.95

b) Average over 8281 values computed inside the shaded area of the can with keyhole cross section (Fig. 2b)

	Method	Rich. $k = 2$	Rich. $k = 3$	Rich. $k = 4$	Power law
error	potentials	2.08E-8	1.37E-8	1.26E-8	8.02E-10
	fields	8.42E-7	5.53E-7	5.08E-7	1.68E-8
gain	potentials	1.24	1.87	2.04	32.1
	fields	1.36	2.07	2.26	68.1

truncated cone and a spherical sector, which are needed as basic elements in the simulation of electrostatic lenses. When a potential V is applied to them, a charge distribution is established on their exterior surfaces in such a way that the enclosed volume is kept at the applied potential and, consequently, the electric field is null. To compute the charge distribution numerically, the electrode surface is divided into many small subareas or boundary elements. Each subarea is then assumed to carry a uniform charge density, which allows reducing the integral equation for the conductor to a set of algebraic equations [4]. Once the approximate charge distribution is known, the potential and field can be evaluated at any point of interest. These values will be compared to the exact ones to get an estimation of the error figure.

For the cylindrical can we have analysed the convergence of the errors for a set of calculations with increasing number of boundary elements, at 10201 points located on a square grid in the shaded area. The size of the grid is chosen in such a way that the closest distance of the points to the conductor is twice the smallest width of the elements used in this study. As a measure of the efficiency of each scheme we have computed the arithmetic mean of the absolute errors over the total number of points as well as the quotient between this value for the best set used and the one obtained by extrapolation, which we call the gain by extrapolation. A similar analysis was done for the second geometry over 8281 points inside the shaded area, a region around the axis of symmetry and rather far from the location of the surface elements.

The average errors and the gain of accuracy by extrapolating potentials and fields are given in tables 1a) and b) for both geometries. We observe that, in general, Richardson extrapolation is working the better the higher the order of the interpolating polynomial is, however, the power law algorithm is even superior. This is particularly true in the case of calculations less close to the boundaries - table 1b - with gain factors of 16 for the potentials and 30 for the fields. A detailed view of how element doubling and further power law extrapolation work on the computed potential error may be seen in Fig. 3 for the cylindrical can.

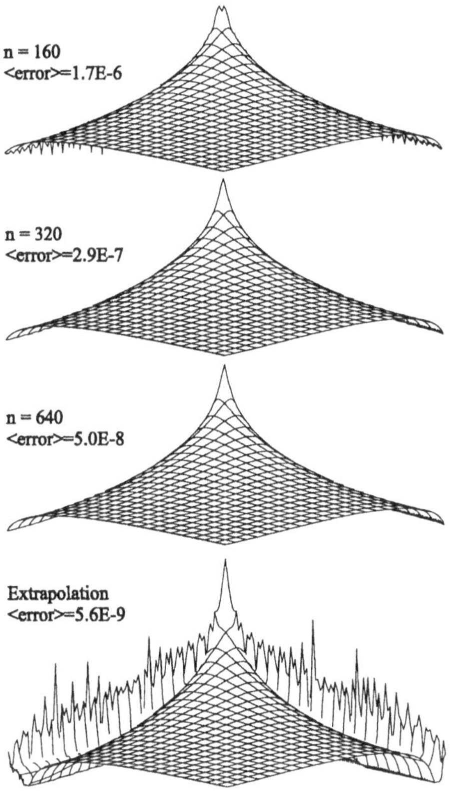

Figure3. Maps of the potential error computed for a sequence of element doubling and for the power law extrapolation: Cylindrical can; applied potential 1 volt

By augmenting the number of elements the error is lowered in general, but it always will be much greater near the corner where the approximation of constant charge density is more critical. The extrapolation map is showing an irregular behaviour and poor efficiency near the conductor, but it is giving a uniform and good improvement of the solutions for the rest of the area. The reason is that if potentials are calculated too close to a boundary the neighbourhood relation generally will be changed for computations with different number of elements and therefore the systematic convergence is spoiled in taking more elements.

For the can with keyhole geometry the results follow the same trend but with some peculiarities as observed in Fig. 4. Here extrapolation is always

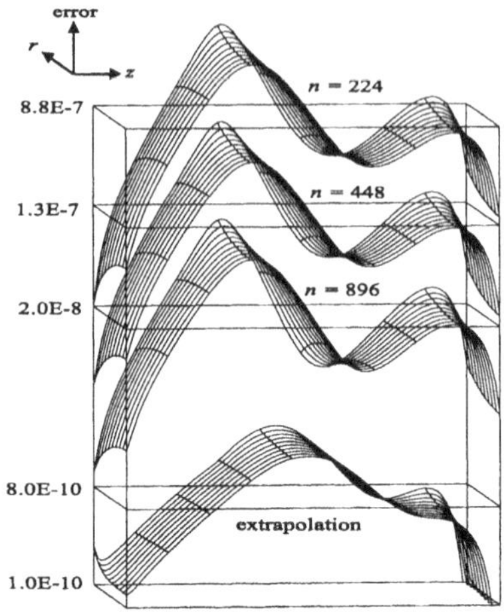

Figure4. Maps of the potential error computed for a sequence of element doubling and for power law extrapolation: Keyhole can; applied potential 1 volt

giving an important gain in accuracy; the greatest errors being found at points relatively near the inward corner formed by the truncated cone and the spherical sector. The power law extrapolation is able to reduce the errors efficiently just in the worst parts.

4 The Cure of Discretisation Artifacts

In order to show the effect of extrapolation on some possible artifacts coming from a rough simulation of the boundary, we have studied the electrostatic field in a spherical condenser replacing the curved boundary by a polygonal trace, so that the original division in spherical sectors is performed here using truncated cones.

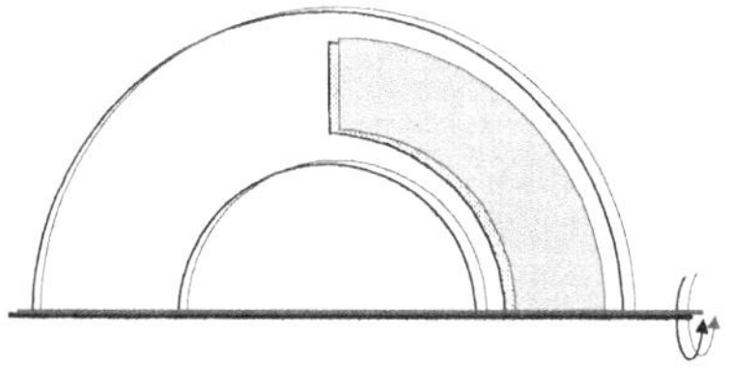

Figure5. Geometry of the spherical condenser. In this study the spherical sectors have been substituted by truncated cones tangent to the central line of each sector. The errors are computed along 4 arcs with radii on equidistant subdivisions between inner and outer sphere, represented by the shaded area

We have computed the deviation of the potential and field values with respect to the analytical solutions, see [2], at four arc lines inside the shaded

Table2. Average error and gain of accuracy of extrapolated solutions for the spherical condenser test. The averaging is done on 940 values

	Method	Rich. $k = 2$	Rich. $k = 3$	Rich. $k = 4$	Power law
error	potentials	6.72E-7	4.28E-7	3.86E-7	5.71E-7
	fields	6.87E-6	4.38E-6	3.94E-6	2.32E-6
gain	potentials	24.9	39.1	43.4	29.3
	fields	4.18	6.55	7.28	12.4

area of Fig. 5. The mean errors and the gain with respect to the most accurate set are listed in table 2.

Here Richardson extrapolation is performing best for the potential solution but the power law is somewhat better in extrapolating the fields.

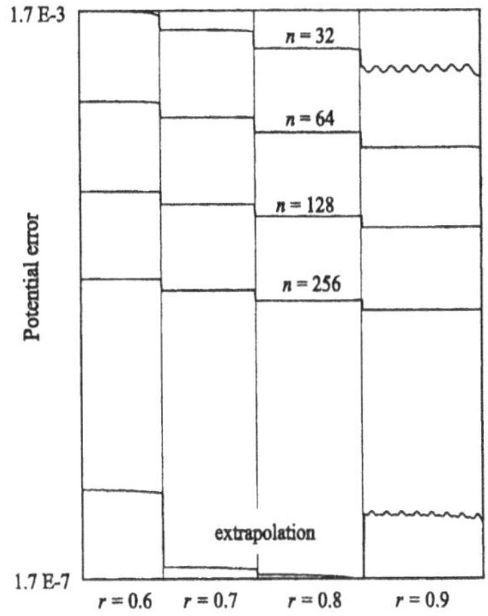

Figure6. Graphs of the logarithm of the potential error inside the spherical condenser computed for a sequence of element doubling, and for Richardson extrapolation ($k = 3, p = 2$) along arcs of r = 0.6, 0.7, 0.8, 0.9. Extrapolation is more efficient at points located at the two central arcs which are less affected by the artefacts introduced by a piecewise linear boundary representation. Applied potentials: outer/ inner sphere 0/1 volt

Figure 6 represents the decrease of the potential error at observation points as the number of elements increases, and the errors of extrapolation by third order polynomials. In points located at radial lines far away from both electrodes more advantage may be taken from extrapolation, even if the error at the outer arc is behaving better for the four sets used. As discussed in the previous example, the closer the points are to the boundary the lesser the efficiency of extrapolation will be.

5 Summary

It has been demonstrated that extrapolation is recovering the accuracy of numerical field and potential calculations, which is lost by discretisation of space and/or insufficient mathematical elaboration of the simulation.

In combination with BEM, the advantage of extrapolation can be used efficiently at points that are not too close to the boundary. This is, however, not a serious restriction for particle optics because trajectories should never come into the vicinity of electrodes to avoid aberrations.

A power law for the extrapolating function has been found superior to Richardson polynomials in most of our investigated cases. Only those polynomials with even powers ($p = 2$) have been found useful.

6 Acknowledgments

One of the authors, GM, is grateful to Spanish Ministerio de Educación y Cultura for the award of a grant supporting a sabbatical stay at J.W. Goethe-Universität, Frankfurt.

References

1. L.F. Richardson, The approximate solution of physical problems involving differential equations using finite differences, with an application to the stress in a masonry dam. Phil. Trans. Roy. Soc., London, Ser. A **210** (1910) 307-357
2. G. Martínez and R. Becker, Accuracy of ray tracing in fields with numerical errors, Optik **111, 3** (2000) 113-118
3. G.I. Marchuk and V.V. Shaidurov, Difference Methods and Their Extrapolations, Springer-Verlag, (1983) Chap. 1
4. G. Martínez and M. Sancho, Integral equation methods for the analysis of electrostatic potentials and electron trajectories, Adv. Electron. And Electron Phys. **81**, Academic Press, (1991) 1-41

Calculation of Electromagnetic Eigenmodes in Complex Structures Using Coupled S-Parameter Calculation*

K. Rothemund, H.-W. Glock, M. Borecky, U. van Rienen

Institut für Allgemeine Elektrotechnik, Universität Rostock, Germany

Abstract Eigenmode calculations are usually carried out with dedicated solvers as e.g. MAFIA's E-module [1]. If eigenmodes of large and complex structures have to be calculated, the application of these solvers is limited even on the most powerful workstations by the available computational resources in terms of memory and computation time. This paper will present a technique based on scattering parameters that allows to split the complete geometry into subsections which can be modeled individually. The broadband S-parameters of each section have to be calculated externally using appropriate solvers. Such runs are significantly smaller problems and may be performed in parallel on different machines. In the paper we present a description which is well suited to calculate the eigenfrequencies of an entire structure and the wave amplitudes needed to determine the internal fields once the S-parameter of the segments are known. This approach is of special advantage if the structure contains segments of identical shape, parts that can be described analytically or components of local complexity, which enforce enhanced grid resolution. Further it establish a mean for frequency selective eigenmode analysis, which is difficult for most eigenmode solvers. We illustrate the procedure with a test example that was calculated both with Coupled S-Parameter Calculation and a direct eigenmode solver. Results of the comparison are presented.

1 Introduction

The analysis of rf components often is based on the knowledge of their eigenmodes. Length and complexity of some structures demand even in modern computational environment a significant amount of computational resources to perform 'direct' eigenmode calculations[1].

This paper presents a method which we call Coupled S-Parameter Calculation (CSC). It allows to calculate the eigenfrequencies and -modes of a complex rf structure by splitting the whole structure into several smaller sections using their broadband scattering parameters. The S-parameters of the sections have to be calculated externally with appropriate solvers like MAFIA-T(ime domain), CST MicroWaveStudio [2], Agilent HFSS [3], etc.

* Work supported by DESY Hamburg

[1] As 'direct' eigenmode calculation we will understand the solution of discretized Helmholtz' equation on the complete solution domain at once.

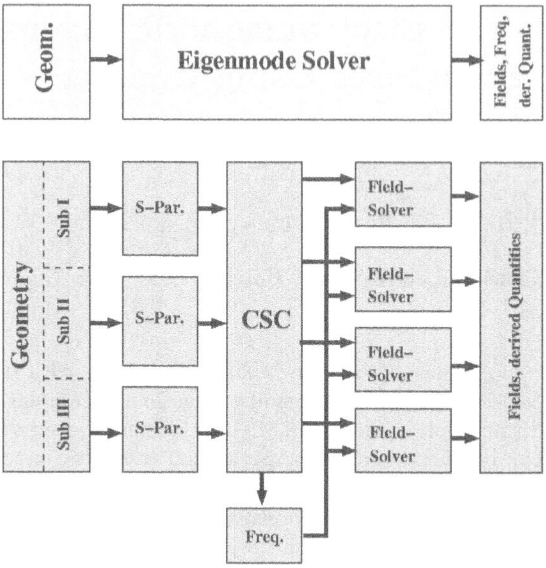

Figure1. Calculation of eigenmodes with direct solver and using the CSC-technique: The geometry is split into subsections and their S-parameters are calculated. The S-parameters are combined using CSC and yield the eigenfrequencies and wave amplitudes needed to determine the eigenfields in the subsections, which are found in separate runs.

The partitioning may be chosen in a way that symmetry properties or repetitions of certain sections can be exploited. Very simple subsections, especially waveguides, may allow for analytical solution, or even measured data can be taken into account. Once the S-parameters are known, CSC determines both the eigenfrequencies of the entire structure and the corresponding complex wave amplitudes at the sections ports. The latter ones are used to calculate the field distribution inside all sections again in external programs like MAFIA-W, HFSS, etc.. These may work either in time domain or in frequency domain by exciting (numerically) waves of the appropriate frequency, amplitude and phase at every port of each subsection.

First in the paper, a general procedure to calculate the coupling between the external ports of any arbitrarily structured system of scattering sections is described. Then resonators which are entirely closed by definition are introduced as a special case without external coupling. It will be shown that in this case the problem is reduced to the repeated solution of low dimensional eigenvalue equations parameterized by frequency. The occurrence of eigenvalue 0 indicates the resonance frequencies; the eigenvectors belonging to eigenvalue 0 of this system are the amplitudes of all the waveguide modes in the system which are needed to calculate the corresponding field distributions.

Finally CSC is demonstrated using an example that also allows for direct eigenmode solution using MAFIA-E 4.20 running on a 2 GB RAM workstation. The results of both methods are compared.

It is our primary aim to perform eigenmode determination in cases, where memory requirements exceed computing capabilities. Therefore we accept a numerical effort that in total may be higher than for a direct calculation (if this could be carried out). CSC performs best - as mentioned above - if there are repeated sections or sections with analytical description, or if the influence of modifications of single sections on the entire system is under consideration.

2 Theory

Signal reflection and transmission between the ports of certain rf-components can be described by scattering-parameters (briefly: S-parameters). All S-parameters of a n_k-port structure can be represented by a $(n_k \times n_k)$-matrix $\mathbf{S}$ where the entry S_{ij} describes the transmission of a signal from port j to port i. Because S_{ij} contains information about phase and amplitude it usually is a complex function of frequency. With $\boldsymbol{a} = (a_1, \ldots, a_{n_k})^T$ and $\boldsymbol{b} = (b_1, \ldots, b_{n_k})^T$ describing the input and output signals, resp., the S-matrix $\mathbf{S}_k$ of the k-th section of a segmented structure (N segments) forms the relation

$$\boldsymbol{b}_k = \mathbf{S}_k(\omega)\, \boldsymbol{a}_k = \begin{pmatrix} S_{11}(\omega) & \cdots & S_{1n_k}(\omega) \\ \vdots & & \vdots \\ S_{n_k1}(\omega) & \cdots & S_{n_kn_k}(\omega) \end{pmatrix}_k \boldsymbol{a}_k \qquad \text{with} \quad k = 1 \ldots N. \quad (1)$$

Neither the S-matrix description is restricted to single moded waveguide ports, nor is it the principle of the procedure under discussion, but every mode has its individual scattering properties and therefore enlarges the size of the matrix system. For technical reasons this will limit the applicability of the method to a certain number of modes under consideration, especially since the number of S-parameter needed grows up quadratically with the mode number. As the effort for most eigenmode solver is increased strongly with higher frequencies due to growing mode density and enhanced mesh requirements for shorter wavelengths this is not a special disadvantage of our approach.

2.1 Open Structures

If $\boldsymbol{a}$ and $\boldsymbol{b}$ simultaneously hold the signals of all the ports, the entries can be arranged in a way that $\mathbf{S}$ becomes a block diagonal matrix

$$\begin{pmatrix} \boldsymbol{b}_1 \\ \vdots \\ \boldsymbol{b}_N \end{pmatrix} = \boldsymbol{b} = \mathbf{S}\boldsymbol{a} = \begin{pmatrix} \mathbf{S}_1 & & 0 \\ & \ddots & \\ 0 & & \mathbf{S}_N \end{pmatrix} \begin{pmatrix} \boldsymbol{a}_1 \\ \vdots \\ \boldsymbol{a}_N \end{pmatrix}. \quad (2)$$

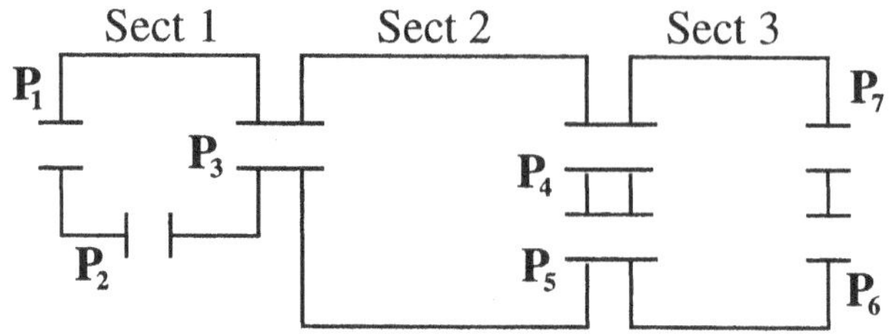

Figure2. A chain of three rf-components with external ports (P_1, P_2, P_6, P_7) and internal ports (P_3, P_4, P_5).

Now we are going to separate signals that are incident, but coming from a neighbouring section in the entire structure, which we well refer to as 'internal' signals, from those 'external' signals that come from outside the structure. With $\boldsymbol{a}_{cop}$ carrying the internal and $\boldsymbol{a}_{inc}$ representing the external incident signals we define a permutation matrix $\mathbf{P}$ as follows:

$$\boldsymbol{a} = \mathbf{P}\begin{pmatrix}\boldsymbol{a}_{cop}\\ \boldsymbol{a}_{inc}\end{pmatrix}. \tag{3}$$

The coupling between the inner ports is described using another permutation matrix $\mathbf{F}$. It is constructed as a map from the signal order of the scattered signals in $\boldsymbol{b}$ to the order of corresponding incoming signals, thus expressing the feedback of outgoing signals of one internal port are incident signals at another port. The inverse permutation $\mathbf{P}^{-1}$ ensures the same order of the scattered signals like in (3) and we get

$$\begin{pmatrix}\boldsymbol{a}_{cop}\\ \boldsymbol{a}_{sct}\end{pmatrix} = \mathbf{P}^{-1}\,\mathbf{F}\,\boldsymbol{b} \tag{4}$$

where $\boldsymbol{a}_{sct}$ represents the scattered (outgoing) signals, which are left untouched by the feedback matrix.

Combining (1) to (4) results in the following matrix vector equation:

$$\begin{pmatrix}\boldsymbol{a}_{cop}\\ \boldsymbol{a}_{sct}\end{pmatrix} = \mathbf{P}^{-1}\mathbf{F}\,\mathbf{S}\,\mathbf{P}\begin{pmatrix}\boldsymbol{a}_{cop}\\ \boldsymbol{a}_{inc}\end{pmatrix}. \tag{5}$$

Setting $\mathbf{G} = \mathbf{P}^{-1}\,\mathbf{F}\,\mathbf{S}\,\mathbf{P}$ leads to

$$\begin{pmatrix}\boldsymbol{a}_{cop}\\ \boldsymbol{a}_{sct}\end{pmatrix} = \mathbf{G}\begin{pmatrix}\boldsymbol{a}_{cop}\\ \boldsymbol{a}_{inc}\end{pmatrix} = \begin{pmatrix}\mathbf{G}_{11} & \mathbf{G}_{12}\\ \mathbf{G}_{21} & \mathbf{G}_{22}\end{pmatrix}\begin{pmatrix}\boldsymbol{a}_{cop}\\ \boldsymbol{a}_{inc}\end{pmatrix}, \tag{6}$$

where the dimensions of the block matrices $\mathbf{G}_{ij}$ correspond to the dimensions of $\boldsymbol{a}_{cop}$, $\boldsymbol{a}_{inc}$ and $\boldsymbol{a}_{sct}$. Therefore (6) can be split into the following system of equations

$$\boldsymbol{a}_{cop} = \mathbf{G}_{11}\boldsymbol{a}_{cop} + \mathbf{G}_{12}\boldsymbol{a}_{inc}, \tag{7}$$

$$\boldsymbol{a}_{sct} = \mathbf{G}_{21}\boldsymbol{a}_{cop} + \mathbf{G}_{22}\boldsymbol{a}_{inc}. \tag{8}$$

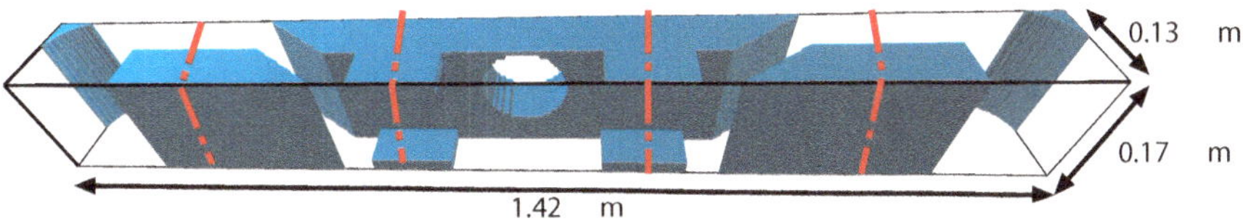

Figure3. Geometry of the test resonator used to compare the direct eigenmode calculation with CSC. $26 \times 46 \times 481 \approx 1.018\ 10^6$ mesh points are used to model the geometry. The segmentation is indicated with grey lines.

Herein $\boldsymbol{a}_{inc}$ and $\boldsymbol{a}_{sct}$ represent the incident and the outgoing waves at the external ports. Thus the coupling between these signals is given by

$$\boldsymbol{a}_{sct} = \left[\mathbf{G}_{21}\ (\mathbf{E} - \mathbf{G}_{11})^{-1}\ \mathbf{G}_{12} + \mathbf{G}_{22}\right] \boldsymbol{a}_{inc} \tag{9}$$

and the overall S-matrix, denoted by $\mathbf{S}^{(tot)}$, can be written as

$$\mathbf{S}^{(tot)} = \mathbf{G}_{21}\ (\mathbf{E} - \mathbf{G}_{11})^{-1}\ \mathbf{G}_{12} + \mathbf{G}_{22} \tag{10}$$

where $\mathbf{E}$ represents the unity matrix.

2.2 Resonators

In the case of a resonator problem there are no open (external) ports. Then $dim(\boldsymbol{a}_{sct}) = dim(\boldsymbol{a}_{inc}) = 0$ holds and only the coupling between the internal ports remains. Simultaneously the block matrices $\mathbf{G}_{12}$, $\mathbf{G}_{21}$ and $\mathbf{G}_{22}$ in (6) vanish. Equation (7) reduces to

$$[\mathbf{G}_{11}(\omega_0) - \mathbf{E}]\, \boldsymbol{a}_{cop} = \mathbf{0} \tag{11}$$

which has to be fulfilled. This is only valid for discrete frequencies ω_0 which are the resonant frequencies aimed for.

In case of resonance the vector $\boldsymbol{a}_{cop}$ contains the amplitudes of the waveguide modes at the location of the internal ports. $\boldsymbol{a}_{cop}$ is found as the eigenspace - usually one vector - of the eigenvalue 0 which is the kernel of the matrix $[\mathbf{G}_{11}(\omega_0) - \mathbf{E}]$. The dimension of this matrix is equal to the total number of modes appearing in the system, which is usually below or in the order of 10^2. Therefore the determination of the matrix kernel is a task best performed from black box-solvers like they are implemented in several math packages.

3 Procedure and Numerical Results

To verify the formulation given above a test geometry was modelled (see Fig. 3) which is split into 5 subsections. The S-parameters of each single subsection were calculated using the MAFIA time domain solver T3 in a frequency range of 1.2 . . . 1.75 GHz.

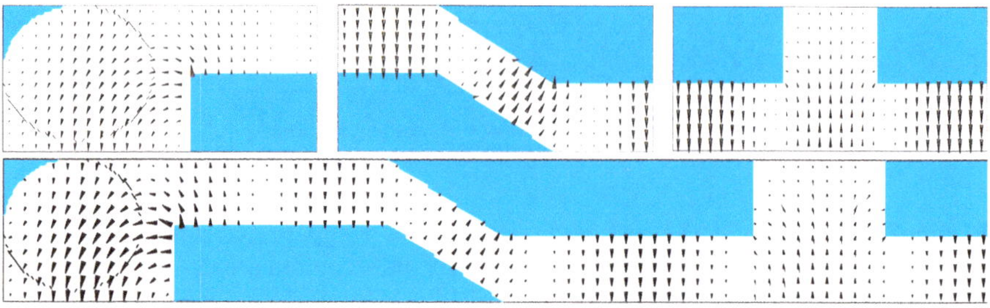

Figure4. Electric field of the $f = 1.44258$ GHz-eigenmode computed by the CSC (upper plot) and directly with MAFIA's E-module (lower plot).

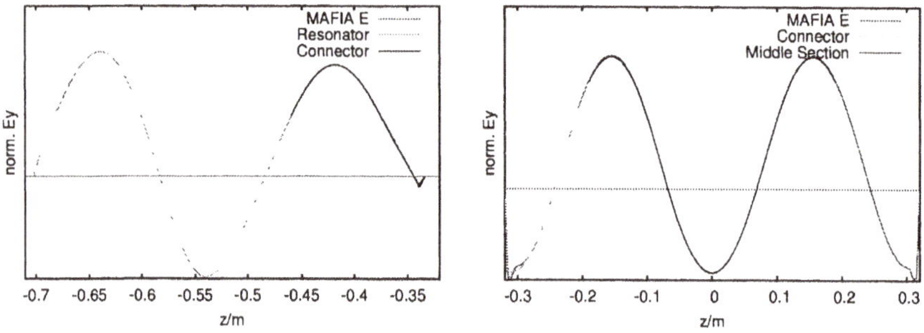

Figure5. E_y-component of the $f = 1.44258$ GHz-eigenmode along some path perpendicular to the port plane calculated by the CSC and directly with MAFIA's E-module.

In order to monitor the amplitudes of the in- and outgoing waves a 2D-mode decomposition at the individual ports was done which is a standard MAFIA routine. The 2D-modes were determined from the port geometries using MAFIA-E. These modes are power-normalized, but it is not predictable,

Table1. Comparison of the eigenfrequencies found by the CSC-technique and by MAFIA's E-module in the frequency range of 1.2 ... 1.75 GHz using the same mesh. The reason for the systematic sign of the deviation is under research.

CSC	MAFIA E-mod.	rel. error
1.21226 GHz	1.21031 GHz	1.61×10^{-3}
1.24006 GHz	1.24002 GHz	2.98×10^{-5}
1.34805 GHz	1.34728 GHz	5.73×10^{-4}
1.38341 GHz	1.38220 GHz	8.73×10^{-4}
1.44250 GHz	1.44268 GHz	-1.26×10^{-4}
1.46394 GHz	1.46387 GHz	4.99×10^{-5}
1.50326 GHz	1.50261 GHz	4.30×10^{-4}
1.53157 GHz	1.53139 GHz	1.19×10^{-4}
1.60484 GHz	1.60329 GHz	9.64×10^{-4}
1.65345 GHz	1.65276 GHz	4.19×10^{-4}
1.68786 GHz	1.68244 GHz	3.21×10^{-3}

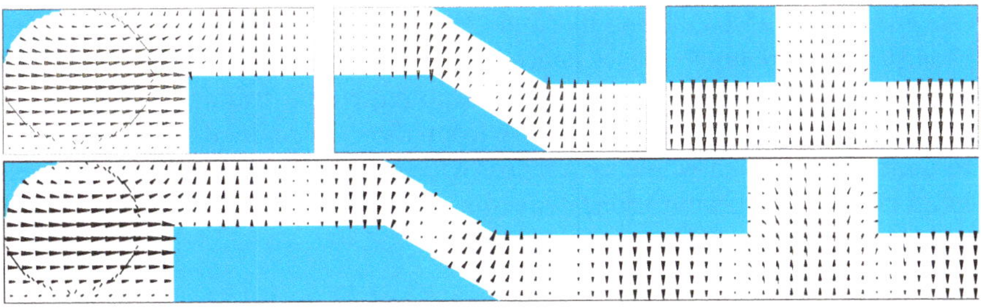

Figure6. The electric field of the $f = 1.50325$ GHz-eigenmode computed by the CSC (upper plots) and directly with MAFIA's E-module (lower plot).

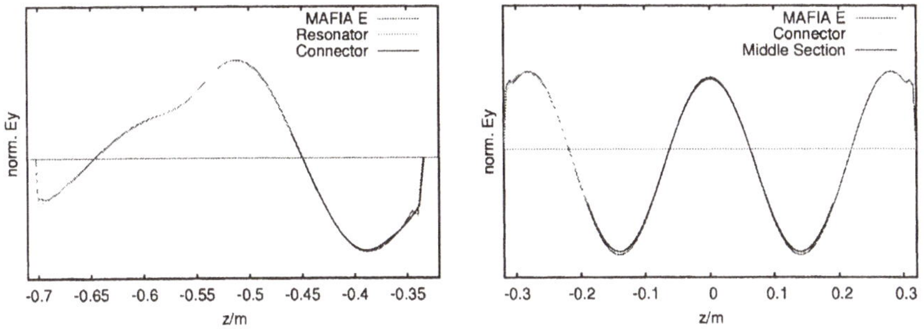

Figure7. E_y along some path in z-direction shown for the eigenmode with frequency ($f = 1.50325$ GHz).

which orientation relative to the coordinate frame occurs. In order to keep orientational consistency an extra routine was implemented providing a uniform mode orientation for all waveguide mode determining runs.

A Mathematica [4] program was used to implement CSC, i.e. to compute the resonant frequencies as well as the amplitudes at each internal port. The field distributions of the eigenmodes in all subsections were computed using MAFIA's frequency domain solver W3 with port input power and phase values found by CSC. This results in the resonant field pattern of the according resonator section. Alternatively the time domain solver T3 can be used to derive the field patterns of the eigenmodes. In this case monochromatic waves of the given frequency are excited at every port with the appropriate amplitudes and phases. The field is monitored after it stabilized to steady state.

For comparison, the eigenmodes of the whole structure were computed with MAFIA's eigenmode solver E. As shown in Tab. 1 the frequencies found by the CSC technique match very well those calculated directly. The field distributions of two eigenmodes are compared in Fig. 4 and 6. Figure 5 and 7 show the y-component of the electric field through the ports along some path perpendicular to the port plane. As can be seen, the field distribution computed by MAFIA's E-module and CSC match extremely good. Deviations near the outer boundaries are caused by interpolation errors.

It is worth to remark that there was no individual scaling of the fields in the subsections. Only the field pattern from 'direct' eigenmode calculation has to be multiplied as a whole by a factor which in principle is a free parameter in *all* eigenmode computations depending on the special normalization of the modes.

4 Conclusion

The presented CSC-technique allows to calculate rf properties of large and/or complex structures by dividing them into smaller subsections and using the S-parameters of the segments. Combination of those yields both the S-matrix of the complete structure for structures with open ports or the eigenfrequencies and the corresponding field distributions in the case of resonator problems. It was shown for a test example that the results of the 'direct' eigenmode calculation and the calculation via CSC, correspond very well.

The advantage of this technique is the possibility to calculate the S-parameters of each subsection in separate runs and with different machines. It benefits from symmetries or repetitions of particular subsections. Further, it easily allows to specify frequency search ranges for the eigenmodes and it eases the study of structure modifications significantly.

References

1. MAFIA V4.20, CST GmbH, Büdinger Straße 2a, D-64289 Darmstadt, Germany.
2. Micro Wave Studio V1.0, CST GmbH, Büdinger Straße 2a, D-64289 Darmstadt, Germany.
3. HP HFSS, Agilent technologies.
4. Mathematica V4.0.1, Wolfram Research Inc, 100 Trade Center Drive, Champaign, IL 61820-7237, USA.
5. T. Itoh, "Numerical Techniques for Microwave and Millimeter-Wave Passive Structures", John Wiley & Sons, 1989.
6. Thomas-Alfred Abele, "Über die Streumatrix allgemein zusammengeschalteter Mehrpole", AE, Vol. 34(6), pp. 262-268, 1960.
7. M. Dohlus and R. Schuhmann and T. Weiland, "Calculation of Frequency Domain Parameters Using 3D Eigensolutions", International Journal of Numerical Modelling, 1999.

Multigrid Solvers for Poisson's Equation in Computational Electromagnetics

G. Pöplau* and U. van Rienen

Institute of General Electrical Engineering, University of Rostock,
Albert-Einstein-Str. 2, D-18051 Rostock, Germany

Abstract Complex real life problems, as they appear with the simulation of electromagnetic fields, demand the construction of efficient and robust solvers for the related equations. In the present paper we investigate multigrid algorithms for the solution of static problems on adaptive discretizations. Two multigrid strategies are compared: algebraic multigrid, which performs the adaption to the discretization automatically and geometric multigrid with semi-coarsening, which has fast convergence if the discretization fits the coarsening strategy.

1 Introduction

The construction of fast and robust solvers for the simulation of electromagnetic phenomena is still an important field of research, because problems get more and more demanding with respect to computing time. Further, the appropriate modeling of practical problems often requires adaptive discretizations. The state-of-the-art for the fast solution of Poisson's equation is the application of multigrid methods. While multigrid algorithms lead to optimal convergence rates for model problems, they often get worse in real life situations.

In the present paper we investigate the application of multigrid algorithms for adaptive grids, which show up with the problem of the tracking of electron beams in linear colliders [10]. Here, the electrostatic potential φ caused by the electrons of the beam is computed in the beam's restframe via Poisson's equation

$$\begin{aligned} -\Delta\varphi &= \frac{\varrho}{\varepsilon_0} && \text{in } \Omega \subset \mathbb{R}^3, \\ \varphi &= 0 && \text{on } \partial\Omega, \end{aligned}$$

from the charge density ϱ with the dielectric constant ε_0. During the tracking process this equation has to be solved for a large number of time steps (~ 1000) with changing distributions of the particles. A good approximation of the distribution of the particles in the beam requires an adaptive discretization with ~ 1 million grid points. It is assumed that the beam transverses a longitudinally (z-direction) homogeneous beam pipe of elliptic or polygonal cross-section (in the (x, y)-plane). The pipe is perfectly conducting which causes the Dirichlet boundary conditions at its wall. The boundary condition

* supported by a research grant from DESY, Hamburg

in longitudinal direction ought to be of open type describing the decay of the electrostatic potential of the beam. Nevertheless, we restrict our considerations in this paper to Dirichlet boundary conditions in order to compare the convergence of the proposed algorithms in the numerical studies without influence of the formulation of the open boundary condition. The computational domain has to be chosen large enough in this case.

The three–dimensional rectangular grid is generated adaptively by a grid–based tracking algorithm [10] depending on the distribution of the particles. This discretization results in an adaptive grid with small subintervals in the centre of the beam and larger subintervals near the boundary. Further, the grids are stretched in longitudinal direction which is caused by the Lorentz transformation [7]. Thus the global aspect ratio h_{max}/h_{min} can be very large (see Fig. 1). The adaptive discretization does not allow the use of the Fast Fourier Transformation as a direct Poisson solver. Further, it is taken into consideration, that a solver for a large number of time steps is needed. In a multigrid iteration, this allows to use the information of the previous time step, which saves a lot of computational work compared to FFT. In the present paper we only investigate the convergence of multigrid algorithms within one time step. The study of the behaviour of these methods used for a large number of time steps is planned for the next future.

2 Adaptive Multigrid Algorithms

For the description of the multigrid method for the above problem let us consider Poisson's equation with Dirichlet boundary conditions as described in the introduction. The original domain Ω will be embedded in a cuboid Q. With the identification

$$\begin{aligned} u &= \varphi && \text{in } \Omega \cup \partial\Omega, \\ f &= \frac{\varrho}{\varepsilon_0} && \text{in } \Omega, \\ f &= 0 && \text{in } Q \setminus \Omega \end{aligned}$$

the following equation has to be solved:

$$\begin{aligned} -\Delta u &= f && \text{in } Q, \\ u &= 0 && \text{on } \partial Q. \end{aligned}$$

The discretization of Poisson's equation by finite differences (seven–point stencil), adapted to the adaptive mesh, leads to a system of equations

$$A_h u_h = f_h,$$

where h denotes the vector of step sizes on the original (fine) grid. The matrix A_h is symmetric and positive definite. The generation of the grids considered in this paper is adapted to the actual distribution of the particles in the beam. Fig. 1 gives an example for such a discretization.

The system of equations above can be solved efficiently by a suitable multigrid algorithm. The general idea of multigrid is the combination of the

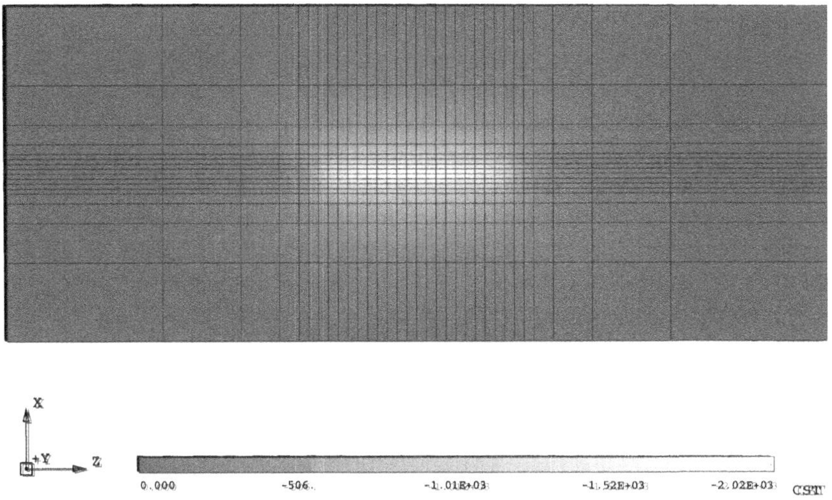

Figure1. Discretization of a beam pipe, longitudinal cross-section ((x, z)-plane). The potential φ of a beam with Gaussian distribution of the charge density ϱ is shown.

approximate solution of the system of equations obtained by a few steps of a relaxation method with a coarse grid correction computed on a coarser grid. The multigrid method is characterized by its fast convergence. In contrast to classical iteration methods (Jacobi, Gauss–Seidel) the convergence speed remains the same when the discretization is refined. Consequently, the computational work is proportional to the number of unknowns in the system of equations. Detailed descriptions of the multigrid method can be found e.g. in [1,11].

The multigrid algorithm works on a certain number of grids becoming coarser and coarser. In the case of an adaptive discretization, as in the tracking context, the strategy for the construction of the coarser grids is a crucial point with respect to retaining the efficiency of the multigrid algorithm. There are two possibilities: First, the geometrical method, that is the removal of grid lines and, second, the algebraic method which operates on the matrix entries. Especially on rectangular grids as in the tracking context the advantages of the geometrical strategy are often the fast performance and simple implementation.

The *geometric multigrid scheme* for the tracking procedure is performed with a *semi-coarsening* technique. The semi-coarsening is carried out for each coordinate axis in the following way: Two neighbouring steps h_1 and h_2 remain also in the next coarser grid as long as either $h_1 \geq 2h_{min}$ or $h_2 \geq 2h_{min}$, where h_{min} denotes the overall minimal step size of the corresponding fine level. The objective of this strategy is to obtain a decreasing aspect ratio.

Conversely, it is most efficient to construct the discretization of the original problem such that it is equidistant at least on the coarsest level.

In our tracking context only the fine grid is given and the automatic construction of the coarser grids could bo too expensive in implementation. In this case the *algebraic multigrid* method (AMG) [8,12] can be an attractive alternative. As already mentioned above the algebraic multigrid algorithm uses the matrix entries for the construction of the coarser levels and the grid transfer operators. The effect of the geometric multigrid with semi-coarsening is similar to the algebraic method, which distinguishes between strongly and weakly coupled variables for the selection of the coarse grid points. The advantage of AMG is that the adaption to the adaptive grid is done automatically. On the other hand the setup phase in AMG, i.e. the construction of the coarser levels and the computation of the grid transfer operators, takes a lot of time. We have to point out that we used the AMG approach of Ruge and Stüben and applied their AMG–code [9] for our numerical experiments.

3 Numerical Results

Numerical experiments have been performed with the data of a 10 MeV beam in a cylindrical pipe. The charge density ϱ has Gaussian distribution in the longitudinal and equidistribution in the transversal direction.

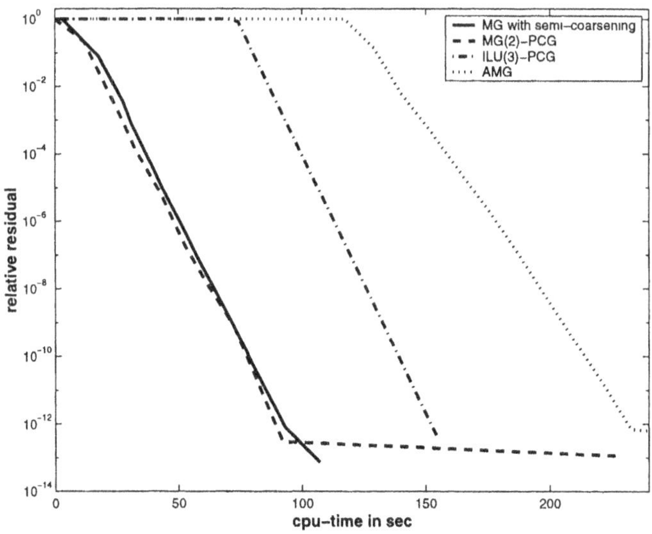

Figure2. Convergence of several algorithms (MG: multigrid, PCG: preconditioned conjugate gradients, AMG: algebraic multigrid) for a discretization on a 65x65x65–mesh with an equidistant mesh spacing for each coordinate direction.

Table1. Numerical results obtained for several multigrid strategies on a SUN Ultra Sparc 1 with 167 MHz to get a relative residual less than 10^{-10} in the maximum norm.

Grid	No. of levels	Method	No. of iterations	cpu–time in sec.
65x65x65		MG	16	79.8
equidistant for	6	MG(2)–PCG	7	82.8
each coordinate		AMG	10	222.0
17x17x33		MG	17	3.65
adaptive mesh	4	MG(2)–PCG	7	3.64
		AMG	12	10.7
33x33x65		MG	17	35.0
adaptive mesh	5	MG(2)–PCG	7	37.6
		AMG	17	146.0

Geometric and algebraic multigrid are tested and compared to a conjugate gradient method with ILU(3)–preconditioner (ILU(3)–PCG) as it is implemented in the software package MAFIA [5]. Furthermore the geometric multigrid algorithm (MG) was applied as preconditioner for the conjugate gradient method [3,4]. The geometric multigrid algorithm uses a V–cycle and has the following components [1,11]: one pre–smoothing step with red–black Gauss–Seidel relaxation and one post–smoothing step black–red Gauss-Seidel relaxation, full–weighting restriction, trilinear interpolation and semi–coarsening for the construction of the coarser levels. The choice of these components ensures that the multigrid preconditioner for the conjugate gradient method is symmetric and positive definite [4]. We used the multigrid preconditioned conjugate gradient method (MG(2)–PCG) with 2 V–cycles per cg iteration. The algebraic multigrid method is applied as black box solver with the AMG code of Ruge and Stüben [9].

First the numerical results on a 65x65x65–grid (=274 625 mesh points) are given (see also [6]). This grid has an equidistant mesh spacing for each separate coordinate axis. Only the large aspect ratio with $h_z \approx 25h_x \approx 25h_y$ poses a difficulty. Thus standard coarsening is performed and semi–coarsening is only applied in the following sense: do not coarsen the mesh in the z–direction if the step size is at least two times larger than the step size of the transversal directions. Fig. 2 shows that the best results are obtained with the geometric multigrid method. The multigrid preconditioned conjugate gradient method does not result in a significant improvement, because the geometric multigrid algorithm already has a good convergence rate. This numerical result coincides with the theoretical considerations in [4]. Neglecting the setup phase, the convergence speed of AMG has nearly the same slope. The ILU(3)–preconditioned conjugate gradient method needs a considerable amount of work for constructing the preconditioner which is rather costly for the ILU(3)–preconditioner.

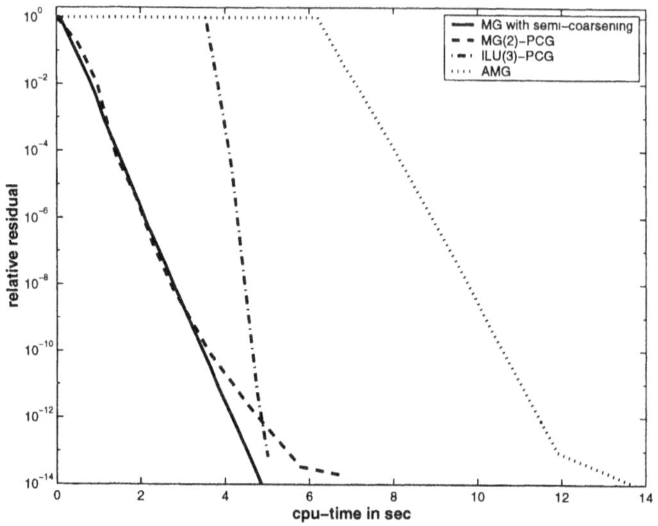

Figure3. Convergence of several algorithms for an adaptive discretization on a 17x17x33–mesh.

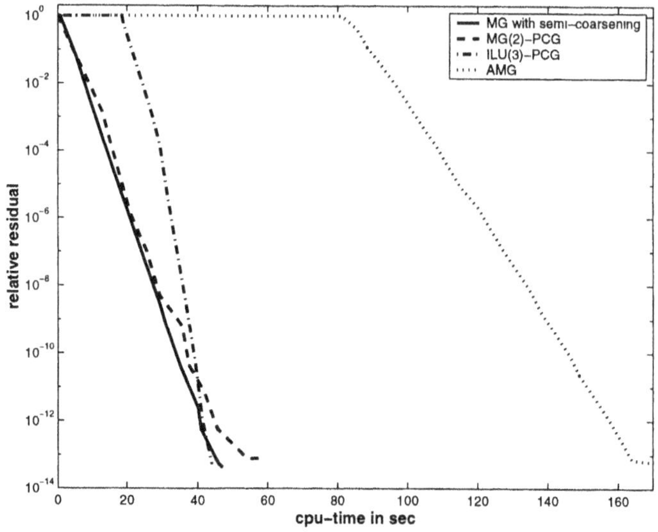

Figure4. Convergence of several algorithms for an adaptive discretization on a 33x33x65–mesh.

The situation does not change on adaptive grids, considering the multigrid algorithms. The proposed multigrid methods have been investigated for two adaptive discretizations. The related grids have 17x17x33 (=9 537) and 33x33x65 (=70 785) mesh points, respectively. Fig. 1 shows the grid for the first example in a longitudinal cross-section. For the sake of symmetry the y-direction is discretized in the same way as the x-direction. The second grid is twice as fine as the first one. In both cases the aspect ratio is $h_{max}/h_{min} = 32$.

The numerical results for the adaptive discretizations are given in Fig. 3 and Fig. 4. It turns out that the convergence behaviour of the geometric multigrid algorithms are nearly the same as in the case shown in Fig. 2. Only in the last third the convergence curve of the multigrid preconditioned conjugate gradient method becomes worse. After the setup phase, the convergence of AMG is not better than for the geometric method.

Table 1 summerizes the numerical results for the different multigrid strategies. For the MG(2)-PCG the number of cg iterations is given (7 cg iterations=18 multigrid iterations). The cpu-time for AMG includes the setup phase. It can be seen that the convergence speed for the geometric multigrid methods remains the same on finer discretizations which is ensured by the semi-coarsening strategy.

For both examples the ILU(3)-preconditioned conjugate gradient algorithm converges very fast, because the eigenvalues of the matrix A_h form clusters for the adaptive discretization. This property leads to a faster convergence than in the equidistant case [2]. But at least it can not beat the geometric multigrid algorithm.

4 Conclusions

Tracking simulation of particle beams is a crucial task for the design of high current or low emittance particle accelerators. The related algorithm [10,7] requires a fast Poisson solver robust with respect to the adaptive discretization of the computational domain. In this context two multigrid strategies have been investigated in this paper. The best numerical results are obtained with the geometric multigrid algorithm with semi-coarsening. It works stable for the adaptive grid, that is the computing time increases equivalent to the number of unknowns.

The AMG code of Ruge and Stüben could be an alternative to the geometric multigrid method if the adaptive grid given from the application does not allow an easy automatic construction of the coarser levels. The disadvantage of the long setup phase can probably be overcome with the recent new development in AMG [12] which is not yet available for common use.

The choice of the solution method for the tracking problem will also depend on its behaviour in the simulation of the particle motion over a large number of time steps. Investigations in that field are planned for the next future.

5 Acknowledgements

The authors want to thank the anonymous referees and Prof. Langer for a lot of helpful comments and J. Ruge and K. Stüben for making the AMG-code (AMG1R5) available to us.

References

1. W. Hackbusch, Multi-Grid Methods and Applications. Springer-Verlag, Berlin, 1985.
2. W. Hackbusch, Iterative Lösung großer schwachbesetzter Gleichungssysteme. B. G. Teubner, Stuttgart, 1991.
3. M. Jung, U. Langer, Applications of multilevel methods to practical problems. Surv. Math. Ind. **1** (1991), 217–257.
4. M. Jung, U. Langer, A. Meyer, W. Queck, M. Schneider, Multigrid preconditioners and their applications. In the Proceedings of the 3rd GDR Multigrid Seminar held at Biesenthal, 1989, 11–52.
5. MAFIA Version 4.0, CST GmbH, Büdinger Str. 2a, D-64289 Darmstadt, Germany.
6. G. Pöplau, U. van Rienen, Multigrid algorithms for the tracking of electron beams. In Multigrid Methods VI, (E. Dick, K. Riemslagh, J. Vierendeels, eds.), LNSCE, Vol. 14, Springer-Verlag, Berlin, 2000, 214–220.
7. G. Pöplau, U. van Rienen, J. Staats, T. Weiland, Fast algorithms for the tracking of electron beams. In Proceedings of the 7th European Particle Accelerator Conference (EPAC 2000), Vienna, 2000 (to appear).
8. J. Ruge, K. Stüben, Algebraic Multigrid (AMG). In Multigrid Methods, (S.F. McCormick, ed.), SIAM, Frontiers in Applied Mathematics, Philadelphia, 1987.
9. J. Ruge, K. Stüben, AMG1R5, release 1.5 Oct. 1990.
10. J. Staats, T. Weiland, S. Kostial, A. Richter, Tracking of electron beams with numerically determined space charge forces. In Proceedings of the 1999 Particle Accelerator Conference PAC'99, New York, 2740-2742.
11. K. Stüben, U. Trottenberg, Multigrid Methods: Fundamental Algorithms, Model Problem Analysis and Applications. GMD-Studien 96, St. Augustin, 1984.
12. K. Stüben, Algebraic Multigrid (AMG): An Introduction with Applications. GMD Report 53, St. Augustin, 1999.

Part II

Circuit Analysis

Model Design Criteria for Integrated Circuits to Have a Unique Solution and Good Numerical Properties*

C. Tischendorf

Institute of Applied Mathematics, Humboldt-University of Berlin

Abstract The numerical simulation of integrated circuits represents an essential part for cheap and time effective circuit design. It provides all interesting signals, voltages, currents and resulting data. New developments in the production engineering demand a higher abstraction level for circuit element models which may cause difficulties for the numerical analysis. Well-known problems are higher index configurations of the resulting differential algebraic systems (DAEs) and inconsistent initial values. Due to the large dimension of many circuits, the localization of problematic circuit configurations becomes difficult. Here, we exploit the special structure of equation systems arising from the modified nodal analysis (MNA) - one of the most used modeling technique that enables an automatic generation of the network equations and preserves the circuit structure. Assuming certain modeling criteria one can calculate the index, guarantee low index systems and guarantee unique solvability. Additionally, convergence and stability (weak instability in certain cases) of standard integration methods for the time domain analysis is guaranteed. The model criteria are circuit element related and can be simply checked by the model designer. Furthermore, we propose some rules for the model design in order to obtain model equations with a unique solution and a low DAE index.

Key words: Circuit simulation; differential-algebraic equation; DAE; index; modified nodal analysis; MNA; structural properties; model design; unique solvability; consistent initial values; perturbation; BDF; topological criteria.

1 Introduction

Present circuit design is strongly based on computer simulations. Depending on the interesting information, the circuits are treated with different kinds of analyses. The transient analysis belongs to the basic ones and is the subject matter of this paper. The task is to analyze the time domain behavior of the network voltages and currents depending on the input signals.

Present-day circuits are often integrated ones of enormous size. The development of electronic circuits has progressed to a state where millions of transistors can be put on a chip. The numerical simulation of such circuits requires modeling techniques which allow an automatic generation of network equations and need only a minimal number of independent network variables. The

* invited paper

modified nodal analysis allows such a modeling and is adopted successfully in numerous programs (e.g. SPICE[1] and TITAN[2]). The resulting systems represent differential-algebraic equations (DAEs).

A straightforward approach (BDF methods) for the numerical solution of such DAEs suggested by Gear [11] has been successfully used for numerous circuits and "is still among the best for this purpose" (see [24]). But, the more and more complex circuit models demand a serious treatment with the fundamental question whether we may believe the numerical results. Do the circuit models provide unique solutions? Does the numerical solution converge to the real solution? Has the numerical solution the same qualitative behavior as the real solution? Of special interests are answers to these problems basing on criteria that can may be checked during the model design. Some answers, particularly for unique solvability, are known for certain kind of networks (see e.g. [3], [19], [16], [2]), [10]).

Beside to the investigations of special circuit systems (usually by electrical engineers), general differential-algebraic equations were widely investigated (usually by mathematicians) during the last 20 years (see e.g. [13], [1], [15], [17]). The results cover among other things questions of unique solvability, the feasibility of numerical methods as well as stability questions. However, most of the results suppose a certain structure and depend strongly on the index of the DAE. The aspect, that one is usually confronted with strong nonlinear connections of circuit elements as diodes and transistors, led to the popular opinion that DAEs arising from circuit simulation do not have a special structure and their index is in general unknown. But, the investigations in [18], [23], [9], and [5] have shown that the network DAEs are well structured (depending on the network topology), the relevant subspaces may be described by suitable projections and the index may be checked by topological criteria.

The main task of this paper is to combine well-known results for DAEs with the structural knowledge of the circuit systems. We start (Section 2) with a description of the network analysis which is the basis for the transient analysis. In Section 3, we expose the correlation between the network topology and mathematical properties of corresponding network matrices. Afterwards, we may present unique solvability as well as feasibility and stability of numerical methods for very general networks satisfying certain criteria that may be checked locally. Finally, we give in Section 7 some suggestions for model designers in which way they can guaranty that the models fulfill the required criteria. Consequently, following the presented rules for the model design, the model equation systems provide always unique solutions and the numerical methods behave properly.

---

[1] originally developed at Berkeley in the 70s and early 80s; commercialized versions: HSPICE, PSPICE, et.al.

[2] developed by SIEMENS / Infineon Technologies

2 Network Analysis

We consider lumped directed electric circuits containing nonlinear, time-variant resistances, capacitances, inductances, voltage sources and current sources. Usually circuit simulation tools are based on these kinds of network elements. For two-terminal (one-port) lumped elements, the current through the element and the voltage across it are well-defined quantities. For lumped elements with more than two terminals, the current entering any terminal and the voltage across any pair of terminals are well defined at all times (cf. [4]). Hence, general n-terminal elements are completely described by $(n-1)$ branch currents entering the $(n-1)$ terminals and the $(n-1)$ branch voltages across each of these $(n-1)$ terminals and the reference terminal n. We denote by j the vector of all branch currents and by v the vector of all branch voltages. Then, the network is described entirely by Kirchhoff's Current Law (KCL)

$$Aj = 0, \tag{1}$$

by Kirchhoff's Voltage Law (KVL)

$$v = A^{\mathrm{T}} e, \tag{2}$$

and the voltage-current relations of all network elements

$$h(\frac{\mathrm{d}v}{\mathrm{d}t}, \frac{\mathrm{d}j}{\mathrm{d}t}, v, j, t) = 0. \tag{3}$$

Here, $A \in \mathbb{R}^{n \times b}$ represents the incidence matrix for the directed network with n nodes and b branches. The entries of the function h depend strongly on the kind of network variables. In case of linear time-invariant resistors, capacitors and inductors, respectively, we are confronted with equations of the form

$$j = Gv, \quad j = C\frac{\mathrm{d}v}{\mathrm{d}t}, \quad v = L\frac{\mathrm{d}j}{\mathrm{d}t}$$

for resistances, capacitances and inductances, respectively, where j is the current and v is the voltage of the corresponding branch. Independent sources may be described by typically time-dependent waveforms

$$v = v_s(t), \qquad j = i_s(t)$$

for voltage and current sources. For general (also nonlinear) n-terminal network elements, again we want to distinguish between resistors, capacitors and inductors. Correspondingly, the element related equations look like

$$j_k = r_k(v_1, ..., v_{n-1}, t), \quad j_k = \frac{\mathrm{d}q_k}{\mathrm{d}t}(v_1, ..., v_{n-1}, t), \quad v_k = \frac{\mathrm{d}\phi_k}{\mathrm{d}t}(j_1, ..., j_{n-1}, t)$$

for resistances, capacitances and inductances, respectively when j_k represents the current entering the terminal k and v_l describes the voltage across the pair of terminals $\{l, n\}$ (for $k, l = 1, ..., n-1$). Note that q_k represent the charges and ϕ_k describe the fluxes. For the following, we want to introduce the conductance, the capacitance and the inductance matrix, resp.,

$$G(v,t) := \frac{\mathrm{d}r}{\mathrm{d}v}(v,t), \quad C(v,t) := \frac{\mathrm{d}q}{\mathrm{d}v}(v,t), \quad L(j,t) := \frac{\mathrm{d}\phi}{\mathrm{d}j}(j,t)$$

for n-terminal elements, where $j = (j_1, ..., j_{n-1})$ and $v = (v_1, ..., v_{n-1})$. Controlled sources may be described by functions of the type

$$v = v_s(v_c, j_c, t), \quad j = i_s(v_c, j_c, t)$$

where j_c and v_c represent the vectors of the controlling currents and voltages, respectively.

Remark 1. More complex network elements like transistors may be described as a combination of dynamical elements (capacitors and inductors) and static elements (resistors, controlled sources). The following example illustrates this fact.

Example 2. We want to consider a MOSFET model (see Figure 1, cf. [21], [23]) describing the branch currents as follows (considering the source terminal as reference terminal):

$$\begin{aligned} j_G &= C_{GS}\dot{v}_{GS} + C_{GD}(\dot{v}_{GS} - \dot{v}_{DS}), \\ j_D &= -C_{GD}(\dot{v}_{GS} - \dot{v}_{DS}) - C_{BD}(\dot{v}_{BS} - \dot{v}_{DS}) \\ &\quad + d(v_{BS} - v_{DS}) + i(v_{GS}, v_{DS}, v_{BS}) + G v_{DS}, \\ j_B &= C_{BS}\dot{v}_{BS} + C_{BD}(\dot{v}_{BS} - \dot{v}_{DS}) - d(v_{BS}) - d(v_{BS} - v_{DS}). \end{aligned}$$

The indices G, D, B, and S represent the terminals gate, drain, bulk, and

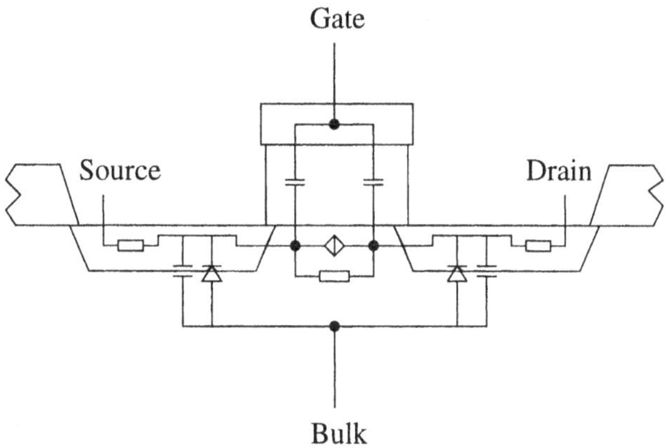

Figure1. MOSFET model using capacitances, resistances, and a voltage controlled current source

source, respectively. The dynamical behavior is reflected by the capacitances C_{GS}, C_{GD}, C_{BD}, and C_{BS}. The corresponding capacitance matrix looks like

$$C(v_{GS}, v_{DS}, v_{BS}) = \begin{pmatrix} C_{GS} + C_{GD} & -C_{GD} & 0 \\ -C_{GD} & C_{GD} + C_{BD} & -C_{BD} \\ 0 & -C_{BD} & C_{BS} + C_{BD} \end{pmatrix}.$$

Furthermore G represents the conductance of the isolation layer between source and drain. The gate controlled current between source and drain is modeled by the voltage controlled current source $i(v_{GS}, v_{DS}, v_{BS})$. Finally, the static current from bulk to source and from bulk to drain, respectively, is described by nonlinear resistances (diodes) $d(v_{BS})$ and $d(v_{BD})$. This yields to a conductance matrix

$$G(v_{GS}, v_{DS}, v_{BS}) = \begin{pmatrix} 0 & 0 & 0 \\ 0 & G - d'(v_{BS} - v_{DS}) & d'(v_{BS} - v_{DS}) \\ 0 & d'(v_{BS} - v_{DS}) & -d'(v_{BS}) - d'(v_{BS} - v_{DS}) \end{pmatrix}.$$

The system (1)-(3) reflects a system of differential-algebraic equations with the unknown network variables j, v, and the nodal potentials e. The dimension of this system is $2b+n-1$. Usually, the number of branches is quite large with respect to the number of nodes. For cheap computation, a reduction to a system with $n-1$ equations with only the nodal potentials as variables is preferable. This is always possible if the network does not contain elements for which the branch voltages are controlled (inductances and voltage sources). In general, the so called modified nodal analysis (MNA) leads as follows to a system with $n - 1 + \bar{b}$ equations where $\bar{b}$ denotes the number of branches for which the voltages are controlled.

1. Replace the branch voltages v in (3) by nodal potentials e using (2) and remove all equations of (2).
2. Replace all currents j_k in (1) of branches for which the currents are controlled (capacitances, resistances, current sources) using the corresponding network element relations from (3) and remove the chosen element relations.

If we split the incidence matrix A into the element-related incidence matrices

$$A = (A_C, A_L, A_R, A_V, A_I),$$

where A_C, A_L, A_R, A_V, and A_I describe the branch-current relations for capacitive, inductive, resistive branches, branches of voltage sources and current sources, respectively, then the resulting system has the form

$$A_C \frac{dq_C(A_C^T e, t)}{dt} + A_R r(A_R^T e, t) + A_L j_L + A_V j_V + A_I i(A^T e, \frac{dq(A_C^T e, t)}{dt}, j_L, j_V, t) = 0, \tag{4}$$

$$\frac{d\phi_L(j_L, t)}{dt} - A_L^T e = 0, \tag{5}$$

$$A_V^T e - v(A^T e, \frac{dq(A_C^T e, t)}{dt}, j_L, j_V, t) = 0. \tag{6}$$

Here, the vector of unknowns consists of the node potentials e (excepting the datum node), the branch currents j_L of inductances and the branch currents

j_V of voltage sources. Note that the vectors $A_C^T e$, $A_L^T e$, $A_R^T e$ and $A_V^T e$ describe the branch voltages for the capacitive, inductive, resistive and voltage source branches, respectively.

Remark 3. Due to the fact that the currents through resistances are functions of the branch voltages, we do not include them separately as controlling functions. Of course, if the network does not contain controlled sources, then the source functions reduce to functions $i(t)$ and $v(t)$ which depend on time only.

The modified nodal analysis is a commonly used method for network analysis and realized in packages like SPICE[3] and TITAN[4]. The conventional realization rewrites the system (4)-(6) as a differential equation in (e, j_L, j_V):[5]

$$A_C C(A_C^T e, t) A_C^T \frac{de}{dt} + A_C q'_{Ct}(A_C^T e, t) + A_R r(A_R^T e, t)$$

$$+ A_L j_L + A_V j_V + A_I i(A^T e, A_C^T \frac{de}{dt}, j_L, j_V, t) = 0, \quad (7)$$

$$L(j_L, t)\frac{dj_L}{dt} + \phi'_{Lt}(j_L, t) - A_L^T e = 0, \quad (8)$$

$$A_V^T e - v(A^T e, A_C^T \frac{de}{dt}, j_L, j_V, t) = 0. \quad (9)$$

In this paper, we want to concentrate on the charge-oriented realization (used in TITAN), where the charges q as well as the fluxes ϕ are included in the vector of unknowns and the original voltage-charge as well as the current-flux

[3] originally developed at Berkeley in the 70s and early 80s, and commercialized versions (e.g. HSPICE, PSPICE)

[4] developed by SIEMENS / Infineon Technologies

[5] Note that we have

$$\frac{dq_C(A_C^T e, t)}{dt} = C(A_C^T e, t) A_C^T \frac{de}{dt} + q'_{Ct}(A_C^T e, t).$$

Therefore, $i(A^T e, \frac{dq_C(A_C^T e,t)}{dt}, j_L, j_V, t) = i_*(A^T e, A_C^T \frac{de}{dt}, j_L, j_V, t)$ for a suitable function i_*. An analogous relation is valid for the controlled voltage-sources. For simplicity, we drop the index $*$.

equations are added to the system. The resulting systems reads (cf. [14]):

$$A_C\frac{\mathrm{d}q}{\mathrm{d}t} + A_R r(A_R^T e, t) + A_L j_L + A_V j_V$$
$$+ A_I i(A^T e, \frac{\mathrm{d}q}{\mathrm{d}t}, j_L, j_V, t) = 0, \qquad (10)$$
$$\frac{\mathrm{d}\phi}{\mathrm{d}t} - A_L^T e = 0, \qquad (11)$$
$$A_V^T e - v(A^T e, \frac{\mathrm{d}q}{\mathrm{d}t}, j_L, j_V, t) = 0, \qquad (12)$$
$$q - q_C(A_C^T e, t) = 0, \qquad (13)$$
$$\phi - \phi_L(j_L, t) = 0, \qquad (14)$$

with continuously differentiable charge and flux functions $q(t)$ and $\phi(t)$, respectively.

Remark 4. The charge oriented system (10)-(14) represents a DAE system of the type

$$\bar{A}x' + \bar{g}(x,t) = 0 \qquad (15)$$

with a constant leading coefficient matrix $\bar{A}$, whereas the conventional system (7)-(9) is a DAE system of the type

$$\bar{\bar{A}}(x,t)x' + \bar{\bar{g}}(x,t) = 0$$

with an x- and t-dependent coefficient matrix $\bar{\bar{A}}(x,t)$.

Remark 5. If we apply implicit methods like the BDF method for solving the network equations then the solution of nonlinear equations is required. Usually, this is done by Newton-type methods. Considering the conventional system (7)-(9), one has to differentiate the capacitance matrix $C(v,t)$ and the inductance matrix $L(j,t)$ with respect to v and j, respectively. Contrarily, this is not necessary for the charge oriented system (10)-(14). This is an important fact since the models often have only low smoothness properties.

Remark 6. The simple form of the equations (13) and (14) makes it possible to reduce the nonlinear equation systems (resulting by the application of numerical methods to (10)-(14)) to a system of the same dimension as the system (7)-(9) has.

3 Topological Properties of the Splitted Incidence Matrix

The splitted incidence matrix $A = (A_C, A_L, A_R, A_V, A_I)$ corresponding to certain branches provides useful structural information for lumped circuits. Mathematical properties of the incidence matrix A are strongly related to certain position relations of the network elements. Here, we want to cite such ones that are relevant for the index of the simulated circuit system.

Lemma 7. *Given a lumped circuit with capacitances, inductances, resistances, voltage sources and current sources. Then, the following relations are satisfied for the (reduced) incidence matrix $A = (A_C A_L A_R A_V A_I)$.*

1. *The matrix $(A_C A_L A_R A_V)$ has full row rank if and only if the network does not contain a cutset*[6] *of current sources.*
2. *The matrix A_V has full column rank if and only if the network does not contain a loop*[7] *of voltage sources.*
3. *The matrix $(A_C A_R A_V)$ has full row rank if and only if the circuit does not contain a cutset consisting of inductances and current sources (cf. Figure 2).*
4. *The matrix $Q_C^T A_V$ has full column rank for any projector Q_C onto $ker A_C^T$ if and only if the circuit does not contain a loop consisting of capacitances and voltage sources with at least one voltage source (cf. Figure 2).*

Proof. We only show property *1.* because the correctness of the other properties can be checked in a similar manner and were proven in detail in [9]. We consider the subnetwork graph G_s which contains all network elements of the originally network G excluding the current sources. Then, G_s is connected if and only if $(A_C A_L A_R A_V)$ does not contain a zero row. All rows belonging to nodes of G_s are linear independent since, for each connected subgraph of G_s, the corresponding incidence matrix has full row rank. Consequently, A_s has full row rank if and only if G_s is connected. This is obviously the case if and only if, G does not contain a cutset of current sources only. □

Remark 8. Well modeled circuits satisfy always the properties *1.* and *2.* of Lemma 7 since they satisfy the Kirchhoff's laws.

The special cutsets and loops considered in item *3.* and *4.* of Lemma 7 play an important role in the following sections. Therefore we define:

Definition 9. An L-I cutset is a cutset consisting of inductances and current sources.

Definition 10. A C-V loop is a loop consisting of capacitances and voltage sources with at least one voltage source.

---

[6] Remember, a subgraph G_c of a network is a cutset if the network decomposes after removing G_c and any branch of G_c connects the decomposed parts of the network.

[7] Remember, a subgraph G_l of a network is a loop if G_l is connected and each node of G_l combines exactly two branches of G_l.

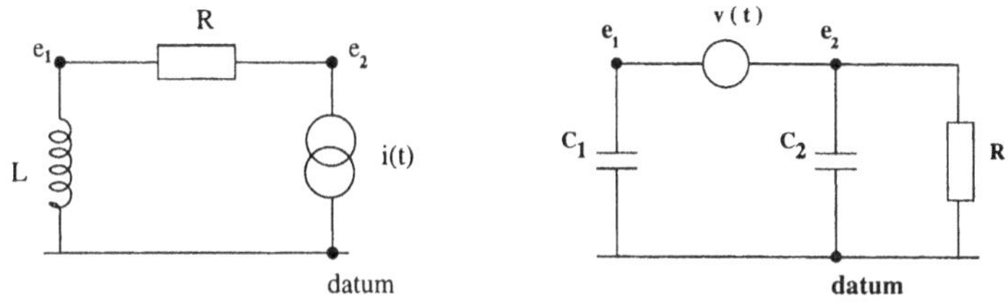

Figure2. Example for an L–I cutset (left) and a C–V loop (right)

4 Index and Solvability of Network Equation Systems

The question of solvability of nonlinear differential-algebraic systems is strongly connected to the question of the index of a DAE. A variety of solution properties, as e.g. smoothness and stability, depends on the index[8]. Therefore, it is advisable to study the index of the network equation systems before deriving some general solvability statements for nonlinear circuit systems.
A long experience in circuit simulation has shown that L-I cutsets as well as C-V loops may lead to difficulties in the transient analysis. It turns out that these network configurations lead to higher index configurations. Combining these configurations using controlled sources sufficiently often in a certain way one obtains even systems with arbitrarily high index (see Figure 3). Nevertheless, the most practical circuit configurations do not exceed the

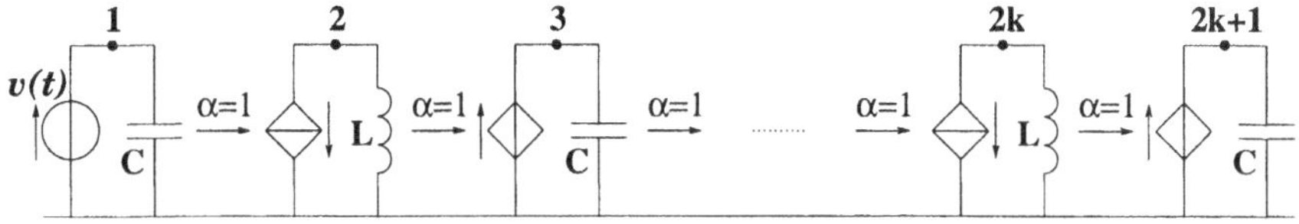

Figure3. The network equation system obtained by MNA has an index $2k+2$ since $e_1 = v(t)$, $e_{2i+1} = e_{2i} = L^i C^i v^{(2i)}(t)$ for $i = 1, ..., k$, $j_{2i-1} = j_{2i} = L^{i-1} C^i v^{(2i-1)}$ for $i = 1, ..., k$ and $j_{2k+1} = L^k C^{k+1} v^{(2k+1)}$, where e_i the nodal potential at node i and j_i the current through the i-th source.

index 2. For networks with passive resistances, capacitances and inductances and without controlled sources, it was shown in [23]. This result could be generalized to networks containing ideal transformers and gyrators (see [20])

[8] The index categorizes DAEs with respect to their numerical behavior. The notion is not unique in the literature. We consider here the tractability index ([13], [17]) that coincides with the differential index (see [1]) for the network equation systems (cf. [9]).

and for networks containing controlled sources satisfying the Assumptions 11 and 12 (see [9]). More precisely, the next theorem (Theorem 13) is satisfied.

Assumption 11. *Any controlled voltage source (CVS) satisfies the following conditions:*

1. *It does not form a part of a C-V loop.*
2. *The controlling voltages are voltages of capacitances or independent voltage sources.*
3. *The controlling currents are currents of inductances or independent current sources.*

Assumption 12. *Any controlled current source (CCS) satisfies the following conditions:*

1. *It does not form a part of an L-I cutset.*
2. *The controlling voltages are voltages of capacitances, independent voltage sources or resistances.*
3. *The controlling currents are currents of inductances, independent current sources, resistances or voltage sources that do not form a part of a C-V loop.*
4. *If the nodes of the CCS are not connected by a C-path (path containing only capacitances) then currents of voltage sources are not allowed as controlling currents.*
5. *If the nodes of the CCS are not connected by a CV-path (path containing only capacitances and voltage sources), then voltages and currents of resistances (whose nodes are not connected by a CV-path) are not allowed as controlling voltages and currents, respectively.*

Theorem 13. *Let a circuit network without loops of voltages sources and without cutsets of current sources be given. Furthermore, the matrices $G(v,t)$ for all resistances which nodes are not connected by a CV-path, the matrices $C(v,t)$ for all capacitances and the matrices $L(j,t)$ for all inductances are assumed to be positive definite*[9]*. Additionally, all controlled sources satisfy the Assumptions 11 and 12. Then, under sufficient smoothness conditions, the DAE (10)-(14) combined with a consistent initial value has a unique solution on compact time intervals. It has the index 1 if it does not contain C-V loops and L-I cutsets. Otherwise, it has the index 2.*

Remark 14. The theorem includes, for the first time, simple topological criteria for calculating the index of a network model. In contrary to the algorithm presented in [20], this approach provides a very cheap and trustworthy index determination and has been successfully implemented as an index monitor in the professional code TITAN (cf. [7]).

[9] For capacitances and inductances with affine characteristics, the positive definiteness implies that they are strictly locally passive (cf. [10]).

Remark 15. Regarding Lemma 7, the assumptions for the controlled voltage sources may be expressed as follows:

$$\bar{Q}_{V-C}v(\cdot) = \bar{Q}_{V-C}v(t),$$
$$v(\cdot) = v(A_C^T e, j_L, t),$$

where $\bar{Q}_{V-C}$ is a projector onto $\ker Q_C^T A_V$. Analogously, the assumptions for the controlled current sources may be described as:

$$Q_{CRV}A_I i(\cdot) = Q_{CRV}A_I i(t),$$
$$i(\cdot) = i(A_C^T e, A_V^T e, A_R^T e, j_L, \bar{P}_{V-C} j_V, t),$$
$$Q_C^T A_I i'_{j_V}(\cdot) = 0,$$
$$Q_{V-C}^T Q_C^T A_I i'_v(\cdot) A_R^T Q_C Q_{V-C} = 0$$

where Q_{CRV} is a projector onto $\ker(A_C\, A_R\, A_V)^T$, $\bar{P}_{V-C} = I - \bar{Q}_{V-C}$, and Q_{V-C} is a projector onto $\ker A_V^T Q_C$.

Remark 16. Taking into account Remark 15, it becomes clear that Theorem 13 keeps true if, for the controlled sources, the controlling voltages and currents can be expressed in terms of the voltages and currents listed in the Assumptions 11 and 12. For instance, the voltage of a resistance, which nodes are connected by a C-path, may always be a controlling one.

Remark 17. The smoothness conditions can be specified in the following way: All network functions including the capacitance, inductance and resistance matrices should be continuous. As already mentioned before, the charge functions $q(v,t)$ of all capacitances and the flux functions $\phi(j,t)$ of all inductances are assumed to be continuous differentiable. The input signals $v(t)$ of voltage sources of C-V loops and the input signals $i(t)$ of current sources of L-I cutsets have to be continuously differentiable. Furthermore, the charges $q(v,t)$ of capacitances of C-V loops and the fluxes $\phi(j,t)$ of inductances of L-I cutsets have to be twice continuously differentiable. Then, the solution (q, ϕ, e, j_L, j_V) is continuous with smooth charges q, smooth fluxes ϕ, smooth capacitive nodal potentials[10] $P_C c$ (with $P_C = I - Q_C$) and smooth inductive currents j_L.

Remark 18. An initial value of the system (10)-(14) is consistent (cf. [9]) if and only if it satisfies the explicit constraints

$$Q_C^T[A_R r(A_R^T e, t) + A_L j_L + A_V j_V + A_I i(\cdot)] = 0, \tag{16}$$
$$A_V^T e - v(\cdot) = 0, \tag{17}$$
$$q = q_C(A_C^T e, t), \tag{18}$$
$$\phi = \phi_L(j_L, t) \tag{19}$$

[10] Nodes that are connected with the datum node by a C-path

as well as the hidden constraints

$$\bar{Q}_{V-C}^T A_V^T H_1^{-1}(\cdot) P_C^T [A_C q_C{}'_t(A_C^T e,t) + A_R r(A_R^T e,t) + A_L j_L + A_V j_V + A_I i(\cdot)] + \bar{Q}_{V-C}^T v'(t) = 0, \quad (20)$$

$$Q_{CRV}^T [A_L L^{-1}(\cdot)(A_L^T e - \phi_L{}'_t(j_L,t)) + A_I i'(t)] = 0 \quad (21)$$

for $H_1(A_C^T e,t) = A_C C(A_C^T e,t) A_C^T + Q_C^T Q_C$. If one wants to compute consistent initial values, one only needs to calculate a DC-operation point for fulfilling the explicit constraints and solving additionally a linear system for satisfying the hidden constraints. One may develop such a linear system because of the special structure (22) of the circuit systems (see the proof of Theorem 13). For the details we refer to [6] and [5] (pages 103ff.).

Proof. (Theorem 13) The statement about the index was shown in [9] (see Theorem 4.2 and Corollary 4.3). This proof is rather long and uses projector techniques. The statement about the unique solvability is well-known for index-1 systems (see e.g. [13]). The most solvability results for index-2 systems are not applicable since the considered DAE has not a Hessenberg structure. But, as shown in [5] (cf. Corollary 3.2.8.), the DAE (10)-(14) has the special form

$$\bar{A}x' + \bar{b}(Ux,t) + BTx = 0 \quad (22)$$

for $x = (q, \phi, e, j_L, j_V)$, for the complementary projectors[11]

$$T = \begin{pmatrix} 0 & 0 & 0 & 0 & 0 \\ 0 & 0 & 0 & 0 & 0 \\ 0 & 0 & Q_{CRV} & 0 & 0 \\ 0 & 0 & 0 & 0 & 0 \\ 0 & 0 & 0 & 0 & \bar{Q}_{V-C} \end{pmatrix}, \quad U = \begin{pmatrix} I & 0 & 0 & 0 & 0 \\ 0 & I & 0 & 0 & 0 \\ 0 & 0 & P_{CRV} & 0 & 0 \\ 0 & 0 & 0 & I & 0 \\ 0 & 0 & 0 & 0 & \bar{P}_{V-C} \end{pmatrix}$$

and for the matrices

$$\bar{A} = \begin{pmatrix} A_C & 0 & 0 & 0 & 0 \\ 0 & I & 0 & 0 & 0 \\ 0 & 0 & 0 & 0 & 0 \\ 0 & 0 & 0 & 0 & 0 \\ 0 & 0 & 0 & 0 & 0 \end{pmatrix}, \quad BT = \begin{pmatrix} 0 & 0 & 0 & 0 & A_V \bar{Q}_{V-C} \\ 0 & 0 & -A_L^T Q_{CRV} & 0 & 0 \\ 0 & 0 & 0 & 0 & 0 \\ 0 & 0 & 0 & 0 & 0 \\ 0 & 0 & 0 & 0 & 0 \end{pmatrix}.$$

Let us point out here that the projector T represents a projector onto the so called index-2 components. This becomes clear if one regards that $\bar{Q}_{V-C} = 0$ if and only if the network does not contain C-V loops and $Q_{CRV} = 0$ if and only if the network does not contain L-I cutsets. Both facts are simple conclusions from Lemma 7. The index-2 components are of importance when solving the network equation systems (10)-(14) numerically since they are effected by instability effects. We will present the details for that in Section 6.

Now we know from [5] (cf. Theorem 2.4.6 and pages 98ff.) that x is a solution of (22) if and only if the initial value x_0 is consistent (cf. Remark 18) for (22) and x is a solution of the index-1 system

$$\big(\bar{A} + W_1(Ux,t)(K_{W_1} W_0 \bar{b})'_x(Ux,t)\big)\, x'(t) + \bar{b}(Ux,t) + BTx - (\hat{W}_1 \bar{b})(Ux,t) + W_1(Ux,t)(K_{W_1} W_0 \bar{b})'_t(Ux,t) = 0 \quad (23)$$

[11] Note that $P_{CRV} = I - Q_{CRV}$ and $\bar{P}_{V-C} = I - \bar{Q}_{V-C}$.

for

$$W_1(\cdot) = \begin{pmatrix} Q_{CRV}^T & 0 & 0 & 0 & Q_{CRV}^T A_L L^{-1}(\cdot) \\ 0 & 0 & 0 & 0 & 0 \\ 0 & 0 & \bar{Q}_{V-C}^T & \bar{Q}_{V-C}^T A_V^T H_1^{-1}(\cdot) A_C & 0 \\ 0 & 0 & 0 & 0 & 0 \\ 0 & 0 & 0 & 0 & 0 \end{pmatrix}$$

with $H_1(A_C^T e, t) = A_C C(A_C^T e, t) A_C^T + Q_C^T Q_C$ and for

$$K_{W_1} W_0 = \begin{pmatrix} Q_{CRV}^T & 0 & 0 & 0 & 0 \\ 0 & 0 & 0 & 0 & 0 \\ 0 & 0 & \bar{Q}_{V-C}^T & 0 & 0 \\ 0 & 0 & 0 & I & 0 \\ 0 & 0 & 0 & 0 & I \end{pmatrix}, \qquad \hat{W}_1 = \begin{pmatrix} Q_{CRV}^T & 0 & 0 & 0 & 0 \\ 0 & 0 & 0 & 0 & 0 \\ 0 & 0 & \bar{Q}_{V-C}^T & 0 & 0 \\ 0 & 0 & 0 & 0 & 0 \\ 0 & 0 & 0 & 0 & 0 \end{pmatrix}.$$

Using the fact that

$$\ker\left(\bar{A} + W_1(Ux, t)(K_{W_1} W_0 \bar{b})'_x(Ux, t)\right) = \ker \bar{A}$$

is constant and that index-1 systems having a leading coefficient matrix with a constant nullspace are uniquely solvable, one can immediately follow that the system (10)-(14) is uniquely solvable. The required smoothness conditions are discussed in Remark 17. □

5 Perturbation Analysis

Our aim is to derive some statements about the feasibility, convergence and stability of numerical methods applied to the circuit equations systems. All the problems are closely connected to the question about the solution behavior of slightly perturbed network systems. The following theorem gives a detailed information about the influence of small perturbations.

Theorem 19. *Let the assumptions of Theorem 13 be satisfied and a solution* $x_* := (q_*, \phi_*, e_*, j_{L*}, j_{V*})$ *of (10)-(14) be given.*

(i) Then, the perturbed initial value problem

$$A_C q' + A_R r(A_R^T e) + A_L j_L + A_V j_V + A_I i(\cdot) = \delta_e, \tag{24}$$

$$\phi_L' - A_L^T e = \delta_L, \tag{25}$$

$$A_V^T e - v = \delta_V, \tag{26}$$

$$q - q_C(A_C^T e, t) = 0, \tag{27}$$

$$\phi - \phi_L(j_L, t) = 0 \tag{28}$$

combined with a consistent initial value $x_0 = (q_0, \phi_0, e_0, j_{L0}, j_{V0})$ *are uniquely solvable on compact time intervals supposing* $|x_0 - x_*(t_0)|$ *as well as the perturbations* $\|\delta\|_\infty$ *(for* $\delta := (\delta_e, \delta_L, \delta_V)$*),* $\|Q_{CRV}^T \delta_e'\|_\infty$, $\|\bar{Q}_{V-C}^T \delta_V'\|_\infty$ *are sufficiently small.*

(ii) For the solution $x = (q, \phi, e, j_L, j_V)$ *of the perturbed system (24)-(28), the inequality*

$$\|x - x_*\|_\infty \leq K(\|\delta\|_\infty + \|Q_{CRV}^T \delta_e'\|_\infty + \|\bar{Q}_{V-C}^T \delta_V'\|_\infty + |x_0 - x_*(t_0)|) \tag{29}$$

is true for a constant $K > 0$.

Remark 20. Solving (27) and (28) requires only function evaluations in suitable implementations. Therefore, we do not consider perturbations for (27) and (28).

Remark 21. For the characterization and computation of consistent initial values see Remark 18.

Remark 22. The theorem implies immediately that the perturbation index (cf. [15]) of the network equation systems (10)-(14) does not exceed 2 and coincides with the differential as well as the tractability index.

Proof. We want to apply Theorem 3.1 and Theorem 3.12 from [22]. Regarding Remark 3 on page 51 in [22], it is sufficient to show that the space $N \cap S(\cdot)$ of index-2 components does not depend on x, where

$$N = \ker\bar{A} \quad \text{and} \quad S(\cdot) = \{z : \bar{b}'_x(\cdot)z \in \text{im } \bar{A}\}.$$

Referring again to [5] (pages 98ff.), we obtain via certain matrix manipulations that

$$N \cap S(\cdot) = \{0\} \times \{0\} \times \text{im } Q_{CRV} \times \{0\} \times \text{im } \bar{Q}_{V-C}$$

is even constant.

Following the proof of Theorem 3.12 in [22] and considering that Tx appears only linear in (22), it turns out that besides the perturbation δ itself the derivative of the part $PQ_1(\cdot)G_2^{-1}(\cdot)\delta$ has an influence onto the solution of the perturbed system whereas

$$P = \begin{pmatrix} \bar{P}_C & 0\,0\,0\,0 \\ 0 & I\,0\,0\,0 \\ 0 & 0\,0\,0\,0 \\ 0 & 0\,0\,0\,0 \\ 0 & 0\,0\,0\,0 \end{pmatrix}$$

is a projector along $\ker\bar{A}$,

$$Q_1 = \begin{pmatrix} \tilde{P}_C & 0 & 0\,0\,0 \\ 0 & \tilde{P}_L & 0\,0\,0 \\ 0 & Q_{CRV}H_5^{-1}(\cdot)Q_{CRV}^T A_L L^{-1}(\cdot) & 0\,0\,0 \\ 0 & 0 & 0\,0\,0 \\ -\bar{Q}_{V-C}H_6^{-1}\bar{Q}_{V-C}^T A_V^T H_1^{-1}(\cdot)A_C & 0 & 0\,0\,0 \end{pmatrix}$$

is a projector onto the nullspace of $A_1(x,t) := \bar{A} + (\bar{b}'_x(Ux,t)U + BT)(I-P)$ along $S_1 = \{z : \ (\bar{b}'_x(Ux,t)U + BT)Pz \in \text{im } A_1\}$, and

$$G_2 = A_1 + (\bar{b}'_x(Ux,t)U + BT)PQ_1$$

$$= \begin{pmatrix} A_C & 0 & A_R G(\cdot)A_R^T + A_I \frac{\mathrm{d}i}{\mathrm{d}e}(\cdot)P_{CRV} & A_L^T + A_I \frac{\mathrm{d}i}{\mathrm{d}\jmath_L}(\cdot) & A_V + A_I \frac{\mathrm{d}i}{\mathrm{d}\jmath_V}(\cdot)\bar{P}_{V-C} \\ 0 & I & -A_L^T & 0 & 0 \\ 0 & 0 & A_V^T - \frac{\mathrm{d}v}{\mathrm{d}e}(\cdot)A_C^T & -\frac{\mathrm{d}v}{\mathrm{d}\jmath_L}(\cdot) & 0 \\ \bar{P}_C + \bar{Q}_C & 0 & -C(\cdot)A_C^T & 0 & 0 \\ 0 & \tilde{P}_L & 0 & -L(\cdot) & 0 \end{pmatrix}$$

is nonsingular. Here, we used the notations

$$\tilde{P}_C = \bar{P}_C A_C^T \bar{H}_1^{-1} A_V \bar{Q}_{V-C} H_6^{-1}(\cdot)\bar{Q}_{V-C}^T A_V^T H_1^{-1}(\cdot)A_C$$
$$\tilde{P}_L = A_L^T Q_{CRV} H_5^{-1}(\cdot)Q_{CRV}^T A_L L^{-1}(\cdot)$$

as well as the notations

$$\begin{aligned} H_1(A_C^T e, t) &= A_C C(A_C^T e, t) A_C^T + Q_C^T Q_C \\ \bar{H}_1 &= A_C A_C^T + Q_C^T Q_C \\ H_5(j_L, t) &= A_{CRV}^T A_L L^{-1}(j_L, t) A_L^T Q_{CRV} + P_{CRV}^T P_{CRV} \\ H_6(A_C^T e, t) &= \bar{Q}_{V-C}^T A_V^T H_1^{-1}(A_C^T e, t) A_V \bar{Q}_{V-C} + \bar{P}_{V-C}^T \bar{P}_{V-C} \end{aligned}$$

from [9] whereas $\bar{Q}_C = I - \bar{P}_C$ is a projector onto $\ker A_C$ and $P_{CRV} = I - Q_{CRV}$ and $\bar{P}_{V-C} = I - \bar{Q}_{V-C}$. This leads to

$$PQ_1(\cdot)G_2^{-1}(\cdot) = \begin{pmatrix} 0 & 0 & \bar{P}_C A_C^T \bar{H}_1^{-1} A_V \bar{Q}_{V-C} H_6^{-1}(\cdot) \bar{Q}_{V-C}^T & \bar{P}_C & 0 \\ A_L^T Q_{CRV} H_5^{-1}(\cdot) Q_{CRV}^T & 0 & 0 & 0 & \tilde{P}_L \\ 0 & 0 & 0 & 0 & 0 \\ 0 & 0 & 0 & 0 & 0 \\ 0 & 0 & 0 & 0 & 0 \end{pmatrix}$$

which implies the assertion of the theorem. □

Remark 23. Following the proof of Theorem 3.12 in [22] and considering the special structure (22) of the network equation systems, it turns out that the component Ux may be determined independently from the index-2 component Tx[12]. This allows a more detailed estimation than given in (29). There are constants K_1 and K_2 such that

$$\begin{aligned} &\|Ux - Ux_*\|_\infty \leq K_1(\|\delta\|_\infty + |Ux_0 - Ux_*(t_0)|), \\ &\|Tx - Tx_*\|_\infty \leq K_2(\|\delta\|_\infty + \|Q_{CRV}^T \delta_e'\|_\infty + \|\bar{Q}_{V-C}^T \delta_V'\|_\infty + |Ux_0 - Ux_*(t_0)|). \end{aligned}$$

This means that the derivatives of the perturbations have only an influence onto the index-2 components Tx which correspond to $Q_{CRV}e$ and $\bar{Q}_{V-C}j_V$.

6 Feasibility, Convergence and (In)stability of the BDF Methods

We assume again the assumptions of Theorem 13 to be fulfilled and we consider a partition π of a compact interval $[t_0, t_N]$ with the following properties.

$$\begin{aligned} &\pi : t_0 < t_1 < \cdots < t_N, \qquad (30) \\ &h_{min} \leq t_n - t_{n-1} \leq h_{max}, \quad h_{min} > 0, \quad n \geq 1, \\ &\kappa_1 \leq \frac{h_{n-1}}{h_n} \leq \kappa_2, \quad n \geq 1, \end{aligned}$$

where κ_1 and κ_2 are suitable constants such that the BDF methods are stable for explicit ordinary differential equations (cf. [12], [13]).
Then, the BDF method applied to network equation systems (15) may be formulated as

$$\bar{A}\frac{1}{h_n}\sum_{i=0}^{k} \alpha_{ni} x_{n-i} + \bar{g}(x_n, t_n) = \delta_n, \quad n \geq k. \qquad (31)$$

[12] For the definition of T and U see the proof of Theorem 13 on page 190.

Here, δ_n describes the perturbations on the n-th step for $n \geq k$, which is caused by numerical computations including the errors arising from solving the nonlinear equations (e.g. with a Newton-like method). As usually, we denote the stepsize for the n-th step by h_n, i.e., $h_n = t_n - t_{n-1}$. Moreover, we introduce

$$\tilde{x}_n := x_n - x_*(t_n), \quad n \geq 0,$$

$$\tau_n := \bar{A}\frac{1}{h_n}\sum_{i=0}^{k} \alpha_{ni} x_*(t_{n-i}) + \bar{g}(x_*(t_n), t_n), \quad n \geq k.$$

The term τ_n defined in this way represents the local error of the BDF of order k on the n-th step, and it is of order $O(h_n^k)$ if Px_* is sufficiently smooth. This becomes obvious if we regard that

$$\tau_n = \frac{1}{h_n}\sum_{i=0}^{k} \alpha_{ni}\bar{A}x_*(t_{n-i}) - \bar{A}x_*'(t_n)$$

since $x_*(t_n)$ is a solution of the system (15). The next theorem gives us an information about the feasibility of the BDF methods for network equation systems. Furthermore, it provides some stability statements.

Theorem 24. *Let the assumptions of Theorem 13 be fulfilled. Supposed there is a constant $K > 0$ such that the starting values satisfy the relation*

$$\|x_n - x_*(t_n)\| \leq K h_n, \quad n < k, \tag{32}$$

then it holds that:

(i) There are constants $\vartheta > 0$ and $r > 0$ such that the BDF with

$$\|\delta_n\| \leq \vartheta \text{ for } n \geq k \quad \text{and} \quad \frac{\|Q_{CRV}^T \delta_{en}\| + \|\bar{Q}_{V-C}^T \delta_{Vn}\|}{h_n} \leq \vartheta \text{ for } n \geq 0,$$

is feasible for all partitions (30) with sufficiently small stepsizes, i.e., the nonlinear equations are solvable with $x_n \in B(x_*(t_n), r)$.

(ii) Supposed there is a constant $K_1 > 0$ with

$$\|\delta_n\| \leq K_1 h_n \text{ for } n \geq k,$$
$$\|Q_{CRV}^T \delta_{en}\| + \|\bar{Q}_{V-C}^T \delta_{Vn}\| \leq K_1 h_n^2 \text{ for } n \geq 0,$$

then we find a constant $K_2 > 0$ such that the following error estimation holds:

$$\max_{n\geq k} \|x_*(t_n) - x_n\| \leq K_2 \Big[\max_{n<k} \|Ux_*(t_n) - Ux_n\| + \max_{n\geq k} \|\delta_n - \tau_n\| + \max_{n\geq 0} \frac{\|Q_{CRV}^T \delta_{en}\| + \|\bar{Q}_{V-C}^T \delta_{Vn}\|}{h_n} \Big].$$

Proof. Since $N \cap S(\cdot)$ is constant and the errors $Q_{CRV}^T \delta_e$ as well as $\bar{Q}_{V-C}^T \delta_V$ reflect the errors in those constraints which have to be differentiated (in order to obtain the hidden constraints), we may apply Theorem 4.2 of [22] and obtain the assertion. For the details, follow the proof of Theorem 19. □

Remark 25. Theorem 24 shows that the BDF methods applied to network equation systems are stable if the circuit does not contain L-I cutsets and C-V loops. In general, they are is weekly unstable.

Corollary 26. *Suppose the errors in the initial values are $O(h^k)$ as well as the errors in the Newton iteration satisfy $O(h^k)$ for all equations and $O(h^{k+1})$ for the equations*

$$Q_{CRV}^T(A_L j_L + A_I i(t)) = 0, \tag{33}$$

$$\bar{Q}_{V-C}^T(A_V^T e - v(t)) = 0. \tag{34}$$

Then, the k-step BDF method applied to network equation systems (10)-(14) is convergent and globally accurate to $O(h^k)$.

This is a simple conclusion from Theorem 24 if one regards that the errors $Q_{CRV}^T \delta_e$ and $\bar{Q}_{V-C}^T \delta_V$ correspond to the errors solving the equations (33) and (34) respectively.

Remark 27. Since, the equations (33) and (34) are linear, they are always solved quite accurate by using Newtons method. This explains that the BDF method often works well in praxis independently of the index.

7 Consequences for the Model Design

For all the results in the sections, the circuits have to satisfy some local criteria. On the one hand they have to fulfill the positive definiteness criteria, i.e., $C(v,t)$ is pos. def. for all capacitances, $L(j,t)$ is pos. def. for all inductances, $G(v,t)$ is pos. def. for all resistances whose nodes are not connected by CV-paths, and, on the other hand, the position criteria for controlled sources (see the Assumptions 11 and 12).
For linear time-invariant capacitances, inductances, and resistances, the criteria are obviously fulfilled. For capacitances and inductances with affine characteristics, the positive definiteness means that they are strictly locally passive (cf. [10]). More complex models like MOSFETs should always be constructed in such a way that the capacitive part is positive definite. All active elements should be considered as controlled sources.
The position criteria for controlled sources may always be fulfilled by introducing some regularizing/parasitic capacitances, inductances and resistances. The following rules are one (but not the only one) possibility to guaranty this.

If a controlled voltage source

- belongs to a C-V loop then include a resistance or an inductance into the loop;
- is voltage controlled and not controlled by
  - a capacitance or
  - an independent voltage source

then put a capacitance in parallel to the controlling element;
- is current controlled and not controlled by
  - an inductance or
  - an independent current source

  then put an inductance in series to the controlling element.

If a controlled current source

- belongs to an L-I cutset then put a resistance or a capacitance in parallel to the cutset;
- is voltage controlled and not controlled by
  - a capacitance,
  - an independent voltage source or
  - a resistance

  then put a capacitance in parallel to the controlling element;
- is current controlled and not controlled by
  - an inductance,
  - an independent current source,
  - a resistance or
  - a voltage source that does not form a part of a C-V loop

  then put an inductance in series to the controlling element;
- is controlled by the current of a voltage source and its nodes are not connected by a C-path then put an inductance in series to the VS;
- is controlled by the current of a resistance (whose nodes are not connected by a CV-path) and its nodes are not connected by a CV-path then put an inductance in series to the controlling resistance.

8 Conclusion

Exploitation of the special structure of network equations is beneficial for applying recent results of DAE theory to circuit simulation problems of industrial relevance and complexity. The relevant DAE subspaces of the circuit equations are simply to describe if one has determined all C-V loops and all L-I cutsets of the network. Using this information, it is possible to develop very efficient and reliable methods for

- calculating the DAE index of even very large circuits;
- identifying critical circuit configurations and providing suggestions for their regularization;
- computing consistent initial values;
- implementing a clean handling of user given initial conditions;
- calculating the transient solution by BDF methods.

A key issue of the approaches presented here is to combine *global topological* criteria – like the existence of C–V loops – with *local numerical* criteria – like the positive definiteness of the device stamps – thus combining the speed of graph oriented methods with the generality and reliability of numerical checks. The presented criteria give a suggestion how to construct future device and circuit models in order to avoid numerical problems due to too high DAE index as far as possible.

Acknowledgments

I am indebted to D. Estévez Schwarz for many fruitful discussions and to U. Feldmann for encouragement and suggestions. I would like to thank R. März and R. Lamour for valuable comments.

References

1. Brenan, K.E., Campbell, S.L., Petzold, L.R.: The Numerical Solution of Initial Value Problems in Ordinary Differential-Algebraic Equations. North Holland Publishing Co. (1989)
2. Chua, L.O., Desoer, Ch.A., Kuh, E.S.: Linear and nonlinear Circuits. McGraw-Hill Book Co., Singapore (1987)
3. Chua, L.O., Lin, P.-M. Computer-Aided Analysis of Electronic Circuits. Prentice Hall, Englewood Cliffs (1975)
4. Desoer, C.A., Kuh, E.S.: Basic Circuit Theory. McGraw-Hill, Singapore (1969)
5. Estévez Schwarz, D.: Consistent initialization for index-2 differential algebraic equations and its application to circuit simulation. Humboldt-Univ. Berlin, PhD Thesis (2000), published electronically under http://dochost.rz.hu-berlin.de/dissertationen/
6. Estévez Schwarz, D.: A step-by-step approach to compute a consistent initialization for the MNA. Submitted to Int. J. Circ. Theor. Appl.
7. Estévez Schwarz, D., Feldmann, U., März, R., Sturtzel, S., Tischendorf, C.: Finding Beneficial DAE Structures in Circuit Simulation. Institut für Mathematik, Humboldt-Univ. zu Berlin **00–7** (2000)
8. Estévez Schwarz, D., Lamour, R.: The computation of Consistent Initial Values for Nonlinear Index–2 Differential-Algebraic Equations. Institut für Mathematik, Humboldt-Univ. zu Berlin **99–13** (1999), to appear in Numerical Algorithms
9. Estévez Schwarz, D., Tischendorf, C.: Structural analysis of electric circuits and consequences for MNA. Int. J. Circ. Theor. Appl. **(28)** (2000) 131–162
10. Fosséprez, M.: Non-linear Circuits: Qualitative analysis of non-linear, non-reciprocal circuits John Wiley & Sons, Chichester (1992)
11. Gear, C.W.: Simultaneous Numerical Solution of differential-algebraic equations. IEEE Trans. Circuit Theory **CT-18 (1)** (1971) 89–95
12. Grigorieff, R.D.: Stability of multistep methods on variable grids. Numer. Math. **42** (1983) 359–377
13. Griepentrog, E., März, R.: Differential-algebraic equations and their numerical treatment. Teubner-Texte zur Mathematik **88** BSB B.G. Teubner Verlagsgesellschaft, Leipzig (1986)
14. Günther, M., Feldmann, U.: CAD based electric modeling in industry. Part I: Mathematical structure and index of network equations. Surv. Math. Ind. **8** (1999) 97–129
15. Hairer, E., Wanner, G.: Solving ordinary differential equations II: Stiff and differential-algebraic problems. Springer Series in Computational Mathematics **14** Springer–Verlag, Berlin, Heidelberg (1991)
16. Hasler, M.: Non-linear non-reciprocal resistive circuits with a structurally unique solution. Int. J. Cir. Theor. Appl. **14** (1986) 237-262

17. März, R.: Numerical methods for differential-algebraic equations. Acta Numerica (1992) 141–198
18. März, R., Tischendorf, C. Recent results in solving index 2 differential algebraic equations in circuit simulation IAM J. Sci. Stat. Comput. **18 (1)** (1997) 139–159
19. Nishi, T., Chua, L.O.: Topological criteria for non-linear resistive circuits containing controlled sources to have a unique solution. IEEE Trans. Circuits and Systems **CAS-31** (1984) 722-741
20. Reißig, G.: The index of the standard circuit equations of passive RLCTG-networks does not exceed 2. Proc. ISCAS'98 **3** (1998) 419–422
21. Shichman, H., Hodges, D.A.: Insulated-gate field-effect transistor switching circuits. IEEE J. Solid State Circuits **SC-3** (1998) 285-289
22. Tischendorf, C.: Solution of index-2 differential algebraic equations and its application in circuit simulation. Humboldt-Univ. Berlin, PhD Thesis (1996)
23. Tischendorf, C.: Topological index calculation of DAEs in circuit simulation. Surv. Math. Ind. **8(3-4)** (1999) 187–199
24. Vlach, J., Singhal, K.: Computer methods for circuit analysis and design. Van Nostrand Reinhold, New York (1994)

CHORAL — A One Step Method as Numerical Low Pass Filter in Electrical Network Analysis*

M. Günther[1], P. Rentrop[1], and U. Feldmann[2]

[1] Universität Karlsruhe (TH), Fachbereich Mathematik, Institut für Wissenschaftliches Rechnen und Mathematische Modellbildung (IWRMM), Engesserstr. 6, D-76128 Karlsruhe
[2] Infineon Technologies, Otto-Hahn-Ring 6, D-81739 München

Abstract Circuit simulation packages generate the network equations automatically. In time domain analysis this results in a system of differential-algebraic equations, which is solved numerically by BDF schemes and/or the trapezoidal rule. CHORAL, a charge-oriented Rosenbrock-Wanner method, has been developed as an alternative approach for digital circuits. By its successful implementation into TITAN, Infineon Technologies' circuit simulator, a second integration scheme is available for the first time. Results for benchmarks and industrial circuits show that CHORAL is competitive with the standard ansatz. A careful analysis shows that CHORAL can be interpreted as a numerical (non-ideal) low pass filter with all its beneficial properties: oscillations of physical significance are preserved, but highly oscillatory perturbations are damped out very rapidly.

1 Introduction

In circuit simulators the time domain analysis mode computes the time-dependent (transient) behavior of electrical signals responding to time varying input signals. The numerical circuit analysis is based on the network approach, see Hoschek et al. [14], which has proven to be a powerful tool in computer-aided analysis of various technical systems. A typical data flow of a circuit simulator like TITAN (Infineon Technologies) is outlined in Fig. 1.

- A network description of the circuit is generated automatically in computer-aided electronics-design systems (CAE) from designer's drafts or fabrication data files. To end up with a network model of the circuit, semiconductor devices like MOS transistors are replaced by network companion models which are built up of basic elements [6]. This is shown in Fig. 2. As an example we use the NAND-gate circuit presented in Fig. 2 which produces a low output signal V_{out} if both input signals V_{in1}, V_{in2} are higher than a certain threshold value, and a high output signal else, such realizing a logical *NOT-AND* operation of the input signals.

* invited paper

- Using Modified Nodal Analysis (MNA), an input processor translates this graphical – or a corresponding textual network description – into a data format representing the mathematical model of the system. The mathematical network equations are based on the application of basic physical laws like Kirchhoff's equations onto the network and insertion of the characteristic equations for the network elements.
- The network equations are solved in the analysis kernel by numerical integration schemes adapted to the type of the network equations.
- Finally, the simulation results are printed into an output file, which can be analyzed by a graphical postprocessor.

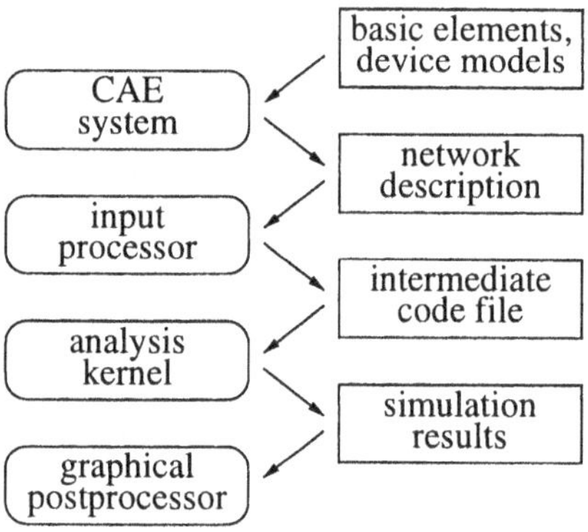

Figure1. Data flow of circuit simulation

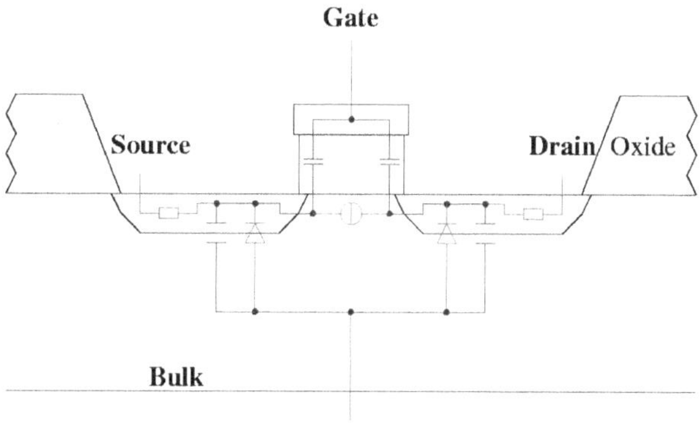

Figure2. Cross section of an MOS-transistor with associated companion network model

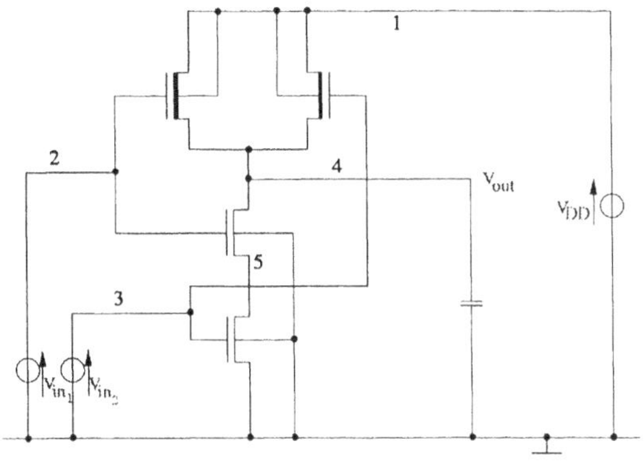

Figure3. Nand-gate circuit

The core of any circuit simulator is its analysis kernel which consists of the generation and numerical solution of the network equations, and model routines describing the characteristic relations of the basic elements. This will be discussed in the next section. CHORAL, an alternative to the usual BDF approach for numerical integration, and its implementation into existing simulation packages are described in Sect. 3. After discussing first numerical results we will see in Sect. 5 by inspecting two examples that CHORAL shows useful properties for oscillatory circuits and has favourable numerical damping qualities. To explain these results, the harmonic oscillator is introduced as a model equation. The corresponding analysis is performed in Sect. 7. We close our discussion with some concluding remarks.

2 The analysis kernel of a circuit simulator

As an alternative to conventional schemes, which describe dynamic effects by capacitors and inductors, a charge/flux oriented formulation has become a standard to build up the network equations, since it guarantees charge conservation during simulation. The automatic modeling approach preserves the topological structure of the network and therefore does not aim at state systems with a minimal set of unknowns. Hence an initial-value problem of differential-algebraic equations (DAEs) is generated, which reads

$$\mathcal{F}(\dot{q}, x, t) = A \cdot \underbrace{\begin{pmatrix} \dot{q}_1(x,t) \\ \dot{q}_2(x,t) \end{pmatrix}}_{=: \dot{q}(x,t)} + \underbrace{\begin{pmatrix} f_1(x) \\ f_2(x) \end{pmatrix}}_{=: f(x)} + \underbrace{\begin{pmatrix} s_1(t) \\ s_2(t) \end{pmatrix}}_{=: s(t)} = 0 \quad \text{on} \quad t \in [0,T] \qquad (1)$$

with consistent initial values $x(0) = x_0$. The system (1) can be described as follows: the vector $x \in \mathbb{R}^n$ of unknowns consists of all node potentials and branch currents through voltage-controlling elements. A denotes an incidence

Table1. Network description of nand gate

network description	explanations
`$$$$ CMOS NAND gate`	title line
`cload 4 0 100f`	capacitor between nodes 4 and ground (0)
`vdd 1 0 dc 5`	constant voltage source, node 1/0, 5V
`vin1 2 0 pwl( V A L U E S )`	piecewise linear time dependent
`vin2 3 0 pwl( V A L U E S )`	input voltage sources
`mpkan1 1 2 4 1 pmos l=2u w=20u`	PMOS transistor
`mpkan2 1 3 4 1 pmos l=2u w=20u`	PMOS transistor
`menh1 4 2 5 0 nmos l=2u w=20u`	NMOS transistor
`menh2 5 3 0 0 nmos l=2u w=20u`	NMOS transistor
`.model nmos nmos1 vto=0.8`	model for NMOS transistor
`.model pmos pmos1 vto=-0.8`	model for PMOS transistor
`.tran 1n 90n`	time domain analysis until 90ns
`.output tran v(*) i(vdd)`	output: all voltage and vdd-current
`.save nand`	name of output file
`* hmin hmax tol h`	stepsizes and tolerance
`.end`	end of input file

network element	notation	nodes	parameters
capacitor	C$xxxxxxx$	N+ N-	value in Farad
indep. voltage source	V$xxxxxxx$	N+ N-	value in Volt
MOS-transistor	M$xxxxxxx$	ND NG NS NB	model reference, length, width in meter

Table2. NAND-gate circuit

matrix, which describes the circuit topology; so its entries are only 0, +1 and −1. In general, A has *not* full row rank. The terminal charges q_1 and branch-fluxes q_2 of all energy-storing elements are nonlinear functions of branch voltages and currents. f_1 collects the currents, f_2 describes the voltage drop of the voltage-controlling elements. s_1 (s_2) contains the contributions of the independent current (voltage) sources.

The properties of the network equations depend not only on the type of the circuit and its parameters, but additionally on its modeling assumptions [6,7]. The network equations include characteristic time constants of several orders of magnitude (stiff equations) and suffer from poor smoothness properties of transistor model equations.

Since SPICE2 [17] most circuit simulators solve the network equations either with the trapezoidal rule or with *backward differentiation formulas* (BDF), which are shortly described here: for a time step h from t_{k-1} to $t_k = t_{k-1} + h$ the derivative $\dot{q}(x(t_k), t_k)$ in (1) is replaced by a backward difference (BDF) operator $\rho\, q_k$, which is defined by

$$\rho\, q_k := \frac{\alpha_0}{h} q(x_k, t_k) + \frac{1}{h} \sum_{i=1}^{l} \alpha_i q(x_{k-i}, t_{k-i})$$

with real coefficients α_i. The numerical solution of the DAE system (1) is thus reduced to the solution of a system of nonlinear equations

$$\mathcal{F}(\rho q_k, x_k, t_k) = 0.$$

Due to the structure of the nonlinear equations the Jacobian (iteration matrix) for Newton's method is

$$\frac{\alpha_0}{h} \cdot \mathcal{F}_{\dot{x}} + \mathcal{F}_x \quad \text{with} \quad \mathcal{F}_{\dot{x}} = A \cdot \left. \frac{\partial q(x,t)}{\partial x} \right|_{(x,t):=(x_k,t_k)}, \quad \mathcal{F}_x = \left. \frac{\partial f(x)}{\partial x} \right|_{x:=x_k}.$$

This system has a unique solution for any index, at least theoretically, if $\{\mathcal{F}_{\dot{x}}, \mathcal{F}_x\}$ is a regular matrix pencil.

BDF methods have been successfully adapted to charge-oriented network problems up to index-2 systems [15]. A bunch of *numerical tricks* is used to make BDF more powerful. Some of them are

- A warning for numerical index problems of the step size control can be given by an *index monitor*. It controls the step size predictions caused by index anomalies [19]. Usually, the implementation of such a monitor yields no difficult problems.
- Additional capacitors are used to regularize the circuit. This may lead to oscillations of high frequency [3].
- The TR-BDF approach tries to combine the large timestep of the trapezoidal rule (TR) – which is $\approx 40\%$ larger than that of BDF order 2 for getting same accuracy – and its property to conserve energy with the (moderate) damping properties of BDF [2].
- To gain efficiency, modified Netwon methods are used: the Jacobian evaluation and its LR decomposition (which may be expensive for large circuits) are reused for subsequent iterations in the same time step.
- Parts of a circuit are latent during some steps; this behavior is exploited by bypass strategies or even by *multi rate methods* [9]. Only the active part of a circuit is integrated accurately using small step sizes. The latent parts can be treated with larger steps.
- Due to the design, circuits possess a natural *partition*. Different areas of the chip are weakly coupled. Therefore, the large systems may be handled in a multilevel process by solving smaller subsystems with different schemes and/or in parallel on different CPUs.

In professional packages the conventional methods have achieved a high degree of maturity, and have proven to be efficient and very robust in an extremely large variety of applications. Nevertheless there is some motivation to look at alternative schemes also from an industrial point of view:

- The BDF methods are applicable to much more general classes of non-linear DAEs; can methods be superior, which are definitely constructed for the special linear-implicit nonlinear form (1) of the circuit equations?
- In the charge/flux oriented form of conventional codes, time step control is difficult, since charge/flux tolerances are not of interest for the user, and extra effort is necessary to derive charge/flux tolerances from the desirable user given node voltage or current tolerances.
  Are there methods with a more natural embedding of time step control even in charge oriented formulation?
- What about semi-implicit methods in circuit simulation?

Facing the extremly large variety of applications in circuit simulation, we do not believe that there is one single integration scheme which matches all requirements perfectly. However, all of the questions raised above can be answered in some positive sense by modified Rosenbrock Wanner schemes, as will be seen in the following sections.

3 Implementation of CHORAL in TITAN

Based on the class of embedded Rosenbrock-Wanner schemes, which have been used successfully for solving classical network equations [18], charge-oriented schemes [4,5] have been developed for charge-oriented network equations of type (1) up to index two. The numerical approximation x_k after one time step from t_{k-1} to $t_k = t_{k-1} + h$, together with an embedded approximation $\widehat{x}_k$ of lower order for error control and time step prediction, is given by

$$x_k = x_{k-1} + \sum_{i=1}^{s} d_i \kappa_i, \quad \widehat{x}_k = x_{k-1} + \sum_{i=1}^{s} \widehat{d}_i \kappa_i$$

with the increments κ_i computed by the linear system

$$\left(\frac{1/\gamma}{h}\mathcal{F}_{\dot{x}} + \mathcal{F}_x\right)\kappa_i = \frac{1/\gamma}{h} A\left(q(x_{k-1}) - q(a_i)\right) -$$
$$-\sum_{j=1}^{i} \widetilde{\beta}_{ij}\left[f(a_j) + s(t_{k-1} + h\alpha_j)\right] - \sum_{j=1}^{i-1} \widetilde{\beta}_{ij} \frac{\partial f}{\partial x}(x_{k-1})\kappa_j - h\widetilde{\gamma}_i \dot{s}(t_0)$$

using the internal stages

$$a_i := x_{k-1} + \sum_{j=1}^{i-1} \sigma_{ij}\kappa_j \, .$$

The coefficients of the method are free to fulfil order conditions and guarantee A- and L-stability, respectively. See [5] for more details. On account of the low smoothness properties of transistor models, as well as of usually moderate accuracy demands required by the user, an embedded method of order (2)3 seems to be most suitable. The corresponding scheme, CHORAL, has four stages and only three function evaluations due to $a_2 = a_3$. To avoid a constant term in the error estimate due to inconsistent initial values, both methods are chosen to be stiffly accurate [12]. The corresponding coefficient set of CHORAL with $\widetilde{\beta}_{ij} := \beta_{ij}/\gamma$ and $\widetilde{\tau}_i := \tau_i/\gamma$ is given in Tab. 3.

Table3. Coefficients for CHORAL

$\gamma = 0.5728160624821349$	$\beta_{11} = \gamma$
$d_1 = \widehat{d}_1 = \sigma_{21} = \sigma_{31} = \sigma_{41} = 1/\gamma$	$\beta_{21} = -2.0302139317498051$
$d_2 = \widehat{d}_2 = \sigma_{32} = \sigma_{42} = 0$	$\beta_{22} = \gamma$
$d_3 = \widehat{d}_3 = \sigma_{43} = 1.0$	$\beta_{31} = 0.2707896390839690$
$d_4 = 1.0$	$\beta_{32} = 0.1563942984338961$
$\alpha_1 = \widehat{d}_4 = 0$	$\beta_{33} = \gamma$
$\alpha_2 = \alpha_3 = \alpha_4 = 1.0$	$\beta_{41} = 2/3$
$\tau_1 = 0.3281182414375370$	$\beta_{42} = 0.08757666432971973$
$\tau_2 = -2.57057612180719$	$\beta_{43} = -0.3270593934785213$
$\tau_3 = -0.229210360916031$	$\beta_{44} = \gamma$
$\tau_4 = 1/6$	

Since the usual error estimate (see, e. g., [19]) for stiffly-accurate embedded ROW methods is used, a reliable error control and step size selection are offered, which are based on node potentials. No additional techniques are necessary to recompute errors from charges into node potentials.

The method is compatible with circuit simulation packages: model evaluation, direct sparse matrix solvers and parallel simulation already implemented in the numerical integration kernel can be used efficiently. Its implementation into the analysis kernel of TITAN is discussed in [13].

Before discussing first numerical results for CHORAL, we have to outline one main difference between CHORAL and TR-BDF: the use of the Jacobian information. Remember that the extreme sparseness of highly integrated circuits is reflected in the Jacobian, with two important consequences.

As we have seen before, semiconductor devices are replaced by so-called companion models to simulate integrated circuits numerically, i.e. electric circuits, which describe the input-output behavior of devices. To reflect important effects of second order, such models consist of complicated controlled sources with highly nonlinear characteristics and may involve hundreds of technical parameters [16]. Hence function evaluations, the computation of branch currents and node voltages of companion models at a given time point, are highly time consuming.

Since the linear algebra costs are comparatively low due to the use of sophisticated direct sparse solvers, the major portion of CPU time is in standard applications caused by the `LOAD` part, the evaluation of functions and

Jacobians. For example, typical values for critical paths of dynamic memories are: 80 % of the total CPU time is spent for the LOAD part, and only 8 % for the numerical solution of the nonlinear systems [2].

Secondly, it is only about 1.4 times more expensive to evaluate both right-hand side and Jacobian than only the right-hand side by its own. This surprising but practically confirmed result is typical for highly integrated circuits if element stamps are used to evaluate the time-discretized models for network elements [2] — in other words, when right-hand side and Jacobian are assembled element by element.

By its definition, TR-BDF could make use of this cheap Jacobian information, if one applies full instead of modified Newton. This is of course done in practice, as far as the problem size is not so large that the cost for the linear algebra becomes dominant. First steps on how to make efficient use of this cheap Jacobian information in CHORAL are discussed in [8,10] for ROW schemes applied to ODE systems: the cheap Jacobian can be used at each stage to introduce additional method coefficients which allow for a given order of convergence with less stages and/or function evaluations per time step.

4 First numerical results

By the successful implementation of CHORAL into TITAN, a second integration scheme is available in an industrial circuit simulation package for the first time. This opens the possibility to gain experience with numerous real life problems. CHORAL has been tested against the BDF approach in TITAN using a bunch of industrial applications: from standard benchmark examples in circuit simulation (LC oscillator and MOS ring oscillator for oscillatory circuits on the one hand, and a 16 bit adder for logical units on the other hand) to real up-to-date industrial applications: critical paths of dynamic memories (DRAMs) of 1 and 16 MBit, and an arithmetic logical unit, the core of a CPU. In these circuits we had to face with 0 up to 13000 transistors, and with 3 up to more than 30 000 network equations.

Table4. Numerical results: CHORAL versus BDF within TITAN

Circuit	# transistors	# equations	CPU time	
			CHORAL	BDF
LC oscillator	0	3	0.57s	0.33s
MOS ring oscillator	134	73	30.13s	27.61s
16 Bit adder	544	283	2m41.32s	2m30.1s
1 MBit DRAM	2005	1211	10m16.18s	8m29.15s
16 Mbit DRAM	5208	3500	23m37.18s	12m5.11s
arithmetic logical unit	13005	32639	97m31.64	82m21.03s

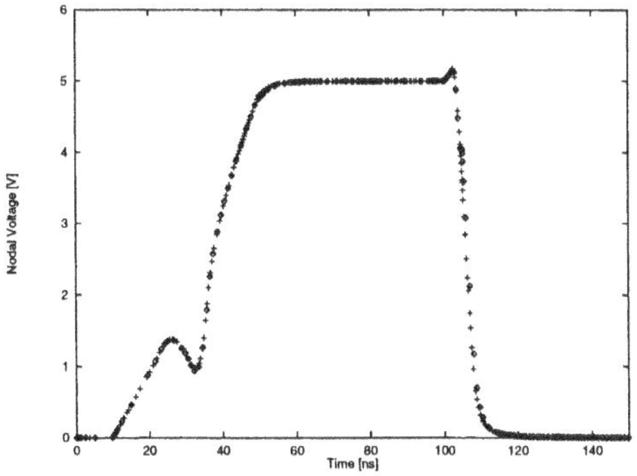

Figure4. One output nodal voltage for the 16 Bit adder: Integration steps of CHORAL (◇) vs. BDF (+).

In order to make results comparable, the CHORAL tolerances were adjusted such that the waveforms of an MOS ringoscillator match with those obtained using the conventional method using standard tolerances. Although the BDF-approach of TITAN had been tuned for more than ten years, CHORAL is comparable to BDF already after a first test implementation, both with respect to computational costs in CPU time (see the corresponding results in Tab. 4 for an HP work station C200 with RISC-processor), and accuracy. One reason for the efficiency of CHORAL seems to be the step size and error control that allow large step sizes by only a few failures of the step size predictions, see Fig. 4.

In contrast to TR-BDF, CHORAL shows an additional merit: excitations and oscillations with physical significance are tracked, but perturbations are damped. This behavior will be carefully examined in the next section.

5 Oscillatory circuits and numerical damping

Dealing with oscillatory behavior, we have to distinguish between two types of oscillations. The first type is given by oscillations of physical significance which reflect the oscillatory property of the mathematical model and the circuit, hopefully, and should be preserved during numerical integration. The LC oscillator shown in Fig. 5 (left side) can serve as a basic example. This linear circuit consists of one capacitance $C = 4\,\mathrm{pF}$ and inductance $L = 1\,\mathrm{nH}$ in parallel driven by an initial current source $I_0 = 6\,\mathrm{A}$. Numerical approximations obtained by CHORAL and BDF2, the BDF scheme of order two, are given in Fig. 5 (right side) for the branch current through the inductor. The current oscillates with the amplitude given by I_0 and frequency $\omega = 1/\sqrt{LC}$, which corresponds to a period of $T = 2\pi/\omega \approx 0.4\,\mathrm{nsec}$. While an error becomes visible both in phase and amplitude for BDF2, both phase and amplitude are preserved by CHORAL.

The second type is given by highly oscillatory perturbations, which may stem from numerical noise caused by step size and error control or by an inappropriate semidiscretization of a PDE model with respect to space. A third cause are discontinuities of the solution, which might be caused by non smooth transistor models or input stimuli. No problems should occur if integration is stopped at these points to make a restart with consistent initial values. Here algorithms for consistent initialization based on structural arguments which have been derived for BDF schemes can be used efficiently for CHORAL, too [1]. However, if integration is not stopped at these points, one may have to deal with inconsistent initial values. When using multistep methods, at least timestep control will run into difficulties here. One example for such a system is given by the operational amplifier circuit. For the trapezoidal rule and CHORAL, for example, this jump in the solution causes a highly oscillatory perturbation with a frequency in the range of the actual step size. The corresponding numerical results are given in Fig. 6. Due to its energy conserving property, the trapezoidal rule has to track this highly oscillatory perturbation, which may yield the impression that the circuit becomes unstable. In contrast, the perturbation is damped immediately by CHORAL, and only a few steps are necessary to capture the smooth solution in the 'middle line'.

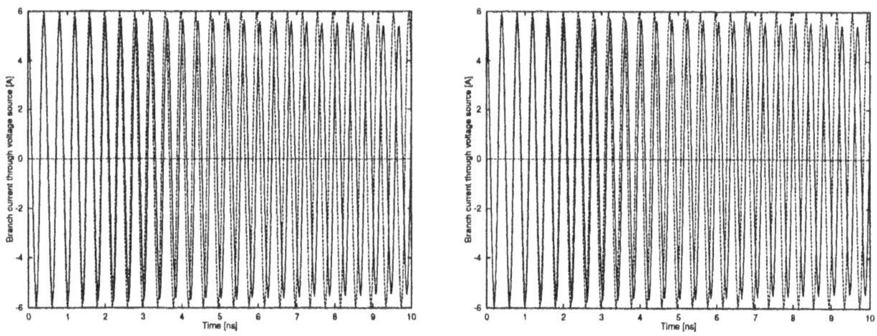

Figure5. LC oscillator (left) and simulation results (right) for BDF2 (—) and CHORAL (- - -)

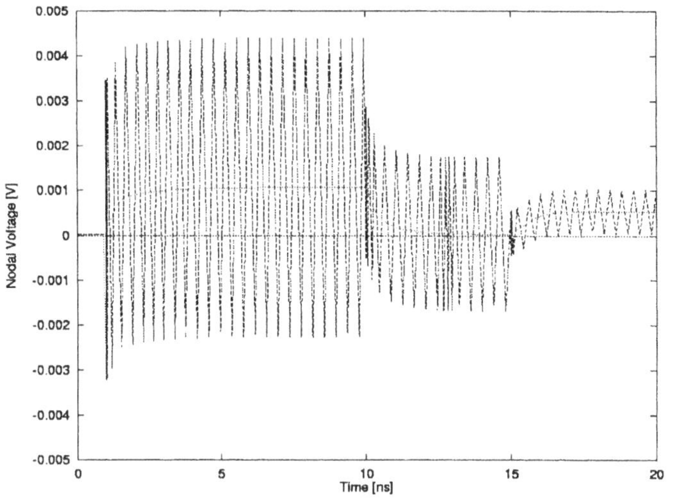

Figure6. Operational amplifier circuit: Simulation results for trapezoidal rule (- - -) and CHORAL ($\cdots$)

CHORAL preserves well the properties of oscillatory circuits: physical oscillations are tracked, but artificial oscillations of high frequency caused by numerical noise are damped out very rapidly. Hence CHORAL combines the advantages of TR-BDF within one single method.

6 Model equation: harmonic oscillator

To explain these results for both physical and artificial oscillations, we investigate the model equation

$$\ddot{x} + \omega^2 x = 0 \tag{2}$$

of a harmonic oscillator with frequency ω over one period $[0, T := 2\pi/\omega]$. With initial values $x(0) = x_0$, $\dot{x}(0) = \dot{x}_0$, the solution reads

$$x(t) = r \cdot \operatorname{Re}\exp(i\varphi) \tag{3}$$

with

$$r := \sqrt{x_0^2 + \dot{x}_0^2/\omega^2}, \quad \varphi := \omega t - \arctan(\dot{x}_0/(\omega x_0)).$$

Note that the LC oscillator discussed above corresponds to a harmonic oscillator with frequency $\omega = 1/\sqrt{LC}$.

This model equation is the special case $\varphi \equiv 0$ of the system

$$\ddot{x} = -\omega^2(x - \varphi) + \ddot{\varphi} = 0, \quad x(0) = \varphi(0) + \xi, \quad \dot{x}(0) = \dot{\varphi}(0) + \gamma$$

with given smooth function φ introduced by van der Houwen and Sommeijer [21] and used by Simeon [20] to investigate oscillatory perturbations of smooth solutions and their numerical treatment within elastic multibody systems.

The results obtained on model equation (2) with initial values $(x(0),\, \dot{x}(0))^\top = (1,\, 0)^\top$ for BDF2 and the trapezoidal rule, the integration schemes TR-BDF is based on, and CHORAL are given in Figs. 7 and 8. For each method one period was resolved with *sample rate* $n = 1, 2, \ldots, 1000$ steps of equidistant step size $h = T/n$.

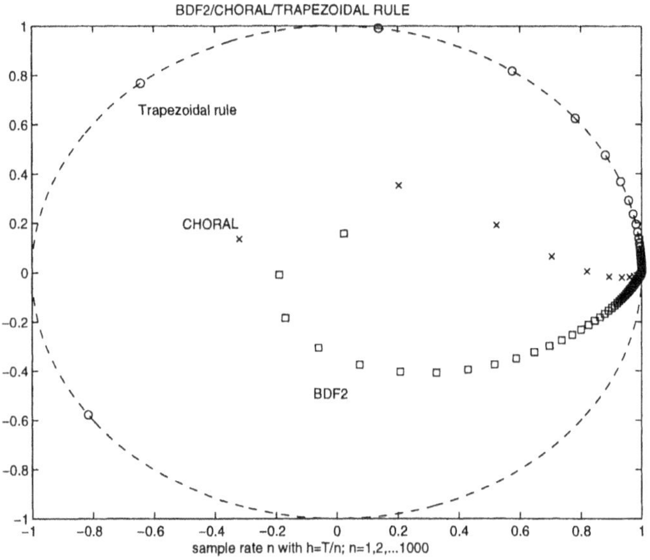

Figure7. Numerical approximation of BDF2 ($\square$), trapezoidal rule ($\bigcirc$) and CHORAL ($\times$) on model equation (2) after one period $T = 2\pi/\omega$. The results are plotted for step sizes $h = T/n$ $(n = 1, 2, \ldots, 1000)$ in phase space $x = r\exp(i\varphi)$.

Comparing the numerical approximations with the exact solution $(x(T),\, \dot{x}(T))^\top = (1,\, 0)^\top$ after one period, we see the following: due to its energy conserving property, no amplitude error occurs for the trapezoidal rule independent of n; however, for small sample rates one has to deal with rather large phase errors. BDF2 acts even worse: in addition to a phase error, one has to deal with amplitude errors, if one period is sampled too roughly. CHORAL, however, has only slight amplitude and phase errors even for rather small sample rates.

These results become more visible, if we zoom into the results for $n = 10, 11, \ldots, 20$. As a rule-of-thumb in circuit simulation, one has to sample one oscillation with approximately 10–20 points to get results which are accurate enough. Thus oscillations of physical significance which are approximated numerically using sample rates in the range of 10–20 yield rather large phase errors (trapezoidal rule) or both amplitude and phase errors (BDF2). CHORAL, however, is highlighted by only slight errors in phase and amplitude.

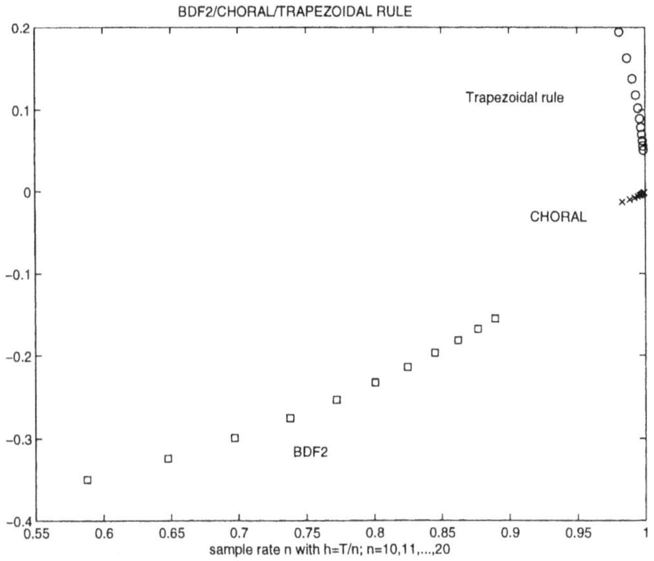

Figure8. Zoom into numerical approximation of BDF2 (□), trapezoidal rule (○) and CHORAL (×) on model equation (2) in phase space with sample rates $n = 10, 11, \ldots, 20$.

7 Analysis of one-step methods

These good properties of CHORAL applied to oscillatory circuits can be explained by investigating the model equation in more detail. Besides that, this analysis can illustrate its excellent damping properties as well. As a first step, we scale and rewrite (2) as an ODE system of first order. With $y := [x, \dot{x}/\omega]^\top$ we have

$$\dot{y} = Jy, \quad y(0) = \begin{pmatrix} x_0 \\ \dot{x}_0/\omega \end{pmatrix}, \tag{4}$$

where

$$J = \begin{pmatrix} 0 & \omega \\ -\omega & 0 \end{pmatrix}.$$

For one-step methods of Runge-Kutta and Rosenbrock-Wanner type such as trapezoidal rule and CHORAL, the numerical solution y_n^h after one period with n equidistant steps of size $h = T/n$ reads

$$y_n^h = [R(hJ)]^n \begin{pmatrix} x_0 \\ \dot{x}_0/\omega \end{pmatrix}.$$

The stability matrix $R(hJ)$ has eigenvalues $R(\pm i\omega h)$ which are given by evaluating the scalar stability function $R(z)$ for imaginary arguments $z = \pm i\omega h$. Furthermore, its eigenvectors are $(1, i)^\top$ and $(1, -i)^\top$, and thus it holds

$$[R(hJ)]^n = U \begin{pmatrix} R(z)^n & 0 \\ 0 & R(-z)^n \end{pmatrix} U^{-1}, \quad U = \begin{pmatrix} 1 & 1 \\ i & -i \end{pmatrix}.$$

Therefore the numerical properties of a one-step method applied to the model equation is fixed by its stability function along the imaginary axis:

$$y_n^h = U \begin{pmatrix} R(z)^n & 0 \\ 0 & R(-z)^n \end{pmatrix} U^{-1} \begin{pmatrix} x_0 \\ \dot{x}_0/\omega \end{pmatrix},$$

see Fig. 9. Note that we have

$$\lim_{z \to 0} R(z) = 1$$

for convergent methods, and

$$\lim_{z \to \pm\infty} R(z) = 0,$$

for L-stable methods [12]. Thus there is a range of small step sizes where $|R(z)|$ is close to one and information is almost preserved, and another range where $|R(z)|$ tends to zero and strong damping prevails.

Depending on the type of oscillation, we demand different properties:

- *Oscillations of physical significance.* These should be preserved. Assuming a sample rate of 10–20 steps for oscillations of physical significance, we demand $|R(z)| \approx 1$ in the range of $|z| \in [0.1\pi, 0.2\pi]$.
  Having a look at Fig. 9, we see that this demand is fulfilled by all methods but the implicit Euler scheme.
- *Perturbations.* Such oscillations of high frequency, either numerical noise caused by time step and error control, inconsistent initial values or by an inappropriate semidiscretization of a PDE model, should be damped as much and soon as possible. Hence a slight damping — independent of the order of the method — should already occur for $|z|$ larger than $\approx 0.5\pi$, which corresponds to 4 sampling points for each cycle, and $|R(z)| \approx 0$ for highly oscillatory signals, i.e. $|z| > 100$.
  Except the trapezoidal rule, which is not L-stable, all methods show good damping properties for highly oscillatory signals ($|z| > 100$). But only the implicit Euler scheme and CHORAL damp already for $|z| > 0.5\pi$ significantly.

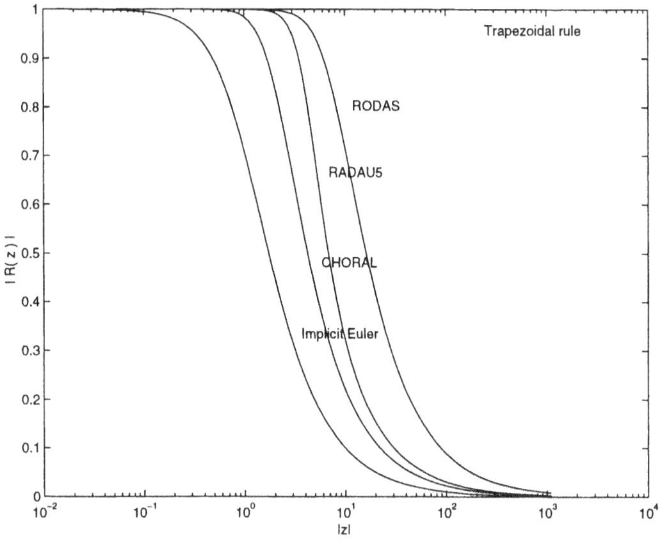

Figure9. Decay of stability functions along the imaginary axis (from left to right: implicit Euler, CHORAL, RADAU5, RODAS, and trapezoidal rule with $|R(z)| \equiv 1$).

Summing up, CHORAL shows all the desired properties of a (non-ideal) numerical low pass filter: physical oscillations of low frequency are preserved, but highly oscillatory perturbations are efficiently damped.

The corresponding analysis for multi-step methods can be found in the appendix.

8 Conclusion

Charge oriented ROW methods like CHORAL are promising alternatives for conventional multistep methods in industrial circuit simulation packages. Starting from an existing test implementation in TITAN, it is now the part of industry to improve efficiency and robustness, and to gain knowledge about the circuit classes for which the method can be recommended, and for which classes improvements are necessary. This may lead to the development of a bunch of different integration schemes tailored to different classes of circuits, and its implementation in circuit simulation packages is a desirable aim for future research.

References

1. Estévez Schwarz, D., Feldmann, U., März, R., Sturtzel, S., Tischendorf, C.: Finding benificial DAE structures in circuit simulation. Submitted for publication.
2. Feldmann, U.; Wever, U.; Zheng, Q.; Schultz, R.; Wriedt, H.: Algorithms for modern circuit simulation. AEÜ 46, 274–285 (1992).
3. Feldmann, U., Günther, M.: Some remarks about regularization of circuit equations. Proc. ISTET'99, Sept. 6.-9. 1999 (Magdeburg), 343–348.
4. Günther, M.: Ladungsorientierte Rosenbrock-Wanner-Methoden zur numerischen Simulation digitaler Schaltungen. VDI Verlag, Düsseldorf, 1995.
5. Günther, M.: Simulating digital circuits numerically – a charge-oriented ROW approach. Num. Math. 79, 203–212 (1998).
6. Günther, M.; Feldmann, U.: CAD based electric circuit modeling I: mathematical structure and index of network equations. Surv. Math. Ind. 8 97–129 (1999).
7. Günther, M.; Feldmann, U.: CAD based electric circuit modeling II: impact of network structure and parameters. Surv. Math. Ind. 8, 131–157 (1999).
8. Günther, M.; Hoschek, M.: ROW methods adapted to electric circuit simulation packages. J. Comp. Appl. Math. 82, 159-170 (1997).
9. Günther, M.; Hoschek, M.; Rentrop, P.: Differential-Algebraic Equations in Electric Circuit Simulation. Int. J. Electron. Commun. (AE) 54, 101-107 (2000).
10. Günther,M.; Hoschek, M.; Weiner, R.:. ROW methods adapted to a cheap Jacobian. To appear in Appl. Numer. Math.
11. Günther,M.; Kværnø, A.; Rentrop, P.: Multirate partitioned Runge-Kutta methods. To appear in BIT.
12. Hairer, E.; Wanner, G.: Solving ordinary differential equations II. Stiff and differential-algebraic problems. Springer-Verlag, Berlin, 1991.
13. Hoschek, M.: Einschrittverfahren zur numerischen Simulation elektrischer Schaltungen. VDI Verlag, Düsseldorf, 1999.
14. Hoschek, M.; Rentrop, P.; Wagner, Y.: Network approach and differential-algebraic systems in technical applications. Surv. Math. Ind. 9, 49–76 (1999).
15. März, R., Tischendorf, C.: Recent results in solving index 2 different algebraic equations in circuit simulation. SIAM. J. Sci. Comp. 18(1), 139–159 (1997).
16. G. Massobrio and P. Antognetti. Semiconductor device modelling with SPICE. McGraw-Hill, New York, 1993.
17. Nagel, W.: SPICE 2 – a computer program to simulate semiconductor circuits. Dissertation. Berkeley, CA: UC Berkeley, 1975.
18. Rentrop, P.: ROW-type methods for the integration of electric circuits. In Bank, R. et al. (Eds): Mathematical modelling and simulation of electrical circuits and semiconductor devices, Basel, Birkhäuser Verlag, 59–71 (1990)
19. Rentrop, P.; Roche, M.; Steinebach, G.: The application of Rosenbrock–Wanner type methods with stepsize control in differential–algebraic equations. Num. Math. 55, 545–563 (1989).
20. Simeon, B.: Order reduction of stiff solvers at elastic multibody systems. Appl. Numer. Math. 28, 459-475 (1998)
21. van der Houwen, P.J., Sommeijer, B.P.: Explicit Runge-Kutta-Nyström methods with reduced phase errors for computing oscillating solutions. SIAM J. Numer. Anal. 24, 595–617 (1987).

Appendix: Analysis of multi-step methods

It remains to compute the numerical approximation of the model equation obtained by multi-step methods. Since only BDF schemes of low order are used in circuit simulation, we restrict the analysis to BDF2 in the following.

Applying BDF2 to the model equation (4) with equidistant steps of size $h = T/n$ leads to

$$\left(I - \frac{2}{3}hJ\right) y_{j+2}^h - \frac{4}{3}y_{j+1}^h + \frac{1}{3}y_j^h = 0 \tag{5}$$

with y_j^h denoting the numerical approximation of y at $t_j = j \cdot h$.

By multiplication of (5) with $[(I - 2hJ/3) \cdot U]^{-1}$ the system can be written in decoupled form

$$\bar{y}_{j+2}^h - \frac{4}{3}D\bar{y}_{j+1}^h + \frac{1}{3}D\bar{y}_j^h = 0, \tag{6}$$

where

$$\bar{y}_j^h := U^{-1}y_j^h, \quad U = \begin{pmatrix} 1 & 1 \\ i & -i \end{pmatrix}, \quad D = \begin{pmatrix} \lambda_1 & 0 \\ 0 & \lambda_2 \end{pmatrix}, \quad \lambda_{1/2} = \frac{1 \pm 2ih\omega/3}{1 + (2h\omega/3)^2}.$$

This defines two difference equations with characteristic polynomials

$$\psi_1 = z^2 - \frac{4}{3}\lambda_1 z + \frac{1}{3}\lambda_1, \quad \psi_2 = w^2 - \frac{4}{3}\lambda_2 w + \frac{1}{3}\lambda_2,$$

with roots

$$z_{1/2} = \frac{2}{3}\lambda_1 \pm \sqrt{(2\lambda_1/3)^2 - \lambda_1/3}, \quad w_{1/2} = \frac{2}{3}\lambda_2 \pm \sqrt{(2\lambda_2/3)^2 - \lambda_2/3}.$$

Thus the general solution after one period reads

$$y_n^h = U\bar{y}_n^h = U \begin{pmatrix} k_1 & k_2 & 0 & 0 \\ 0 & 0 & l_1 & l_2 \end{pmatrix} \begin{pmatrix} z_1^n \\ z_2^n \\ w_1^n \\ w_2^n \end{pmatrix},$$

where the constants k_1, k_2, l_1 and l_2 are determined by the initial values and the starting procedure. Using $y(0) = (1, 0)^\top$ and the implicit Euler scheme as starter, for example, yields the linear system

$$\begin{pmatrix} 1 & 1 & 1 & 1 \\ i & i & -i & -i \\ z_1 & z_2 & w_1 & w_2 \\ iz_1 & iz_2 & -iw_1 & -iw_2 \end{pmatrix} \cdot \begin{pmatrix} k_1 \\ k_2 \\ l_1 \\ l_2 \end{pmatrix} = \begin{pmatrix} 1 \\ 0 \\ 1/[1 + (\omega h)^2] \\ -h\omega/[1 + (\omega h)^2] \end{pmatrix}.$$

Periodic Steady-State Analysis of Free-running Oscillators

S.H.M.J. Houben, J.M. Maubach

Eindhoven University of Technology

1 Introduction

A common problem in the simulation of electric circuits for RF (Radio Frequency) applications is finding a periodic steady-state (PSS) of such a circuit. Several approaches exist for solving this problem. For non-autonomous circuits, i.e. circuits that are driven by an input source with an a priori known period T, many methods exist (see [1], [2], [3]). However, when dealing with autonomous circuits, the situation is less satisfactory. For an autonomous circuit, the period T becomes an additional unknown, which makes the resulting system under-determined. A common solution method is harmonic balance, a frequency-domain method (see [4], [5]). Harmonic balance performs well if the waveform to be computed contains mostly low harmonics, but it becomes very expensive if a large number of harmonics is present. Therefore, there has been much interest in hybrid (see [6]) and pure time-domain methods (see [4]), such as shooting or finite difference. However, convergence of these methods is often problematic; typically there is only convergence when the initial guess for the period T_0 is already very close to the actual solution T^*.

In this paper, two novel methods will be presented. The first, called Poincaré-map method, has very strong convergence properties, but converges only linearly for many real-world circuits. The second, the accelerated Poincaré-map method, converges super-linearly when the Poincaré-map method converges linearly, but has somewhat weaker convergence properties. Some numerical results comparing both methods will be presented.

The algorithms discussed in this paper are described in more detail in the upcoming paper [7]. This paper also contains more numerical experiments and a discussion of how differential-algebraic equations (DAE) can be handled. In this paper, we restrict ourselves to ordinary differential equations (ODE).

2 Periodic steady-state

Definition 1. Consider an autonomous ordinary differential equation (ODE) of the form:

$$\frac{d\mathbf{x}}{dt} = \mathbf{f}(\mathbf{x}), \quad \mathbf{x} \in \mathbf{R}^n. \tag{1}$$

A function $\mathbf{x} : \mathbf{R} \to \mathbf{R}^n$ is called a *periodic steady-state* (PSS) of (1) if:

1. $\mathbf{x}$ is a solution to (1).
2. $\mathbf{x}$ is periodic, i.e. there is a $T > 0$ such that for all $t \in \mathbf{R}$, $\mathbf{x}(t) = \mathbf{x}(t+T)$.

Note that according to this definition, a stationary solution, i.e. a solution of the form $\mathbf{x}(t) \equiv \mathbf{x}_0$, is also a PSS.

Definition 2. The *limit cycle* $C(\mathbf{x})$ of a PSS $\mathbf{x}$ is the range of the function $\mathbf{x}(t)$, i.e.

$$C(\mathbf{x}) = \{\mathbf{x}(t) \mid t \in \mathbf{R}\}. \tag{2}$$

A set C is called a limit cycle of (1) if there is a PSS $\mathbf{x}$ of (1) so that $C = C(\mathbf{x})$.

Definition 3. A periodic steady-state $\mathbf{x}$ is called *stable*[1] if there is a $\delta > 0$ so that the following holds: For every solution $\mathbf{x}^*$ to (1) which has the property that

$$\exists_{\tau_1 > 0} ||\mathbf{x}^*(0) - \mathbf{x}(\tau_1)|| < \delta, \tag{3}$$

there exists a $\tau_2 > 0$ so that

$$\lim_{t \to \infty} ||\mathbf{x}^*(t)(0) - \mathbf{x}(t + \tau_2)|| = 0 \tag{4}$$

A limit cycle is called stable when one of its periodic steady-states is stable; for an ODE of the form (1), this implies that **all** of its periodic steady-states are stable.

In this paper, we will concentrate on methods for finding a stable periodic steady-state. Periodic steady-states that are not stable are not interesting for the IC designer, since they do not correspond to any physical behaviour of the modelled circuit. In fact, we want to actively avoid non-stable periodic steady-states for this reason.

3 Autonomous oscillating circuits

The circuits in which we are interested are so-called *autonomous* or *free-running* oscillators. Such oscillators have the property that they do not have any time-dependent input signals. This implies that they can be mathematically described by an *autonomous* ordinary differential equation or differential-algebraic equation. In this paper, we restrict ourselves to circuits that can be described with an autonomous ODE of the form (1). An example of such an oscillator is given in figure 1. The equations describing this particular circuit are:

$$\frac{dv}{dt} = \frac{1}{C} i, \qquad \frac{di}{dt} = \frac{-1}{L} v. \tag{5}$$

[1] some authors prefer the term *strongly stable*

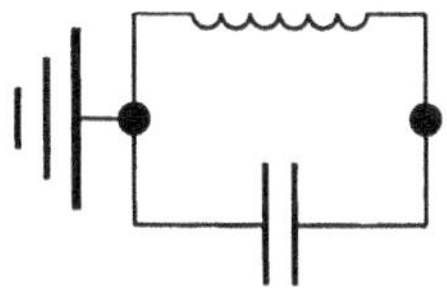

Figure1. The LC-ring is a very simple free-running oscillator. It consists of a capacitor and an inductor.

Note that these equations are linear. For initial conditions $v(0) = v_0$, $i(0) = i_0$, the solution to (5) is given by:

$$v(t) = v_0 \cos \omega t + \sqrt{\frac{L}{C}} i_0 \sin \omega t, \tag{6a}$$

$$i(t) = i_0 \cos \omega t - \sqrt{\frac{C}{L}} v_0 \sin \omega t, \tag{6b}$$

where $\omega = 1/\sqrt{CL}$. For this problem, it isn't difficult to find a PSS, since *every* solution $\mathbf{x}$ of (5) is periodic and hence a PSS. However, none of these PSS is stable.

A nonlinear example of a free-running oscillator is given by the following equations:

$$\frac{dx}{dt} = y + h(\sqrt{x^2 + y^2})x, \tag{7a}$$

$$\frac{dy}{dt} = -x + h(\sqrt{x^2 + y^2})y, \tag{7b}$$

The function h is chosen so that:

1. h continuous and differentiable.
2. $h(0) > 0$.
3. there are several points $r_k > 0$ so that $h(r_k) = 0$.

Possible choices of h include $h(r) = \cos r$, and $h(r) = \varepsilon(1 - r)$.

The problem (7) has the following properties:

1. It has at least one PSS solution, namely the stationary state with $\mathbf{x} = \mathbf{0}$. However, this solution is unstable.
2. For every $r_k > 0$ with $h(r_k) = 0$, we have that the circle described by $x^2 + y^2 = r_k^2$ is a limit cycle. Moreover, if $h'(r_k) < 0$, then the limit cycle is stable.
3. As $h'(r_k) \to 0$ from below, the limit cycle $x^2 + y^2 = r_k^2$ becomes a weaker and weaker attractor for nearby solutions of (5).

As we will see later, finding a stable PSS becomes more difficult when the PSS behaves only as a weak attractor, i.e. convergence towards the PSS is very slow.

4 The Poincaré-map method

The Poincaré-map method is based on the following observation: starting sufficiently close to a stable limit cycle $\mathcal{C}$, a transient simulation will eventually converge towards $\mathcal{C}$. After all, this is implied in the definition of a stable limit cycle. There are, however, two disadvantages to this approach:

1. We have to find a way to detect if we have approached the PSS close enough. If T is known, a "running window" can be used, i.e. the value $\mathbf{x}(t)$ at the current integration time t is compared to the value at $\mathbf{x}(t-T)$. However, T is an unknown in the autonomous case.
2. Convergence will be linear at best, which means that excessive computing time is needed to arrive at the solution.

The first problem will be addressed in this section, leading to the (unaccelerated) Poincaré-map method. The second problem will be addressed in the next section when considering the Accelerated Poincaré-map method.

The length of the period can be estimated by looking for periodic recurring features in the computed circuit behaviour. A possible recurring feature is the point at which a specific condition (the so-called *switch condition*) is satisfied. This is equivalent to carrying out a Poincaré-map iteration. The

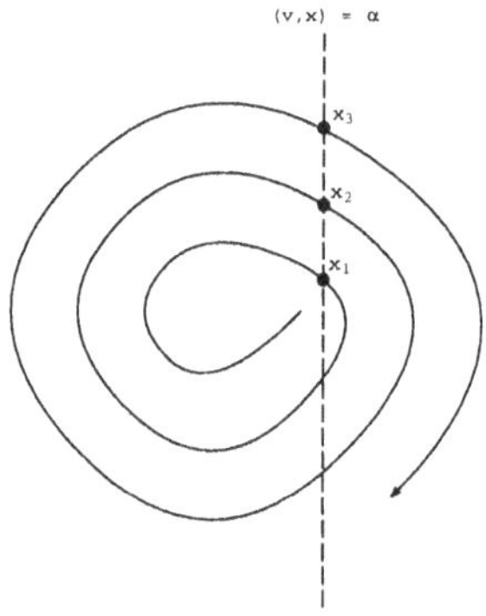

Figure2. The trajectory of a solution $\mathbf{x}(t)$ to the Initial Value Problem. The points $\mathbf{x}_n$ are chosen so that they satisfy $(\mathbf{x}, \mathbf{v}) = \alpha$, for some given $\mathbf{v}$ and α.

switch condition has to be chosen in such a way that the solution becomes locally unique. Moreover, the switch condition has to be satisfied at some point during the periodic steady state. In [7], some heuristics for finding a suitable switch condition are given.

The unaccelerated Poincaré-map method can now be described as follows.

Algorithm 1. *Provide the algorithm with the following inputs: an initial state* $\mathbf{x}_0$*, a switch condition of the form* $(\mathbf{v}, \mathbf{x}(t)) = \alpha$*, and a tolerance* $\varepsilon > 0$.

The algorithm will iteratively produce approximations for the period T and for a point on the periodic waveform $\mathbf{x}^*$.

1. *Set $i \leftarrow 0$ and $t_0 \leftarrow 0$.*
2. *Starting with $t = t_i$, $\mathbf{x}(t_i) = \mathbf{x}_i$, integrate (1) until $(\mathbf{v}, \mathbf{x}(t)) = \alpha$ and $d(\mathbf{v}, \mathbf{x}(t))/dt > 0$.*
3. *Set $\mathbf{x}_{i+1} \leftarrow \mathbf{x}(t)$ and $t_{i+1} \leftarrow t$.*
4. *Compute $\delta_i := \|\mathbf{x}_{i+1} - \mathbf{x}_i\|$. If $\delta_i > \varepsilon$, set $i \leftarrow i+1$ and proceed to step 2. If $\delta_i \leq \varepsilon$, proceed to step 5.*
5. *Set $T \leftarrow t_{i+1} - t_i$ and $\mathbf{x}^* \leftarrow \mathbf{x}_{i+1}$. Done.*

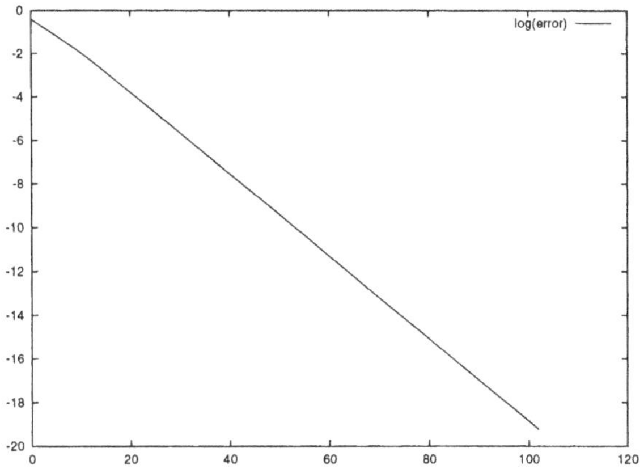

Figure3. log(error) after each iteration for the Poincaré-map method applied to (7) with $h(r) := \varepsilon(1-r), \varepsilon = 3 \cdot 10^{-2}$.

This method seems promising for two reasons:

1. It has rather good convergence properties.
2. It is simple to implement in an existing simulator, since it can essentially be considered as a post-processing step to an ordinary transient simulation.

5 The accelerated Poincaré-map method

The Poincaré-map method essentially leads us to find the fixed point of a function $F : \mathbf{R}^n \to \mathbf{R}^n$. This function F can be formally defined as:

$$F(\mathbf{x}_0) := \mathbf{x}(T), \tag{8}$$

where $\mathbf{x}(t)$ is the solution of (1) with $\mathbf{x}(0) = \mathbf{x}_0$, and T is the smallest $t > 0$ such that $(\mathbf{v}, \mathbf{x}(t)) = \alpha$ and $d(\mathbf{v}, \mathbf{x}(t))/dt > 0$. Given $\mathbf{x}_0$, the vector $F(\mathbf{x})$ can

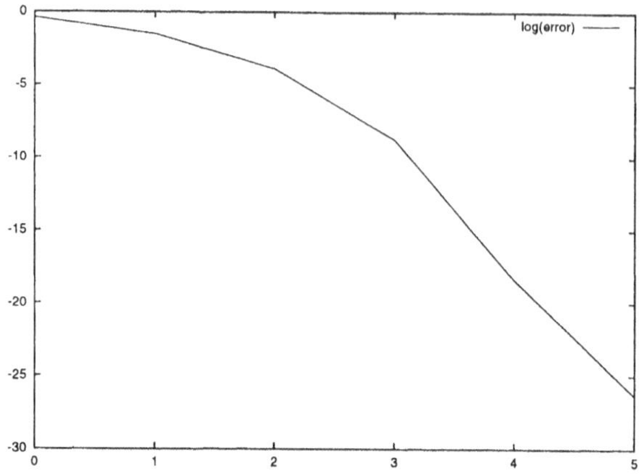

Figure4. log(error) after each outer loop iteration for the accelerated Poincaré-map method applied to (7) with $h(r) := \varepsilon(1-r), \varepsilon = 3 \cdot 10^{-2}$.

effectively be computed by using Algorithm 4, i.e. by applying the ordinary Poincaré-map method.

The successive approximations of the Poincaré-map method satisfy the recursion:

$$\mathbf{x}_{n+1} = F(\mathbf{x}_n) \tag{9}$$

Note that the n-th iteration $\mathbf{x}_n$ does *not* include the period T. Suppose that this sequence converges linearly to some fixed point $\mathbf{x}^*$ of F. As said in the previous section, convergence might be slow. Hence we are interested in accelerating convergence using an acceleration method.

An acceleration method operates on the first k vectors of a sequence $\{\mathbf{x}_n\}$, and produces an approximation $\mathbf{y}$ to the limit of $\{\mathbf{x}_n\}$. This approximation can then be used to restart (9) and generate the beginning of a new sequence $\mathbf{y}_0, \mathbf{y}_1, \mathbf{y}_2, \ldots$. Again, the acceleration method can be applied to this new sequence, resulting in a new approximation $\mathbf{z}$ of the limit. The idea is that the sequence $\mathbf{x}, \mathbf{y}, \mathbf{z}, \ldots$ converges much faster to the limit of $\{\mathbf{x}_n\}$ than the sequence $\{\mathbf{x}_n\}$ itself. Typically, if $\{\mathbf{x}_n\}$ converges linearly, then $\{\mathbf{x}, \mathbf{y}, \mathbf{z}, \ldots\}$ converges super-linearly.

We applied the accelerated Poincaré-map method based on the well-known minimal polynomial extrapolation (MPE) method. Rather than describing MPE here in detail, the reader is referred to [8]. With MPE, we obtain a super-linear converging sequence , provided that the original sequence produced has in the limit linear convergence. This is typically the case for periodic circuits, and it is in particular the case for our test problems.

6 Numerical results

In this section we compare the ordinary Poincaré-map method and the accelerated Poincaré-map method. They have been applied to problem (7), where h was chosen as $h(r) := \varepsilon(1 - r)$. A switch condition of the form $y = 0$ was taken. For all $\varepsilon > 0$, the resulting problem has exactly one stable limit cycle, namely the unit circle. The parameter $\varepsilon > 0$ affects the speed of convergence towards the limit cycle; as ε approaches 0, speed of convergence also goes to 0.

The number of iterations needed for decreasing values of ε is shown in Table 1. From this, it is easy to see that the unaccelerated Poincaré-map

Table1. The numbers of iterations (i.e. the number of evaluations of F) needed by both methods for decreasing values of ε

ε	$1 \cdot 10^{0}$	$1 \cdot 10^{-1}$	$3 \cdot 10^{-2}$	$1 \cdot 10^{-2}$	$3 \cdot 10^{-3}$	$1 \cdot 10^{-3}$
Unaccel. Poincaré	6	35	104	290	899	2516
Accel. Poincaré	7	11	17	17	20	23

method becomes impractical when ε approaches 0. On the other hand, the accelerated Poincaré-map method performs well even for very small values of ε.

For $\varepsilon = 3 \cdot 10^{-2}$, the errors after each iteration for both methods have been plotted in Figures 3 and 4. Note that in Figure 4, only the error after each outer loop iteration has been plotted, whereas the number in Table 1 indicates the total number of iterations.

From Figures 3 and 4, it is clear that the unaccelerated Poincaré-map method gives linear convergence, whereas accelerated Poincaré gives super-linear convergence for this test problem.

7 Conclusions

The following conclusions can be drawn:

1. Unaccelerated Poincaré is impractical for finding the PSS, because of its slow convergence.
2. Unaccelerated Poincaré probably is a good way to generate an initial approximation. After that, we might switch to the accelerated Poincaré method.
3. Accelerated Poincaré gives super-linear convergence towards the solution in all the test cases.
4. Both methods are simple to implement in existing simulators, since they can be implemented as a post-processing step to an ordinary transient simulation. Implementation details for both methods are given in [7].

References

1. T. J. Aprille and T.N. Trick, Steady State Analysis of Nonlinear Circuits with Periodic Inputs, Proceedings IEEE 1972 Vol.60 No.1 pages:108-114
2. Stig Skelboe, Time-Domain Steady-State Analysis of Nonlinear Electrical Systems, Proceedings of the IEEE 1982 Vol. 70 No.10 pages:1210-1228
3. Telichevesky, R. and Kundert, K. and Elfadel, I. and White, J., Fast Simulation Algorithms for RF Circuits, Proceedings of the IEEE 1996 Custom Integrated Circuits Conference 1996 pages:437-444
4. Ken Kundert, Simulation Methods for RF Integrated Circuits, Proceedings of ICCAD'97 1997
5. E.J.W. ter Maten, Numerical methods for frequency domain analysis of electronic circuits, Survey on Mathematics for Industry 1999 Vol.8 pages: 171-185
6. Semlyen, A. and Medina, A., Computation of the Periodic Steady State in Systems with Nonlinear Components using a Hybrid Time and Frequency Domain Method, IEEE Transactions on Power Systems 1995 Vol.10 No.3 pages:1498-1504
7. Stephan H.M.J. Houben and Joseph M. Maubach, Periodic Steady-State Analysis of Free-running Oscillators, pre-print 2000
8. David A. Smith and William F. Ford and Avram Sidi, Extrapolation Methods for Vector Sequences, SIAM Review, 1987 Vol.29 No.2 pages: 199-233

Efficient Analysis of Oscillatory Circuits

R. Neubert[1] and A. Schwarz[2]

[1] Infineon Technologies AG, MP PTS, 81730 München, Germany
[2] Technische Universität München, Zentrum Mathematik, 80290 München, Germany

Abstract Circuit simulation is a standard task in the computer-aided design of electronic circuits. The generation of the circuit equations leads to differential-algebraic equations (DAEs). With respect to the analysis of oscillatory circuits, periodic solutions are of major interest. For highly oscillatory circuits, direct methods are used to obtain the limit cycle efficiently. An efficient shooting method for solving the associated boundary-value problem (BVP) of the differential-algebraic network equations is presented. Also, the applicability of a shooting method with implicitly used sensitivity is demonstrated.

1 Introduction

For modeling, electrical circuits are treated as networks consisting of nodes, elements and their interconnections. The model equations are derived from the element equations and Kirchhoff's laws. The charge/flux-oriented modified nodal analysis (MNA) (cf. e.g. [4]) leads to a system of quasi-linear differential-algebraic equations of large dimension of the form

$$A \cdot \dot{q}\,(x(t)) + f(x(t), t) = 0\,. \tag{1}$$

The vector $x(t)$ consists of the node potentials and the branch currents of voltage controlling elements, $q(x)$ is the vector of the electrical charges and the magnetic fluxes, A is the (usually singular) incidence matrix that describes the network topology, and $f(x,t)$ includes the static linear and non-linear parts of the circuit and the time dependent input signals. In general, the equations are of index 1 or index 2. The special structure is analyzed in detail in [2,4,11].

Oscillatory circuits present a special task in the field of circuit simulation. The periodic oscillation is important e.g. for radio technology, analog-to-digital conversion and as clock signals in microprocessors. The analysis of the oscillatory circuits includes the computation of periodic steady-state solutions, their stability, periodic noise and distortion analysis. This article deals exclusively with the determination of the periodic solution. To avoid the expansive transient integration of numerous periods, direct methods are utilized to approximate the limit cycles. The most common methods are harmonic balance and shooting.

The method of harmonic balance approximates the periodic solution using a Fourier-polynomial and solves for the coefficients by means of a Galerkin approach (cf. e.g. [1]). Harmonic balance has been shown to be efficient for a wide range of circuits, but for very large circuits or for circuits having pulse like excitations (which force the usage of many Fourier coefficients within the approximation), the method may run into efficiency or accuracy problems. An alternative approach is to formulate and solve a periodic BVP using a shooting method. The periodic boundary condition reads

$$x(0) - x(\tau) = 0\,, \tag{2}$$

where τ is the period of the input and output signal. If the system (1) is autonomous, the period is unknown. For autonomous systems, the BVP can be transformed to the interval $[0,1]$ and the period is then treated as an additional variable. Besides the periodicity condition, a phase condition must be specified. In this paper, the non-autonomous case is treated exclusively.

In [7] a shooting method is adapted to DAEs by adding regularizing constraints. In Sec. 2 an efficient alternative, suitable for large networks, is presented. For sake of convenience only DAEs of index 1 are treated even though the concept applies as well to systems of index 2. In Sec. 3 the applicability of another efficient version which is based on implicitly used sensitivities is discussed. Finally, in Sec. 4 numerical results are given.

2 An Efficient Shooting Method for DAEs

Define the function $\varphi(t;x_0) : \mathbb{R} \to \mathbb{R}^n$ as the solution of the DAE-system (1) with the consistent initial value $\varphi(0;x_0) = x_0$. This initial value must be determined so that the condition of periodicity (2) is fulfilled.

The problem of finding the unknown initial state x_0 is formulated as the minimization of the nonlinear residual function

$$\min_{x_0 \in \mathbb{R}^n} \|\varphi(\tau;x_0) - x_0\|_2^2\,, \tag{3}$$

where the residual vanishes for the initial state of a periodical solution.

A Gauss-Newton method is used to solve (3). It consists of a sequence of linear approximations of the residual function. For the actual approximation of the initial state x_0^i, a correction Δx^i is determined as the solution of the linearization of the argument of the minimization problem: $x_0^{i+1} = x_0^i + \Delta x^i$, $i = 0, 1, \ldots$ and Δx^i is the solution of

$$\min_{\Delta x^i \in \mathbb{R}^n} \| \underbrace{(X(\tau) - X(0))}_{=:\,S_X} \Delta x^i + \underbrace{\varphi(\tau;x_0^i) - \varphi(0;x_0^i)}_{=:\,r} \|_2^2\,. \tag{4}$$

The matrix $X(t) = \frac{\partial \varphi}{\partial x_0}(t)$ is the sensitivity at time t. For DAEs, the iteration matrix S_X is singular, in which case (4) has no unique solution Δx^i. This property is exploited in the following to improve the efficiency of the method. To enlarge the convergence region a damped Gauss-Newton method is used in practice (cf. [10]).

Differentiating (1) with respect to the initial value of the solution $\varphi(t;x_0)$, it can be seen, that $X(t)$ is the solution of the so called variational problem

$$F_{\dot{x}}(\dot{\varphi},\varphi,t)\,\dot{X}(t)+F_x(\dot{\varphi},\varphi,t)\,X(t)=0\,, \tag{5}$$

with $F_{\dot{x}} := A\,q_x(\varphi(t))$ and $F_x := A\frac{d}{dt}q_x(\varphi(t)) + f_x(\varphi(t),t)$, where $q_x = \frac{d}{dx}q$ and $f_x = \frac{d}{dx}f$. The matrix pencil $(F_{\dot{x}},F_x)$ should be regular. The initial conditions can be given by $P_{\mathrm{can}}(0)(X(0)-I)=0$, where P_{can} is the canonical projector along $N(t)$ onto the subspace

$$S(t) := \{z \in \mathbb{R}^n : F_x\, z \in im\, F_{\dot{x}}\}, \qquad N(t) := \ker F_{\dot{x}}\,.$$

All consistent initial conditions belong to $S(0)$.

As shown in [3], $X(t)$ can be expressed as

$$X(t) = K(t)P_{\mathrm{can}}(0)\,,$$

where $K(t)$ is a nonsingular matrix. Using $P_{\mathrm{can}}(0)^2 = P_{\mathrm{can}}(0)$ and

$$S_X = (K(\tau)-K(0))P_{\mathrm{can}}(0) = S_X P_{\mathrm{can}}(0)$$

the minimization problem (4) can be written (omitting the iteration index i)

$$\min_{\Delta x\in\mathbb{R}^n}\|S_X P_{\mathrm{can}}(0)\Delta x - r\|^2 = \min_{\Delta x\in S(0)}\|S_X\Delta x - r\|^2\,. \tag{6}$$

Without increasing the minimum, the solution can be selected out of the image space of the canonical projector $P_{\mathrm{can}}(0)$, $im\,P_{\mathrm{can}}(0) = S(0)$.

To get the solution $\Delta x \in S(0)$ of (6), consider the matrix $V \in \mathbb{R}^{n\times k}$ whose columns form an orthonormal basis of $S(0)$. Then $V\,V^T$ is a projector onto $S(0)$. Since $\Delta x \in im\,V$, $\Delta x = VV^T\Delta x = V\Delta\alpha$ holds, with $\Delta\alpha \in \mathbb{R}^k$. This leads to

$$\begin{aligned}\min_{\Delta x\in\mathbb{R}^n}\|S_X\Delta x - r\|^2 &= \min_{\Delta x\in S(0)}\|S_X\,VV^T\Delta x - r\|^2\\ &= \min_{\Delta\alpha\in\mathbb{R}^k}\|S_X\,V\Delta\alpha - r\|^2, \quad \Delta x = V\Delta\alpha\,. \end{aligned} \tag{7}$$

Using the abbreviation $X_\alpha(t) := X(t)\cdot V$ the minimization problem (7) reads

$$\min_{\Delta\alpha\in\mathbb{R}^k}\|(X_\alpha(\tau)-V)\Delta\alpha - r\|^2, \qquad \Delta x = V\,\Delta\alpha\,.$$

The minimization problem in the space $S(0)$ does not require the full square sensitivity matrix, but rather the matrix $X_\alpha \in \mathbb{R}^{n\times k}$ suffices. The correction Δx lies in $S(0)$, the tangent space of the constrained manifold at the current point x_0^i. With it, the new iterative x^{i+1} fulfills the explicit algebraic constraints up to the first order. For the index 1 case, these are the consistency conditions.

To determine complete consistent values for x^{i+1}, a projection onto the constrained manifold can be performed.

2.1 Approximation of the Sensitivity

The sensitivity is approximated by the integration of X_α is the solution of (5) multiplied by V from the right, i.e. the solution with the initial values

$$X_\alpha(0) = X(0)V = V\,.$$

For the integration (5) is written in the form

$$A\,\frac{d}{dt}(q_x(\varphi(t))\,X(t)) + f_x(\varphi(t),t)\,X(t) = 0\,. \tag{8}$$

In general a combination of low order BDF-methods and the trapezoidal rule is used to integrate the network equation (1). This is also true for the integration of (8). The step size and the order of the integration method for the sensitivity is regulated by the integration of the trajectory $\varphi(t)$, i.e. the solution of (1). Using the discretization of a BDF-method

$$\frac{d}{dt}(q_x(t_m)X(t_m)) = \sum_{i=0}^{k} \alpha_i\, q_x(t_{m-i})X(t_{m-i})\,,$$

where α_i are the coefficients and k is the order, we arrive at a linear equation for the unknown $X(t_m)$:

$$(\alpha_0\, A\, q_x(t_m) + f_x(t_m))\, X(t_m) = -A\sum_{i=1}^{k} \alpha_i\, q_x(t_{m-i})\, X(t_{m-i})\,. \tag{9}$$

Because the same step size and order are used for the integration of the sensitivity as for the integration of the trajectory, the matrix of (9) is known already. It results from solving the nonlinear equation of the implicit discretization of the network equations. During the calculation of the trajectory the sparse matrices are saved. In this way the matrix X_α can be computed efficiently.

2.2 Calculation of the Basis *V*

Determining the basis V means calculating k linear independent consistent initial values of (5). A general way to get the basis is to solve a generalized eigenvalue problem. Inserting the particular solution $X(t) = v\, e^{\lambda t}$ and $\dot{X}(t) = \lambda\, v\, e^{\lambda t}$ in (5) leads to the equation

$$\lambda\, F_{\dot{x}}\, v + F_x\, v = 0$$

at the initial time $t = 0$. $V = (v_1, ..., v_m)$ with $v_i, i = 1, ..., m$ are the linear independent eigen- and mainvectors for the eigenvalues λ_i.

In the case of the network equations the basis is also available from the topology of the network. Especially for high dimensional problems this may be cheaper than solving the generalized eigenvalue problem. The topological investigations in [2] result in a projector W along $im\, F_{\dot{x}}$. With it, k linear independent consistent initial values can be calculated. In both cases V must be orthonormalized.

3 A Shooting Method with Implicitly used Sensitivity

In [5] a shooting method with implicitly used sensitivity is presented. The correction for an initial guess of the solution is calculated out of

$$(X(\tau) - I)\Delta x^i = -(\varphi(\tau; x_0^i) - x_0^i)\,. \tag{10}$$

This linear equation is solved, using an iterative method, like e.g. GMRES [8]. Instead of building the sensitivity matrix $X(\tau)$ explicitly, these methods only require matrix-vector products $X(\tau)\cdot q$, $q \in \mathbb{R}^n$. The product is realized by integrating (5), since $z(\tau) := X(\tau)\cdot q$ is the solution of

$$F_{\dot{x}}\,\dot{z} + F_x\,z = 0\,,\; z(0) = q\,. \tag{11}$$

The calculation of $z(\tau)$ has the same complexity as the explicit computation of one column of $X(\tau)$. The method is efficient, as in general much less then n steps are needed to approximate Δx^i to sufficient accuracy (cf. [5]).

On the other hand, the aspects of treating DAEs are not considered yet. In general, the initial values q for (11) used in the iterative solving of (10) are *not* consistent. Nevertheless the following Lemma holds.

Lemma. *For the network equations of index 1 resp. 2, resulting from the charge/flux-oriented MNA, the above described method is applicable, if only the integration of (11) is started using one resp. two implicit Euler steps.*

Proof. Let $z_i := z(t_i)$ and $h_i := t_i - t_{i-1}$, $i = 1, 2$. For the network equations $N(t) \equiv N$ and $N \cap S(t)$ are constant. Let Q be a projector onto N and $P = I - Q$. In the index 1 case $\mathbb{R}^n = N \oplus S(t)$ holds and with it $q = Qq + Pq$. Using $z_0 = q$ and applying an implicit Euler step to (11) yields:

$$(\frac{1}{h_1}\,F_{\dot{x}}(t_1) + F_x(t_1))\,z_1 = \frac{1}{h_1}\,F_{\dot{x}}(t_1)q = \frac{1}{h_1}\,F_{\dot{x}}(t_1)(Qq + Pq) = \frac{1}{h_1}\,F_{\dot{x}}(t_1)Pq\,.$$

Consequently no inconsistent parts of the initial value q contribute to $z(\tau)$.

For problems of index 2 the property $\mathbb{R}^n = N \oplus S(t)$ gets lost. In addition to the purely explicit constraints some so called hidden constraints arise. The corresponding parts of q lie in $N \cap S(t)$. Following [2], let W be a projector along $im\,F_{\dot{x}}$, especially $WF_{\dot{x}} = 0$. Discretization of (11) using the implicit Euler scheme leads to:

$$F_{\dot{x}}(t_i)\frac{z_i - z_{i-1}}{h_i} + F_x(t_i)z_i = 0\,, i = 1, 2\,,\; z_0 = q\,.$$

As in the index 1 case, z_1 fulfills all explicit constraints ($W\,F_{\dot{x}} = 0 \Rightarrow W\,F_x(t_1)\,z_1 = 0$), but in general not the hidden constraints. However, a consistent z_1^c exists, such that $z_1^c = z_1 + \hat{z}_1$, $\hat{z}_1 \in N \cap S(t)$ (cf. [2]). Let T be a projector onto $N \cap S(t)$. The following holds

$$F_{\dot{x}}(t_2)z_1 = F_{\dot{x}}(t_2)(z_1^c - \hat{z}_1) = F_{\dot{x}}(t_2)P(z_1^c - T\hat{z}_1) = F_{\dot{x}}(t_2)z_1^c\,,$$

since $F_{\dot{x}}P = F_{\dot{x}}$ and $PT = 0$. Consequently, after the second Euler step, all inconsistent parts of q are again damped out and do not contribute to $z(\tau)$. □

Remark: The usage of implicit initial conditions as described above is suitable to integrate (11) resp. to solve (10) for the correction Δx_0. On the other hand, the so obtained sensitivity information is *not* sufficient for a stability analysis, since the sensitivity results at $t = \tau$ must be investigated *with respect to the defacto consistent* initial values and not with respect to some arbitrary starting vector whose inconsistent parts are just damped out because of the applied numerical discretization (cf. [9]).

4 Numerical Results

Both methods described in Sec. 2 and Sec. 3 have been implemented in the industrial circuit simulator TITAN of Infineon Technologies AG [1]. They have been tested on various examples. Both methods perform very well though for large circuits the second method is slightly faster and less memory consuming. Here the results for two circuits are presented. The first considered circuit is a mixer, as shown schematically in Fig. 1. The multi-tone system

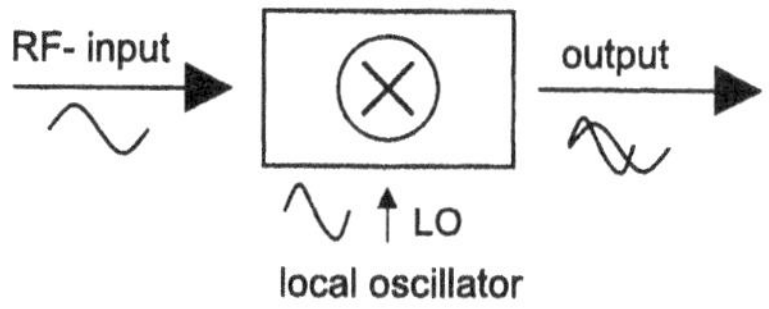

Figure1. Scheme of a mixer circuit

is clocked by a local oscillator of frequency 900 MHz. The output signal is the superposition of a radio frequent input signal, e.g. of frequency 950 MHz and the local oscillator. In this example the mixer is down converting, i.e. the output signal is of frequency 50 MHz. Figure 2 shows the time and phase diagram of the output signal.

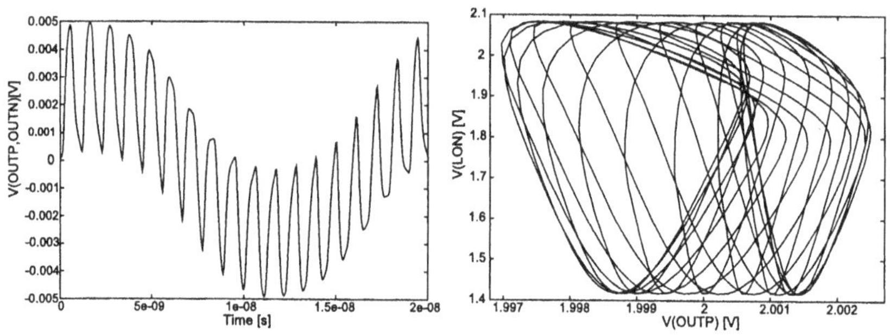

Figure2. Time and phase diagram of a low-noise, highly linear 0.25 μm CMOS-mixer

The second example illustrates the differences between the harmonic balance (HB) and the shooting method. For these purposes a commercially available simple operational amplifier μA 709 circuit is considered (cf. Fig. 3).

First, sinusoidal input is applied. The output signal obtained by the method of harmonic balance with nine calculated frequencies nearly coincides with the solution obtained by the shooting method. With 17 frequencies the solutions are identical (cf. Fig. 4, left diagram). The calculation time for both methods is the same. Second, the circuit is excited by a pulse like input signal. In this case the method of harmonic balance needs about 25 frequencies for a solution of satisfying accuracy (cf. Fig. 4, right diagram). For the shooting method the pulse like input does not cause any problems. As expected in this case, it is superior to harmonic balance with respect to run time and accuracy.

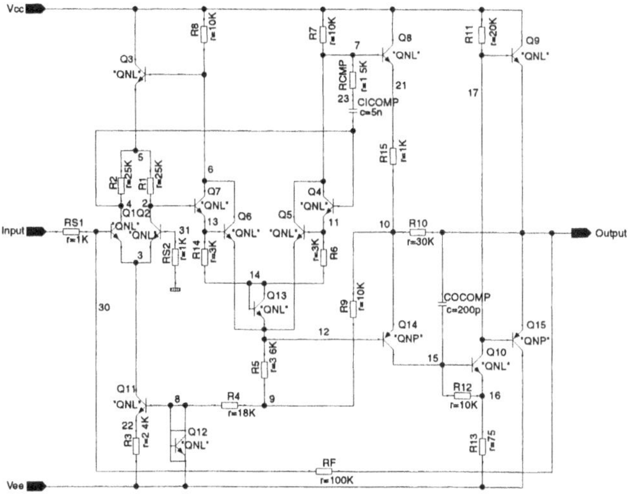

Figure3. Schematic of an operational amplifier μA 709

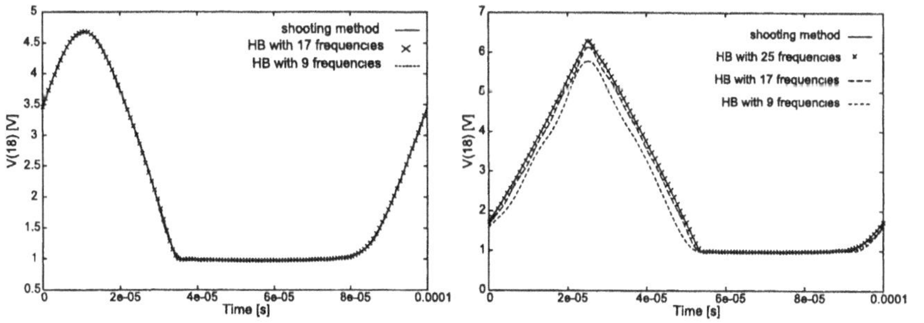

Figure4. Output signal of the operational amplifier μA 709, sinusoidal (left) and pulse like (right) input signal

5 Conclusions

Two efficient shooting methods for the determination of the limit cycle of oscillatory circuits described by DAEs have been presented. The methods have been tested on various examples. Both methods perform well.

Acknowledgements

The authors would like to thank D. Estévez Schwarz for valuable comments and remarks. This research has been supported by the Federal BMBF under grant 03BU7TM1/5. The authors bear responsibility for the contents of this publication.

References

1. Feldmann, U. and Wever, U. and Zheng, Q. and Schultz, R. and Wriedt, H., Algorithms for modern circuit simulation, Archiv für Elektronik und Übertragungstechnik 1992 Vol.46 No.4 pages:274-285
2. Estévez Schwarz, D., Consistent initialization for index-2 differential algebraic equations and its application to circuit simulation, Humboldt-Universität 2000 Berlin
3. E. Griepentrog and R. März, Differential–Algebraic Equations and their Numerical Treatment, Teubner Leipzig 1996
4. Günther, M. and Feldmann, U., CAD-based electric-circuit modeling in industry I.Mathematical structure and index of network equations, Surveys on Mathematics for Industry 1999 Vol.8 pages:97-129
5. Telichevesky, R. and Kundert, K. and White, J., Efficient Steady-State Analysis based on Matrix-Free Krylov-Subspace-Methods, Proceedings of the 32rd Design Automation Conference 1995 pages: 480-484
6. Telichevsky, R. and Kundert, K. and Elfadel, I. and White, J., Fast Simulation Algorithms for RF Circuits, Proceedings of the IEEE Custom Integrated Circuits Conference 1996 pagees:437-444
7. Lamour, R., A shooting method for fully implicit index-2 differential algebraic equations, SIAM Journal on Scientific Computing 1997 Vol.18 No.1 94-114
8. Saad, Y., Iterative Methods for Sparse Matrices, PWS publishing company 1996 Boston
9. Selting, P. and Zheng, Q., Numerical stability analysis of oscillating integrated circuits, Journal Comp. and Applied Mathematics 1997 Vol.82 pages:367-378
10. Stoer, J. and Bulirsch, R., Introduction to Numerical Analysis, Springer Verlag 1993 New York
11. Tischendorf, C., Topological index calculation of DAEs in circuit simulation, Surveys on Mathematics for Industry 1999 Vol.8 pages: 187-199

Robust Limit Cycle Calculations of Oscillators

S. Lampe[1], H.G. Brachtendorf[1], E.J.W. ter Maten[2], S.P. Onneweer[2], R. Laur[1]

[1] Institute for Electromagnetic Theory and Microelectronics, University of Bremen, 28334 Bremen, Germany, E-mail: lampe, brachtd, laur@item.uni-bremen.de
[2] Philips Research Laboratories, ED&T / Analogue Simulation, 5656 AA Eindhoven, The Netherlands, E-mail: jan.ter.maten, siep.onneweer@philips.com

Abstract Standard approaches for limit cycle calculations of autonomous circuits exhibit poor convergence behavior in practice. By introducing an additional periodic probe voltage source, we can reformulate the system of autonomous differential algebraic equations (DAEs) as a system of non-autonomous DAEs with the constraint, that the current through the source has to be zero for the limit cycle. A one or a two stage approach now leads to a greater convergence domain, but in practice still additional techniques are necessary to improve robustness. The range of convergence towards the limit cycle for the initial probe amplitude can be expanded drastically by employing the affine invariance technique as damping strategy to Newton's method.

1 Introduction

Using modified nodal analysis, electronic circuits can be described mathematically by a system of N (nonlinear) differential algebraic equations (DAEs)

$$f(x,t) = i(x,t) + \frac{\mathrm{d}q(x)}{\mathrm{d}t} = 0, \tag{1}$$

where $x \in \mathbb{R}^N$ is the vector of unknown node voltages and branch currents of inductors, $i : \mathbb{R}^{N\times 1} \to \mathbb{R}^N$ is the vector of conductive contributions to the circuit equations and $q : \mathbb{R}^N \to \mathbb{R}^N$ is the vector of charges and fluxes. We require that i, q are sufficiently smooth functions and that the partial derivatives with respect to x are available. We further assume that i_x is regular for all t, otherwise there is no unique solution of the DAE. In the autonomous case, equation (1) reads as follows

$$f(x) = i(x) + \frac{\mathrm{d}q(x)}{\mathrm{d}t} = 0. \tag{2}$$

We assume that (2) has at least one non-trivial periodic solution $x(t) = x(t+T)$ with an a priori unknown period T and that the DAEs do not exhibit any chaotic behavior. Additionally, we assume that (2) is uniquely solvable for any consistent set of initial conditions x_0 at t_0. Under these assumptions the calculation of limit cycles of (2) can be written as the solution of the boundary value problem

$$
\begin{aligned}
f(x) &= i(x) + \frac{\mathrm{d}q(x)}{\mathrm{d}t} = 0 \\
x(0) &= x(T).
\end{aligned} \tag{3}
$$

If $x(t)$ is a solution of (3), $x(t+\Delta t)$, $\Delta t \in \mathbb{R}$ is a solution as well. As we will see in the following section, this continuity of solutions leads to ill-posed discretized systems, which e.g. can be avoided by an additional constraint on the solution.

Standard techniques for solving periodic boundary value problems of type (3) are shooting techniques, finite-difference methods and Ritz-Galerkin techniques. Harmonic Balance is a Ritz-Galerkin method employing trigonometric basis functions. In what follows we restrict ourselves to the Harmonic Balance technique, which is a reliable technique for simulating steady state responses of electronic systems.

In the following section we discuss algorithms for limit cycle calculations. Section 3 deals with strategies for increasing the range of convergence of the algorithms for limit cycle calculations. Examples that demonstrate the drastic improvement of convergence behavior by application of the affine invariance technique to Newton's method conclude this paper.

2 Limit Cycle Calculations

In the following we use the complex form of the discrete Fourier transform for ease of presentation, resulting into $2K+1$ harmonics of the signal from $-K\omega, ..., K\omega$. The system size can be reduced roughly by a factor of two, when employing a sine/cosine representation of the waveforms.

In the sections below, lower case characters represent quantities in the time domain (waveforms), upper case letters quantities in the frequency domain (spectra).

2.1 The Standard Approach

Application of Harmonic Balance to the boundary value problem (3) leads to a system of $N(2K+1)$ nonlinear algebraic equations

$$F(X,\omega) = I(X) + j\,\Omega(\omega)\,Q(X) = 0, \tag{4}$$

where $\omega = \frac{2\pi}{T}$, $j = \sqrt{-1}$, $F : \mathbb{C}^{N(2K+1)} \times \mathbb{R} \to \mathbb{C}^{N(2K+1)}$. Ω represents the time derivative in the frequency domain

$$\Omega = [\Omega_{nm}], \qquad \Omega_{nm} = \begin{cases} \Omega_{nn} & \text{if} \quad n = m \\ 0 & \text{if} \quad n \neq m \end{cases} \tag{5}$$

$$\Omega_{nn} = \omega \operatorname{diag}\{-K, \ldots, -1, 0, 1, \ldots, K\}, \qquad n = 1, 2, ..., N.$$

Nonlinear equations of type (4) are generally solved by (damped) Newton type techniques, which require the partial derivatives of the constitutive element relations. Let ν be the counter of the Newton procedure, then we get

$$\left[J_F(X^{(\nu)},\omega^{(\nu)}) \;\; \frac{\partial F(X^{(\nu)},\omega^{(\nu)})}{\partial\omega} \right] \begin{bmatrix} \Delta X^{(\nu+1)} \\ \Delta\omega^{(\nu+1)} \end{bmatrix} = -\left[F(X^{(\nu)},\omega^{(\nu)}) \right],$$

$$\text{where } J_F = \partial I(X) \,/\, \partial X + j\,\Omega(\omega)\,\partial Q(X) \,/\, \partial X = G(X) + j\,\Omega(\omega)\, C(X)$$

$$\text{and } \frac{\partial F(X,\omega)}{\partial\omega} = \frac{j}{\omega}\,\Omega(\omega)\,Q(X). \tag{6}$$

(4) is an under-determined system, because the additional angular frequency ω is an unknown. This reflects the property of autonomous systems having a continuum of phase-shifted solutions. There are two well-known techniques to overcome this problem: Fix the phase of one harmonic, i.e. $\mathrm{Re}\{(e_l^m)^T X\} = 0$ where e_l^m is the l-th canonical vector of dimension $m := N\,(2K+1)$ or solve the under-determined system (6) by a Moore-Penrose pseudoinverse [4]. The latter technique unfortunately needs a singular value decomposition (SVD) or QR-decomposition which both require $O(m^3)$ operations. Furthermore, SVD and QR do not exploit the sparsity of the system and are therefore not applicable for industrial relevant circuits. The advantage of the Moore-Penrose inverse on the other hand is that it calculates always the solution of (6) with the least Euclidian norm. The standard approach for limit cycle calculations of oscillators can be found e.g. in [6].

2.2 The Two Stage Approach

By introducing an additional periodic probe voltage source $b(t) = b(t+T)$ to the node n of the oscillator circuit, $1 \le n \le N$, we can reformulate the autonomous DAE (3) with unknown periodic steady state as a non-autonomous DAE (1) with additional constraints [13]. The probe voltage source injects a current added to the list of unknowns, $x_{N+1}(t)$. From (2) we obtain the non-autonomous DAE

$$f(x) = i(x) + \frac{\mathrm{d}q(x)}{\mathrm{d}t} + e_n^N\, x_{N+1} = 0$$

$$x_n - b(t) - 0, \tag{7}$$

where e_n^N is the n-th canonical vector of dimension N. Equation (7) is now a non-autonomous DAE of the form (1) and existing software for solving driven Harmonic Balance problems can be re-used in solving (7) [12,8]. The autonomous steady state solution of (3) is obtained if and only if $x_{N+1}(t) \equiv 0$. In the frequency domain the following equations are obtained, using Kronecker's $\otimes$ product

$$F(X) = I(X) + j\,\Omega\,Q(X) + e_n^N \otimes X_{N+1}$$

$$X_n - B = 0. \tag{8}$$

The steady state condition is obtained if and only if the constraint

$$X_{N+1}(B,\omega) = 0 \tag{9}$$

is met. Note that (9) is an under-determined system $\mathbb{C}^{2K+1} \times \mathbb{R} \to \mathbb{C}^{2K+1}$, because the non-autonomous system (8) is assumed to be time-invariant. Equation (8) is solved, keeping stimulus B and ω fixed until convergence is reached, by a damped Newton method in a first step. In the next step the approximate $(B^{(\nu)}, \omega^{(\nu)})$ of (9) will be updated. The algorithm proceeds with the first step until convergence of the outer loop is achieved. The under-determined system (9) can be solved either by a Moore-Penrose pseudoinverse, but now applied to a much smaller system than that given by (6), or by gauging the problem by fixing the phase of the k-th harmonic of the stimulus, i.e. ${e_k^K}^T B - {e_{-k}^K}^T B = 0$.
This two stage approach can be modified in a natural way. Assume that there is a frequency selective resistance in series with the source $b(t)$ which is zero for the fundamental waveform of $b(t)$ and infinite otherwise. We assume furthermore a natural ordering of the frequencies, i.e. $-K \leq k \leq K$ and let

$$D = \operatorname{diag}(0, \ldots, 0, 1, 0, 1, 0, \ldots, 0) \in \mathbb{C}^{(2K+1)\times(2K+1)},$$

D selects the fundamental harmonic of the system. In a straightforward way the algorithm with the frequency dependent element can be written in the frequency domain as

$$\begin{aligned} I(X) + j\Omega Q(X) + e_n^N \otimes (D X_{N+1}) &= 0 \\ D(X_n - B) + (E - D) X_{N+1} &= 0 \end{aligned} \tag{10}$$

with the constraint

$$D\, X_{N+1}(D\, B, \omega) + (E - D)\, B = 0, \tag{11}$$

where E represents the identity matrix.
Roughly speaking, voltage source and source current consist only of a fundamental waveform, all higher harmonics are set to zero. This variant has some natural advantages because it is less error prone to detect a node of the underlying autonomous circuit to which the source can be attached. This is due to the fact that with such an additional source it is less likely to create a system with inconsistent initial conditions. This can be explained as follows: Suppose, one creates, while using the equations (8) and (9), a loop of ideal voltage sources and inductors. Because an inductor is simply a short for DC, one creates a system with a singular Jacobian and the source currents tend to infinity. The under-determined system of equations (11) may be solved by either a Moore-Penrose inverse or by gauging the phase of the fundamental harmonic of the stimulus.

2.3 The One Stage Approach

The one stage approach [12] might be considered as a combination of the standard and the two stage approach because it is a one stage method like the standard approach and it retains the idea of a probe voltage source with a frequency selective element in series like the two stage algorithm. Using the frequency selecting matrix D introduced above, the algorithm comes down to this:

$$
\begin{aligned}
I(X)+j\Omega Q(X)+e_n^N\otimes(DX_{N+1}) &= 0\\
D\,(X_n-B)+(E-D)\,X_{N+1} &= 0\\
D\,X_{N+1}(D\,B,\omega)+(E-D)\,B &= 0,
\end{aligned}
\tag{12}
$$

where $(X,\, X_{N+1},\, B,\, \omega)$ are the unknowns. Eq. (12) is still under-determined. Authors in [12] gauge this system by fixing the phase of the fundamental harmonic. Furthermore, to avoid convergence towards the trivial solution a deflation technique [3] by dividing by $\|B\|_2$ has been used in [12]. Eq. (12) is obviously very similar to the two stage algorithm (10,11). The difference lies in the fact that in (10) the inner system of equations is solved in a first step for a fixed stimulus $b(t)$ and in a second step the stimulus $b(t)$ will be upgraded. The one stage approach is doing both in one step. Note that block-Gaussian elimination may represent the two stages.

3 Damping Strategies

For all methods mentioned above Newton's method either fails to converge or converges towards the trivial unstable operating point, if the start-values are not close enough to the correct solution.

To increase the convergence range of the algorithms we applied damping to Newtons's method by introducing a damping factor α:

$$
\begin{aligned}
\Delta x^{(\nu+1)} &= -J_F(x^{(\nu)})^{-1}\cdot F(x^{(\nu)})\\
x^{(\nu+1)} &= x^{(\nu)}+\alpha_\nu\cdot\Delta x^{(\nu+1)}, \quad 0<\alpha_\nu\le 1, \quad \alpha_\nu\in\mathbb{R}.
\end{aligned}
\tag{13}
$$

Let x^* be the exact solution. Because x^* is unknown, the convergence criterion

$$
\left\|x^{(\nu+1)}-x^*\right\| \le \left\|x^{(\nu)}-x^*\right\| \tag{14}
$$

cannot be verified. For this reason we have to substitute the "true" convergence criterion.

The steepest descent method accepts an α if and only if the residual norm of the $(\nu+1)$-st iteration is smaller than the residual of the ν-th iteration, namely

$$
\left\|F(x^{(\nu)})\right\|^2 - \left\|F(x^{(\nu+1)})\right\|^2 \ge 0. \tag{15}
$$

A more sophisticated descent technique has been proposed by Schwetlick et al [11], where the α is chosen in such a way, that the residuals of the $(\nu+1)$-st and ν-th iteration fulfill the requirement

$$
\left\|F(x^{(\nu)})\right\|^2 - \left\|F(x^{(\nu+1)})\right\|^2 \ge \delta\cdot\left[\left\|F(x^{(\nu)})\right\|^2 - A\right], \tag{16}
$$

$$
\text{where } A=\left\|F(x^{(\nu)})+J_F(x^{(\nu)})\cdot(x^{(\nu+1)}-x^{(\nu)})\right\|^2.
$$

The "true" descent on the left hand side of (16) has to be at least δ times the predicted descent, which we get through a linearization of $F(x^{(\nu+1)})$ at point $x^{(\nu)}$.
The convergence criterion of the affine invariance technique published by Deuflhard et al [10,?] is as follows:

$$\left\| J_F(x^{(\nu)})^{-1} \cdot F(x^{(\nu+1)}) \right\|^2 \leq \left\| J_F(x^{(\nu)})^{-1} \cdot F(x^{(\nu)}) \right\|^2 . \qquad (17)$$

The name affine invariance stems from the fact, that the solution $F(x) = 0$ is invariant under the affine transformation $F \to G := W \cdot F$ where W is an arbitrary regular matrix. Authors in [10,5] choose an iteration-dependent $W^{(\nu)} := J_F(x^{(\nu)})^{-1}$, i.e. instead of monitoring the residual, the affine invariance technique monitors the update vector Δx of two Newton iterations: the ν-th update vector is $\Delta x^{(\nu)} = J_F(x^{(\nu)})^{-1}\, F(x^{(\nu)})$ and the $(\nu+1)$-st iteration is approximately given by $\Delta x^{(\nu+1)} \approx J_F(x^{(\nu)})^{-1}\, F(x^{(\nu+1)})$. In case of convergence, the series of the Δx converges to zero. Therefore, an approximate estimate of the correct update $\Delta x^{(\nu+1)} = J_F(x^{(\nu+1)})^{-1} \cdot F(x^{(\nu+1)})$ is given by $\Delta x^{(\nu+1)} \approx J_F(x^{(\nu)})^{-1} \cdot F(x^{(\nu+1)})$, hence we get (17). If the convergence condition (17) is not fulfilled, a new damping parameter α_ν has to be chosen. For a sophisticated choice of the parameter α_ν we refer the reader to the papers [10,5].
Alternatively, the global homotopy method

$$f(x(\lambda), \lambda) = F(x) - (1 - \lambda)\, F(x_0), \quad 0 \leq \lambda \leq 1$$

where x_0 is any initial guess, has been tested for the algorithms.
In order to avoid convergence towards the trivial solution, a deflation technique [3] by dividing $F(X)$ by $\|B\|_2$ has been used for the different algorithms.

4 Examples

In this section, the simulation results of a VHF-oscillator (VHF), a symmetric oscillator (SYM) and a voltage controlled oscillator (VCO) are given. The circuits were simulated with the fast one stage approach (OSA) and the robust two stage approach (TSA), both stand alone and in conjunction with damping strategies mentioned in section 3. The initialisation of the start-values was done by an AC-sweep (Kurokawa condition [7]). For the probe voltage source we used as amplitude $U = 0.01V$, $0.1V \leq U \leq 1.5V$ (stepsize = 0.1V) and $2.0V \leq U \leq 5.0V$ (stepsize = 1.0V). Table 1 shows the simulation results.

- VHF: The one stage approach converges to the limit cycle for $U = 0.1V$, $0.3V \leq U \leq 0.7V$ within 8 iterations in average. The two stage algorithm is able to find the limit cycle within a range of the stimulus $0.3V \leq U \leq 0.9V$ within 18 total inner Newton iterations in average. The steepest descent, modified steepest descent and the global homotopy method (section 3) did not increase the convergence range of the algorithms, whereas the affine invariance technique applied to the one stage

Table1. Simulation results

Circuit	Algorithm	Damping	Convergence Range	Av. Iter.
VHF	OSA	/	$U = 0.1V, 0.3V \leq U \leq 0.7V$	8
VHF	OSA	Affine Inv.	$0.1V \leq U \leq 0.9V$	14
VHF	TSA	/	$0.3V \leq U \leq 0.9V$	18
VHF	TSA	Affine Inv.	$0.01V \leq U \leq 1.2V$	37
SYM	OSA	/	$0.1V \leq U \leq 0.3V$	9
SYM	OSA	Affine Inv.	$0.01V \leq U \leq 0.6V$	28
SYM	TSA	/	$0.1V \leq U \leq 0.4V$	21
SYM	TSA	Affine Inv.	$0.01V \leq U \leq 0.7V$	65
VCO	OSA	/	$0.2V \leq U \leq 1.0V, U = 2.0V$	8
VCO	OSA	Affine Inv.	$0.01V \leq U \leq 1.0V, U = 2.0V$	9
VCO	TSA	/	$U = 0.6V, 1.0V \leq U \leq 2.0V$	24
VCO	TSA	Affine Inv.	$0.01V \leq U \leq 3.0V, U = 5.0V$	42

approach leads to a convergence range of $0.1V \leq U \leq 0.9V$ demanding 14 iterations in average. The convergence domain of the two stage algorithm increases to $0.01V \leq U \leq 1.2V$ when applying the affine invariance technique to the inner and outer loop including deflation to the outer loop, with an average amount of 37 inner Newton iterations.

- SYM: The affine invariance technique is able to increase the range of convergence of the one stage approach from $0.1V \leq U \leq 0.3V$ up to $0.01V \leq U \leq 0.6V$ and of the two stage approach from $0.1V \leq U \leq 0.4V$ up to $0.01V \leq U \leq 0.7V$. The steepest descent, modified steepest descent and the global homotopy method did not increase the convergence range of the algorithms.

- VCO: Also for this circuit the steepest descent, modified steepest descent and the global homotopy method did not increase the range of convergence of the algorithms. But the application of the affine invariance technique increases the convergence range of the one and the two stage algorithm significantely.

5 Conclusion

The one stage algorithm requires roughly 1/3 of the Newton iterations compared to the two stage technique. The small range of convergence towards the limit cycle for the initial probe amplitude U can be expanded drastically by employing the affine invariance technique. The affine invariance shows superior properties compared with (modified) steepest descent techniques and homotopy methods, requiring only slightly more overhead. In all circuits under test the affine invariance method applied to the two stage algorithm exhibits a wider range of convergence for the initial probe amplitude than the one stage method in conjunction with affine invariance.

References

1. H.G. Brachtendorf, *Simulation des eingeschwungenen Zustands elektronischer Schaltungen*, Shaker, Aachen, 1994.
2. H.G. Brachtendorf, G. Welsch, R. Laur, A. Bunse-Gerstner, "Numerical steady state analysis of electronic circuits driven by multi-tone signals", Electrical Engineering 79, pp. 103-112, 1996.
3. K.M. Brown, W.B. Gearhart,"Deflation Techniques for the Calculation of Further Solutions of a Nonlinear System", Numerische Mathematik, Vol. 16, pp. 334-342, 1971.
4. P. Deuflhard, "Computation of periodic solutions of nonlinear ODE's", BIT, Vol. 24, pp. 456-466, 1984.
5. P. Deuflhard, A. Hohmann, *Numerische Mathematik I: Eine algorithmisch orientierte Einführung*, de Gruyter, New York, 1993
6. K.S. Kundert, J.K. White, A. Sangiovanni-Vincentelli, "Steady-state methods for simulating analog circuits", Kluwer Academic Publ., Boston, 1990.
7. K. Kurokawa, "Some basic characteristics for broadband negative resistance oscillators", Bell. Syst. Tech. J., Vol. 48, pp. 1937-1955, 1969.
8. E.J.W. ter Maten, "Numerical methods for frequency domain analysis of electronic circuits", Surv. Meth. Ind., Vol. 8, pp. 171-185, 1999.
9. E. Ngoya, A. Suárez, R. Sommet, R. Quéré, "Steady State Analysis of Free or Forced Oscillators by Harmonic Balance and Stability Investigation of Periodic and Quasi-Periodic Regimes", Int. J. Microwave and Millimeter-Wave CAD, Vol. 5, No. 3, pp. 210-223, 1995.
10. U. Nowak, L. Weimann, "A Family of Newton Codes for Systems of Highly Nonlinear Equations", Technical Report TR-91-10, Konrad-Zuse-Zentrum, Berlin, 1991.
11. H. Schwetlick, H. Kretzschmar, *Numerische Verfahren für Naturwissenschaftler und Ingenieure*, Fachbuchverlag Leipzig, Leipzig, 1991.
12. M.F. Sevat, T.J. Engelen, J.C.H. van Gerwen, E.J.W. ter Maten, "Harmonic Balance Algorithm in Pstar", Philips Electronics N. V. 1998.
13. G. Welsch, *Analyse des eingeschwungenen Zustands autonomer und nichtautonomer elektronischer Schaltungen*, Shaker, Aachen, 1998.

Multirate ROW Methods of Mixed Type for Circuit Simulation

A. Bartel *

Centre for Scientific Computing and Mathematical Modelling, University of Karlsruhe, Engesser Straße 6, D-76128 Karlsruhe, Germany.

Abstract The simulation of integrated circuits demands an increasing amount of computational resources. Due to latency, the numerical integration based on multirate methods might be another denouement. These methods are tailored to coupled settings by the adaption of the numerical effort to the corresponding activity level. Furthermore, a refined modelling is more and more pursued. Therefore coupled systems arise and multirate is topic of many applications.
Following the idea of multirate step sizes, a flexible multirate ROW method is presented, which bases on both the original and the generalised type. The scope of the method are stiff initial value problems founded on ordinary differential equations with the Inverter-Chain serving as a benchmark.

1 Introduction to Multirate

Multirate settings comprise of systems which behave on largely differing time scales. This induces an inherent partition on the set of state variables, which can be exploited on several levels and with a bunch of possible approaches. We focus on partitionings of the whole system of equations and multirate step size algorithms. Rewriting our initial value problem $\mathbf{y}' = \mathbf{f}(\mathbf{y})$, $\mathbf{y}(0) = \mathbf{y}_0 \in \mathbb{R}^n$, which is just autonmous for simlicity of notation, in partitioned form introduces the following notation

$$\begin{matrix}\text{latent} \\ \text{active}\end{matrix} \quad \begin{pmatrix}\dot{\mathbf{y}}_L \\ \dot{\mathbf{y}}_A\end{pmatrix} = \begin{pmatrix}\mathbf{f}_L(\ \mathbf{y}_L\ , \boxed{\mathbf{y}_A}) \\ \mathbf{f}_A(\boxed{\mathbf{y}_L}, \ \mathbf{y}_A\ )\end{pmatrix} \qquad \begin{matrix}\mathbf{y}_L(t_0) = \mathbf{y}_{L,0} \\ \mathbf{y}_A(t_0) = \mathbf{y}_{A,0}\ ,\end{matrix} \tag{1}$$

where $\mathbf{y_L}$ and $\mathbf{y_A}$ denote the latent and active components of the state variables, respectively. Of course, one can think of an arbitrary number of parts, but considering a simple depiction and the more overhead, this is oppressed here. The partition might be viewed through the reading glass of hierarchical modelling such as encountered in circuit analysis. Moreover, the partitioned formulation invokes occurrences of the variables as couplings, which are boxed in the above equation and throughout the paper. Gear and Wells investigated a multirate step size algorithm, tailored to (1), which was based on Backward Differentiation Formulae (BDF) [5]. Later, Günther, Kværnø

* Thanks to M. Günther and P. Rentrop for many fruitful discussions and their guidance.

and Rentrop developed multirating for one step methods of Runge-Kutta (RK) and Rosenbrock-Wanner-type (ROW) [8,10], while Engstler and Lubich formulated multiorder methods based on Richardson Extrapolation and high order RK-schemes [4,3]. The latter kind is well suited for astro-physical problems, where explicit integrators are fine.

The multirate step size approach bases on the usage of inherent step sizes, which mirror the activity level of the specified units. In our case of eq. 1 we will have two integrations: (A) the latent part will be integrated with *macro step size* H and (B) the active part with *micro step size* h. Now, the crucial question is, how to realise the coupling numerically.

2 Idea of Mixed ROW

The various kinds of multirate address the couplings differently. The first one-step based multirate ROW methods (Günther, Rentrop [8] – *original type*) investigated couplings done by extrapolation formulae. This results in a two step algorithm, where the active and non-active parts are totally decoupling over the so-called macro step, which is the time grid for synchronisation.

Later, Kværnø and Rentrop [10] proposed the incremental formulation of the coupling variables, which returns us to a genuine one-step method: *generalised multirate*. This type couples the first micro step of the active components to the macro step of the latent components – so-called *compound step*, but decouples all later micro steps, computationally. Within that conception, a W-type[1] formula was developped [2], where the Jacobian is lagged over the macro step. This is computationally efficient, but had the consequence of a significantly increased number of micro steps for the nonlinear, stiff test problem (inverter chain benchmark). Since in circuit simulation the load part is most time consuming, one needs to minimise the right-hand side evaluations, this is to provide updated Jacobians, while the number of micro steps returns to the single rate algorithm scale. The easiest extension is found in the mixed type multirate formulation. Here, we want to extract the positive features of each type, by using an incremental formulation for a compound step; for the later micro steps we deduce the couplings by a dense output formula (cf. [11]) of the non active components. This allows: (1) coupling of active to latent in the interior of the macro step, (2) complete ROW scheme (cf. [2]), (3) step size adjustment in the later micros without macro step update (see below).

3 Formulae of Mixed ROW

To employ the compact Kronecker notation in the formulation of method, let $A = (a_{ij}) \in \mathbb{R}^{n\times m}$, $B = (b_{kl}) \in \mathbb{R}^{p\times q}$, then the *Kronecker product* is defined by

[1] W-methods [13] allow the usage of an approximated Jacobian.

$$A \otimes B := (c_{st}) \in \mathbb{R}^{np \times mq}, \text{ with } c_{p(i-1)+k,\, q(j-1)+l} = a_{ij} \cdot b_{kl},$$

where the employed matrices will consist of method defining coefficients and identity operators of associated spaces. Moreover, we need 'multi'-evaluations, which we denote by the corresponding capital letters:

$$\mathbf{F}_*\left(\left(\mathbf{y}_L^{(i)}\right)_{i=1}^{s_*}, \left(\mathbf{y}_A^{(i)}\right)_{i=1}^{s_*}\right) := \left(\mathbf{f}_*\left(\mathbf{y}_L^{(i)}, \mathbf{y}_A^{(i)}\right)\right)_{i=1}^{s_*}, \tag{2}$$

where $*$ varies over $\{A, L\}$. One-step methods give the numerical approximation in weighted sums of increments – for simplicity of notation, the micro step size is assumed to be constant ($H = m\,h,\ m \in \mathbb{N}$) over each macro step. Thus we have

$$\mathbf{y}_L^H\big(t_0 + H\big) = \mathbf{y}_{L,0} + \big(\mathbf{b}_L \otimes \mathcal{I}_L\big)^\top \mathbf{k}_L \tag{3}$$

$$\mathbf{y}_A^h\big(t_0 + (\lambda+1)\,h\big) = \mathbf{y}_{A,\lambda} + \big(\mathbf{b}_A \otimes \mathcal{I}_A\big)^\top \mathbf{k}_A^\lambda, \tag{4}$$

where $\lambda = 0, \ldots, m-1$ denotes any internal micro step, $\mathcal{I}_L$ and $\mathcal{I}_A$ represent the identity in the space of latent and active variables, respectively, and $\mathbf{b}_* \in \mathbb{R}^{s_*}$ (s_*: no. of stages). Before we fix the increments, we look at the dense output formula for (3): It is obtained by introducing an additional parameter, θ say, solely for the weights, such that the solution at intermediate timelevels $t_0 + \theta\,h$ is given by

$$\mathbf{y}_L^{H,\,\mathrm{dense}}(\theta) := \mathbf{y}_0 + \big(\mathbf{b}_L(\theta) \otimes \mathcal{I}_L\big)^\top \mathbf{k}_L\,,$$

where θ varies over $[0,1]$. Similar to (2), we have a 'multi'-evaluation $\mathbf{Y}^{H,\,\mathrm{dense}}$ for the dense output formula. Now, the increments $\mathbf{k}$ are defined by

compound step

$$\begin{aligned}\mathbf{k}_L = {}& H\,\mathbf{F}_L\Big(\mathbb{1}_{s_L} \otimes \mathbf{y}_{L,0} + (\mathcal{A}_L \otimes \mathcal{I}_L)\,\mathbf{k}_L,\ \boxed{\mathbb{1}_{s_L} \otimes \mathbf{y}_{A,0} + m(\mathcal{D}_L \otimes \mathcal{I}_A)\,\mathbf{k}_A^0}\Big)\\ &+ H\left(\mathcal{G}_L \otimes \left.\frac{\partial \mathbf{f}_L}{\partial \mathbf{y}_L}\right|_{\mathbf{y}_0}\right)\mathbf{k}_L + m\,H\left(\mathcal{N}_L \otimes \left.\frac{\partial \mathbf{f}_L}{\partial \mathbf{y}_A}\right|_{\mathbf{y}_0}\right)\mathbf{k}_A^0\end{aligned} \tag{5}$$

$$\begin{aligned}\mathbf{k}_A^0 = {}& h\,\mathbf{F}_A\Big(\boxed{\mathbb{1}_{s_A} \otimes \mathbf{y}_{L,0} + 1/m(\mathcal{D}_A \otimes \mathcal{I}_L)\,\mathbf{k}_L},\ \mathbb{1}_{s_A} \otimes \mathbf{y}_{A,0} + (\mathcal{A}_A \otimes \mathcal{I}_A)\,\mathbf{k}_A^0\Big)\\ &+ h\left(\mathcal{G}_A \otimes \left.\frac{\partial \mathbf{f}_A}{\partial \mathbf{y}_A}\right|_{\mathbf{y}_0}\right)\mathbf{k}_A^0 + h/m\left(\mathcal{N}_A \otimes \left.\frac{\partial \mathbf{f}_A}{\partial \mathbf{y}_L}\right|_{\mathbf{y}_0}\right)\mathbf{k}_L\,.\end{aligned} \tag{6}$$

later micro steps for $\lambda = 1, \ldots, m-1$

$$\begin{aligned}\mathbf{k}_A^\lambda = {}& h\,\mathbf{F}_A\left(\boxed{\mathbf{Y}_L^{H,\,\mathrm{dense}}\left(\frac{\lambda\mathbb{1} + \mathcal{A}_A\mathbb{1}}{m}\right)},\ \mathbb{1}_{s_A} \otimes \mathbf{y}_{A,\lambda} + (\mathcal{A}_A \otimes \mathcal{I}_A)\,\mathbf{k}_A^\lambda\right)\\ &+ h\left(\mathcal{G}_A \otimes \left.\frac{\partial \mathbf{f}_A}{\partial \mathbf{y}_A}\right|_{\mathbf{y}_\lambda}\right)\mathbf{k}_A^\lambda\end{aligned} \tag{7}$$

where the Jacobian in the later micro steps is evaluated at

$$\mathbf{y}_\lambda^\top = \big(\mathbf{y}_L^{H,\,\mathrm{dense}}(\lambda/m)^\top,\ \mathbf{y}_{A,\lambda}^{h\,\top}\big). \tag{8}$$

Note, that coefficients used for computing active variables are flagged by A and coefficients for latent parts by L. ($\mathcal{A}_*, \mathcal{G}_* \in \mathbb{R}^{s_* \times s_*}$ are quadratic, while $\mathcal{D}_L, \mathcal{N}_L \in \mathbb{R}^{s_L \times s_A}$, and $\mathcal{D}_A, \mathcal{N}_A \in \mathbb{R}^{s_A \times s_L}$.)

In addtion we remark that the compound step (eq. 5, 6) is equivalent to the compound step defined by the W-method [2]; furthermore, the later micros do not explicitly depend on the previously chosen m, such that, step size control is allowed within all later micro steps. Actually, we may regard the second coupling (latent to the active) to be a given function: i.e. the dense output formula. Finally, there is no constraint, why to use the same coefficients in the compound step and in the later micro steps. Thus we can even substitute (7) by

$$\mathbf{k}_A^\lambda = h\mathbf{F}_A\left(\boxed{\mathbf{Y}_L^{H,\,\text{dense}}\left(\frac{\lambda\mathbb{1} + \mathcal{A}\mathbb{1}}{m}\right)},\ \mathbb{1}_{s_A} \otimes \mathbf{y}_{A,\lambda} + (\mathcal{A} \otimes \mathcal{I}_A)\mathbf{k}_A^\lambda\right) + h\left(\mathcal{G} \otimes \left.\frac{\partial \mathbf{f}_A}{\partial \mathbf{y}_A}\right|_{\mathbf{y}_\lambda}\right)\mathbf{k}_A^\lambda \tag{9}$$

using its own set of coefficients, non flagged this time. This 'multimethod' multirate approach comprises the possibility to do multi-order. But note, since we want to do couplings via dense output, we can only reduce the order in the later micros, which is not advisable, since the computed components are active and hence will change a lot. But there is another possibility, namely to swap parent and embedded solution. This will be discussed below.

4 Order Conditions

We consider the 'multimethod' approach for the active increments (9), since it is most general and allows a straight forward deduction of the order conditions: in this case we have the direct sum of the order conditions for the compound step and for the later micros. The compound step can by analysed using P-theory, which yields the order equations for a P-ROW method (cf. [2,?]), by the same proof. Furthermore, since the coupling is performed by a given function, these steps are nothing else but steps of a MROW method [8]. Therefore the order conditions of MROW apply for these steps. Finally, the coupling function is the dense output formula for the latent part, hence those equations accomplish the set of order conditions. Table 4 presents all of them for an order three method, sorted by the corresponding rooted tree. Note, that fat and light nodes represent the active and latent partition, respectively. Furthermore, the parameter θ varies over $[0, 1]$. Next, we want to deduce an instant of the mixed type multirate method.

5 Coefficient Set

To keep computational cost minimal, while aiming a scheme of order 2(3), we can employ a three staged scheme for the compound step, but have to

Table1. Order equations mixed multimethod ROW-multirate. Notation: $\mathcal{B} = \mathcal{A} + \mathcal{G}$, $\mathcal{E} = \mathcal{D} + \mathcal{N}$.

	LATENT			ACTIVE			
order	tree		eq. no.	eq. no.	compound	tree	later micros
1	∘	$\mathbf{b}_L^\top(\theta)\mathbb{1} = \theta$	1.1L	1.1A	$\mathbf{b}_A^\top\mathbb{1} = 1$	•	$\mathbf{b}^\top\mathbb{1} = 1$
2		$\mathbf{b}_L^\top(\theta)\mathcal{B}_L\mathbb{1} = \theta/2$	2.1L	2.1A	$\mathbf{b}_A^\top\mathcal{E}_A = 1/2$		$\mathbf{b}^\top\mathcal{A}\mathbb{1} = 1/2$
		$\mathbf{b}_L^\top(\theta)\mathcal{E}^L = \theta/2$	2.2L	2.2A	$\mathbf{b}_A^\top\mathcal{B}_A\mathbb{1} = 1/2$		$\mathbf{b}^\top\mathcal{B}\mathbb{1} = 1/2$
3		$\mathbf{b}_L^\top\mathrm{diag}(\mathcal{A}_L\mathbb{1})\mathcal{A}_L\mathbb{1} = 1/3$	3.1L	3.1A	$\mathbf{b}_A^\top\mathrm{diag}(\mathcal{D}_A\mathbb{1})\mathcal{D}_A\mathbb{1} = 1/3$		
		$\mathbf{b}_L^\top\mathrm{diag}(\mathcal{A}_L\mathbb{1})\mathcal{D}_L\mathbb{1} = 1/3$	3.2L	3.2A	$\mathbf{b}_A^\top\mathrm{diag}(\mathcal{A}_A\mathbb{1})\mathcal{D}_A\mathbb{1} = 1/3$		
		$\mathbf{b}_L^\top\mathrm{diag}(\mathcal{D}_L\mathbb{1})\mathcal{D}_L\mathbb{1} = 1/3$	3.3L	3.3A	$\mathbf{b}_A^\top\mathrm{diag}(\mathcal{A}_A\mathbb{1})\mathcal{A}_A\mathbb{1} = 1/3$		$\mathbf{b}^\top\mathrm{diag}(\mathcal{A}\mathbb{1})\mathcal{A}\mathbb{1} = 1/3$ (rows 3.1A–3.5A)
		$\mathbf{b}_L^\top\mathcal{B}_L\mathcal{B}^L\mathbb{1} = 1/6$	3.4L	3.4A	$\mathbf{b}_A^\top\mathcal{E}^A\mathcal{B}^L\mathbb{1} = 1/(6m)$		
		$\mathbf{b}_L^\top\mathcal{B}_L\mathcal{E}_L\mathbb{1} = 1/6$	3.5L	3.5A	$\mathbf{b}_A^\top\mathcal{E}^A\mathcal{E}^L\mathbb{1} = 1/(6m)$		
		$\mathbf{b}_L^\top\mathcal{E}_L\mathcal{E}_A\mathbb{1} = m/6$	3.6L	3.6A	$\mathbf{b}_A^\top\mathcal{B}^A\mathcal{E}^A\mathbb{1} = 1/6$		$\mathbf{b}^\top(\mathcal{B}\mathbb{1})\mathcal{A}\mathbb{1} = 1/6$
		$\mathbf{b}_L^\top\mathcal{E}_L\mathcal{B}_A\mathbb{1} = m/6$	3.7L	3.7A	$\mathbf{b}_A^\top\mathcal{B}^A\mathcal{B}^A\mathbb{1} = 1/6$		$\mathbf{b}^\top(\mathcal{B}\mathbb{1})\mathcal{B}\mathbb{1} = 1/6$

use four stages in the later micro steps [1]. Additionally, we get some second vectors of weights $\mathbf{d}$, which have to define a solution of different order. Since the order conditions are a composite, we may reuse tuned coefficient sets. Especially, MROW [8] provides a small leading error. In addition, it bases the numerical solution on the order 3 approximation, which in our case of the later micro steps is rather nice, since active behaviour is tracked and error control is largely performed on the macro step scale. On the other hand, the coefficient set MWrow (cf. [1]) can be reused for the compound step by skipping the some coefficients (which correspond to coupling computation in the later micro steps), but providing dense output for latent part. This yields a working coefficient set, which is given in Table 2.

Table 2. Coefficient set - Mixed ROW.

Compound step (involves simplifying conditions: $\widehat{\mathbf{b}} := \mathbf{b}_L = \mathbf{b}_A$, $\widehat{\mathbf{d}} := \mathbf{d}_L = \mathbf{d}_A$, $\widehat{\mathcal{A}} := \mathcal{A}_L = \mathcal{A}_A = \mathcal{D}_A = \mathcal{D}_L$, and $\widehat{\mathcal{G}} := \mathcal{G}_L = \mathcal{G}_A$)

$$\widehat{\mathbf{b}} = \begin{pmatrix} 1/6 \\ 2/3 \\ 1/6 \end{pmatrix} \quad \widehat{\mathbf{b}}(\theta) = \begin{pmatrix} -(5/6+2\gamma)\,\theta^2 + (1+2\gamma)\,\theta \\ (2/3+2\gamma)\,\theta^2 - 2\gamma\,\theta \\ \theta^2/6 \end{pmatrix} \quad \widehat{\mathbf{d}} = \begin{pmatrix} 0 \\ 1 \\ 0 \end{pmatrix}$$

$$\widehat{\mathcal{A}} = \begin{pmatrix} 0 & & \\ 1/2 & 0 & \\ 0 & 1 & 0 \end{pmatrix} \quad \mathcal{N}_A = \begin{pmatrix} \gamma & & \\ -\gamma & \gamma & \\ \frac{1-2/m+8\gamma-8\gamma^2}{1-2\gamma} & \frac{2/m-1-10\gamma+12\gamma^2}{1-2\gamma} & \gamma \end{pmatrix}$$

$$\widehat{\mathcal{G}} = \begin{pmatrix} \gamma & & \\ -\gamma & \gamma & \\ 1+2\gamma & -4\gamma-1 & \gamma \end{pmatrix} \quad \mathcal{N}_L = \begin{pmatrix} \gamma & & \\ -\gamma & \gamma & \\ \frac{1-2m+8\gamma-8\gamma^2}{1-2\gamma} & \frac{2m-1-10\gamma+12\gamma^2}{1-2\gamma} & \gamma \end{pmatrix}$$

and $\gamma = (2+\sqrt{2})/2$

Later micro steps

$$\mathbf{b}^\top = \begin{pmatrix} 1 & 1 & 2 & 2 \end{pmatrix}/6 \quad \mathbf{d}^\top = \begin{pmatrix} 0.33257135 & 0.33257135 & 0.33485731 & 0 \end{pmatrix}$$

$$\mathcal{A} = \begin{pmatrix} 0 & & & \\ 1 & 0 & & \\ 1/8 & 3/8 & 0 & \\ 1/8 & 3/8 & 0 & 0 \end{pmatrix} \quad \mathcal{G} = \begin{pmatrix} 0.395 & & & \\ -1.57583400 & 0.395 & & \\ 0.38546938 & 0 & 0.395 & \\ 0.117508005 & -0.72736757 & -0.23026486 & 0.395 \end{pmatrix}$$

6 Numerical Test

In our MATLAB test environment we model the well known inverter chain benchmark by a piecewise quadratic characteristics f for the MOSFETs (see [10]). This yields the following set of equations for n inverters: ($U_{\text{op}} = 5\text{V}$)

$$\begin{aligned} \dot{u}_1 &= U_{\text{op}} - u_1 - \Upsilon f(u_{\text{in}}, u_1, 0), \\ \dot{u}_k &= U_{\text{op}} - u_k - \Upsilon f(u_{k-1}, u_k, 0) \quad \text{for } k = 2, \dots, n\,. \end{aligned} \tag{10}$$

The stiffness is more or less controlled by Υ. Thus we assign $\Upsilon = 100$ to obtain a stiff setup. Supposing n is even, the fed-in signal u_{in} is merely transmitted, i.e. delayed, by the circuit (plus some smoothing effects). The input was defined as the following polyline signal:

$$(0\text{ns}, 0\text{V}),\ (5\text{ns}, 0\text{V}),\ (10\text{ns}, 5\text{V}),\ (15\text{ns}, 5\text{V}),\ (17\text{ns}, 0\text{V}),\ (70\text{ns}, 0\text{V})\,.$$

Furthermore, a dynamic partitioning is provided by an inspection of the right-hand sides (rhs) of (10): Given a positive real number $\varepsilon_{\text{actlevel}}$, u_i is active, if $|\text{rhs}_i| \geq \varepsilon_{\text{actlevel}}$.

Numerical results for the stiff benchmark, comprising $n = 200$ inverters, are summarised in Table 3. As reference we compare the multirate results with the MATLAB build-in routine odes23s [12], which corresponds to the first row of that table (sr = single rate).

Table 3. Stiff inverter chain test.

mode	err scal	$\varepsilon_{\text{actlevel}}$	macro steps	micro steps	# G ops	RHS-eval	ratio eval	remark
sr (ode23s)	-	-	-	2049	35.21	819 600	-	(+rej stp)
mr (mxrow)	off	0.5	248(16)	3289(111)	4.35	327 418	2.50	slight delay
	on	0.7	135(4)	2573(8)	2.36	178 114	4.60	(delay)
	on	0.05	289(11)	3364(47)	5.60	527 054	1.56	

Since the load part in the simulation of circuits is most costly [6], speed up is measured in the ratio of RHS-evaluations (column 'ratio-eval', Table 3), while the MATLAB operations (G ops=10^9 ops) shall give a mere indication. Thus a fair speed up of 2.50 is obtained, but we note that the number of micro steps is increased compared to the single rate case. Of course, the larger the latency, i.e. the longer the inverter chain, the less dominant will be this effect. In addition, tests with an error scaling of the form $\mathbf{err} = 2\,(\mathcal{I} - h\gamma\mathbf{f_y})^{-1}(\mathbf{y}^1 - \mathbf{y}^2)$ were investigated; $\mathbf{y}^1$ and $\mathbf{y}^2$ denote numerical approximations of different order. Here larger speed ups are possible at the cost of a delay. However, the number of micro steps mirrors much better the number of steps in the single rate algorithm. – The effect of the delay can be decreased by tuning the activation level (see Table 3).

Finally, we remark that the success of the internal micro steps was viewed by inspecting the LU-decomposition of the system matrix. If U was 'close' to singular, the interior micro step was updated by halving the micro step size. Thus we really use the controlling advantage of the mixed ROW method.

7 Conclusions

Having developed the consistency theory of mixed multirate ROW, tests on the inverter chain were quite successful. Comparing the results to the MPRK algorithm [9], mixed ROW needs only about a sixth of the number of micro steps. This is of course due to the implicit computation within mixed ROW.

A desirable extension of our algorithm will be a multirate method based on *ROW methods adapted to cheap Jacobian* [7], which allows to save RHS evaluations by the usage of 'internal' Jacobians. This is especially nice for circuit simulation, where the Jacobian is 'cheap' available. Last, general circuit equations are DAEs, therefore developments in the direction of the quasi-linear implicit structure of charge-flux-oriented network equations are necessary.

Acknowledgements

Thanks go to M. Günther and P. Rentrop for many fruitful discussions and their guidance. Moreover, this work is part of the BMBF project "Modifizierte ROW-Methoden in der elektrischen Schaltkreissimulation".

References

1. Bartel, A.: Generalised Multirate – Two ROW-type Versions for Circuit Simulation. *Unclassified NatLab Report* No. 2000/804. Philips Electronics (2000).
2. Bartel, A., Günther, M.: Circuit Simulation with Generalised Multirate. Submitted to *J. Comp. Appl. Math.*
3. Engstler, Ch., Lubich, Ch.: Multirate extrapolation methods for differential equations with different time scales. *Computing* **58** (1997), pp. 173–185.
4. Engstler, Ch. Lubich Ch.: MUR8: A Multirate Extension of the Eighth-order Dormand-Prince Method. *Appl. Numer. Math.* **25** (1997), pp. 185–192.
5. Gear, C.W., Wells, D.R.: Multirate Linear Multistep Methods. *BIT* **24** (1984), pp. 484–502.
6. Günther M., Hoschek, M.: ROW methods adapted to electric circuit simulation packages. *J. Comp. Appl. Math.* **82** (1997), pp. 159–170.
7. Günther, M., Hoschek, M., Weiner, R.: ROW methods adapted to cheap Jacobian. To appear in *Appl. Num. Math.*
8. Günther, M., Rentrop P.: Multirate ROW methods and latency of electric circuits. *Applied Numerical Mathematics* **13** (1993), pp. 83–102.
9. Günther, M., Kværnø, A., Rentrop, P.: Multirate Partitioned Runge-Kutta Methods. Submitted to BIT.
10. Kværnø, A., Rentrop, P.: Low Order Multirate Runge-Kutta Methods in Electric Circuit Simulation. Preprint Nr. 99/1, IWRMM, University of Karlsruhe.

11. Ostermann, A.: Continuous Extensions of Rosenbrock-Type Methods. *Computing* **44** (1990), pp. 59–68.
12. Shampine, L.F., Reichelt, M.W.: The MATLAB ode suite. *SIAM J. of Scient. Comp.*, **18** (1997), pp. 1–22.
13. Steihaug, T., Wolfbrandt, A.: An attempt to avoid exact Jacobian and nonlinear equations in the numerical solution of stiff ordinary equations. *Math. Comp.*, **5** (1989), pp. 151–160.

Neural Networks for Device and Circuit Modelling

P.B.L. Meijer

Philips Research Laboratories, Eindhoven, The Netherlands
E-mail: Peter.B.L.Meijer@philips.com

Abstract The standard backpropagation theory for static feedforward neural networks can be generalized to include continuous dynamic effects like delays and phase shifts. The resulting non-quasistatic feedforward neural models can represent a wide class of nonlinear and dynamic systems, including arbitrary nonlinear static systems and arbitrary quasi-static systems as well as arbitrary lumped linear dynamic systems. When feedback connections are allowed, this extends to arbitrary nonlinear dynamic systems corresponding to equations of the general form $\boldsymbol{f}(\boldsymbol{x}, \dot{\boldsymbol{x}}, t) = \mathbf{0}$. Extensions of learning algorithms to include combinations of time domain and frequency domain optimization lead to a semi-automatic modelling path from behaviour to simulation models. Model generators have been implemented for a range of existing analog circuit simulators, including support for the VHDL-AMS and Verilog-AMS language standards.

1 Introduction

With the continuing advances in digital technology, now with deep-submicron devices and wires switching at such high frequencies that RF effects appear, it becomes increasingly hard to maintain a clear separation between the digital abstraction from 1's and 0's upward and the underlying physical world which is characterized by analog behaviours. Together with the rapidly increasing complexity of digital and mixed-signal designs, this calls for methodologies that further support reuse as well as the abstraction from analog behaviour at the device level to intermediate levels of abstraction. Our aim is to find simplified simulation models while sufficiently preserving the analog functional behaviour in order to allow for further testing and validation of designs and design rules through simulation. New language standards like VHDL-AMS and Verilog-AMS help to co-simulate analog models in a digital environment, but the task of finding efficient and sufficiently accurate analog and mixed-signal models often remains a daunting one.

The methodology as outlined in this paper may on the one hand assist in quickly obtaining accurate analog simulation models where no suitable or efficient physical device models are available (yet), while on the other hand it may find applications in macro-modelling for mixed-level mixed-signal simulation where structural models already do exist in the form of netlists of interconnected transistors and other devices, but not or not yet in the form of more efficient simplified functional models.

The proposed approach is based on a generalization of feedforward neural networks, also known as multilayer perceptron (MLP) networks [6]. We add time differentiation to the connecting weights and use nonlinear transfer functions and time integration in the neuron bodies to arrive at a modelling formalism capable of representing the time-dependent nonlinear behaviour of a very wide class of electronic circuits. Learning of dynamic nonlinear multivariate behaviour can be done with combinations of time domain and frequency domain data. However, since the approach is still ultimately rooted in continuous optimization algorithms for minimizing errors between supplied behavioural data and model, it does not on its own take away the intrinsic problems associated with all known continuous nonlinear optimization methods, such as a possibly slow convergence (when far from the optimum point) and the risk of being trapped in some local minimum. These problems, if they occur, can be circumvented or alleviated by using suitable templates for initializing the neural network topology and parameters. This hybrid approach will be outlined and illustrated in sections 3 and 4.

2 Neural Network Equations

The neural networks used in this paper are defined in this section. A detailed motivation for the various specific choices can be found in [4]. Layers are counted starting with the input layer as layer 0, such that a network with output layer K involves a total of $K+1$ layers. Layer k by definition contains N_k neurons. A vector notation with bold font is used to denote information on all neurons in a particular layer. A neural network has a vector of inputs $\boldsymbol{x}^{(0)}$ and a vector of outputs $\boldsymbol{x}^{(K)}$.

The differential equation for the output, or excitation, y_{ik} of one particular neuron i in layer $k > 0$ is given by

$$\tau_{2,ik} \, \frac{\mathrm{d}^2 y_{ik}}{\mathrm{d}t^2} + \tau_{1,ik} \, \frac{\mathrm{d}y_{ik}}{\mathrm{d}t} + y_{ik} = \mathcal{F}^{(ik)}(s_{ik}, \delta_{ik}) \tag{1}$$

with timing parameters $\tau_{1,ik}$ and $\tau_{2,ik}$, and the source term $\mathcal{F}^{(ik)}$ a (generally nonlinear) function having an optional transition parameter δ_{ik}. The weighted sum s_{ik} of results from the preceding layer is further defined as

$$\begin{aligned} s_{ik} &\triangleq \boldsymbol{w}_{ik} \cdot \boldsymbol{y}_{k-1} - \theta_{ik} + \boldsymbol{v}_{ik} \cdot \frac{\mathrm{d}\boldsymbol{y}_{k-1}}{\mathrm{d}t} \\ &= \sum_{j=1}^{N_{k-1}} w_{ijk} \, y_{j,k-1} - \theta_{ik} + \sum_{j=1}^{N_{k-1}} v_{ijk} \, \frac{\mathrm{d}y_{j,k-1}}{\mathrm{d}t} \end{aligned} \tag{2}$$

for $k > 1$, involving weighting parameters w_{ijk} and v_{ijk}, an offset parameter θ_{ik}, and similarly for the neuron layer $k = 1$ connected to the network inputs

$x_j^{(0)}$

$$s_{ik} \triangleq \boldsymbol{w}_{ik} \cdot \boldsymbol{x}^{(0)} - \theta_{ik} + \boldsymbol{v}_{ik} \cdot \frac{\mathrm{d}\boldsymbol{x}^{(0)}}{\mathrm{d}t}$$
$$= \sum_{j=1}^{N_0} w_{ij,0}\, x_j^{(0)} - \theta_{i,0} + \sum_{j=1}^{N_0} v_{ij,0} \frac{\mathrm{d}x_j^{(0)}}{\mathrm{d}t} \tag{3}$$

which is analogous to having a dummy neuron layer $k = 0$ with enforced neuron j outputs $y_{j,0} \equiv x_j^{(0)}$, or in vector notation $\boldsymbol{y}_0 \equiv \boldsymbol{x}^{(0)}$.

Finally, to allow for arbitrary network output ranges in case of bounded functions $\mathcal{F}^{(ik)}$, a linear scaling transformation is added to the output stage

$$x_i^{(K)} = \alpha_i\, y_{iK} + \beta_i \tag{4}$$

yielding a network output vector $\boldsymbol{x}^{(K)}$.

The above function $\mathcal{F}^{(ik)}$ is for neuron i in layer k applied to the weighted sum s_{ik} of neuron outputs $y_{j,k-1}$ in the preceding layer $k-1$. The optional transition parameter δ_{ik} may be used to set an appropriate scale of change in qualitative transitions in function behaviour, as is common to semiconductor device modelling. For example, choosing the following function $\mathcal{F}_2$ as

$$\mathcal{F}^{(ik)}(s_{ik}, \delta_{ik}) = \mathcal{F}_2(s_{ik}, \delta_{ik}) \triangleq \frac{1}{\delta_{ik}^2} \ln \frac{\cosh \frac{\delta_{ik}^2 (s_{ik}+1)}{2}}{\cosh \frac{\delta_{ik}^2 (s_{ik}-1)}{2}} \tag{5}$$

lets the parameter δ_{ik} be used to optimize the transition between a nearly linear region and two asymptotically exponential tails. The function $\mathcal{F}_2$ is illustrated in Fig. 1.

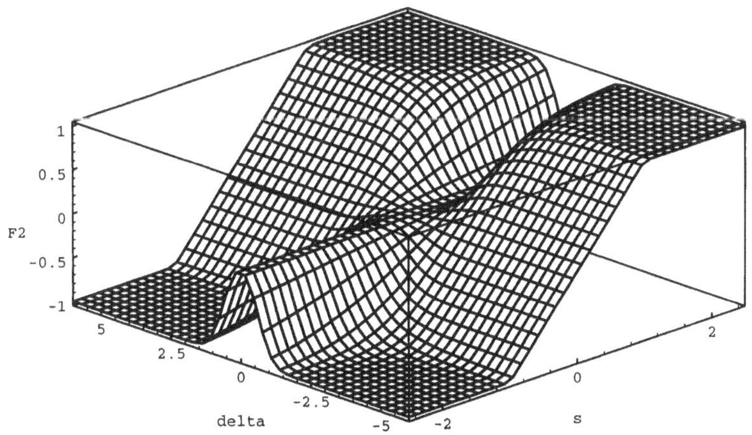

Figure1. Neuron nonlinearity $\mathcal{F}_2(s_{ik}, \delta_{ik})$.

Again referring to [4], it can be shown that the above neural network equations can represent any lumped linear dynamic system, and can arbitrarily closely approximate any multivariate static (DC) or quasistatic model—such as typically used to model the DC and capacitive currents of MOSFET transistors. Under very weak conditions, this approximation property applies even when the $\mathcal{F}^{(ik)}$ are the same for all neurons, while requiring only three layers for static models [1,2,3] and at most four layers for quasistatic models.

Moreover, when allowing for external feedback connections from the outputs of the dynamic feedforward neural network as defined above back to the network inputs, it can be shown[1] that the resulting class of neural networks can represent arbitrary nonlinear dynamic systems corresponding to equations of the general form $\boldsymbol{f}(\boldsymbol{x}, \dot{\boldsymbol{x}}, t) = \mathbf{0}$, thus covering a very broad class of multidimensional nonlinear non-quasistatic systems [4,5]. This includes systems having multiple DC solutions, which the above-defined feedforward neural networks cannot represent without these additional feedback connections.

3 Initialization Templates and Learning Phase

Apart from using one and the same nonlinearity for all neurons, one may also select an appropriate function $\mathcal{F}^{(ik)}$ for each neuron individually, for instance by using initializing templates that capture any available a priori knowledge about the approximate behaviour or structure of the device or circuit to be modelled. Furthermore, once known, the set of poles and zeros of any linear transfer function can be mapped exactly and constructively to corresponding linear(ized) neural (sub)networks [4]. Univariate transcendental functions, such as sine, square root and exponential, are readily associated with corresponding instances of $\mathcal{F}^{(ik)}$, while the (bivariate) multiplication of two sub-expressions is easily mapped to a neural (sub)network through a linear combination of scalar squaring functions as in $xy = \frac{1}{4}[(x+y)^2 - (x-y)^2]$, using two instances of $\mathcal{F}^{(ik)}$ for the squaring operation. One can apply these basic mappings to create templates manually, but one can also devise a general parser for automatically mapping differential-algebraic systems of equations onto exactly equivalent dynamic neural networks.

[1] The existence proof builds on the theorems for static models [1,2,3,4]. By defining a companion function $\boldsymbol{F}(\boldsymbol{x}, \dot{\boldsymbol{x}}, t)$ we may arbitrarily closely approximate the *static* (w.r.t. the direct arguments $\boldsymbol{x}$, $\dot{\boldsymbol{x}}$ and t) function expression $\boldsymbol{f}(\boldsymbol{x}, \dot{\boldsymbol{x}}, t) + \boldsymbol{x}$, by making use of the time differentiation available in the input connections to the neural network and adding time as an additional input, the latter either directly or via a time dependent vector function $\boldsymbol{u} = \boldsymbol{u}(t)$. Next closing the feedback connections for $\boldsymbol{x}$ to obtain $\boldsymbol{f}(\boldsymbol{x}, \dot{\boldsymbol{x}}, t) + \boldsymbol{x} \equiv \boldsymbol{x}$ completes the implicit system equivalent to the state equation $\boldsymbol{f}(\boldsymbol{x}, \dot{\boldsymbol{x}}, t) = \mathbf{0}$ that was to be modelled. Any separate output equations of the general form $\boldsymbol{y} = \boldsymbol{G}(\boldsymbol{x}, \boldsymbol{u}, \dot{\boldsymbol{u}})$, can similarly be accounted for and included in the neural network.

After the optional initialization phase, where topologies and parameters are set through a template containing a priori modelling knowledge, the actual optimization ("learning") starts. This involves a continuous optimization algorithm for minimizing a cost function for the model error w.r.t. the supplied behavioural data. This data in turn can be a mixture of DC, time domain and frequency domain data. The latter may include scatter parameters obtained from small signal AC analysis or measurements. Many optimization algorithms, such as conjugate gradient and BFGS, need the gradient of the cost function, and this implies the use of DC sensitivity, transient sensitivity and AC sensitivity. A basic time integration algorithm such as Backward Euler can be used for the discretized transient analysis and transient sensitivity. Higher order algorithms may be applied for greater simulation efficiency at the expense of increased algorithmic complexity.

4 Using A Priori Knowledge - An Example

To illustrate the use of a priori knowledge in macro-modelling a complex circuit by means of dynamic neural networks, we will consider the folding part of a folding AD converter designed within Philips. This circuit block has one input and 32 outputs. Within the operating range of interest, the DC output of each of these outputs is approximately sinusoidal as a function of the input. Different outputs differ only by subsequent phase steps of $\pi/32$. Therefore, we can make use of $\sin(x+\phi_i) = c_{1i} sin(x) + c_{2i} sin(\pi/2 - x)$ with $c_{1i} = \cos(\phi_i)$, $c_{2i} = \sin(\phi_i)$, to write all the DC outputs as linear combinations of just two instances of sine functions. Next, the neural network of choice was correspondingly initialized as a 1-2-32 network, with two instances of $\mathcal{F}^{(i1)}(s_{i1}, \delta_{i1}) = \sin(2\pi s_{i1})$ in the single "hidden" layer $k = 1$, and the identity functions $\mathcal{F}^{(i2)}(s_{i2}, \delta_{i2}) = s_{i2}$ for the 32 outputs in layer $k = 2$. The neural network topology is illustrated in Fig. 2.

The initial values for $\tau_{1,ik}$ were initially all identical and estimated from graphical inspection of the high-frequency roll-off of the transistor-level circuit behaviour. The initial values for $\tau_{2,ik}$ were based on those for $\tau_{1,ik}$ while ensuring a small initial quality factor $Q \lll 1$ for each neuron. The details of setting proper values for the weights and offsets will be omitted here. Through subsequent optimization with our neural modelling software, the timing parameters were readily adapted for a much better fit to the actual circuit response curves.

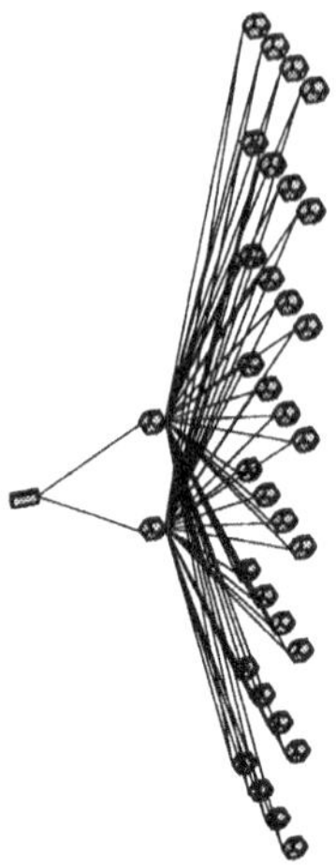

Figure2. Neural network topology for 1-2-32 network.

The final result of neural network training is illustrated in Fig. 3 where model behaviour and circuit behaviour ("Target output") are shown for one of the 32 outputs (all outputs gave comparable results), and using a sinusoidal time-dependent input signal—which had also been used in the training phase[2]. The higher frequencies in the output signal are the result of the input signal amplitude spanning several periods of the approximately sinusoidal DC response. The model accurately represents the high-frequency roll-off, which causes the non-constant envelope in the output signal, with smaller excursions at the higher frequencies.

The original circuit block contained about five hundred transistors, normally simulated using advanced analog CAD device models, while the neural model now only needs two sine functions and 32 simple linear combinations of these, plus low-order linear differential equations to account for the dynamic behaviour. The computational complexity was therefore lowered by several orders of magnitude, and depending on the simulator used, simulation speedups of two or three orders of magnitude were obtained for simulating the model versus the circuit block. The differential-algebraic equations and parameters of any neural model can be automatically mapped to the syntax supported by the preferred analog circuit simulator. For example, the neuron body as specified in Eq. (1) can be rewritten and translated into corresponding VHDL-AMS code like

[2] In general one should run a cross-validation with data that was not used in the training in order to reduce the risk of accidental over-fitting.

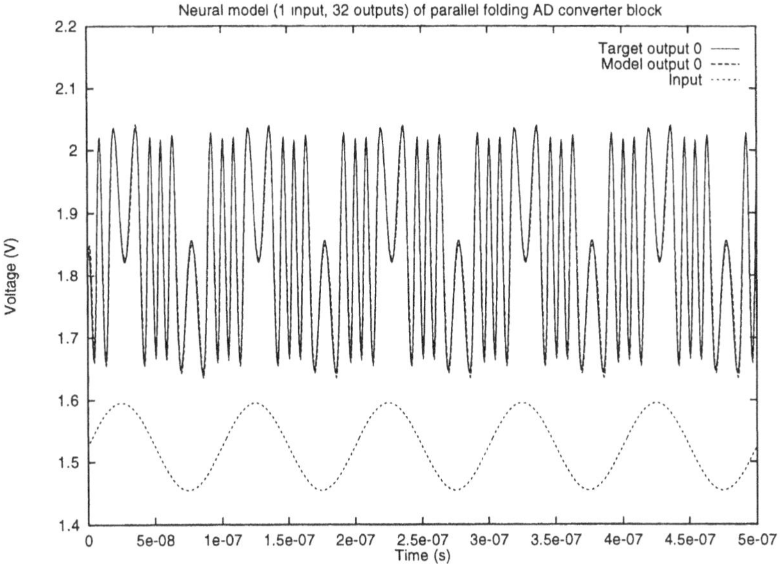

Figure3. Simulation with original circuit and neural model.

```
PACKAGE BODY soma_pkg IS

   USE IEEE.MATH_REAL.all;

   -- DC nonlinearity selection and function evaluation
   FUNCTION somadc(ftype: INTEGER; s, delta: REAL) RETURN REAL IS
      VARIABLE f: REAL;
   BEGIN
     CASE ftype IS
     WHEN 0 => f := s;
     WHEN 5 => f := sin(math_2_pi*s);
     WHEN OTHERS => RETURN 0.0;
     END CASE;
     RETURN f;
   END somadc
END soma_pkg;

ENTITY dynsoma IS
   GENERIC (ftype: INTEGER; delta, tau1, tau2: REAL);
   PORT (TERMINAL INA, OUTA, REF: ELECTRICAL);
END ENTITY dynsoma;

ARCHITECTURE soma OF dynsoma IS
   TERMINAL AUX: ELECTRICAL;
   QUANTITY s ACROSS INA TO REF;                -- Weighted sum s
   QUANTITY x ACROSS AUX TO REF;                -- Auxiliary variable
   QUANTITY y ACROSS iout THROUGH OUTA TO REF; -- Neuron output y
BEGIN
   x == tau1 * y'dot;
   y == somadc(ftype,s,delta) - x - tau2/tau1 * x'dot;
END ARCHITECTURE soma;
```

where the function type ftype was used to specify what function $\mathcal{F}$ should be used in the folding AD converter model. A different description is used for model instances with zero-valued τ_1. Similarly, model code can be generated for Verilog-AMS or any other sufficiently rich simulation or programming language. Consistency among models in different simulation languages is always ensured through the automatic mapping by model generators, which all use the same network topology and parameter set.

5 Conclusions

Dynamic neural networks can be used for a wide range of device and circuit modelling applications. The generalized formalism allows for a hybrid modelling approach where existing knowledge can be incorporated before starting the general optimization or learning phase. This approach can help to trade off the respective strengths and weaknesses of physical modelling and automatic black-box behavioural modelling.

References

1. K.-I. Funahashi, "On the Approximate Realization of Continuous Mappings by Neural Networks," *Neural Networks*, Vol. 2, pp. 183-192, 1989.
2. K. Hornik, M. Stinchcombe and H. White, "Multilayer Feedforward Networks are Universal Approximators," *Neural Networks*, Vol. 2, pp. 359-366, 1989.
3. M. Leshno, V. Y. Lin, A. Pinkus and S. Schocken, "Multilayer Feedforward Networks With a Nonpolynomial Activation Function Can Approximate Any Function," *Neural Networks*, Vol. 6, pp. 861-867, 1993.
4. P. B. L. Meijer, "Neural Network Applications in Device and Circuit Modelling for Circuit Simulation," *Ph.D. thesis*, Eindhoven University of Technology, May 2, 1996.
5. P. B. L. Meijer, "Signal generator for modelling dynamical system behaviour,", *U.S. patent*, No. 5790757, August 4, 1998.
6. D. E. Rumelhart and J. L. McClelland, Eds., *Parallel Distributed Processing, Explorations in the Microstructure of Cognition*. Vols. 1 and 2. Cambridge, MA: MIT Press, 1986.

Sub-micrometer Bipolar Transistor Modeling Using Neural Networks

A. Plebe[1], A.M. Anile[1], and S. Rinaudo[2]

[1] University of Catania Department of Mathematics and Informatics V.le Andrea Doria, 8 I-95125 Catania, Italy
[2] ST Microelectronics Str. Primosole, 50 I-95121 Catania, Italy

Abstract An approach based on Artificial Neural Networks (ANN) for constructing models of high speed bipolar transistors is described. This method is proposed as an alternative for physical modeling for circuit simulation, when high frequency and small device size make classical models very complex or even unreliable. In the ANN here adopted, neurons are represented in terms of continuous-time differential equations, allowing the immediate application inside conventional circuit simulators. The most difficult task in this approach is the network training from the measurements on the real device, and the usual learning rules for ANN's easily lead to poor approximation or unacceptable slowness. A generative method has been developed, where a subset of the network parameters is trained inside an auxiliary static network, using measurements at fixed DC bias. The complete network is trained on the full set of measurements using this subset as a starting point. The learning rule is a combination of global optimization followed by a quasi-Newton conjugate-gradient iterative process.

1 Introduction

Until recently physical modeling has been the prevailing approach to semiconductor modeling for circuit simulation, Gummel-Poon being the choice for Bipolar Junction Transistors (BJT) at high frequencies [1]. The knowledge of the underlying physical principles was sufficient for deriving effective models, where each parameter has a clear physical meaning. The effort required to derive the parameters was still acceptable, since it was basically reduced to a number of one-dimensional phenomena analyses, and neglecting several complex side-effects was not causing a significant loss in accuracy.

This situation is gradually changing, with the current size of BJT made available by the microelectronics manufacturing capabilities. As long as the ratio of surface to volume increases for the smallest devices, many boundary effects should be taken into account, leading to complex multidimensional analysis of the phenomena. Reliable models require now a high development cost. Furthermore, precise physical models should now rely on a number of largely empirical fit parameters.

An adopted alternative is the use of so called Table Models [9] [11], where interpolation techniques are used to fit behavior of BJT in every possible working condition, from a set of measured data. One of the drawbacks of table models is the difficulty in imposing a smoothness better then C^1, which may limit the performance of the integration scheme of the circuit simulation where the model is used.

The method here investigated is based on ANN, which can be seen again as a fitting technique, with the advantage of C^∞ output functions. ANN approaches have gained widespread attention over the past two decades, with a variety of applications, including dynamic systems identification, control and simulation. However, the traditional schemes used for dynamic ANN, like tapped-delays feed-forward networks, feedback and self-recurrent networks, are not immediately usable for models to be embedded in circuit simulators. Simulators require a representation in terms of continuous-time equations, since time discretisation will be an internal process of the simulator itself.

Recently Meijer [5] introduced a new neural architecture where neurons are instances of second order differential equations, corresponding to the type **dynamic** in Table 1, and the non-linear neural function is expressed by a combination of typical semiconductor junction exponential non-linearity, as in (4).

A difficulty in training this type of ANN on BJT data arises because of the large number of parameters for each neuron. On the other hand, the total number of neurons in the network cannot be limited due to the complex static behavior of the BJT, as well as the large variation of dynamic behavior inside the DC bias space. The strategy here proposed, is based on a mixed architecture, where simple static neurons are combined with dynamic neurons, now extended to allow variable dynamics when DC bias point changes. This last type of neuron has a larger set of parameter, but the major advantage is the possibility to learn its parameters using an auxiliary simplified network architecture, with the method described in section 3. Still training is not a trivial task, and even the most refined back-propagation learning rules (*Quick-Prop* [2], *Delta-Bar-Delta* [4], Levenberg-Marquardt [10]) are not sufficient for learning real BJT behavior.

A quasi-Newton conjugate-gradient method has been used, where the objective function is the square error cumulated over the entire training set. The speed of this method is counterweighted by the sensitivity on the initial parameter vector. To overcome this problem, the first guess is obtained by an other optimization method, based on the simplex geometry, which performs a global search over the parameter space. This method will be described in section 2.1, and the former in 2.2.

2 Extended Dynamic Neural Networks

The model of artificial Dynamic Neural Networks (DNN) here proposed is based on four different types of neurons, as illustrated in Table 1, where y_i is the value at a generic neuron, and x_j the neurons in the layer below. The number of parameters for each neuron is given by a constant for each

Table1. Node types in the DNN, with equations, number of parameters, and graphic symbol

type	equation	$\#_{\text{par}}$	symbol
static	$y_i = \mathcal{F}\left(\sum_{j=1}^{N} w_{ij}x_j - \theta_i,\ \delta_i\right)$	$N+2$	
dummy	$y_i = \sum_{j=1}^{N} w_{ij}x_j$	N	
sigmoidal	$y_i = \mathcal{S}\left(\sum_{j=1}^{N} w_{ij}x_j - \theta_i\right)$	$N+1$	
dynamic	$\tau_2 \frac{\partial^2 y_i}{\partial t^2} + \tau_1 \frac{\partial y_i}{\partial t} + y_i = \mathcal{F}(a_i, \delta_i)$	$2N+4$	
varying dynamic	$\mathcal{T}_2(o_i, \omega_i)\frac{\partial^2 y_i}{\partial t^2} + \mathcal{T}_1(o_i, z_i, \omega_i, \zeta_i)\frac{\partial y_i}{\partial t} + y_i = \mathcal{F}(a_i, \delta_i)$	$4N+6$	

type, and a term proportional to the size N of the underlying layer. Neural activations a_i, o_i and z_i used in the dynamic nodes are the following:

$$a_i = \sum_{j-1}^{N} w_{ij}x_j + \sum_{j-1}^{N} v_{ij}\frac{\partial x_j}{\partial t} - \theta_i \ , \tag{1}$$

$$o_i = \sum_{j=1}^{N} u_{ij}x_j - v_i \ , \tag{2}$$

$$z_i = \sum_{j=1}^{N} s_{ij}x_j - \sigma_i \ . \tag{3}$$

The non-linear functions are summarized in the following equations, now the neuron index i is omitted, and other local parameters of the functions (δ, ω, ζ)

are made explicit:

$$\mathcal{F}(a,\delta) = \frac{1}{\delta^2} \log \frac{e^{\frac{\delta^2 (a+1)}{2}} + e^{\frac{-\delta^2 (a+1)}{2}}}{e^{\frac{\delta^2 (a-1)}{2}} + e^{\frac{-\delta^2 (a-1)}{2}}} , \tag{4}$$

$$\mathcal{S}(a) = \frac{1}{1+e^{-a}} , \tag{5}$$

$$\mathcal{T}_1(o,z,\omega,\zeta) = \frac{2\zeta(1+e^{-o})}{\omega(1+e^{-z})} , \tag{6}$$

$$\mathcal{T}_2(o,\omega) = \frac{1+2e^{-o}+e^{-2o}}{\omega^2} . \tag{7}$$

There are also four different type of possible connections in the DNN, expressed in the (1), (2),(3), which nature and graphical symbol are summarized in Table 2.

Table2. Kinds of connection between neurons

w proportional connection	——
v derivative connection	········
u natural frequency connection	········
s damping ratio connection	········

2.1 Global Optimization

In the training methods used for DNN, the objective function is computed on the whole set of measurements:

$$\mathcal{E}(\boldsymbol{p}) = \sum_k \sum_i \left(\widehat{\boldsymbol{x}}_{k,i} - \boldsymbol{x}_{k,i}(\boldsymbol{p})\right)^T \left(\widehat{\boldsymbol{x}}_{k,i} - \boldsymbol{x}_{k,i}(\boldsymbol{p})\right) , \tag{8}$$

where k is a set of samples in a series (in time or frequency), i is the sample index, $\widehat{\boldsymbol{x}}$ is the measured output of the BJT and $\boldsymbol{x}(\boldsymbol{p})$ the output of the DNN when using the parameter vector $\boldsymbol{p}$.

The purpose of the global method is to span a broad region of the parameter space, searching for a global rough minimum. The method adopted is the *Controlled Random Search* [8], based on simplex geometry. Being A the desired region of the parameter space to search, a set $\mathcal{P}$ of N points is sampled randomly in A. The method is an iteration over the following three steps:

1. simplex building, i.e., building the set:

$$\mathcal{S} = \{P_1 \dots P_{n+1} : P_i \in \mathcal{P},\ \mathcal{E}(\boldsymbol{p}_1) = \min_{\mathcal{P}}\{\mathcal{E}(\boldsymbol{p})\}\} , \tag{9}$$

where n is the parameter dimension and $\boldsymbol{p}_i$ the vector associated with point P_i;

2. generation of the new point $\widehat{P}$ with vector $\widehat{\boldsymbol{p}} = 2\ E\{\mathcal{S}\} - \boldsymbol{p}_{n+1}$, $E\{\cdot\}$ being the average operation;
3. if $\widehat{P}$ is inside A and condition $\mathcal{E}(\widehat{\boldsymbol{p}}) < \max_{\mathcal{P}}\{\mathcal{E}(\boldsymbol{p})\}$ is satisfied, the new point is insert in $\mathcal{P}$:

$$\mathcal{P} \leftarrow \mathcal{P} \bigcup \{\widehat{P}\} - \left\{P_\bullet : P_\bullet \in \mathcal{P},\ \mathcal{E}(\boldsymbol{p}_\bullet) = \max_{\mathcal{P}}\{\mathcal{E}(\boldsymbol{p})\}\right\} . \tag{10}$$

Although this method is extremely slow, it succeeds in producing a first guess of the parameter for the fast optimization described in the next section.

2.2 Quasi-Newton Optimization

This classical method [6], as its name suggests, searches for a minimum of the objective function quite like the way the Newton method searches for the root of an equation, which in this case is given equating the gradient of the objective function to zero.

Using the objective function defined in (8), the optimization at each step s proceeds on a one-dimensional non-dimensional α.

$$\alpha_s = \arg \min_{0<\alpha<\alpha_M} \{\mathcal{E}(\boldsymbol{p}_s + \alpha \boldsymbol{v}_s)\} , \tag{11}$$

where $\boldsymbol{v}$ is the search direction and α_M the maximum search step. Point and direction at each next step are computed as:

$$\boldsymbol{p}_{s+1} = (\boldsymbol{p}_s + \alpha_s \boldsymbol{v}_s) , \qquad \boldsymbol{v}_{s+1} = -\boldsymbol{H}_{s+1} \left.\frac{\partial \mathcal{E}}{\partial \boldsymbol{p}}\right|_{\boldsymbol{p}_s} ; \tag{12}$$

where $\boldsymbol{H}$ is the inverse Hessian of $\mathcal{E}$:

$$\boldsymbol{H} \approx \begin{bmatrix} \cdots & \cdots & \cdots \\ \cdots & \frac{\partial^2 \mathcal{E}}{\partial p_i \partial p_j} & \cdots \\ \cdots & \cdots & \cdots \end{bmatrix}^{-1} . \tag{13}$$

Actually $\boldsymbol{H}$ is just an approximation (hence the *quasi-* in the method name), using the gradient variation at each step:

$$\nabla \boldsymbol{g}_s = \left.\frac{\partial \mathcal{E}}{\partial \boldsymbol{p}}\right|_{\boldsymbol{p}_s} - \left.\frac{\partial \mathcal{E}}{\partial \boldsymbol{p}}\right|_{\boldsymbol{p}_{s-1}} , \tag{14}$$

$\boldsymbol{H}$ is iteratively corrected, starting from the identity matrix, by:

$$\boldsymbol{H}_{s+1} = \boldsymbol{H}_s - \frac{1}{\nabla \boldsymbol{g}_s^T\ \alpha_s \boldsymbol{v}_s} \left(\boldsymbol{H}_s\ \nabla \boldsymbol{g}_s\ \alpha_s \boldsymbol{v}_s^T + \alpha_s \boldsymbol{v}_s\ \nabla \boldsymbol{g}_s^T\ \boldsymbol{H}_s\right) \tag{15}$$

$$+ \frac{1}{\nabla \boldsymbol{g}_s^T\ \alpha_s \boldsymbol{v}_s} \left(1 + \frac{\nabla \boldsymbol{g}_s^T\ \boldsymbol{H}_s\ \nabla \boldsymbol{g}_s}{\nabla \boldsymbol{g}_s^T\ \alpha_s \boldsymbol{v}_s}\right) \alpha_s \boldsymbol{v}_s\ \alpha_s \boldsymbol{v}_s^T ; \tag{16}$$

full details are available in [3].

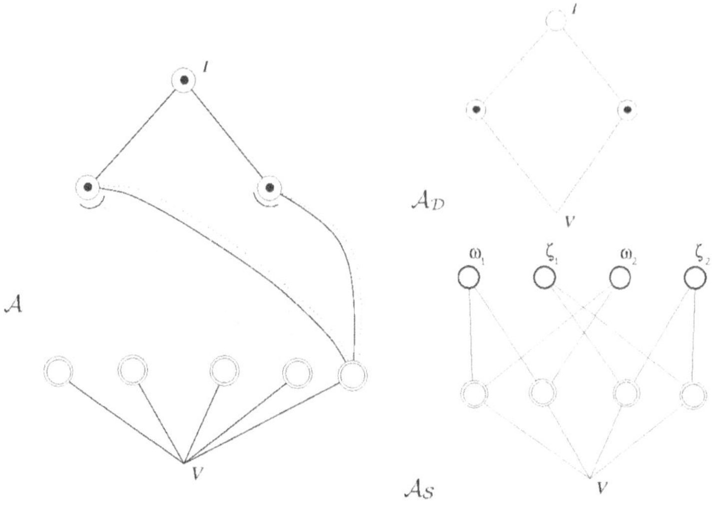

Figure1. An architecture $\mathcal{A}$ with two varying-dynamic neurons and the corresponding auxiliary architectures

3 Generative Method for DNN Models

In the design of a DNN model, a first specification is the *architecture* $\mathcal{A}$, i.e. number and type of neurons in each layer, connections, deemed sufficient for modeling the BJT device, and in general quite a complex architecture is required. The advantage of using varying-dynamic neurons is mainly because it is suitable to be partially trained in a simpler architecture.

Given the final architecture $\mathcal{A}$, the initial training of the parameters ω and ζ of the varying-dynamic neurons is made with two auxiliary architectures, $\mathcal{A}_\mathcal{D}$ and $\mathcal{A}_\mathcal{S}$. The former uses only fixed dynamic neurons, one for each of the original varying-dynamic neurons, while $\mathcal{A}_\mathcal{S}$ is made of all the static neurons of $\mathcal{A}$ connected with the varying-dynamic neurons, being the latter substituted by sigmoidal neurons. For example, from the complete architecture in the left side of Fig. 1, the derived auxiliary architectures are those on the right side. on the right side. Many copies of architecture $\mathcal{A}_\mathcal{D}$ are trained, each at a single DC bias point, generating a set of ω-ζ-type parameters. The architecture $\mathcal{A}_\mathcal{S}$ is trained to compute the dependency of such parameters from the DC conditions, using static neurons. The parameters computed with $\mathcal{A}_\mathcal{S}$ are then mapped back into the complete $\mathcal{A}$, and the remaining parameters are now optimized using the full set of data.

It is necessary to impose certain restriction in order to limit the effect of parameters other then of kind ω-ζ when changing DC bias point, for example the dynamic nodes connected to the output of varying-dynamic in $\mathcal{A}$, are substituted with a dummy node with fixed weights in $\mathcal{A}_\mathcal{D}$, as visible in Fig1.

4 Results

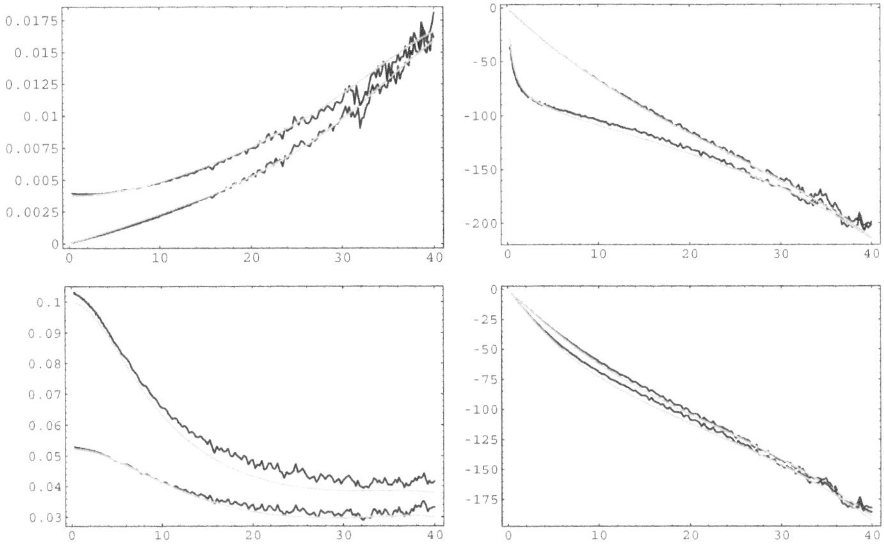

Figure2. Comparison of Y_{21} components generated with the DNN model (light line) and measured on the BJT (black line). Frequency, on the X axes, is in GHz, left graphics are modules (in Ω^{-1}), right ones are phases (in degrees). The top graphics correspond to $V_{BE} = 0.7, 0.8$, bottom ones to $V_{BE} = 0.9, 1.0$.

Results on a BJT device are reported, with details on the Y_{21} component, which is the most difficult to model, in range $V_{CE} = 1.0 \cdots 3.0$. The neural architecture used is the same as in Fig. 1, therefore the preliminary training was based on the auxiliary architectures of Fig. 1.

The error was 0.053, in terms of Normalized Root of Mean Square Error ($\mathcal{E}_{21} = \frac{1}{\sigma_{21}} \sqrt{\frac{1}{N} \sum_k \sum_f \left|\widehat{Y_{21}} - Y_{21})\right|^2}$ where σ_{21} is the standard deviation of $|Y_{21}|$ in the original data, and N the total number of points).

In Fig. 2 results of the DNN are compared with the original data, at $V_{CE} = 2.0$, and at four different values of V_{BE}.

5 Conclusions

To cope with the current scenario of high speed sub-micron bipolar devices, and their use in CAD applications, clearly new and more efficient modeling techniques are needed. The neural network strategy presented in this work seems to be a valid candidate. One of the major difficulties in ANN application to dynamic BJT modeling, the training of the internal parameters, has been solved, to a certain extent, by the generative method here described. The proposed method certainly has no claim to ultimately solve the general BJT modeling problem. In the DNN approach the design criteria of the architecture still remain a critical problem, and more work is needed towards an automated solution. However, the promising results achieved so far confirm that spending efforts in this direction is worthwhile.

References

1. D. Celi. BJT model equations. Technical Report V 2.0, UNICAD, 1991.
2. S.E. Fahlman. Faster-learning variations on back-propagation: an empirical study. In D. Touretzky, Hinton G. E., and Sejnowski T. J., editors, *Proc. of the 1988 Connectionist Models Summer School*, San Mateo, CA, 1988. Morgan Kaufmann.
3. P. E. Gill, W. Murray, and M. H. Wright. *Practical Optimization.* Academic Press, London, 1981.
4. R.A. Jacobs. Increased rates of convergence through learning rate adaptation. *Neural Networks*, 1(4):295–307, 1988.
5. B. L. Meijer, Peter. *Neural Network Applications in Device and Subcircuit Modelling for Circuit Simulation.* PhD thesis, Technical University of Eindhoven, 1996.
6. J. Nocedal and S. J. Wright. *Numerical Optimization.* Springer-Verlag, New York, 1999.
7. A. Plebe, A. M. Anile, and S. Rinaudo. Neural networks with higher level architecture for bipolar device modeling. In *Proc. of the Eleventh ECMI Conference*, Palermo, Italy, 2000.
8. W. L. Price. Global optimization algorithms for a cad workstation. *Journal of Optimization Theory and Applications*, 55:133–146, 1987.
9. K.-G. Rauh. A table model for circuit simulation. In *Proc. of the Twelfth European Solid-State Circuit Conference*, pages 211–213, Delft, The Netherlands, 1986.
10. A. J. Shepherd. *Second-Order Methods for Neural Networks.* Springer-Verlag, New York, 1997.
11. T. Shima. Device and circuit simulator integration techniques. In W. L. Engl, editor, *Process and devics modeling*, pages 433–459, Amsterdam, The Netherlands, 1986. North Holland.

Numerical Simulation for Lossy Microwave Transmission Lines Including PML

G. Hebermehl[1], F. Hübner[1], R. Schlundt[1]
T. Tischler[2], H. Zscheile[2], and W. Heinrich[2]

[1] Weierstrass Institute for Applied Analysis and Stochastics, Mohrenstr. 39, D-10117 Berlin, Germany
[2] Ferdinand-Braun-Institut für Höchstfrequenztechnik, Albert-Einstein-Str. 11, D-12489 Berlin, Germany

Abstract Finite-difference analysis of transmission lines including lossy materials and radiation effects leads to a complex eigenvalue problem. A method is presented which preserves sparseness and delivers only the small number of interesting modes out of the complete spectrum. The propagation constants are found solving a sequence of eigenvalue problems of modified matrices with the aid of the shift-and-invert mode of the Arnoldi method. In an additional step non physical Perfectly Matched Layer modes are eliminated.

1 Introduction

Microwave circuits are used in mobile communications, radio links, sensors, and automotive systems.

Basic elements of microwave integrated circuits are their transmission lines, whose propagation behavior has to be determined accurately. The propagation behavior of the transmission lines can be calculated by applying Maxwell's equations to the infinitely long homogeneous transmission line structure and solving an eigenvalue problem [1].

For numerical treatment, the computational domain has to be truncated by electric or magnetic walls or by a so-called absorbing boundary condition simulating open space. A very efficient formulation for the latter case is the Perfectly Matched Layer (PML). These layers consist of an artificial material with complex anisotropic material properties [2]. The PML provide absorbing properties for any frequency, polarization and angle of incidence.

In the presence of losses or absorbing boundary conditions the matrix of the eigenvalue problem becomes complex. The system matrix is sparse and of high order. This requires efficient solvers that preserve sparseness and deliver only the small number of interesting modes out of the complete spectrum.

In earlier time- and memory-consuming methods [1] and [3] the complete set of eigenvalues and of corresponding propagation constants was computed and sorted in order to select the desired propagation constants. In the previous papers [4] (lossless case) and [5] (lossy case including PML) the authors presented a method which avoids the computation of all eigenvalues to find the few required propagation constants. The sparse-storage technique is applied. The computing times increase with the frequency and using PML. A feasible computation is possible for frequencies up to 600 GHz. In order to reduce the numerical effort and to simulate circuits for higher frequencies a new method is presented here.

Using an estimation for the maximum value of the real part and a preset maximum value of the imaginary part the region containing the interesting propagation constants is defined as a rectangle $\hat{F}$. Using conformal mapping relations between the plane of propagation constants and the plane of eigenvalues the rectangle $\hat{F}$ is mapped into an area F which is bounded by parabolas. A method is presented which finds the eigenvalues of this area solving a sequence of eigenvalue problems of modified matrices with the aid of the shift-and-invert mode of the Arnoldi method [6].

Special attention is paid to the so-called Perfectly Matched Layer boundary conditions, which influence the mode spectrum in a significant way [7]. Particularly, the problem of detecting the desired modes out of the mode spectrum is treated.

2 Boundary Value Problem

The structures under investigation can be described as an interconnection of infinitely long transmission lines, which have to be longitudinally homogeneous. The junction, the so-called discontinuity, may have an arbitrary structure. Cross-sectional planes, the so-called ports, are defined on the transmission lines. The whole structure may be surrounded with an enclosure. A three-dimensional boundary value problem can be formulated using the integral form of Maxwell's equations in the frequency domain

$$\oint_{\partial\Omega} \boldsymbol{H} \cdot d\boldsymbol{s} = \int_{\Omega} \jmath\omega[\epsilon]\boldsymbol{E} \cdot d\boldsymbol{\Omega}, \qquad \oint_{\cup\Omega} ([\epsilon]\boldsymbol{E}) \cdot d\boldsymbol{\Omega} = 0, \tag{1}$$

$$\oint_{\partial\Omega} \boldsymbol{E} \cdot d\boldsymbol{s} = -\int_{\Omega} \jmath\omega[\mu]\boldsymbol{H} \cdot d\boldsymbol{\Omega}, \qquad \oint_{\cup\Omega} ([\mu]\boldsymbol{H}) \cdot d\boldsymbol{\Omega} = 0, \tag{2}$$

$$\boldsymbol{D} = [\epsilon]\boldsymbol{E}, \ \ \boldsymbol{B} = [\mu]\boldsymbol{H}, \quad [\epsilon] = \operatorname{diag}(\epsilon_x, \epsilon_y, \epsilon_z), \ \ [\mu] = \operatorname{diag}(\mu_x, \mu_y, \mu_z), \tag{3}$$

in order to compute the electromagnetic field. The PML's are filled with an artificial material with complex anisotropic material properties. Therefore, the quantities are diagonal complex tensors. At the ports p the transverse electric field $\boldsymbol{E}_t(z_p)$ is given by superposing transmission line modes $\boldsymbol{E}_{t,l}(z_p)$. On all other parts of the computation domains surface the tangential electric or magnetic field is assumed to be equal zero.

$$\boldsymbol{E}_t(z_p) = \sum_{l=1}^{m^{(p)}} w_l(z_p)\boldsymbol{E}_{t,l}(z_p), \quad \boldsymbol{E} \times \boldsymbol{n} = 0 \quad \text{or} \quad \boldsymbol{H} \times \boldsymbol{n} = 0. \tag{4}$$

The transverse electric mode fields are the solutions of an eigenvalue problem for the transmission lines. Starting from the Maxwellian equations (1),

(2) each of these equations is solved on a three-dimensional grid. Using the Finite Integration Technique [8] in the frequency domain (FDFD for Finite-Difference method in the Frequency Domain) [9] Equations (1), (2) are transformed into a set of Maxwellian grid equations

$$A^T D_{s/\mu} \boldsymbol{b} = \jmath\omega\epsilon_0\mu_0 D_{A_\epsilon} \boldsymbol{e}, \qquad B D_{A_\epsilon} \boldsymbol{e} = 0, \tag{5}$$

$$A D_s \boldsymbol{e} = -\jmath\omega D_{A_\mu} \boldsymbol{b}, \qquad \tilde{B} D_{A_\mu} \boldsymbol{b} = 0. \tag{6}$$

The vectors $\boldsymbol{e}$ and $\boldsymbol{b}$ contain the components of the electric field intensity and the magnetic flux density of the elementary cells, respectively. The diagonal matrices $D_{s/\mu}$, D_{A_ϵ}, D_s, and D_{A_μ} contain the information on cell dimension and material. A, B, and $\tilde{B}$ are sparse and contain the values 1, -1, and 0 only. Eliminating the components of the magnetic flux density from the two equations of the left-hand side of (5), (6) we get the system of linear algebraic equations

$$(A^T D_{s/\mu} D_{A_\mu}^{-1} A D_s - k_0^2 D_{A_\epsilon})\boldsymbol{e} = 0, \quad k_0 = \omega\sqrt{\epsilon_0\mu_0}, \tag{7}$$

which have to be solved using the boundary conditions. k_0 is the wavenumber in vacuum.

3 Eigenvalue Problem

Because the transmission lines are longitudinally homogeneous any field can be expanded into a sum of so-called modal fields which vary exponentially in the longitudinal direction

$$\boldsymbol{E}(x,y,z) = \underline{\boldsymbol{E}}(x,y)e^{\mp\jmath k_z z}. \tag{8}$$

A substitution of ansatz (8) into the system of linear algebraic equations (7) and the elimination of the longitudinal electric field intensity components by means of the electric-field divergence equation $BD_{A_\epsilon}\boldsymbol{e} = 0$ (see (5)) gives an eigenvalue problem

$$C\underline{e} = \gamma\underline{e}, \quad \gamma = -4\sin^2(hk_z) \approx -4(hk_z)^2 = u + \jmath v. \tag{9}$$

$\underline{e}$ consists of components of the discretized eigenfunctions $\underline{\boldsymbol{E}}$. $2h$ is the length of an elementary cell in z-direction. The sparse $n \times n$ matrix C is in general nonsymmetric complex. We can use the approximation $\sin(x) \approx x$ in (9) if we choose h to be small enough, which is anyway necessary to get small discretization errors. The relation between the propagation constants k_z and the eigenvalues γ is nonlinear, and can be expressed as

$$k_z = \frac{1}{h}\arcsin\left(\frac{\jmath}{2}\sqrt{\gamma}\right) = \frac{\jmath}{2h}\ln\left(\frac{\gamma}{2} + 1 + \sqrt{\frac{\gamma}{2}\left(\frac{\gamma}{2}+2\right)}\right) = \beta - \jmath\alpha. \tag{10}$$

A propagation constant k_z and its corresponding eigenfunction is called a mode. We are interested only in the modes with the smallest magnitude of imaginary part, but possibly with large real part of their propagation constant.

For our method we must limit the search for propagation constants by a maximum value k_f of their real part. A reasonable estimation of this maximum value is derived for the lossy case including PML for inhomogeneously filled waveguides in [5]

$$\Re(k_z) \leq k_f = \omega \Re(\sqrt{\epsilon_m \mu_m}). \tag{11}$$

ϵ_m and μ_m are properties of the material that yields the largest value of the right-hand side of Equation (11). Using a maximum value k_f of the real part and a preset maximum value α_m of the imaginary part of the propagation constant the region containing the interesting propagation constants is defined as a rectangle $\hat{F}$ (see Fig. (2)) bounded by the lines

$$\beta = \pm k_f \quad \text{and} \quad \alpha = \pm \alpha_m. \tag{12}$$

We consider the plane of eigenvalues (γ-plane) and the plane of propagation constants (k_z-plane). The conformal mapping relations between the γ- and the k_z-plane are (see (9) and (10))

$$u = -4h^2(\beta^2 - \alpha^2), \quad v = 8h^2\alpha\beta. \tag{13}$$

Using the inverse of the mapping (13) the lines (12) of the k_z-plane are transformed into the parabolas (see Fig. (1))

$$v = \pm 4hk_f\sqrt{u + 4h^2{k_f}^2} \quad \text{and} \quad v = \pm 4h\alpha_m\sqrt{-u + 4h^2{\alpha_m}^2}, \tag{14}$$

respectively. That means, we have to find all eigenvalues of the region bounded by the parabolas (14).

The eigenvalues are obtained as follows. q points $\hat{P}_k(\beta_k, \alpha_m)$ with

$$\beta_k = \begin{cases} k\frac{k_f}{q}, & \text{if } \frac{k_f}{q} \geq \bar{\beta},\ q \ \text{given},\ k = 1(1)q,\ \text{with } \bar{\beta} = \sqrt{3}\alpha_m, \\ k\bar{\beta}, & \text{otherwise},\ q = \left[\frac{k_f}{\bar{\beta}}\right] + 1,\ k = 1(1)q-1,\ \beta_q = k_f, \end{cases} \tag{15}$$

are defined on the interval $[0, k_f]$ of the line $\alpha = \alpha_m$. q is found empirically (see section 5). The meaning of $\bar{\beta}$ is discussed below. Some simple case differentions are not described here in detail. The points $\hat{P}_k$ are transformed into the points $P_k(u_k, v_k)$ of the γ-plane. They are located on the parabola ((14), right formula). The q circles C_k of the γ-plane

$$(u + m_k)^2 + v^2 = {r_k}^2, \quad r_k = \sqrt{v_k^2 + (m_k - u_k)^2}, \quad k = 1(1)q, \tag{16}$$

with

$$m_1 = 0, \quad m_k = \frac{u_k^2 - u_{k-1}^2 + v_k^2 - v_{k-1}^2}{2|u_k - u_{k-1}|}, \quad k = 2(1)q, \tag{17}$$

centered on the u-axis cover the region bounded by the parabolas (14).

l points Q_i (see section 5) are defined on the periphery of C_k in order to find all eigenvalues, located in the circle C_k. The matrix C is extended by the diagonal matrix Q (see (19)) which consists of the set $\mathcal{E}$ with the l complex elements Q_i. The q eigenvalue problems

$$(\bar{C} - m_k I)\underline{e} = (\gamma - m_k)\underline{e}, \quad k = 1(1)q, \tag{18}$$

$$\text{with} \quad \bar{C} = \begin{pmatrix} Q & \\ & C \end{pmatrix}, \quad Q = diag(Q_1, ..., Q_l), \tag{19}$$

are solved with the aid of the implicitly restarted Arnoldi method. The number m (see section 5) of eigenvalues to be computed on each circle C_k must be l on the first call to the Arnoldi procedure. m is raised by l for so long until at least one value $Q_i \in \mathcal{E}$ was found. Separating the new values on each eigenvalue problem k, we are sure to have found all eigenvalues which are located in the corresponding circles C_k. Applying (13) the circles C_k are transformed into Cassinian curves $\hat{C}_k$

$$(\beta^2 + \alpha^2)^2 - \frac{m_k}{2h^2}(\beta^2 - \alpha^2) = \frac{r_k{}^2}{16h^4} - \frac{m_k{}^2}{16h^4}, \tag{20}$$

which cover the rectangle $\hat{F}$ containing all desired propagation constants. The Cassinian curves $\hat{C}_k$, $k = 2(1)q$, consist of two separated ovals (see Figure 2), if $r_k < m_k$. Using $\bar{\beta}$ as minimum distance between the origin and $\hat{P}_1$ (see (15)) other shapes of Cassinian curves, which would lead to higher execution times, are avoided.

The dimension n of the matrix C is too large to use a direct method. We need an algorithm that computes just a few selected eigenvalues and eigenvectors of a sparse nonsymmetric matrix. A state-of-the-art iterative algorithm for such problems is the Arnoldi method. In general the Arnoldi method does not converge using the regular mode for our eigenvalue problem. Thus, the invert mode with shifting (see (18)) is applied with the solution of systems of linear algebraic equations looking for eigenvalues of largest magnitude. We use the combined unifrontal/multifrontal method [10] for the solution of the partly ill-conditioned nonsymmetric complex linear algebraic equations.

4 Detecting PML Modes

We use the PML in order to calculate the eigen modes of open waveguide structures. Without the PML the finite difference computational domain has to be surrounded by either electric or magnetic walls. In this case we would have a huge number of box modes, which are not an intrinsic property of the waveguide. Since the total number of modes is constant, the PML cannot remove any mode but only shift them into other parts of the eigenvalue spectrum. We want to distinguish these PML-modes from the desired ones. The PML-modes are characterized by their high power concentration in the PML area. Thus, to eliminate the PML-modes we calculate the magnitude of the power flow of each computed mode in the PML ($P^{(P)}$), in the waveguide region ($P^{(W)}$), and in the total computational domain (P):

$$P = P^{(P)} + P^{(W)} = \int\limits_{\Omega^{(P)}} (\boldsymbol{E}_t \times \boldsymbol{H}^*_{t,m}) \cdot d\boldsymbol{\Omega} + \int\limits_{\Omega^{(W)}} (\boldsymbol{E}_t \times \boldsymbol{H}^*_{t,m}) \cdot d\boldsymbol{\Omega}\,. \tag{21}$$

A mode is specified as PML-mode if

$$r^{(P)} = \frac{P^{(P)}}{P} > \xi \,, \tag{22}$$

with values $\xi = 0.2, \ldots, 0.6$, found empirically. This criterion works for a wide frequency range. In the case of very high frequencies, where the physical modes are strongly radiating, the relation $r^{(P)}$ of the physical modes and the PML modes approach each other and the criterion may accept even some PML modes. In this case, one has to consult additional information, e. g. field density plots, in order to decide which eigenvalues belong to the desired guided modes.

5 Numerical Example

As an example we have calculated the modes of a coplanar waveguide with finite substrate height, and finite ground metallization according to Figure (4 a), for the frequencies $f = 10(10)800$ GHz. As the waveguide is symmetrical, only its right-hand half must be discretized. The computational domain consists of the cross section of the waveguide, the ground metallization and the PML regions. The PML's are bounded with electric walls, while the symmetry plan is formed as a magnetic wall. The structure is subdivided into $n_{xy} = n_x n_y$ elementary cells. A graded mesh of $n_x = 89$, $n_y = 109$ cells, including 12-cell PML regions is used in the given example, so the dimension of the eigenvalue problem is $n = 2n_x n_y - n_b = 17955$. n_b is determined by the number of cells filled with perfectly conducting material. $q = 4$ and $l = 15$ (see section 3), found empirically, were used for this example. The number m of modes to be computed on each eigenvalue problem k increases with the frequency.

The Figures (1) and (2) refer to the frequency $f = 300$ GHz. Three modes were identified as guided wave modes, according to (22) with $\xi = 0.35$. The corresponding propagation constants are characterized by a " + ". To go into detail these modes are the parallel plate line mode (PPL), the coplanar waveguide mode (CPW), and a higher order mode (HM10) that is caused by the finite substrate height. In Figure (4 b) a field pattern plot of the CPW mode is given, which illustrates the leakage due to lateral radiation of this mode. Figure (3) shows the effective permittivity of the different physical modes over the whole frequency range. The number of propagating modes increases with the frequency.

It should be mentioned, that the PML modes can be well separated for frequencies up to 500 GHz using criterion (22). For higher frequencies some PML modes with $r^{(P)} < 0.35$ arise and one has to separate them by checking the field distribution.

Concerning the sequencing technique for the eigenvalue search, one can say that, especially for high frequencies, the numerical effort could be reduced considerably in comparison to the method presented in [5].

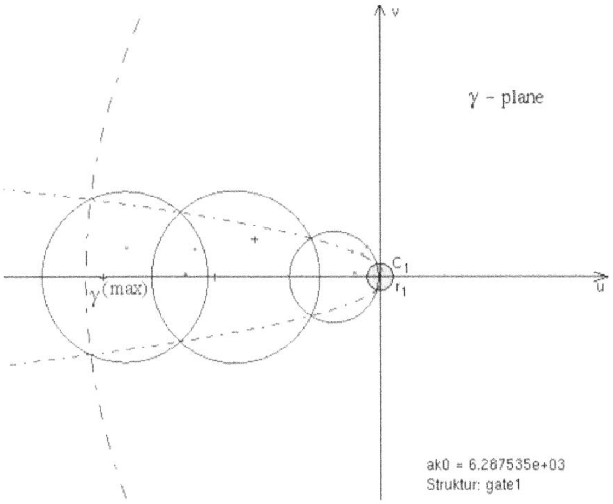

Figure1. γ-plane

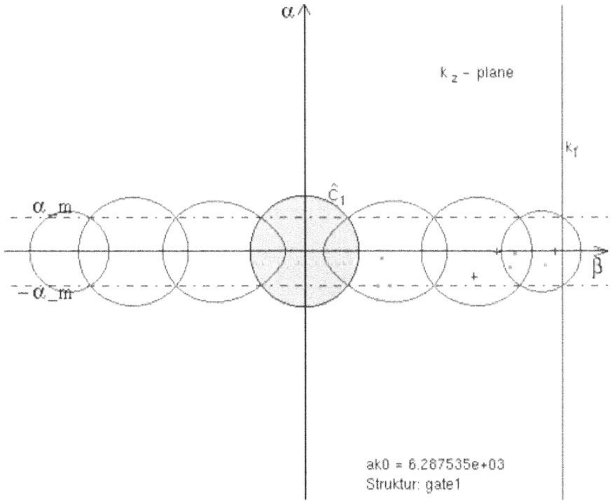

Figure2. k_z-plane

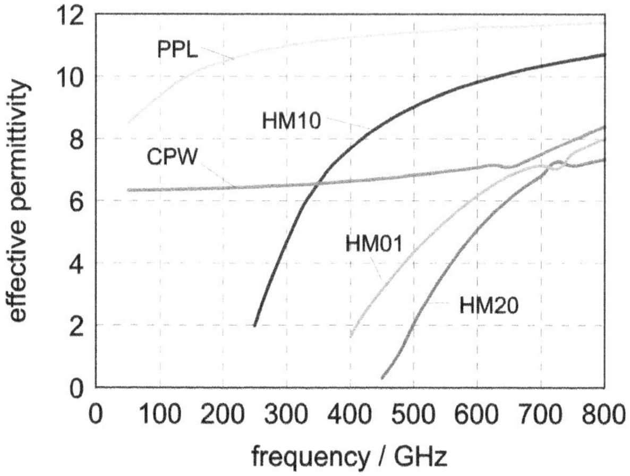

Figure3. Propagating modes of a coplanar waveguide

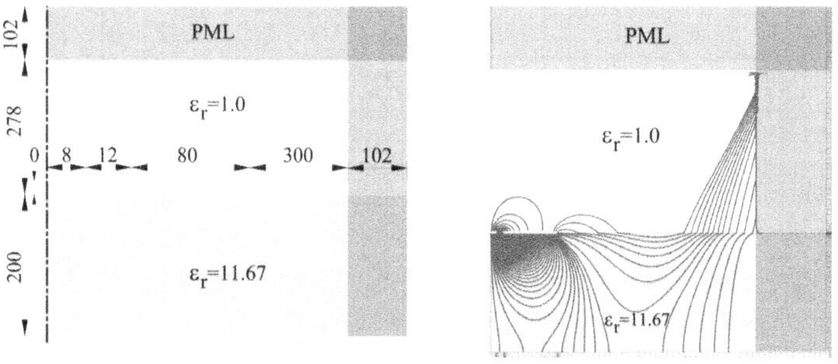

(a) Geometry of the CPW (all parameters in μm) (b) E-Field pattern of the CPW- Mode at f = 300 GHz

Figure4. Lateral radiation of a coplanar waveguide

References

1. Christ, A., Hartnagel, H. L.: Three-Dimensional Finite-Difference Method for the Analysis of Microwave-Device Embedding. IEEE Transactions on Microwave Theory and Techniques **35** (1987) 688–696
2. Sacks, Z. S., Kingsland, D. M., Lee, R., Lee, J.-F.: A Perfectly Matched Anisotropic Absorber for Use as an Absorbing Boundary Condition. IEEE Transactions on Antennas and Propagation **43** (1995) 1460–1463
3. Klingbeil, H., Beilenhoff, K., Hartnagel, H. L.: FDFD Full-Wave Analysis and Modeling of Dielectric and Metallic Losses of CPW Short Circuits. IEEE Transactions on Microwave Theory and Techniques **44** (1996) 485–487
4. Hebermehl, G., Schlundt, R., Zscheile, H., Heinrich, W.: Eigen Mode Solver for Microwave Transmission Lines. The International Journal for Computation and Mathematics in Electrical and Electronic Engineering **16**, (1997) 108–122
5. Hebermehl, G., Hübner, F.-K., Schlundt, R., Tischler, T., Zscheile, H., Heinrich, W.: On the Computation of Eigen Modes for Lossy Microwave Transmission Lines Including Perfectly Matched Layer Boundary Conditions. The International Journal for Computation and Mathematics in Electrical and Electronic Engineering **20** (2000) (to appear)
6. Lehoucq, R. B.: Analysis and Implementation of an Implicitly Restarted Arnoldi Iteration. Rice University, Department of Computational and Applied Mathematics, Technical Report **13** (1995) 1-135
7. Tischler, T., Heinrich, W.: The Perfectly Matched Layer as Lateral Boundary in Finite-Difference Transmission-Line Analysis. 2000 Int. Microwave Symp. Digest **1** (2000) 121-124
8. Weiland, T.: A Discretization Method for the Solution of Maxwell's Equations for Six-Component Fields. Electronics and Communication (AEÜ) **31** (1977) 116–120
9. Beilenhoff, K., Heinrich, W., Hartnagel, H. L.: Improved Finite-Difference Formulation in Frequency Domain for Three-Dimensional Scattering Problems. IEEE Transactions on Microwave Theory and Techniques **40** (1992) 540-546
10. Davis, T. A., Duff, I. S.: A Combined Unifrontal/Multifrontal Method for Unsymmetric Sparse Matrices. University of Florida, Technical Report **16** (1997) 1–18

Electromagnetic Simulation of Real-Life Circuits: A Mobile Phone Power Amplifier Module and an FM Radio Tuner PCB

J. Niehof[1], N.J. Pulsford[1,2], I. Lodema[2], and J. Meeuwis[2]

[1] Philips Research Laboratories, Prof. Holstlaan 4, 5656 AA Eindhoven, The Netherlands, email: jan.niehof@philips.com
[2] Philips Semiconductors, Hurksestraat 19, 5600 MD Eindhoven, The Netherlands

Abstract The effects of the physical implementation of the layout on the overall circuit behaviour is analysed for the output stage of a mobile phone power amplifier module and an FM radio tuner. For this purpose an equivalent circuit model is generated using a 2.5D BEM electromagnetic field solver. The simulation results are compared with measurements. The results show that for the design of the output stage of the power amplifier layout effect simulation is crucial, while for the PCB regular circuit simulation is not enough to predict the actual circuit behaviour accurately above 100 MHz. The simulations including the EM layout effects show a good agreement with the measurements. Layout effect simulation can be used to optimise the circuit layout and reduce the design cycle.

1 Introduction

The demands set by the electronics consumer market are increasing continuously. In order to be competitive in such a market electronic designs need to fulfil demands such as time to market, first time right and legal standards for interference. Furthermore there is the pressure to reduce size and cost. A fast growing segment of the electronics consumer market is the market for wireless applications (e.g. mobile phones, in-house wireless systems). All of these products contain high frequency parts. The overall performance of such circuitry is influenced by its physical implementation. The influence of the interconnect and packaging on the circuit behaviour is increasing with the ongoing trend towards higher frequencies and miniaturisation. Such high-speed circuitry cannot be developed without taking the influence of the layout into account. On the other hand, layout effects are even used to achieve the desired circuit behaviour. The development of today's complex electronic products requires accurate computer simulations. For radio frequency (RF) design in particular, many parts of the physical system, such as IC (integrated circuit) packages, PCB (printed circuit board) and printed components (and also the coupling between them) can only be simulated accurately using electromagnetic field analysis. For this purpose the EMC and layout effect tool Fasterix has been developed [1]. Fasterix offers the possibility of taking the influence of the physical implementation of a circuit on the overall circuit behaviour into account in circuit simulations. More specifically Fasterix can be used for any type of structure built as interconnect structures stacked in layers. Fasterix is used for the design of all kinds of circuits where the layout plays an important role in the overall circuit behaviour [2,3,4]. This paper gives the results of the application of Fasterix for integration of the output match circuit of a power amplifier module and for the FM tuner printed circuit as shown in Figure 1. For the power amplifier layout the physical implementation of the layout determines the desired circuit behaviour. The tuner PCB is analysed using Fasterix in order to study the effect of the physical layout (including the printed coil, refer to Figure 1) on the circuit behaviour. Specifically the signal behaviour of the RF input stage is studied.

2 Fasterix

Fasterix generates an equivalent circuit model for the physical interconnect structure. The equivalent circuit topology and its component values are extracted using a 2.5D BEM (boundary element method) calculation.

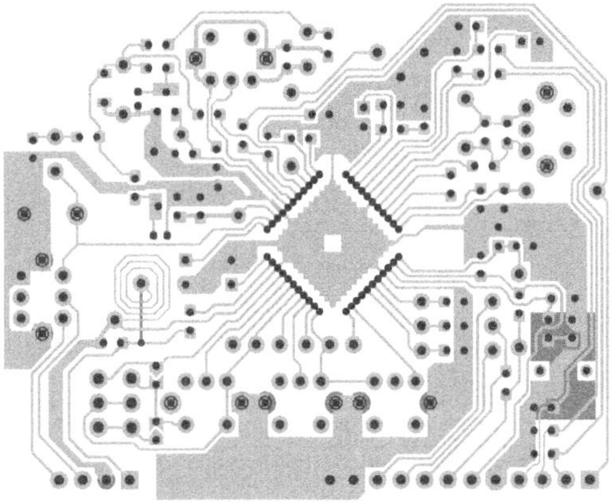

Figure1. Layout of the FM radio tuner PCB.

In all electromagnetic field solvers a geometrical mesh is created, and numerical methods are then used to find solutions for the fields over this mesh. Typically the field analysis is performed for each required time or frequency sample. Because of the geometrical complexity of structures in many real applications it is necessary to create a fine mesh to achieve acceptable accuracy. This leads to very large matrices and concomitant large memory use and long computation times. This problem of large matrix size again occurs in circuit analysis, when combining the results of the electromagnetic field analysis with circuit simulation by extracting a lumped partial element equivalent circuit.

The method used in Fasterix is also based on the equivalent circuit concept, but employs a compressed equivalent circuit which has a much smaller number of nodes, typically reducing the circuit analysis computations by many orders of magnitude [5]. Using a circuit model is extremely efficient compared to electromagnetic field models, this because it removes the need to recompute the field distribution associated with every device every time the model is used. This information is essentially stored in the component values. But in general for conductor/insulator structures the topology and component values are difficult to obtain. Fasterix automatically provides an equivalent circuit model for arbitrarily shaped structures. The metallisation structures that can be modelled by Fasterix are required to be passive and linear. No assumptions are made concerning the spatial charge or current density distribution. The equivalent circuit topology is directly derived from a quasi-static solution of the Maxwell equations. The size (number of circuit nodes) of the equivalent circuit is determined by a user-specified maximum frequency for which the model has to be valid. The most important feature is that the number of components used to characterise the electromagnetic structure depends on the electrical size (i.e. ratio between feature size and wavelength), not geometrical complexity. This allows the electromagnetic analysis of for instance complete PCBs, such as shown in Figure 1, whereas other 2.5D planar EM simulators are restricted to much smaller structures (i.e. small with respect to the minimum wavelength considered). The generated equivalent circuit basically is an RLC- netlist. Fasterix can therefore be used to generate SPICE compatible models for any type of layered metallisation structure. These models may then be combined with models for transistors and other active components to perform both frequency-domain and non-linear time-domain analysis of the complete (interconnect system + components) circuit, using SPICE-type circuit simulators. For the analysis all available circuit simulator functionality can be used. Each component or connector pin in the layout is a circuit node in the equivalent circuit model. Therefore at every pin signals can be measured or applied.

3 Mobile Phone Power Amplifier Module

A current, conventional RF power amplifier module for mobile telephones produced by Philips Semiconductors consists of a hybrid assembly of silicon IC's and SMD 0402 resistors and capacitors on a 8 x 13.5 mm^2 ceramic substrate. The substrate contains a single layer of thick film metallisation for the signal and supply interconnect. The high SMD passive component density and the fact that these hybrid modules typically have a large spread in matching characteristics (due to the spread in the SMD passive components and the spread in the placement of these components), makes parts of the module attractive for some form of passive component integration on silicon. Therefore a new design of the output match is made where the passive components are integrated. In this paper the application of Fasterix for the integration of one part of the module - the output match circuit - is described using the passive integration technology on high ohmic silicon. The conventional hybrid design serves as the basis, this design is analysed using Fasterix simulations. With all the characteristics of this design known, the passive integration layout is designed and optimised using Fasterix simulations.

3.1 Hybrid Power Amplifier Module

The amplifier is a three-stage UHF module in a SOT482B leadless package with a plastic cover. The module consists of one bipolar monolithic integrated circuit and one npn silicon planar transistor die mounted together with matching and bias circuit components on a metallised ceramic substrate. It amplifies an input signal in the frequency range f=880-915MHz from P=1mW at the input to P=2.4W at the output with a power added efficiency η=40%.

The first step in the design process is to use Fasterix simulations to analyse the output match circuit of the hybrid module. These analyses help to define the circuit parameters (Z, S matrices) for the passive integration design in the next step. The Fasterix input includes the thick metal interconnect, the ceramic substrate with ground plane, the substrate vias and the seven bondwires in the electromagnetic simulation. The bondwires are in fact 3D elements in the Fasterix simulation. The 3D trajectory of the bondwires is taken into account in the solution process, therefore the generated equivalent circuit also contains the EM behaviour of the bondwires (interaction between the bondwires, and interaction bondwires with the rest of the layout). Discrete models for 0402 SMD capacitors are added at the circuit simulation level. The mesh generated by Fasterix for the output match metallisation is shown in the left of Figure 2. This mesh consists of 3627 elements. The generation of the equivalent circuit takes 50 min. CPU on an HP9000-800 workstation. This equivalent circuit consists of 113 circuit nodes. Simulating the equivalent circuit model including the models for the discrete SMD components, in a SPICE-like circuit simulator, to determine the impedance levels and the attenuation of the output match circuit, takes a few seconds.

3.2 Passive Integration Power Amplifier Module

In the passive integration technology on Si, resistor, capacitor and inductor networks are formed on an oxidised silicon substrate [6,7]. The passive integration circuit for the power amplifier output match is designed with physical layout simulations using Fasterix. The meshing for the passive integration die on the ceramic module substrate is shown in Figure 2 (right). This mesh consists of 4403 elements. The generation of the equivalent circuit takes 55 min. CPU on an HP9000-800 workstation. This equivalent circuit consists of 82 circuit nodes. Using Fasterix, the passive integration structure was designed to have the same electrical behaviour as the discrete design of the output match circuit. The bondwires both between the active and passive dies and to the ceramic substrate, in total twenty, are included in the simulation. There are 6 (L,C) matching sections included in the network. The passive integration IC

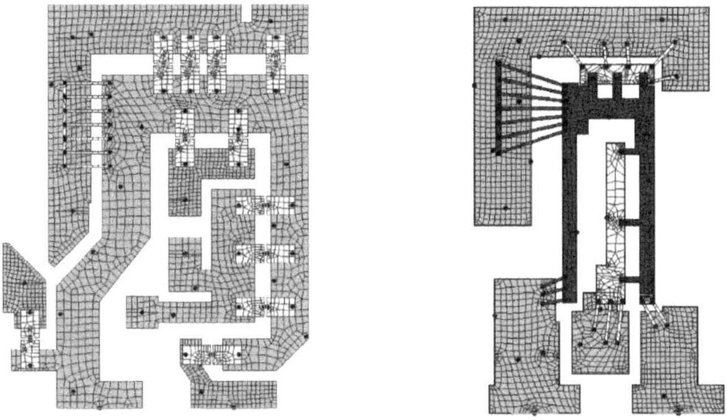

Figure2. Power amplifier output match: hybrid (left) and passive integration (right).

is mounted and wirebonded onto the ceramic substrate carrier alongside the booster IC. The passive integration circuit matches the low load impedance of this IC ($Z \approx 1.2 + 0.45j\ \Omega$) to $Z = 50\ \Omega$ at the module output. The rest of the module is identical to the conventional power amplifier. The passive integration output match occupies 50% area of the discrete output match of the conventional module.

3.3 Results

The power amplifier performance of the modules with passive integration IC has been measured. A comparison of the typical performance characteristics of the conventional hybrid module with that of the passive integration circuit shows that the ouput power and efficiency of the passive integration module is on a similar level to a conventional hybrid module - i.e. the impedance match circuit is working. The area occupied by the passive integration implementation is about 50% of the area used for the conventional hybrid implementation. Layout effect simulation is essential in the design of circuits where the circuit behaviour is completely determined by the layout. Fasterix has succesfully been applied in crucial steps in the design of a passive integration structure. First it was used to characterise the reference design. Then it was used to design the passive integration structures, resulting in first-time-right silicon occupying ~50% of the area used by the hybrid match.

4 FM Radio Tuner PCB

To illustrate the capabilities of Fasterix for a PCB with a complex geometry, the FM radio tuner PCB as shown in Figure 1 is simulated.

The design (layout and components) is transferred to Fasterix for analysis. The most important parameter in the Fasterix simulation flow is the maximum frequency for which the equivalent model has to be valid. After setting the material properties, such as the conductivity of the metallisation and the dielectric constant of the PCB material, the structure is meshed. The resulting mesh consists of 9355 elements.

4.1 Generation of the Full EM Model for the PCB

The next step is to generate the equivalent circuit. The solution to the Maxwell equations on the mesh is compressed into an equivalent circuit model consisting of 711 circuit nodes. The equivalent model generated here is valid in the frequency range from DC up to the maximum frequency, here set to 1 GHz. The CPU simulation times for the steps in Fasterix running on an HP9000-800 workstation are given in Table 1. The equivalent circuit is written in a SPICE compatible format. The size of the equivalent circuit netlist file is 3.8 Mb (ASCII format).

Table1. Fasterix simulation times.

Step	CPU time
Layout	22 s
Mesh	20 s
Equivalent circuit	12 hrs 37 min

For the purpose of this paper all EM effects were calculated for the complete PCB, in order to give an impression of the size and complexity of structures Fasterix can handle. A much more efficient way to simulate this of course is to focus on a sub-section of the complete layout. When focussing on the RF input tracks and some neighbouring interconnect, very fast (in the order of minutes) simulation iterations can be performed. This can be used for instance to optimise the layout to meet certain specifications. When the complete layout needs to be taken into account, both the simulation time and the size of the equivalent circuit netlist file can also be reduced when setting Fasterix to only consider interaction between points within a specified radius.

4.2 Complete PCB Simulation

In order to perform a simulation of the behaviour of the tuner circuit including the effect of the PCB layout, the equivalent circuit needs to be completed by attaching the simulation models, which describe the behaviour of the components on the PCB. The IC circuit model is combined with the package model for its leadframe. This package model is also generated with Fasterix.

The IC circuit model for the FM radio tuner IC is connected to the package model. This combination is then connected to the PCB equivalent circuit model. Also, the models for the other component have to be connected to the PCB model. At this point the circuit simulator input file is complete. In this paper only the RF input section is examined, but since an equivalent circuit has been generated for the complete PCB, it is easily possible to also study other sections of the tuner PCB. Other analyses that can be performed are: simulations to check on the RF content in the audio outputs, the RF content on the supply/bus connections or, in general, to check the circuit behaviour against legally required standards. The PCB layout can be optimised to fulfil these standards, saving design/prototype iterations and thereby shortening the design cycle.

4.3 Simulation Results

The circuit simulation to calculate a 500-points frequency domain curve took 19 minutes and 13 seconds CPU time on an HP9000-800 machine. Figure 3 shows 3 curves: the ideal simulation result (circuit simulation only, no PCB-layout/interconnect effects, the printed coil is replaced by an ideal inductor), the Fasterix simulation result and the measurement.

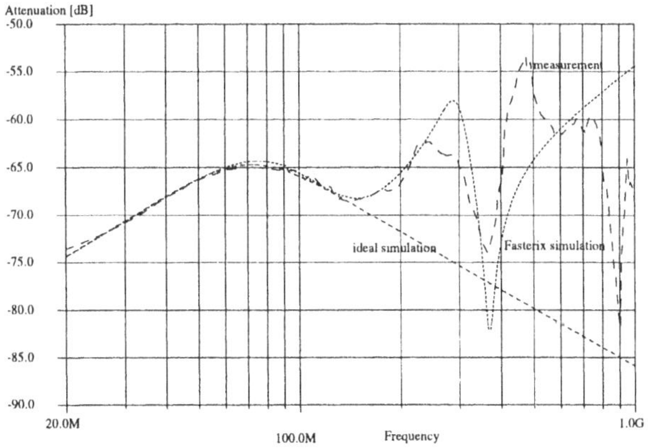

Figure3. Curves at the RF-input connector.

For low frequencies, up to about 100 MHz, the ideal simulation and the measurement agree very well. However, for higher frequencies the mismatch between ideal simulation and measurement becomes significant. The function of the input stage is to prevent frequencies above the FM-band from entering the tuner circuit. The measurement and the Fasterix simulation result in Figure 3 both show that signals in the range 200 - 350 MHz (e.g. TV signals) are not suppressed at all. This clearly undesired effect is introduced by the layout.

5 Conclusions

The electromagnetic layout effect simulation tool Fasterix is capable of generating a compact SPICE compatible model for physical layered metallisation structures. The match between measurement and simulation is very good. Layout effect simulation is essential in the design of circuits where the circuit behaviour is completely determined by the layout. Fasterix has succesfully been applied in crucial steps in the design of a passive integration structure. The output match of an RF power amplifier has been redesigned by the integration of discrete passive components onto high ohmic silicon. This has resulted in first-time-right silicon with a reduction in the module size without any performance loss. The capabilities of Fasterix for complicated physcial structures are demonstrated by the simulations on the FM tuner PCB. Unwanted effects caused by the layout can be identified at a very early stage, before any physical implementation of the PCB is made. Layout effect simulations can be used to check the circuit behaviour against legally required standards. The PCB layout can be optimised to fulfil these standards, saving design iterations and thereby shortening the design cycle.

References

1. R. du Cloux, G.P.J.F.M. Maas, A.J.H. Wachters, *"Fasterix, an environment for PCB simulation"*, Proc. Int. Symp. on EMC. , pp.213-218, March 9-11, 1993.
2. R.F. Milsom, *"Efficient SPICE-Compatible Electromagnetic Model of Arbitrarily-shaped Integrated Passive Structure"*, IEEE Trans. MTT, Vol. 49, No. 7, pp.1033-1045, July 1999.
3. R.F. Milsom, N.J. Pulsford, A. Boogaard and M. de Samber, *"EM Simulation of MCM-D Implementation of PA with Integrated Passive Components"*, Proc. RAWCON 99, IEEE Radio and Wireless Conference, 99EX292, pp.255-258, Denver Colorado USA, 1-4 Aug. 1999.
4. N.J. Pulsford, J.T.M. van Beek, M.H.W.M. van Delden, A. Boogaard and R.F. Milsom, *"Passive Integration on Si for RF Circuits in Wireless Applications"*, Proc. IEEE MTT-S, International Microwave Symposium, Anaheim California USA, June 1999.
5. R.F. Milsom, K.J. Scott, P.R. Simons, *"Reduced equivalent circuit model for PCB"*, Philips Journal of Research, Vol. 48, Nos 1-2, pp.9-35, 1994.
6. M. de Samber and L. Tegelaers, *"Technology for Passive Integration"*, Philips J. Research, Vol. 51, No.3, pp.389-410, 1998.
7. M. de Samber, N. Pulsford, M. van Delden and R. Milsom, *"Low Complexity MCM-D Technology for High Frequency Applications"*, Proc. 1998 Int. Conf. Multichip Modules and High Density Packaging, Denver, 1998.

Error Indicator and Adaptive Meshing for Hierarchical FEM-Simulation of Waveguides

M. Bludszuweit and M. Kasper

Technical University Hamburg Harburg

Abstract The Finite Element Method is applied to solve the full vectorial wave equation for z-invariant dielectric waveguides. A mixed set of scalar and vectorial basis functions is employed for the proper modeling of interface and boundary conditions. In order to improve convergence rate basis functions up to third order are proposed and thoroughly tested. For the modal analysis a simple and efficient error indicator is proposed which uses the hierarchical field expansion. Based on this error indicator an h-adaptive mesh refinement is developed and tested. Numerical results show the convergence behaviour and the efficiency for FE-element types of different order and different refinement strategies.

1 Introduction

Microwave and optical waveguide technology has rapidly developed over the last years. In integrated optics commercially available software is mostly based on the Finite Difference Method. However, the Finite Element (FE) Method (FEM) has gained increasingly recognition over the last years. This is attributed to the superior resolution of complex geometries of boundaries or interfaces and proper field modeling. The local mesh density can be chosen to match the required accuracy and hence solve the problem with a minimum of computational effort. The occurrence of spurious modes in early FEM led to an extensive research about their origin. From this resulted an thorough understanding of the FE solution of the Maxwell equations and the development of vector FE basis functions.

A common drawback of FEM software is the high (user) effort in generating a suitable mesh. To overcome this problem a robust mesh generator and an error indicator are needed to perform adaptive mesh refinement. A further area of significance is the order of convergence of the FEM solution. Most simulators in integrated optics work with first or second order. From classical FE analysis it is well known, that higher order shape functions can lead to significant improvements in accuracy for the same computational load.

In the following we treat the eigenvalue problem and introduce suitable hierarchical vector basis functions up to third order. With these hierarchical elements an error indicator is presented and applied to adaptive h-refinement. Numerical results are given.

2 Problem Formulation

Wavepropagation in a z-invariant waveguide is described by the following full-vectorial wave equation [7] :

$$\nabla \times \left(\frac{1}{\mu_r} \nabla \times \mathcal{E} \right) - k_0^2 \varepsilon_r \mathcal{E} = 0 \tag{1}$$

Here we assume ε_r and μ_r to be complex valued scalars and k_0 the wavenumber. The region is bounded by either perfectly electric or magnetic walls. The electric field expresses as $\mathcal{E}(x, y, z) = \mathbf{E}(x, y)e^{-jk_0 n_{eff} z}$. One requirement for an accurate representation is that the interface condition for the electric field at material boundaries is fulfilled. The continuity of the z-component over the cross section and the discontinuity of the transverse field is incorporated in the following ansatz

$$\mathbf{E}(x, y) = \mathbf{E_t}(x, y) + E_z(x, y)\mathbf{e_z} \tag{2}$$

$$\mathbf{E_t} = \sum_{i=1}^{n_Q} \mathbf{Q}_i e_{ti} \qquad \text{and} \qquad E_z = \sum_{i=1}^{n_N} N_i e_{zi} \tag{3}$$

with the vector basis functions $\mathbf{Q}_i$, the scalar basis functions N_i and the unknowns e_{ti} and e_{zi}. Transforming eq. (1) into a variational functional [7], applying standard Galerkin procedure and substituting $jk_0 n_{eff} e_z^\star = e_z$ results in the following general eigenvalue problem:

$$\begin{bmatrix} A^{tt} & 0 \\ 0 & 0 \end{bmatrix} \begin{Bmatrix} e_t \\ e_z^\star \end{Bmatrix} = -k_0^2 n_{eff}^2 \begin{bmatrix} B^{tt} & B^{tz} \\ B^{zt} & B^{zz} \end{bmatrix} \begin{Bmatrix} e_t \\ e_z^\star \end{Bmatrix} \tag{4}$$

where

$$A_{ij}^{tt} = \sum_{e=1}^{M} \int \int_{\Omega} \mu_r^{-1} \left(\nabla_t \times \mathbf{Q}_i \right) \left(\nabla_t \times \mathbf{Q}_j \right) - k_0^2 \varepsilon_r \mathbf{Q}_i \mathbf{Q}_j d\Omega \tag{5}$$

$$B_{ij}^{tt} = \sum_{e=1}^{M} \int \int_{\Omega} \mu_r^{-1} \mathbf{Q}_i \mathbf{Q}_j d\Omega \tag{6}$$

$$B_{ij}^{tz} = \sum_{e=1}^{M} \int \int_{\Omega} \mu_r^{-1} \nabla_t N_i \mathbf{Q}_j d\Omega \qquad B^{zt} = B^{tz^T} \tag{7}$$

$$B_{ij}^{zz} = \sum_{e=1}^{M} \int \int_{\Omega} \mu_r^{-1} \nabla_t N_i \nabla_t N_j - k_0^2 \varepsilon_r N_i N_j d\Omega \tag{8}$$

with the Nabla - operator

$$\nabla = \nabla_t + \mathbf{e_z} \frac{\partial}{\partial z} = \mathbf{e_x} \alpha_x \frac{\partial}{\partial x} + \mathbf{e_y} \alpha_y \frac{\partial}{\partial y} + \mathbf{e_z} \frac{\partial}{\partial z} \tag{9}$$

The parameter α_x and α_y are set to unity, except for absorbing boundaries represented by a PML [2] . The PML-layer is only activated for weakly guiding optical waveguides. In this case the boundary can be placed closer to the

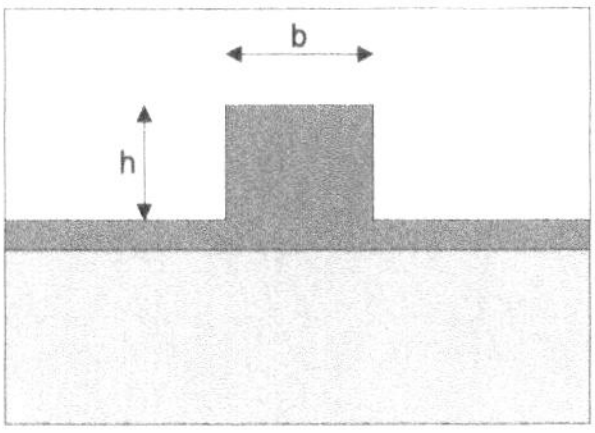

Figure1. Rib waveguide: h=1.1 μm, b=2.0 μm, remaining film thickness: 0.2 μm, refractive index from top to bottom: 1.0, 3.44, 3.34

region of interest and mesh size will be smaller. Usually a small number of the dominant eigenvalues and eigenvectors are of technical importance. Eigenvalues near zero are not of interest for dielectric waveguides.

3 Hierarchical Vector Basis Functions

The choice of the FE basis function has major impact on the performance of the FE software. Nedelec [4] introduced a type of vectorial basis functions suitable for representing electromagnetic fields. For a vector triangular element a polynomial complete expansion of order n requires (n+1)(n+2) degrees of freedom. Nedelec gave a number of constraints on the basis functions to reduce the number of unknown. The constraints of first and second order are given in [5] and and for the cubic case by

$$\frac{\partial^3 Q_x}{\partial x^3} = 0 \quad \text{and} \quad \frac{\partial^3 Q_y}{\partial y^3} = 0 \quad \text{and} \quad 3\frac{\partial^3 Q_x}{\partial x^2 \partial y} + \frac{\partial^3 Q_y}{\partial x^3} = 0$$
$$\text{and} \quad \frac{\partial^3 Q_x}{\partial x \partial y^2} + \frac{\partial^3 Q_y}{\partial x^2 \partial y} = 0 \quad \text{and} \quad 3\frac{\partial^3 Q_y}{\partial x \partial y^2} + \frac{\partial^3 Q_x}{\partial y^3} = 0 \qquad (10)$$

where Q_x and Q_y are the x and y component of $\mathbf{Q}$. With these equations the number of degrees of freedom reduces for a triangular element to n(n+2). The vector basis function can be distinguished in "inner" and "outer" basis functions. "outer" basis functions contribute to the tangential component along one edge, "inner" basis functions only have normal components or are zero along an edge of the triangle. For the linear element this means that each of the 3 DOF are related to one side of the triangle and the tangential component has to be constant along an edge. For the quadratic element there are 2 DOF for each side permitting a linear variation of the tangential component along any edge. 2 DOF are associated with "inner" basis functions. Analogously for the cubic element there are 3 DOF along any edge and 6 "inner" basis functions. A further characteristic useful for the later application of the FE is hierarchy. A class of basis function is called hierarchical, if all basis functions of order n are also contained in the basis functions of order (n+1). The following lists the basis functions for a vector element up to cubic order.

$$
\begin{aligned}
\mathbf{Q_1} &= \nu_1 l_1 \left(\lambda_1 \nabla\lambda_2 - \lambda_2 \nabla\lambda_1\right) \\
\mathbf{Q_2} &= \nu_2 l_2 \left(\lambda_2 \nabla\lambda_3 - \lambda_3 \nabla\lambda_2\right) \\
\mathbf{Q_3} &= \nu_3 l_3 \left(\lambda_3 \nabla\lambda_1 - \lambda_1 \nabla\lambda_3\right)
\end{aligned}
\qquad
\begin{aligned}
\mathbf{Q_4} &= l_1 \left(\lambda_1 - \lambda_2\right)\left(\lambda_1 \nabla\lambda_2 - \lambda_2 \nabla\lambda_1\right) \\
\mathbf{Q_5} &= l_2 \left(\lambda_2 - \lambda_3\right)\left(\lambda_2 \nabla\lambda_3 - \lambda_3 \nabla\lambda_2\right) \\
\mathbf{Q_6} &= l_3 \left(\lambda_3 - \lambda_1\right)\left(\lambda_3 \nabla\lambda_1 - \lambda_1 \nabla\lambda_3\right) \\
\mathbf{Q_7} &= 4 l_1 \lambda_3 \left(\lambda_1 \nabla\lambda_2 - \lambda_2 \nabla\lambda_1\right) \\
\mathbf{Q_8} &= 4 l_2 \lambda_1 \left(\lambda_2 \nabla\lambda_3 - \lambda_3 \nabla\lambda_2\right)
\end{aligned}
$$

$$
\begin{aligned}
\mathbf{Q_9} &= \nu_1 l_1 \left(\lambda_1 - \lambda_2\right)^2 \left(\lambda_1 \nabla\lambda_2 - \lambda_2 \nabla\lambda_1\right) \\
\mathbf{Q_{10}} &= \nu_2 l_2 \left(\lambda_2 - \lambda_3\right)^2 \left(\lambda_2 \nabla\lambda_3 - \lambda_3 \nabla\lambda_2\right) \\
\mathbf{Q_{11}} &= \nu_3 l_3 \left(\lambda_3 - \lambda_1\right)^2 \left(\lambda_3 \nabla\lambda_1 - \lambda_1 \nabla\lambda_3\right) \\
\mathbf{Q_{12}} &= 12 l_1 \left(\lambda_3 - 0.5\right) \lambda_3 \left(\lambda_1 \nabla\lambda_2 - \lambda_2 \nabla\lambda_1\right) \\
\mathbf{Q_{13}} &= 12 l_2 \left(\lambda_1 - 0.5\right) \lambda_1 \left(\lambda_2 \nabla\lambda_3 - \lambda_3 \nabla\lambda_2\right) \\
\mathbf{Q_{14}} &= 12 l_3 \left(\lambda_2 - 0.5\right) \lambda_2 \left(\lambda_3 \nabla\lambda_1 - \lambda_1 \nabla\lambda_3\right) \\
\mathbf{Q_{15}} &= 3 l_2 \left(\lambda_2 \lambda_3 \nabla\lambda_1 - \lambda_3 \lambda_1 \nabla\lambda_2 + \lambda_1 \lambda_2 \nabla\lambda_3\right)
\end{aligned}
\tag{11}
$$

Here λ are the simplex co-ordinates, l_i is the length of side i and ν_i is either $+1$ or -1 depending on the orientation of side i to ensure that adjacent vector functions have the same variation. The linear vector element consists of the basis functions $\mathbf{Q_1}$ to $\mathbf{Q_3}$ and the quadratic one of $\mathbf{Q_1}$ to $\mathbf{Q_8}$. Further details can be found in [1]. The new basis functions of third order $\mathbf{Q_9}$ to $\mathbf{Q_{15}}$ do not show any linear dependency, which is a problem with W_{10}^4 to W_{12}^4 in [1]. Convergence properties of various curl-conforming elements are investigated numerically in [11]. The scalar basis functions N_i for the z-component of the electric field also form a hierarchical group.

$$
\begin{aligned}
N_1 &= \lambda_1 \\
N_2 &= \lambda_2 \\
N_3 &= \lambda_3
\end{aligned}
\qquad
\begin{aligned}
N_4 &= 4\lambda_1\lambda_2 \\
N_5 &= 4\lambda_2\lambda_3 \\
N_6 &= 4\lambda_3\lambda_1
\end{aligned}
\qquad
\begin{aligned}
N_7 &= 10\nu_1\lambda_1\lambda_2\left(\lambda_1 - \lambda_2\right) \\
N_8 &= 10\nu_2\lambda_2\lambda_3\left(\lambda_2 - \lambda_3\right) \\
N_9 &= 10\nu_3\lambda_3\lambda_1\left(\lambda_3 - \lambda_1\right) \\
N_{10} &= 27\lambda_1\lambda_2\lambda_3
\end{aligned}
\tag{12}
$$

For the linear element N_1 to N_3 are used, for the quadratic N_1 to N_6 and for the cubic N_1 to N_{10}. Basis functions $\mathbf{Q_i}$ and N_i together built the basis for a fully vectorial element for z-invariant wave propagation. Basis functions N_i are complete to order n, while basis functions $\mathbf{Q_i}$ are incomplete in the corresponding polynomial order. However, the missing functions to complete the polynomial order are often considered to be nonessential since they belong to the nullspace of the curl operator. The analytical integration of all combination of the basis functions to form the matrices A_{tt} to B_{zz} is cumbersome if done manually. For this reason a software was developed to calculate all element matrices with the symbolic algebra program Maple 6. The output is an optimised Fortran source code. This significantly improves efficiency and reliability of software development.

4 Error Indicator and Mesh Refinement

Despite the high number of publication considering the FE modal analysis of waveguides, the question of an appropriate error indicator has not yet attracted much attention. An ad hoc approach is to use the element energy as an criterion for mesh refinement. This is only a poor indicator since a field variation is worse approximated than a constant field. Lee proposed in [7] to use the duality between the E and H fields. The disadvantage of this approach is, that a generalised eigenvalue problem has to be solved twice. This results in very high numerical cost for the error indicator.

Here an error indicator for the electric field distribution is presented, which only has to solve a single system of linear equation of order N. This idea of using a hierarchical refinement has for instance been presented in [6]. It is assumed that the eigenvalue n_{eff} and eigenvector ϕ_n have already been found for a certain discretisation level n.

$$[A_n]\{\phi_n\} = -k_0^2 n_{eff,n}^2 [B_n]\{\phi_n\} \tag{13}$$

Here ϕ_n stands for the eigenvector of the tangential and z-component of the field. A hierarchical refinement of this discretisation leads to

$$\begin{bmatrix} A_n & A_{n,m} \\ A_{m,n} & A_m \end{bmatrix} \begin{Bmatrix} \psi_n \\ \psi_m \end{Bmatrix} = -k_0^2 n_{eff,m}^2 \begin{bmatrix} B_n & B_{n,m} \\ B_{m,n} & B_m \end{bmatrix} \begin{Bmatrix} \psi_n \\ \psi_m \end{Bmatrix} \tag{14}$$

In order to find the higher order contribution, it is assumed, that

$$n_{eff,n} = n_{eff,m} \qquad \psi_n = \phi_n \tag{15}$$

The second equation from (14) results in

$$\left[A_m + k_0^2 n_{eff,n}^2 B_m\right] \psi_m = \left[-A_{m,n} - k_0^2 n_{eff,n}^2 B_{m,n}\right] \phi_n \tag{16}$$

provided that $\left[A_m + k_0^2 n_{eff,n}^2 B_m\right]$ is a nonsingular matrix. The term

$$\mathbf{E_{err}} = \begin{Bmatrix} 0 \\ \psi_m \end{Bmatrix} \tag{17}$$

is interpreted as an approximation of the deviation of the calculated modal field from the true field distribution. To measure the error in each triangle the energy norm is used:

$$\eta_e = \int\int_{\Omega_e} \varepsilon \mathbf{E_{err}} \mathbf{E_{err}^\star} d\Omega \tag{18}$$

The realised mesh generator, based on [8], [9] and [10], first creates an solid model from geometric input primitives, like rectangles and polygons. The mesh density is than estimated so that the coarsest possible mesh, still having a good triangle quality will be generated. The local mesh density is controlled by an mesh density function, where the nodal values are stored at the triangle vertices and scalar linear basis functions are used for a continuous interpolation. The FE solution process consists of an iterativ solution and mesh refinement loop. In elements with a high local error the mesh density function is decreased proportional to the error criterion. Element size can locally be decreased by arbitrary factors in a single iteration step.

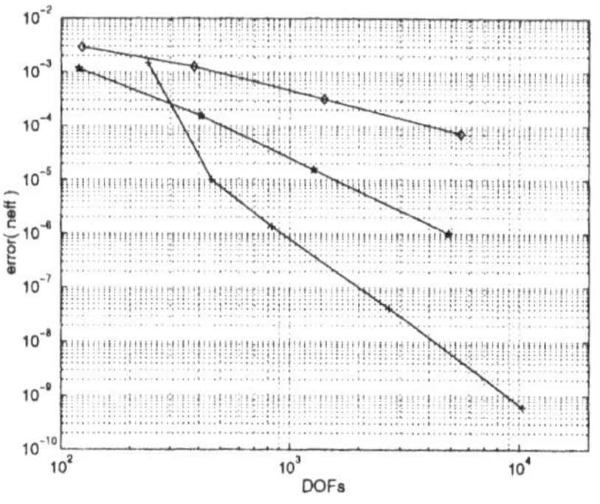

Figure2. Half filled hollow waveguide: error in the propagation constant versus number of degree of freedom for different element order ($\diamond$ 1; $\star$ 2; + 3)

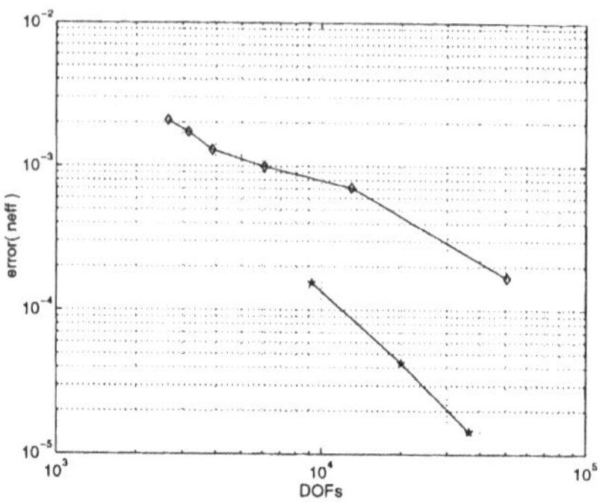

Figure3. Rib waveguide: error in propagation constant for a dielectric waveguide ($\diamond$: order 1; $\star$: order 2)

5 Examples

As typical examples a half filled hollow metallic waveguide and a dielectric optical rib waveguide are chosen. The hollow waveguide has a width of $3\mu m$, a high of $6\mu m$, the lower half is filled with $\varepsilon_r = \sqrt{2}$. The wavenumber is $k_0 = 2\pi/8\mu m$. The third mode was chosen because it has nonzero tangential and longitudinal field components. The dimensions of the rib waveguide are given in fig. 1. To show the convergence behaviour for basis functions of the different order, the mesh density was homogenously increased and the propagation constant of the guided modes are compared with the analytic values. For the rib waveguide the propagation constant of the finest possible discretisation with third order elements was used as a reference value. Fig. 2 and 3 show the convergence rates. The performance of the hierarchical field estimation is presented in fig. 4. The solution ϕ_n of the eigenvalue problem with linear elements is plotted in fig. 4a whereas the approximation of the field ψ by quadratic elements in fig. 4b shows a clearly improved field distribution. To check the performance of the error indicator and the mesh generator an initially coarse mesh was iteratively refined. In fig. 5 the convergence for the half filled hollow waveguide is shown. First and second order elements were tested, both with the energy density and the hierarchical error indicator. The rib waveguide was tested only with the linear element (fig. 6). In all cases the hierarchical error indicator performs better than the energy criterion.

6 Summary

Hierarchical elements formed by scalar and vector basis functions have been proposed and thoroughly tested for polynomial order up to 3. With these results a further extension to truly hp-adaptive analysis is conceivable. An error indicator for the general eigenvalue problem is proposed which requires

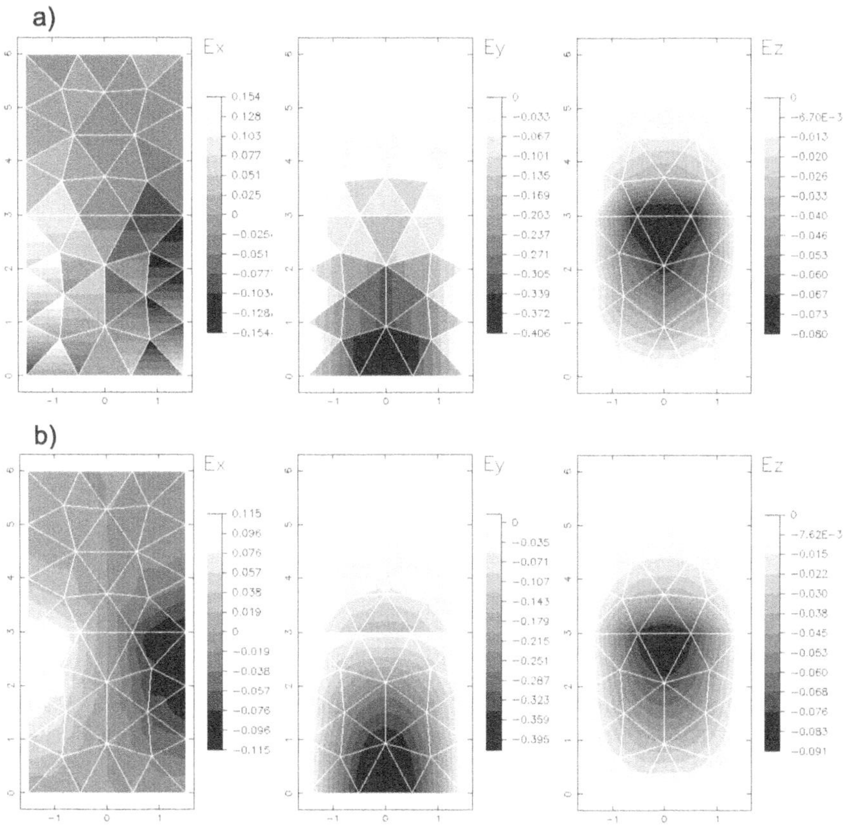

Figure4. Electric field distribution for the third mode in a half filled waveguide: a) original solution with linear basis functions, b) estimated solution by hierarchical quadratic basis functions

only a single solution of a sparse system of linear equation of size N. The accurate prediction of the local field error permits h-adaptive refinement. The results show the expected rate of convergence for different element orders and higher convergence for the adaptive refinement in the number of unknowns. The authors are greatly indebted to the reviewers for their helpful suggestions.

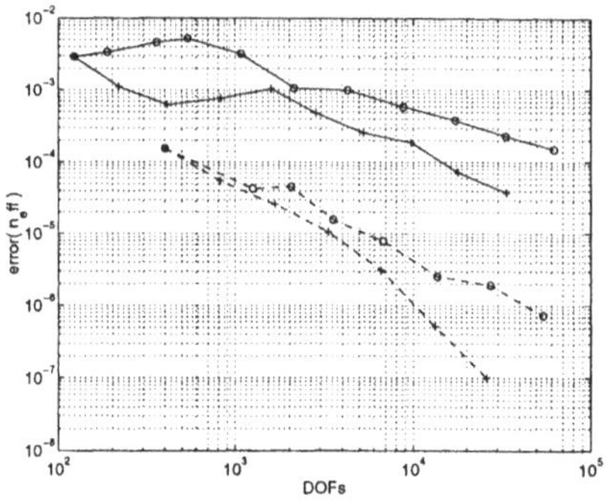

Figure5. Half filled hollow waveguide: error in propagation constant for adaptive mesh refinement , o: element energy and +: hierarchical error indicator , solid line - first order, dotted line - second order

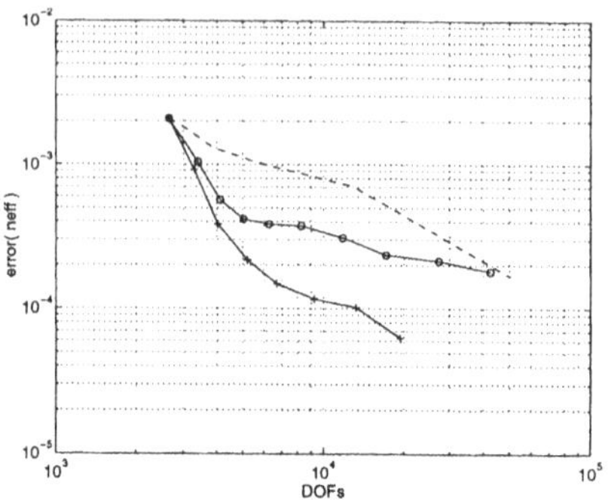

Figure6. Rib waveguide: error in propagation constant for adaptive mesh refinement, o: energy error indicator ; + : hierarchical error indicator , dotted line: homogenous mesh density

References

1. Andersen, L.S., Volakis, J.L.: Development and Application of a Novel Class of Hierarchical Tangential Vector Finite Elements for Electromagnetics. IEEE Trans. Ant. Prop. **47** (1999) 112-120
2. Obbayya, S.S.A., Rahmann, B.M.A., El-Mikati, H.A.: Full-Vectorial Finite-Element Beam Propagation Method for Nonlinear Directional Coupler Devices. IEEE J. Quant. Electr. **36** (2000) 556-562
3. Berenger, J.P.: A perfectly matched layer for the absorption of electromagnetic waves. J. Comput. Phys. **114** (1994) 185-200
4. Nedelec,J.: Mixed finite elements in $\mathbb{R}^3$. Numer. Math. **35** (1980) 315-341
5. Peterson, A.F.: Vector Finite Element Formulation for Scattering from 2D Heterogeneous Bodies. IEEE Trans. Ant. Prop. **43** (1994) 357-365
6. Friberg, P.O.: An Error Indicator for the Generalized Eigenvalue Problem using the Hierarchical Finite Element method. Int. J. Numer. Meth. Eng. **23** (1986) 91-98
7. Lee, J.F., Sun, D.K., Cendes, Z.J.: Full-Wave Analysis of Dielectric Waveguides Using Tangential Vector Finite Elements. IEEE Trans. Micro. Theor. Techn. **39** (1991) 1262-1271
8. Rebay, S.: Efficient Unstructured Mesh Generation by means of Delaunay Triangulation and Bowyer-Watson Alg. . J. Comp. Phys. **106** (1993) 125-38
9. Frey, W.H.: Selective Refinement: A New Strategy for Automatic Node Placement. Int. J. Numer. Meth. Eng. **24** (1987) 2183-2200
10. Anderson, W.K.: A Grid Generation and Flow Solution Method for the Euler Equations on Unstructured Grids. J. Comp. Phys. **110** (1994) 23-38
11. Geuzaine, C., Meys, B., Dular,P., Legros, W.: Convergence of High Order Curl-Conforming Finite Elements. IEEE Trans. Magn. **35** (1999) 1442-1445

A Parallelizable GMRES-type Method for p-cyclic Matrices, with Applications in Circuit Simulation

W. Bomhof[1 2]* and H.A. van der Vorst[1]**

[1] Mathematical Institute, University of Utrecht, Budapestlaan 6, 3584 CD, Utrecht, The Netherlands
[2] Philips Research Laboratories, WAY41, Prof. Holstlaan 4, 5656 AA, Eindhoven, The Netherlands

Abstract In this paper we propose a GMRES-type method for the solution of linear systems with a p-cyclic coefficient matrix. These p-cyclic matrices arise in the periodic steady state simulation of circuits, assuming that the DAE is discretized in the time domain. The method has similarities with existing GMRES approaches for p-cyclic matrices, but in contrast to these methods the method is efficiently parallelizable, even if the p-cyclic matrix has a small block size. However, the serial costs of the method may be somewhat higher. Numerical experiments demonstrate the effectiveness of the method.

1 Introduction

Periodic differential algebraic equations of the form

$$\begin{aligned} \tfrac{d}{dt}q(t,x(t)) + j(t,x(t)) &= 0 \in \mathbb{R}^n \ , \\ x(0) &= x(T) \ , \end{aligned} \tag{1}$$

arise in the periodic steady state analysis of a circuit. The sources of the circuit are included in function $j(t,x(t))$. We assume that all explicitly time-dependent coefficients and sources are periodic with period T. Transient analysis of $\frac{d}{dt}q(t,x(t)) + j(t,x(t)) = 0$ for $t \to \infty$ often converges very slowly to the periodic solution of (1). This leads to an excessive amount of CPU-time for the simulation process. Therefore, it may be more efficient to solve the periodic problem (1) with a method that exploits the periodic structure, for instance, the multiple shooting method or the finite difference method, see [1,5]. We will briefly describe the finite difference method. Time discretization of (1) at the discrete time points $t_0 = 0 < t_1 < \ldots < t_{M-1} < t_M = T$ leads to a system of nonlinear algebraic equations with Mn unknowns: $x_i = x(t_i)$, $i = 1, \ldots, M$. Applying Newton's method to this nonlinear system gives a linear

* e-mail: bomhof@math.uu.nl
** e-mail: vorst@math.uu.nl

system of the form

$$\begin{bmatrix} F_1 & 0 & 0 & -E_M \\ -E_1 & F_2 & 0 & 0 \\ 0 & \ddots & \ddots & 0 \\ 0 & 0 & -E_{M-1} & F_M \end{bmatrix} \begin{bmatrix} y_1 \\ \vdots \\ \vdots \\ y_M \end{bmatrix} = \begin{bmatrix} c_1 \\ \vdots \\ \vdots \\ c_M \end{bmatrix} . \tag{2}$$

For a more detailed derivation of (2) we refer to [5]. Note that both the multiple shooting method and the finite difference method lead to a linear system of the form (2).

The E_i and F_i in (2) are sparse n by n matrices. The matrix in (2) is called an M-cyclic matrix. We will assume that the diagonal blocks F_i are nonsingular; the off-diagonal blocks E_i may be singular. If one y_i is known, the others can be computed by the forward recursion

$$y_j = \begin{cases} F_j^{-1}(c_j + E_{j-1}y_{j-1}) & \text{if } j = i+1, \ldots, M \ , \\ F_j^{-1}(c_j + E_M y_M) & \text{if } j = 1 \ , \\ F_j^{-1}(c_j + E_{j-1}y_{j-1}) & \text{if } j = 2, \ldots, i-1 \ . \end{cases} \tag{3}$$

Block diagonal scaling of (2) with $M = \text{diag}(F_1^{-1}, \ldots, F_M^{-1})$ leads to a linear system that is easier to handle:

$$\begin{bmatrix} I & 0 & 0 & -C_1 \\ -C_2 & I & 0 & 0 \\ 0 & \ddots & \ddots & 0 \\ 0 & 0 & -C_M & I \end{bmatrix} \begin{bmatrix} y_1 \\ \vdots \\ \vdots \\ y_M \end{bmatrix} = \begin{bmatrix} c'_1 \\ \vdots \\ \vdots \\ c'_M \end{bmatrix} , \tag{4}$$

with $c'_i = F_i^{-1}c_i$ and

$$C_i = \begin{cases} F_i^{-1}E_M & \text{if } i = 1 \ , \\ F_i^{-1}E_{i-1} & \text{if } i = 2, \ldots, M \ . \end{cases} \tag{5}$$

The M-cyclic linear system (4) can be reduced by block Gaussian elimination to a p-cyclic one, with $p \le M$. In order to eliminate blocks, we define a partition of the integers $1, \ldots, M$:

$$(1, 2, \ldots, M) = (\underline{q}_1, \ldots, \overline{q}_1, \ldots, \ldots, \underline{q}_p, \ldots, \overline{q}_p) \ . \tag{6}$$

A reduced system is obtained by eliminating the unknowns $y_{\underline{q}_i}, \ldots, y_{\overline{q}_i - 1}$, $i = 1, \ldots, p$ from (4). Hence, the unknowns $y_{\overline{q}_1}$, $y_{\overline{q}_2}$, $\ldots$, $y_{\overline{q}_p}$ are not eliminated. The reduced system is

$$Ax = b \ , \quad \text{with} \quad A = \begin{cases} I - B_1 \ , & \text{if } p = 1 \ , \\ \begin{bmatrix} I & 0 & 0 & -B_1 \\ -B_2 & I & 0 & 0 \\ 0 & \ddots & \ddots & 0 \\ 0 & 0 & -B_p & I \end{bmatrix} , & \text{if } p \ge 2 \ , \end{cases} \tag{7}$$

and

$$B_i = C_{\overline{q}_i} \cdot \ldots \cdot C_{\underline{q}_i} , \ i = 1, \ldots, p \ , \tag{8}$$

$x = \left[x_1^T, \ldots, x_p^T\right]^T$, $b = \left[b_1^T, \ldots, b_p^T\right]^T$. The vectors b_i are defined by the following recursion:

$$\begin{aligned} w_{\underline{q}_i} &= c'_{\underline{q}_i} \ , \\ w_k &= c'_k + C_k w_{k-1}, \ k = \underline{q}_i + 1, \ldots, \overline{q}_i \ , \\ b_i &= w_{\overline{q}_i} \ . \end{aligned} \tag{9}$$

The x_i and y_j are related via $y_{\overline{q}_i} = x_i$, for $i = 1, \ldots, p$. The other y_j, $j = 1, \ldots, M$, can be computed by the forward recursion (3), after (7) has been solved.

Block Gaussian elimination of $x_1, \ldots, x_{p-1}$ in (7) gives a linear system

$$(I - B_p \cdot \ldots \cdot B_1) x_p = \hat{b}_p \ , \tag{10}$$

where $\hat{b}_p$ is defined by the recurrence relation

$$\begin{aligned} \hat{b}_1 &= b_1 \ , \\ \hat{b}_j &= b_j + B_j \hat{b}_{j-1} \quad \text{for } j = 2, \ldots, p \ . \end{aligned}$$

The linear system (10) is equal to the linear system obtained by block Gaussian elimination of $y_1, \ldots, y_{M-1}$ in (4), this is the case $p = 1$ in (7). *LU* factorization of $I - B_p \cdot \ldots \cdot B_1$ in (10) leads to a direct method for solving (2). This method requires $2nM(\text{nnz}(E_i) + \text{nnz}(L_i) + \text{nnz}(U_i))$ flops for explicitly forming $I - B_p \cdot \ldots \cdot B_1$ and $2/3n^3$ flops for the *LU* factorization of $I - B_p \cdot \ldots \cdot B_1$. Here $\text{nnz}(\ \cdot\ )$ is the number of nonzeros of a matrix and $L_i U_i$ is the *LU* factorization of F_i. The other costs are relatively minor. We refer to [1] for other direct methods for (2).

A Krylov subspace method may be attractive for solving (10), because these methods do not require to form $I - B_p \cdot \ldots \cdot B_1$ explicitly. For example, GMRES [4] only needs matrix vector products $(I - B_p \cdot \ldots \cdot B_1)v$. Here B_i is defined by (5) and (8), and sparse *LU* decompositions of the diagonal blocks F_i with partial pivoting are used for F_i^{-1}. The costs of m matrix vector products are $2mM(\text{nnz}(E_i) + \text{nnz}(L_i) + \text{nnz}(U_i))$ flops and the other costs are relatively minor. Usually $m \ll n$, so then there is a gain compared with the direct method.

Instead of applying GMRES to (10), it can be applied to (7). Then each matrix vector product Av requires the matrix vector products $B_i v_i, i = 1, \ldots, p$, which are independent of each other. So, the computation of Av is efficiently parallelizable. Note that the matrix vector product $(I - B_p \cdot \ldots \cdot B_1)v$ is hard to parallelize because of its serial nature. In practice E_i and F_i are often too small for parallelization on a matrix level. Unfortunately, GMRES converges usually much better for (10) than for (7), and even in a parallel environment there is no gain for (7), if $p \neq 1$.

In Sect. 2 we will introduce an efficiently parallelizable iterative method for solving the p-cyclic linear system (7). We will not discuss all the details of the method, a more comprehensive and precise description of the method will appear later in [2]. The method is not restricted to periodic steady state simulation of circuits, because linear systems of the form (2) may arise in the simulation of other periodic phenomena as well, for example, in periodic AC analysis of RF circuits [6].

Independently and simultaneously with our work, Dekker has developed the P-GMRES method [3]. For the special case $p = 2$ our method is equivalent with this P-GMRES method. For $p = 1$ our method is equivalent with the standard GMRES method [4].

Telichevesky, et al., [5] propose to solve (2) with preconditioned GMRES applied to (2). The preconditioner they use is the block lower triangular part of the matrix in (2). In this approach the matrix B_1 of the reduced system (7), with $p = 1$, arises as a block matrix on the diagonal of the (block upper triangular) preconditioned linear system. Therefore, solving (2) by applying our method (or standard GMRES) to the reduced system (7), with $p = 1$, is approximately as efficient as the approach of [5]. Similar to the $p = 1$ case, this method is hard to parallelize because of the serial nature of the matrix-vector recursions.

In Sect. 3 we present numerical experiments for solving linear systems of the form (2) arising from periodic steady state simulation of circuits.

Superscripts are used frequently in this paper. We will use the following convention for it: X^{-1} denotes the inverse of a nonsingular matrix X. The i in X^i is an index number if X is a vector or matrix and $i \geq 0$. Throughout this paper we use $\|\cdot\|$ for the 2-norm of a vector or a matrix.

2 *p*-cyclic GMRES

The GMRES method applied to (7) finds an approximate solution x^m in the Krylov subspace

$$\mathcal{K}^m = \text{span}(b, Ab, A^2, \ldots, A^{m-1}b) \ , \tag{11}$$

with $x^m = \underset{x^m \in \mathcal{K}^m}{\text{argmin}} \|b - Ax^m\|$. For example, $p = 3$ and $m = 4$ leads to

$$\mathcal{K}^4 = \text{span}(\begin{bmatrix} b_1 \\ b_2 \\ b_3 \end{bmatrix}, \begin{bmatrix} B_1 b_3 \\ B_2 b_1 \\ B_3 b_2 \end{bmatrix}, \begin{bmatrix} B_1 B_3 b_2 \\ B_2 B_1 b_3 \\ B_3 B_2 b_1 \end{bmatrix}, \begin{bmatrix} B_1 B_3 B_2 b_1 \\ B_2 B_1 B_3 b_2 \\ B_3 B_2 B_1 b_3 \end{bmatrix}) \ .$$

The idea of the *p*-cyclic GMRES method is to decouple the Krylov subspace (11) into p independent subspaces $\mathcal{K}_i^m$, each with dimension less than or equal to m. We illustrate this idea for the special case $p = 3$ and $m = 4$. The generalization to arbitrary p and m is straightforward. We seek an approximate solution x^4 of (7) in the search space

$$\hat{\mathcal{K}}^4 = \left\{ x^4 = \begin{bmatrix} x_1^4 \\ x_2^4 \\ x_3^4 \end{bmatrix} \middle| \begin{array}{l} x_1^4 \in \mathcal{K}_1^4 = \text{span}(b_1, B_1 b_3, B_1 B_3 b_2, B_1 B_3 B_2 b_1) \\ x_2^4 \in \mathcal{K}_2^4 = \text{span}(b_2, B_2 b_1, B_2 B_1 b_3, B_2 B_1 B_3 b_2) \\ x_3^4 \in \mathcal{K}_3^4 = \text{span}(b_3, B_3 b_2, B_3 B_2 b_1, B_3 B_2 B_1 b_3) \end{array} \right\} \ . \tag{12}$$

Algorithm 1: p-cyclic GMRES

1. Initialization:
   for $i = 1 : p$
   Define $H_i^m = \text{zeros}(m+1, m)$
   $v_i^1 = b_i$
   $[v_i^1, \text{dummy}, u_i^1] = \text{house\_orth}(v_i^1, [\,])$
   $U_i^1 = u_i^1, \; V_i^1 = v_i^1$
   $\delta_i = b_i^T v_i^1$
   end
2. Iterate:
   for $j = 1 : m$

$$\begin{bmatrix} \hat{v}_1^j \\ \hat{v}_2^j \\ \vdots \\ \hat{v}_p^j \end{bmatrix} = \begin{bmatrix} v_p^j \\ v_1^j \\ \vdots \\ v_{p-1}^j \end{bmatrix}$$

   for $i = 1 : p$ (parallelizable for loop)
   $[v_i^{j+1}, H_i^m(1:j+1, j), u_i^{j+1}] = \text{house\_orth}(B_i \hat{v}_i^j, U_i^j)$
   $U_i^{j+1} = [U_i^j \;\; u_i^{j+1}], \; V_i^{j+1} = [V_i^j \;\; v_i^{j+1}]$
   end
   end
3. Finish:
   Solve the least squares problem:

$$\min_{y_1,\ldots,y_p} \left\| \begin{bmatrix} \tilde{I} & 0 & 0 & -H_1^m \\ -H_2^m & \tilde{I} & 0 & 0 \\ 0 & \ddots & \ddots & 0 \\ 0 & 0 & -H_p^m & \tilde{I} \end{bmatrix} \begin{bmatrix} y_1 \\ \vdots \\ \vdots \\ y_p \end{bmatrix} - \begin{bmatrix} \delta_1 e_1 \\ \vdots \\ \vdots \\ \delta_p e_1 \end{bmatrix} \right\|$$

$$x^m = \begin{bmatrix} V_1^m y_1 \\ \vdots \\ V_p^m y_p \end{bmatrix}$$

The approximate solution $x^m \in \hat{\mathcal{K}}^m$ is chosen such that the norm of the residual $\|b - Ax^m\|$ is minimal. It is easily seen that $\mathcal{K}^m \subset \hat{\mathcal{K}}^m$, and we have $\dim(\hat{\mathcal{K}}^m) \leq mp$. Usually the dimension of $\hat{\mathcal{K}}^m$ is p times larger than the dimension of $\mathcal{K}^m$ and therefore we hope that p-cyclic GMRES achieves a much faster convergence than GMRES. Since $\mathcal{K}^m \subset \hat{\mathcal{K}}^m$, the following relation holds for the norm of the residual: $\|r^{m,\, p\text{-cyclic GMRES}}\| \leq \|r^{m,\, \text{GMRES}}\|$.

An orthogonal basis for $\hat{\mathcal{K}}^m$ is desirable, for a practical and accurate method. Fortunately, we only need an orthogonal basis for each $\mathcal{K}_i^m$, $i = 1, \ldots, p$, which immediately leads to an orthogonal basis for $\hat{\mathcal{K}}^m$. In example (12), $B_1 b_3$ has to be orthogonalized with respect to b_1, and $B_1 B_3 b_2$ has to be orthogonalized with respect to b_1 and $B_1 b_3$, etc. Step 1 and step 2 of Alg. 1 construct an orthogonal basis in an Arnoldi-like way. In step 3 the approximate solution with minimal residual norm is computed. For $p = 1$ step 1 and 2 of Alg. 1 reduce to the standard Arnoldi method and Alg. 1

is equivalent to standard GMRES. The Householder method is used for the orthogonalization of the vectors. The function

$$[v_i^{j+1}, H_i^m(1:j+1,j), u_i^{j+1}] = \text{house\_orth}(B_i v_{i\tilde{-}1}^j, U_i^j)$$

of Alg. 1, with $i\tilde{-}1$ defined as

$$i\tilde{-}1 \equiv \begin{cases} i-1 \text{ if } i = 2,\ldots,p \ , \\ p \quad\ \text{ if } i = 1 \ , \end{cases}$$

computes v_i^{j+1} and $H_i^m(1:j+1,j)$ such that

$$B_i v_{i\tilde{-}1}^j = [V_i^j \ \ v_i^{j+1}] H_i^m(1:j+1,j) \ \text{ and } \ [V_i^j \ \ v_i^{j+1}]^T [V_i^j \ \ v_i^{j+1}] = I \ . \quad (13)$$

The Householder vectors are stored in U_i^{j+1}.

Equation (13) can be written in matrix form:

$$B_i V_{i\tilde{-}1}^m = V_i^{m+1} H_i^m \ .$$

Here H_i^m is an $m+1$ by m Hessenberg matrix. This leads to the following block matrix relation:

$$\begin{bmatrix} I & 0 & 0 & -B_1 \\ -B_2 & I & 0 & 0 \\ 0 & \ddots & \ddots & 0 \\ 0 & 0 & -B_p & I \end{bmatrix} \begin{bmatrix} V_1^m & 0 & \ldots & 0 \\ 0 & V_2^m & & \vdots \\ \vdots & & \ddots & 0 \\ 0 & \ldots & 0 & V_p^m \end{bmatrix} =$$

$$\begin{bmatrix} V_1^{m+1} & 0 & \ldots & 0 \\ 0 & V_2^{m+1} & & \vdots \\ \vdots & & \ddots & 0 \\ 0 & \ldots & 0 & V_p^{m+1} \end{bmatrix} \begin{bmatrix} \tilde{I} & 0 & 0 & -H_1^m \\ -H_2^m & \tilde{I} & 0 & 0 \\ 0 & \ddots & \ddots & 0 \\ 0 & 0 & -H_p^m & \tilde{I} \end{bmatrix} ,$$

with the $m+1$ by m matrix $\tilde{I} = \begin{bmatrix} I \\ 0 \ldots 0 \end{bmatrix}$, or shortly

$$AV^m = V^{m+1} H^m \ . \quad (14)$$

It is possible to prove that the subspace $\hat{\mathcal{K}}^m = \text{range}(V^m)$, if and only if $\{H_i^m\}_{j+1,j} \neq 0$, $j = 1,\ldots,m-1$, $i = 1,\ldots,p$ and $b_i \neq 0$, $i = 1,\ldots,p$, but the proof is beyond the scope of this paper (see [2]).

Step 2 is the most expensive part of Alg. 1. Step 2 is efficiently parallelizable on a p processor computer system, because only the $\hat{v}_i^j = v_{i\tilde{-}1}^j$ operations require communication.

In step 3 of Alg. 1 the approximate solution of $Ax = b$ is computed in a similar way as in GMRES. We seek an approximate solution x^m that solves the least squares problem

$$\min_{x^m \in \text{range}(V^m)} \|b - Ax^m\| \ . \quad (15)$$

Since $x^m \in \text{range}(V^m)$, we may substitute $x^m = V^m y$. Now formula (14) can be used to reduce the least squares problem (15) to the much smaller least squares problem of Alg. 1, step 3, see [2]. A common way to solve this reduced least squares problem is by QR factorization of H^m: $Q^m R^m = H^m$, see also [2]. Computing the QR factorization of H^m is relatively inexpensive if a suitable reordering is applied to H^m.

3 Numerical Experiments

In this section we demonstrate the effectiveness of the p-cyclic GMRES method for a number of periodic steady state test problems, arising in circuit simulation. The periodic DAE is solved by the procedure explained in Sect. 1. Table 1 shows the sizes of the test problems (2). The matrices are taken from the first Newton step of the nonlinear system solution process.

Table1. Convergence of p-cyclic GMRES for periodic steady state test problems. M is the number of time points, n is the number of unknowns per time point. The number of iterations needed to satisfy the stopping criterion is displayed in the last 5 columns, for different values of p.

				iterations				
problem	M	n	Mn	$p=1$	$p=2$	$p=4$	$p=8$	$p=16$
pss_1	64	22	1408	5	5	6	7	8
pss_2	128	22	2816	9	10	11	11	11
pss_3	64	54	3456	12	12	12	12	12
pss_4	32	904	28928	1	2	2	4	6
pss_5	128	424	54272	34	36	43	52	67

In Sect. 1 we have described how to reduce (2) to a p-cyclic system. The integer partition (6) is chosen such that $\underline{q}_i - \overline{q}_i =$ constant. This is possible since both p and M are powers of 2 here. Note that in a parallel environment it is desirable to have $\underline{q}_i - \overline{q}_i =$ constant, for a good load balance. We apply p-cyclic GMRES to the reduced p-cyclic system, for different values of p.

The stopping criterion is based on the true residual of (2), which makes a fair comparison possible between the different values of p. The iterative process is stopped if the true residual of (2) is smaller than $10^{-10}\|c\|$, with $c = \left[c_1^T, \ldots, c_M^T\right]^T$.

The convergence results of p-cyclic GMRES are in Table 1. We see that usually the speed of convergence decreases with increasing p, but it decreases only slowly relative to the increase in p (see also Table 1).

The costs of recursion (9), which defines b_i, and the costs of recursion (3), which solves the missing y_i, are not negligible. For both recursions $M - p$ matrix vector products $C_i v$ are needed. Each p-cyclic GMRES iteration costs M matrix vector products $C_i v$. The overhead due to (9) and (3) amounts to approximately $2(M-p)/M \approx 2$ p-cyclic GMRES iterations. The p-cyclic GMRES iterations, as well as recursions (9) and (3) are efficiently parallelizable on a p processor computer system.

Recall that for $p = 1$ the computational costs of our method are approximately equal to the costs of the sequential method proposed in [5]. We conclude that for $p > 1$ our method is somewhat more expensive than the method of [5], but in contrast with this method, our method is efficiently parallelizable.

Acknowledgements

I would like to thank Jan ter Maten, Wil Schilders and Jaap Fijnvandraat (Philips Research) for posing the problem and for being critical on the ideas suggested by the author.

References

1. Ascher, U.M., Mattheij, R.M.M., Russell, R.D.: Numerical solution of boundary value problems for ordinary differential equations. SIAM, Philadelphia, PA, 1995.
2. Bomhof, C.W.: A parallelizable GMRES-type method for p-cyclic matrices. In: PhD thesis, Mathematical Institute, Utrecht University, to appear.
3. K. Dekker, *Parallel GMRES and Domain Decomposition.* Report 00-05, Department of Applied Mathematical Analysis, Delft University of Technology, 2000.
4. Saad, Y., Schultz, M.: GMRES: A generalized minimal residual algorithm for solving nonsymmetric linear systems. SIAM J. Sci. Statist. Comput. **7** (1986), pp. 856-869.
5. Telichevesky, R., Kundert, K., Elfadel, I., White, J.: Fast simulation algorithms for RF circuits. In proc. IEEE 1996 Custom Integrated Circuits Conf., San Diego, pp. 437-444, 1996.
6. Telichevesky, R., Kundert, K., White, J.: Efficient AC and noise analysis of two-tone RF Circuits. In proc. 33rd Design Automation Conference (DAC'96), Las Vegas, pp. 292-297, 1996.

Parallel Algorithms for Solving Linear Equations in VLSI Circuit Simulation

H.K. Dirks, M. Fischer, J. Rüdiger

Lehrstuhl für Allgemeine und Theoretische Elektrotechnik,
Christian-Albrechts-Universität zu Kiel

Abstract Algorithms for parallel execution of Gaussian elimination to solve linear equations arising from VLSI circuit simulation are discussed. Here, elimination trees are used since they embody the dependence of data during the elimination process. They are important means for estimating and controlling parallel solution algorithms. The potential parallelism is increased by reducing the critical path length of the tree. Experimental results rest upon multithreaded methods and compare favourable with theoretical estimates of the speed-up.

1 Introduction

Analog circuit simulation is an indispensable tool in VLSI chip design. Progressive technology results in increasing demands on simulation programs like SPICE and CEDUSA [1], which is used in this study. Circuit simulation requires the numerical solution of a large system of nonlinear differential algebraic equations. First, the differential equations are numerically integrated employing implicit backward difference formulae like the backward Euler formula. For each time step the resulting system of nonlinear algebraic equations is then solved iteratively using a modified Newton-Raphson method. This approach entails the recurrent direct solution of large systems of linear equations $\mathbf{Ax} = \mathbf{b}$. Matrix $\mathbf{A}$ is often structurally symmetric and has a very sparse and irregular pattern of nonzero coefficients, which change numerically during the simulation process. To minimize the number of fill-ins and operations an empirical ordering method (Markowitz) is used. For very large circuits most of the computing time is needed for the direct solution of these linear equations. In order to reduce simulation runtimes, parallel execution on multi-processor architectures can be utilized. A common possibility is to divide the circuit into several subcircuits by topological methods [2]. These subcircuits are treated in parallel, e.g. on a workstation cluster. In this paper a different approach suited for shared memory multi-processor architectures (SMP) is considered, where the total set of arithmetical operations for solving $\mathbf{Ax} = \mathbf{b}$ is partitioned into subsets which are executed in parallel.

2 Equivalent Reordering for Parallel Elimination

In the following **A** is assumed to be a n by n structurally symmetric[1] sparse matrix, which is already filled using the Markowitz reordering method[2]. For solving $\mathbf{Ax} = \mathbf{b}$ by Gaussian elimination, one can define tasks for parallel processing. Therefore a unit task is the elimination of the k−th unknown, i.e. the execution of the i, j double-loop in Fig. 1 with fixed k. Representing these tasks as nodes of a task graph, an elimination tree T is introduced [3] describing the precedence constraints between the unit tasks. The relation

```
for k = 1, ..., n − 1
    for i = k + 1, .., n
        a_ik = a_ik / a_kk
        for j = k + 1, .., n
            a_ij = a_ij − a_ik a_kj
        b_i = b_i − a_ik b_k
```

Figure1. Gaussian Elimination

between matrix structure and elimination tree (ET) is illustrated by an example shown in Fig. 2, where in Fig. 2(a) nonzero elements are depicted as 'X' or digits, respectively. Fig.[3]2(b) presents the corresponding ET. For the following discussion the reader is assumed to be familiar with some basic tree terminology such as parent-child relation, paths, root, subtree, leaf and others [4]. The tree gives an idea of the potential parallelism which can be exploited: A node can be processed after the completion of all its children. Thus, the parallel elimination starts with processing the leaf nodes and it ends at the root.

Relevant measures to discuss parallel methods are explained by means of the example given in Fig. 2. A node weight $\mathbf{Wt}(v)$ is introduced as an estimate of the computational effort for task v. It counts the number of operations, e.g. the elimination of node 4 needs 9 multiplications (flops). The sum $\mathbf{Wt}(T) = 56$ of weights of all nodes in T is some measure of total completion time for the elimination. The weighted height[4] of a node v is given

1 The approach in this paper is restricted to matrices with symmetric pattern. Matrices with nonsymmetric pattern can be treated in a similar manner.

2 Experience shows that in VLSI circuit simulation numerical pivoting is not needed. In contrast to other applications, here the number of fill-ins is kept small using the Markowitz reordering.

3 Fig. 2(c) is used in chapter 2.

4 If all node weights equal one, $\mathbf{Ht}^{Wt}$ and $\mathbf{Dth}^{Wt}$ are used without the superscript Wt.

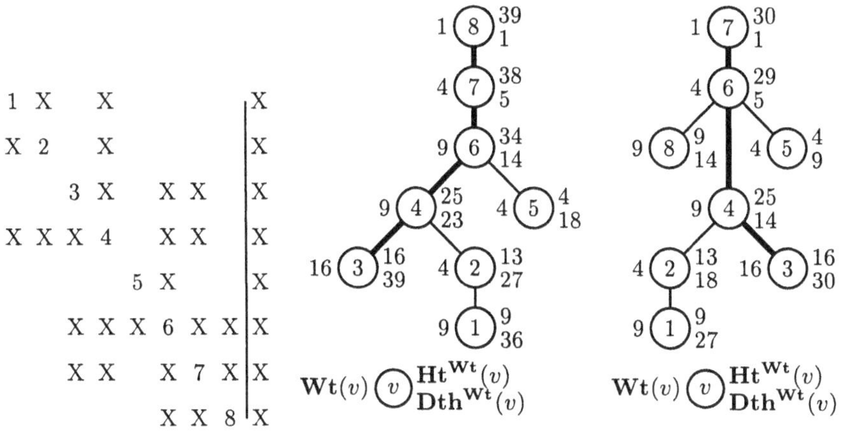

(a) **A**|**b** (b) ET (c) Rotated ET

Figure2. Matrix structure and ETs

by:

$$\mathbf{Ht}^{Wt}(v) = \begin{cases} \mathbf{Wt}(v) & v\text{: leaf ,} \\ \mathbf{Wt}(v) + \max\left(\mathbf{Ht}^{Wt}(w) \mid w\text{: child of } v \text{ in } T\right) & \text{else .} \end{cases} \tag{1}$$

$\mathbf{Ht}^{Wt}(root)$ is the critical path length (CPL), because it predicts the minimum parallel execution time, if any overhead, like synchronization overhead, is neglected and if enough processors are available. The critical path (CP) is highlighted by thicker lines in Fig. 2(b). Obviously, weighted and unweighted CPs differ. The most common measure to characterize parallel execution is the speed-up. The theoretical bound for the achievable speed-up is:

$$\text{Speedup}_{CPL}^{Wt} = \mathbf{Wt}(T)/\mathbf{Ht}^{Wt}(root) \ . \tag{2}$$

The weighted depth of a node v measures the weighted distance between v and *root*:

$$\mathbf{Dth}^{Wt}(v) = \begin{cases} \mathbf{Wt}(v) & v\text{: root ,} \\ \mathbf{Wt}(v) + \mathbf{Dth}^{Wt}\left(\mathbf{parent}(v)\right) & \text{else .} \end{cases} \tag{3}$$

In the following a method for increasing potential parallelism is discussed. For this the adjacency set is needed:

$$\mathbf{Adj}(j) = \{i \mid a_{ij} \neq 0 \wedge i > j\} \tag{4}$$

and $|\mathbf{Adj}(j)|$ denotes the power of the set $\mathbf{Adj}(j)$. The achievable speed-up is typically limited to an unsatisfying factor of about two. To further increase potential parallelism, a modification of Liu's rotation method is proposed. In [3] a heuristic algorithm is introduced consisting of the search and the rotation procedure, which are both performed only once. The former is depicted in Fig. 3, it determines an eligible node x for rotation. Search and Composite_Rotations compute a reordering of the equations causing no additional fill-ins. This leads to an ET with nearly minimum height ($\mathbf{Ht}(root)$). Due

```
x = 0
y = root of tree T
htbar = -1
reduced = 0
WHILE Ht(y) > htbar and y is not a leave
    newdepth = Dth(y) - reduced
    IF newdepth > |Adj(y)|                        line A
        x = y
        reduced = Dth(y) - |Adj(y)|               line B
        htbar = htbar + newdepth - |Adj(y)|       line C
    IF y has more than one child
        m2 = second largest height among
              the children of y
        htbar = max(m2, htbar)
    y = a child of y with the largest height value
```

Figure3. x=Search(T) by Liu

to strongly varying node weights the search algorithm has to be modified such that $\mathbf{Ht}^{Wt}(root)$ is minimized. In addition to the global substitution of $\mathbf{Ht}(y)$ by $\mathbf{Ht}^{Wt}(y)$ and $\mathbf{Dth}(y)$ by $\mathbf{Dth}^{Wt}(y)$, in lines A, B, and C the term $d = |\mathbf{Adj}(y)|$ has to be replaced by:

$$|\mathbf{Adj}(y)| \rightarrow \begin{cases} \sum\limits_{k \in \mathbf{Adj}(y)} \mathbf{Wt}(k) & \text{in line A ,} \\ d(d+1)(2d+1)/6 & \text{in lines B and C ,} \end{cases} \tag{5}$$

where the last replacement is the weighted depth of y in the restructured ET [5,?]. Applying the modified search algorithm to the example introduced in chapter 2, node 4 is selected for rotation. The restructured ET is shown in Fig. 2(c) with CPL being reduced from 39 to 30.

3 Partitioning and Parallel Solving

3.1 Solving with Dynamic Scheduling

The use of a SMP in conjunction with MT-Programming [7] enables an effective parallel dynamic asynchronous strategy. The considered dynamic

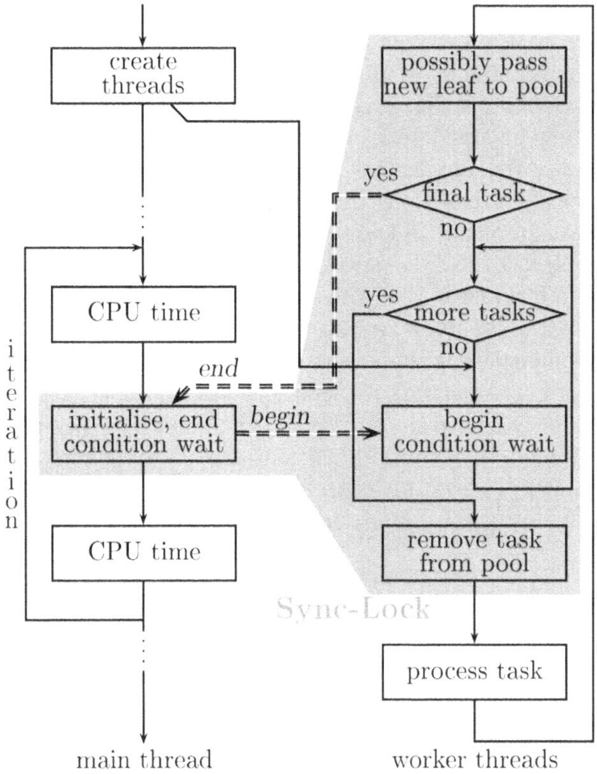

Figure4. Implemented Synchronization Mechanism

scheduling strategy is based on the pool-of-tasks concept. At the beginning of the elimination process all leaves of the ET are part of the pool. Each idle processor can select one of them, remove it from the pool and process the corresponding task. Hereafter, another node of the tree may become a leaf which is immediately passed to the pool. Manipulating the contents of the pool requires a protection realized by a lock mechanism, symbolized by Sync-Lock in Fig. 4, whereas the tasks are carried out unprotectedly. To optimize the sequence of selected tasks a priority list can be introduced, created with respect to $\mathbf{Dth}^{Wt}(v)$. Short experimental completion times justify the decision to determine the sequence in that manner.

To avoid synchronization overhead it is helpful to keep the number of tasks small. Therefore it is an advantage to create groups of nodes (partitions) in such a way that they fit well with the employed scheduling. The partitioning maps the ET onto a so-called partition tree [5].

3.2 Partitioning Rules

Similar to partitioning algorithms used with global synchronization solvers, here the partitioning is also done in a stepwise empirical manner. After each partitioning step grouped nodes are stripped off the elimination tree. Having $\mathbf{Ht}^{Wt}$ in mind, which determines the lower bound of execution time, it makes sense to select partitions along the actual CP. Targeting minimum execution time it is more important to keep the CPL of the partition tree small than to balance partition sizes. This contrasts with other approaches like that in [8]. Entering the ET from the root and branching to subtrees, possible partitions are detected with respect to special rules and a variable threshold. Following these rules, partitions without any advantage for parallel processing, e.g. serial paths of partitions, are avoided. Size and number of suitable subtrees are used to adjust the threshold. The threshold must be increased (decreased) if too many (only a few) initial partitions are found. Among these partitions the p (number of processors) subtrees with the greatest priority:

$$\mathbf{Prio}(T_v) = \mathbf{Wt}(T_v) + \mathbf{Dth}^{Wt}(v) - \mathbf{Wt}(v) \tag{6}$$

are selected and passed to the scheduling list, where T_v is the subtree rooted at node v. This ensures potential parallelism towards the end of the elimination.

4 Results and Conclusion

4.1 Theoretical Results

Because the number of available processors is limited, the CPL is generally not an appropriate measure for performance and the theoretical bound (2) is not always reached. In order to take care of finite p, a new definition of the achievable speed-up $\text{Speedup}_{sched}^{Wt}(p)$ is introduced. For that, $\mathbf{Ht}^{Wt}(root)$ in (2) is replaced by $\mathbf{Ht}_{sched}^{Wt}(root, p)$, which is calculated by a symbolic parallel elimination. Thus, scheduling is modeled by taking into account the number of processors, task dependencies and neglecting any overhead.

By means of three examples, characterized by some data given in Table[5] 1, the success of the described methods is shown. These examples have been provided by NEC Corp., Tokyo and Micronas GmbH, Freiburg. The authors gratefully acknowledge this support. Fig. 5 depicts $\text{Speedup}_{sched}^{Wt}(p)$ over p. The curves are based on three different trees, which are: ET (Markowitz), ET after rotation and partition[6] tree. The advantage of the rotation method can be read off the horizontal lines referring to $\text{Speedup}_{CPL}^{Wt}$. The poor performance of example S3 is caused by a dense lower right corner of the matrix mapping on a path at the root, which covers most of the computing time. Fig. 5 shows the excellent agreement between the results obtained for partition and ETs.

[5] Total number of nonzeros can be derived by adding columns three and four.

[6] Of course, partition tree includes rotation.

Table1. Characteristical Data for Three Industrial Circuits

Circuit	n	entries	fill-ins	flops	population
S1	14793	150729	307804	21610821	0.2095%
S2	21817	302647	209458	7418429	0.1076%
S3	60797	552101	3059690	1177459360	0.0977%

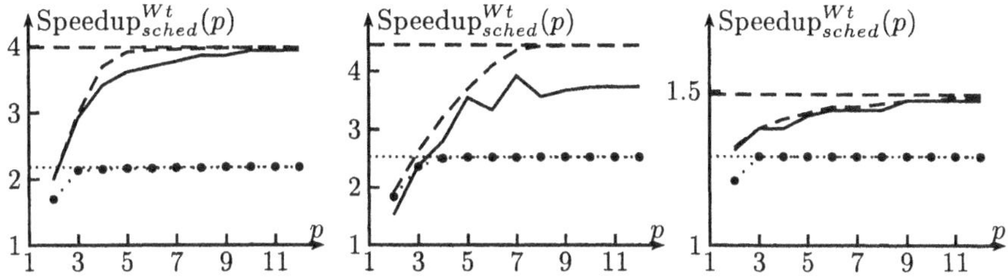

(a) S1 (b) S2 (c) S3

ET Markowitz: • · • ET after Rotation: - - Partition Tree: —

Figure5. Speedup$^{Wt}_{sched}(p)$ for the Three Examples

4.2 Experimental Results

Experimental results presented in this section are based on a modification of CEDUSA's target row directed solver (node weights like Gaussian algorithm) [5]. The scatter-gather technique in conjunction with a compressed row storage scheme is used during the elimination process. Measured data (Speedup$^{Wt}_{meas}(4)$ and execution time $time_{par}(4)$) in Table 2 refer to an E-450[7] with four processors. Measured data exceed theoretical estimates because not all details of the hard- and software are considered. Especially the cache behaviour can cause advantages in parallel applications.

Despite the rotation method the speed-up is typically limited to a factor of four — even for $p > 4$. To further increase speed-up it is necessary to redefine the unit task from chapter 2, e.g. replacing the elimination of the k-th unknown by that of the coefficient a_{ik}. Thus, the new unit task becomes the j-loop for fixed k and i in Fig. 1.

[7] Enterprise 450 (E-450) server with 1.2 GByte RAM (100 MHz), 4 CPUs, (300 MHz, 2 MByte external cache)

Table2. Measured Data Using Four Processors

Circuit	$\text{Speedup}_{sched}^{Wt}(4)$	$\text{Speedup}_{meas}^{Wt}(4)$	$time_{par}(4)/\text{sec}$
S1	3.41	3.58	0.454
S2	2.79	3.28	0.180
S3	1.38	1.75	72.690

References

1. *CEDUSA Reference Guide*, Version 1.10, Institut für Theoretische Elektrotechnik, RWTH Aachen, 1997.
2. B. W. Kerninghan, S. Lin, *An Efficient Heuristic Procedure for Partitioning Graphs*, The Bell System Technical Journal, pp. 291 - 307, 1970.
3. J. W. H. Liu, *Reordering Sparse Matrices for Parallel Elimination*, Parallel Computing, Vol. 11, pp. 73 - 91, 1988.
4. A. V. Aho, J. E. Hopcroft, J. D. Ullman, *Data Structures and Algorithms*, Addison-Wesley, Reading, MA, 1983.
5. J. Rüdiger, *Effizienzabschätzungen und experimentelle Verifikation paralleler Algorithmen für lineare Gleichungssysteme der VLSI-Netzwerksimulation*, Dissertation, CAU Kiel, Shaker Verlag, 2000.
6. K. Hachiya, *Matrix Reordering for Parallel Circuit Simulation*, seminar talk given at CAU Kiel, 1999.
7. *SunOS 5.3 Guide to Multithread Programming*, Sun Microsystems, Inc., Moutain View, California, 1993.
8. A. Schumacher, *Netzwerksimulation von VLSI-Schaltungen auf Parallelrechnern*, Dissertation, RWTH Aachen, Shaker Verlag, 1996.

Investigation of Time Step Control for the Mixed–Level Device/Circuit Simulation of SiGe Bipolar Microwave Power Amplifiers

M. Bartels and B. Meinerzhagen

Institut für Theoretische Elektrotechnik und Mikroelektronik
University of Bremen, Germany

Abstract Important aspects of modern RF power amplifier design require mixed-level device/circuit simulation. In this method, the nonlinear transistor behavior is described by a numerical device simulator which is embedded in a circuit simulator accounting for the surrounding circuit including parasitics.

During transient simulations, time step control limits the time discretization error of the time derivatives of the circuit variables (node potentials and branch currents) and the device variables (carrier densities). The displacement currents at the device terminals are directly connected to the time derivatives of the local electron and hole densities inside the device.

Allowing more discretization error for the time derivatives of the carrier densities typically relaxes the overall time step control and lowers the simulation time. In this study for the special application of RF power amplifier design, we have investigated how much the accuracy of the circuit variables is affected, and which CPU–time advantage can be gained when only the circuit variables are considered for time step control, and consequently the time discretization accuracy within the device simulator is relaxed.

1 Introduction

Power amplifiers are key components within the analog front ends of modern mobile communication devices. For the upcoming mass market applications in the sub–10 GHz regime, the amplifying devices in such circuits will frequently be SiGe bipolar transistors due to their compatibility with silicon technology [1]. In order to assist system designers in optimizing these devices, numerical device models become more and more important, because they offer a more accurate modeling directly based on semiconductor physics, doping profiles, and device structure and therefore a better insight into device–internal behavior, which cannot be obtained from compact models.

A significant drawback in a mixed–level device/circuit simulation environment is the increase in required simulation time introduced by the numerical device model. This issue becomes even more striking, when e. g. an RF PA is simulated in the region of gain compression, where the periodic steady state is often not reached unless several tens of excitation signal periods have passed. It is therefore of major interest that the overall simulation method requires as few time steps as possible without significantly affecting the accuracy of the circuit variables.

2 The Simulation Environment

The mixed–level device/circuit simulation environment consists of the circuit simulator Cedusa [2] and the numerical 2D device simulator Galene III, which considers the drift/diffusion (DD) and the generalized hydrodynamic (HD) model [3], both supplied with Monte Carlo generated transport parameters [4]. The coupling of these simulators features load balancing within a heterogeneous workstation cluster [5] as well as the evaluation of device–internal quantities such as carrier densities [6] through an interface in the circuit simulator.

The BDF time step control in this simulation system has to monitor the individual time discretization errors in either simulator. For example, in principle the accuracy of the carrier density time derivatives at each device internal real space grid point in the device simulator must be monitored to guaranty overall simulation accuracy. Both simulators choose the BDF integration order such that the new time step to be suggested to the simulation framework becomes maximal. The lowest new time step is chosen among all suggested values, thus favoring time integration via the most defensive BDF variant.

3 The Investigated PA

Figures 1 through 3 briefly introduce the investigated RF PA and the bipolar transistor it features as an amplifying device.

The PA is a one–stage (Q_1) amplifier using an additional biasing stage (Q_2). The inductance L_1 represents part of the output matching and packaging parasitics. The input/output powers $P_{\text{in/out}}$ are derived from the transient signals $u/i_{\text{in/out}}$, respectively, via Fourier decomposition to obtain power values of the fundamental and the lowest harmonics.

The transistor itself, a multi–emitter finger NPN–type SiGe heterojunction bipolar transistor (HBT), exhibits a maximum f_{T} (cutoff or unit–current–gain frequency) of about 30 GHz and a maximum $f_{\max}$ (maximum oscillation or unit–power–gain frequency) of about 50 GHz.

The power sweep simulation results in fig. 3 were obtained using a sinusoidal $u_{\text{in}}(t)$ with an amplitude of $U_{\text{in}} = 0.05, 0.1, 0.2, \ldots, 6.4\,\text{V}$. Three particular steps in this row will be subsequently considered (cf. numbers in fig. 3):

1. $U_{\text{in}} = 0.4\,\text{V}$—behavior still linear though gain (G) compression already slightly noticeable,

2. $U_{\text{in}} = 1.6\,\text{V}$—maximum power–added efficiency (PAE) and strong gain compression, and
3. $U_{\text{in}} = 6.4\,\text{V}$—strong gain compression and high harmonic content in output signal, PAE very low again.

4 Results

Simulation Runs. All the results discussed here were obtained with a sinusoidal $u_{\text{in}}(t)$ as described above, at the frequencies $f = 0.95, 1.9, 3.8\,\text{GHz}$, $f = 1.9\,\text{GHz}$ being the intended frequency of operation for this circuit. In order to study the effect of (not) resolving the time dependency of device–internal carrier densities, two simulation runs were performed:

- Run "A"—The BDF methods of both the device *and* the circuit simulator are allowed to suggest a value for the new time step, which ensures a proper time discretization of the semiconductor equations even in cases when the circuit simulator alone would suggest a relatively large new time step. This is the "reference" run.
- Run "B"—Only time step suggestions by the circuit simulator are considered, which in some cases results in a too coarse time discretization of the semiconductor equations, but on the other hand can be expected to significantly decrease the total number of time steps. The network solution obtained from this run should deviate from "A" as little as possible to be a practical alternative.

Before extracting results over one period, always a number of periods was simulated in order to ensure that the periodic steady state was reached with sufficient accuracy. Note that the axis scaling of $0 \leq t/T \leq 1$ in some of the plots refers to the period evaluated. The result of a typical power sweep simulation is shown in figure 3.

Network Solution. The most important stationary figure in bipolar transistor modeling is the collector current, which in the PA observed here is identical with the network variable i_C (cf. fig. 1), and therefore will be the subject of the subsequent investigations. Figures 5 and 6 show details of i_C for different values of U_{in} at $f = 1.9\,\text{GHz}$. Obviously, the overall agreement between runs "A" and "B" is very good, though in cases of gain compression some loss in signal detail becomes visible (cf. especially $U_{\text{in}} = 6.4\,\text{V}$ in fig. 6). Accordingly, the increasing amount of harmonic content with gain compression is confirmed by the Fourier decomposition of i_C in fig. 4.

The number of time steps per period is drastically reduced in the "B" runs by a factor of approximately 4...8, as recorded in table 1. This eases the burden of "steady–state" simulations by the transient time–domain method performed in this work, as summarized in fig. 3.

Table1. Number of time steps per period, n, for different runs

$U_{\text{in}} \rightarrow$	0.4 V			1.6 V			6.4 V		
f	$n^{\text{“A”}}$	$n^{\text{“B”}}$	$n^{\text{“A”}}/n^{\text{“B”}}$	$n^{\text{“A”}}$	$n^{\text{“B”}}$	$n^{\text{“A”}}/n^{\text{“B”}}$	$n^{\text{“A”}}$	$n^{\text{“B”}}$	$n^{\text{“A”}}/n^{\text{“B”}}$
0.95 GHz	707	166	4.3	1722	212	8.1	2313	311	7.4
1.90 GHz	860	139	6.2	2270	284	8.0	2443	299	8.2
3.80 GHz	625	132	4.7	1754	307	5.7	2826	406	7.0

Figures 7, 9, and 11 show i_C from the “A” runs for all cases covered by table 1. The relative error introduced to i_C in the “B” runs by resolving the time dependency of the carrier densities in the numerical device model by the relatively coarse time steps only derived from the network equations is depicted in figures 8, 10, and 12. For each value of U_{in}, the error is related to the peak-to-peak value of i_C from run “A”, $i_{\text{C,pp}}^{\text{“A”}}$, in order to account for the varying signal amplitudes. The error does not exceed 2 % in any of the shown cases, which again confirms that the “B” approach is indeed accurate enough for power amplifier design.

Carrier Density Monitoring. The time dependency of device–internal carrier densities is shown in figures 13 through 18, where the hole density along a one–dimensional cut through the inner transistor is sampled along one period for $U_{\text{in}} = 6.4$ V. The vertical lines subdivide each plot into base, lowly doped collector, and slope of the buried layer, from top to bottom. No significant differences are found comparing the “A” to the “B” runs.

5 Conclusions

For power amplifier design in a mixed–level device/circuit simulation environment resolving the time dependency of the device–internal carrier densities by the relatively coarse time steps sufficient to accurately resolve the variables on the circuit level significantly reduces the number of time steps and consequently reduces computation time by a factor between 4 and 8. The error in the network solution caused by this approach is negligible for the circuit investigated in this study, as long as no information predominately related to higher harmonics is extracted. The enormous gain in computation speed makes periodic steady–state mixed–level device/circuit simulations based on the transient time–domain method feasible even for a power amplifier at high gain compression.

Acknowledgment

This work was supported by the *Bundesministerium für Bildung und Forschung* under contract no. 01 M 2416 C. The authors wish to thank A. Schüppen and S. Gerlach of TEMIC Semiconductor GmbH, Heilbronn, Germany, for their support concerning power amplifier design.

References

1. A. Schüppen, H. Dietrich, U. Seiler, H. von der Ropp, and U. Erben, "A SiGe RF technology for mobile communication systems," *Microwave Engineering Europe*, vol. June, pp. 39–46, 1998.
2. M. Stecher, B. Meinerzhagen, I. Bork, J. M. J. Krücken, P. Maas, and W. L. Engl, "Influence of energy transport related effects on NPN BJT device performance and ECL gate delay analysed by 2D parallel mixed level device/circuit simulation," *IEICE Trans. on Electronics*, vol. E77-C, no. 2, pp. 200–205, 1994.
3. R. Thoma, A. Emunds, B. Meinerzhagen, H. J. Peifer, and W. L. Engl, "Hydrodynamic equations for semiconductors with nonparabolic bandstructures," *IEEE Trans. Electron Devices*, vol. 38, pp. 1343–1352, 1991.
4. B. Neinhüs, S. Decker, P. Graf, F. M. Bufler, and B. Meinerzhagen, "Consistent hydrodynamic and Monte–Carlo simulation of SiGe HBTs based on table models for the relaxation times," *VLSI Design*, vol. 8, pp. 387–391, 1998.
5. S. Scharfenberg, "Mixed–Level Circuit–Device Simulation auf einem heterogenen Workstation–Cluster," Dissertation, RWTH Aachen, Aachen, 1996.
6. T. Kwasniok, "Simulation Integrierter MOS–Schaltungen: Miniaturisierungseffekte und Lebensdauer," Dissertation, RWTH Aachen, Aachen, 1996.

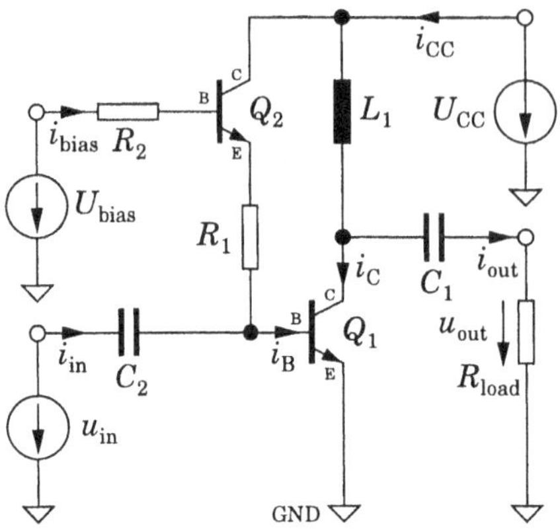

Figure1. Circuit of the investigated RF PA

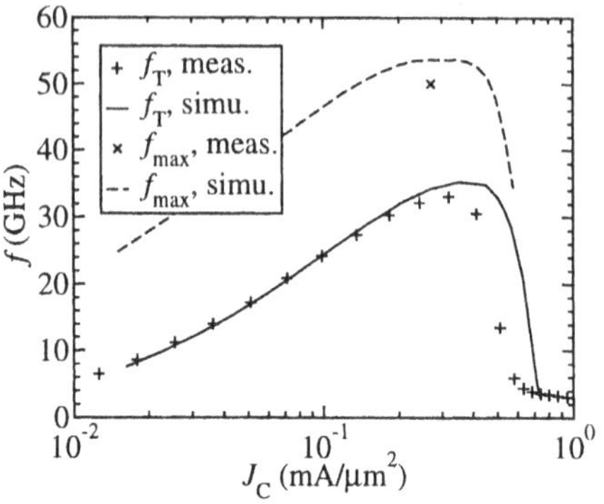

Figure2. f_T and f_{max} at $V_{CE} = 2.8$ V of the HBT used in the PA

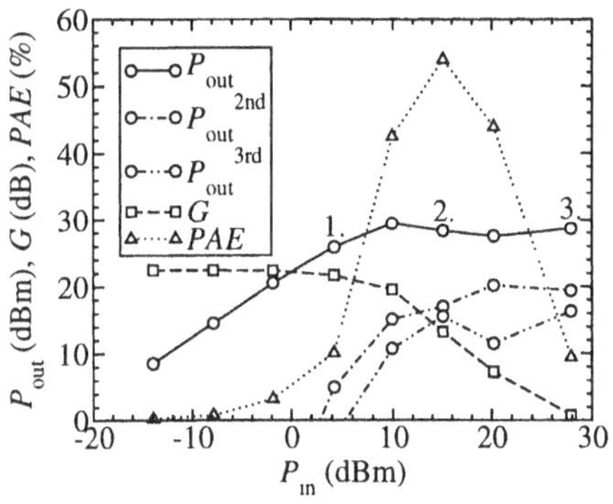

Figure3. Power sweep, $f = 1.9$ GHz, steady state, "B"

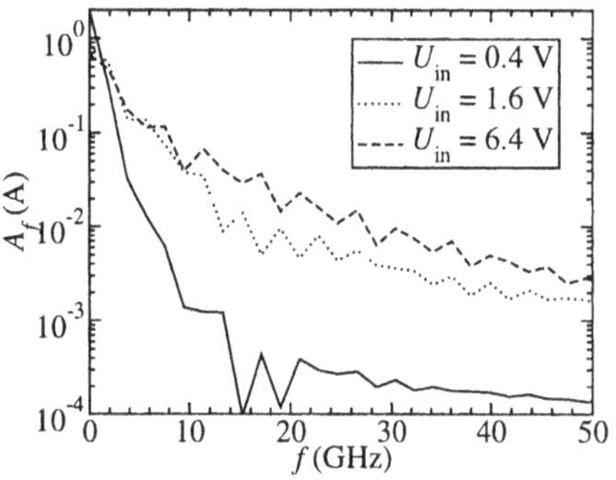

Figure4. Fourier decomposition of i_C, $f = 1.9$ GHz, "A"

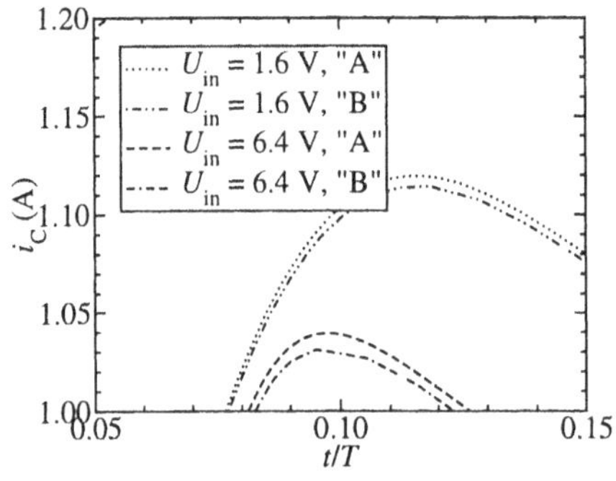

Figure5. Detail of i_C, $f = 1.9$ GHz

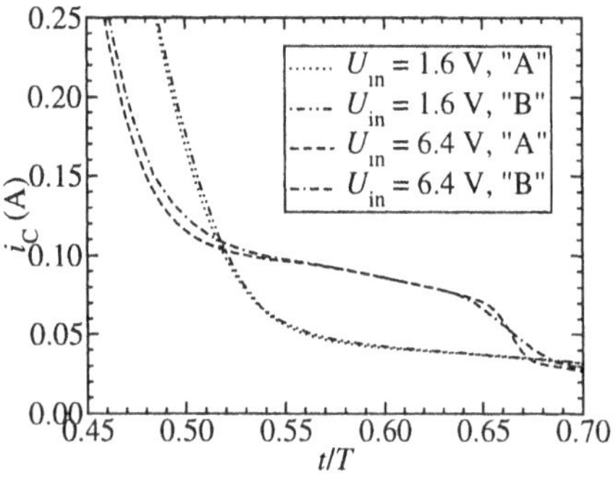

Figure6. Detail of i_C, $f = 1.9$ GHz

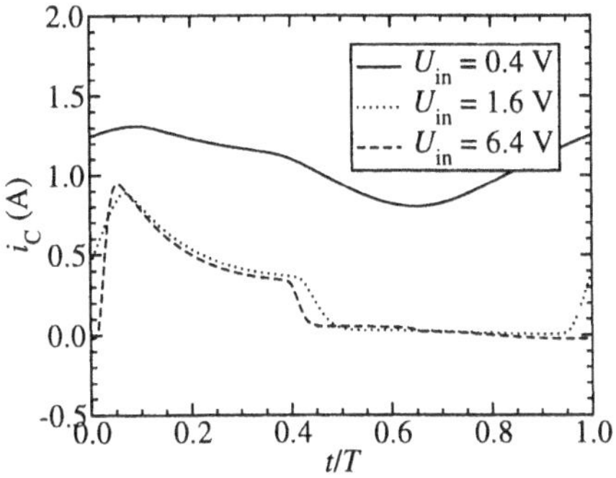

Figure7. i_C, $f = 0.95\,\text{GHz}$, "A"

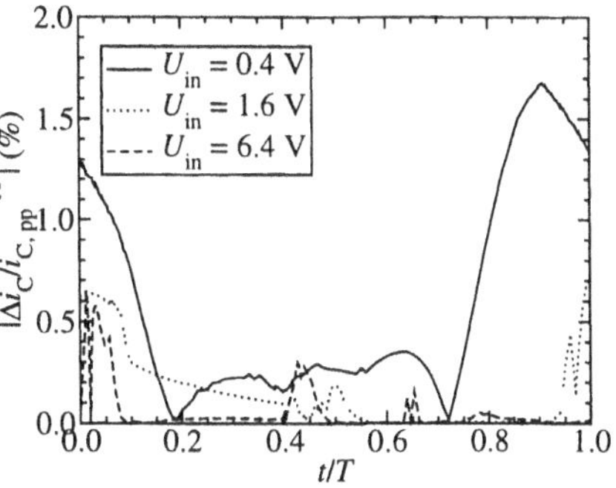

Figure8. Rel. error in i_C due to "B", $f = 0.95\,\text{GHz}$

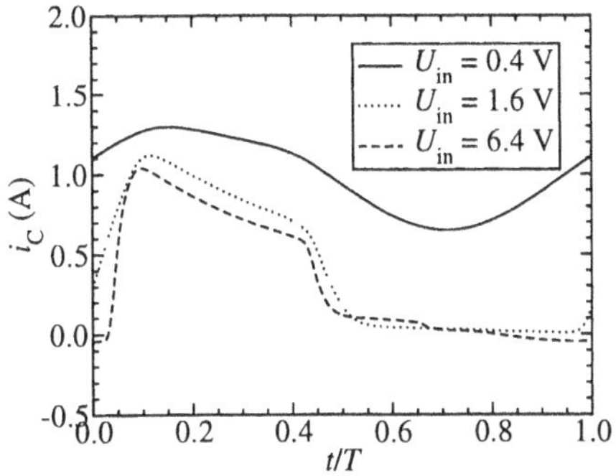

Figure9. i_C, $f = 1.9\,\text{GHz}$, "A"

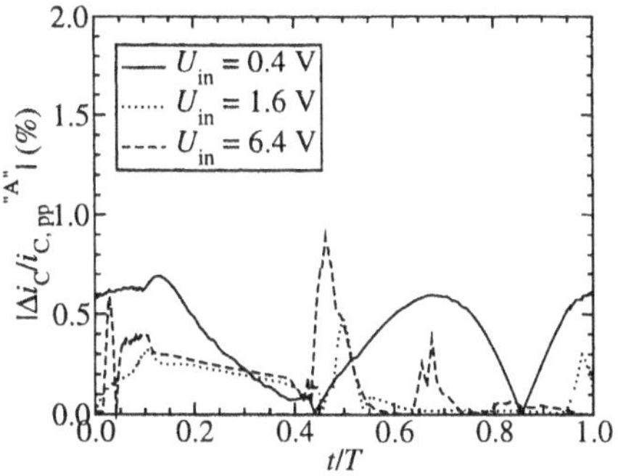

Figure10. Rel. error in i_C due to "B", $f = 1.9\,\text{GHz}$

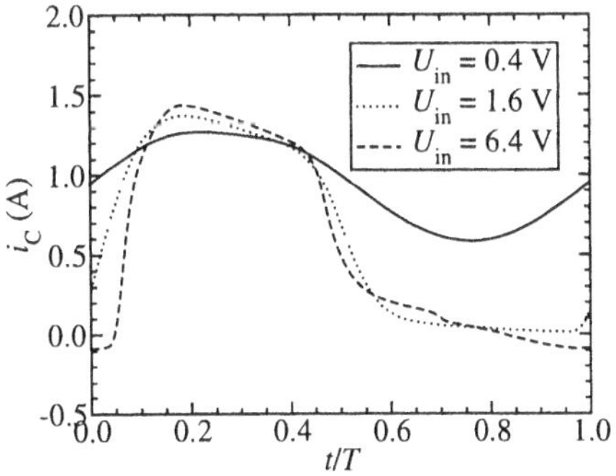

Figure11. i_C, $f = 3.8\,\text{GHz}$, "A"

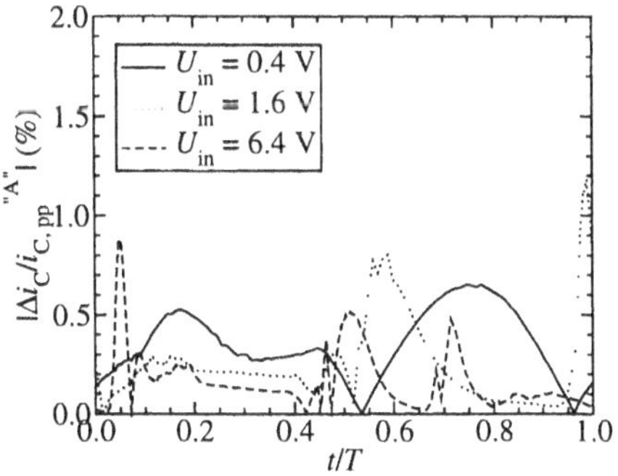

Figure12. Rel. error in i_C due to "B", $f = 3.8\,\text{GHz}$

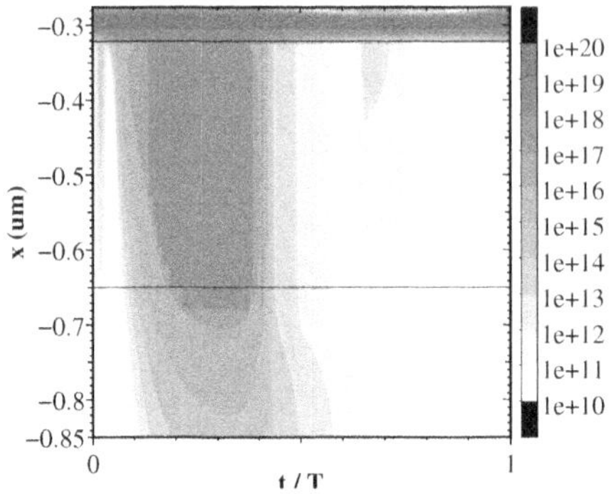

Figure13. p (cm^{-3}), $U_{in} = 6.4\,V$, $f = 0.95\,GHz$, "A"

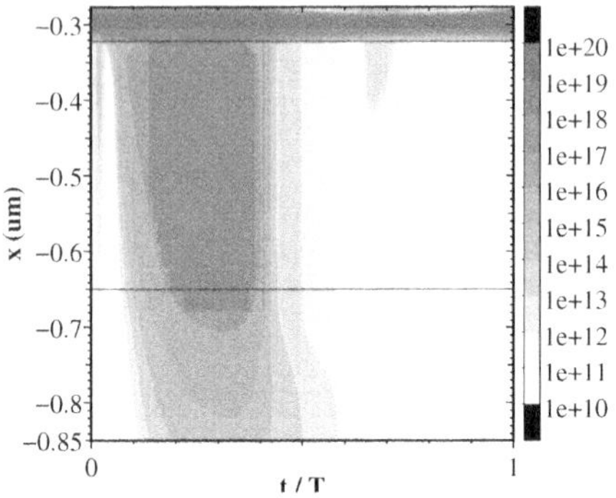

Figure14. p (cm^{-3}), $U_{in} = 6.4\,V$, $f = 0.95\,GHz$, "B"

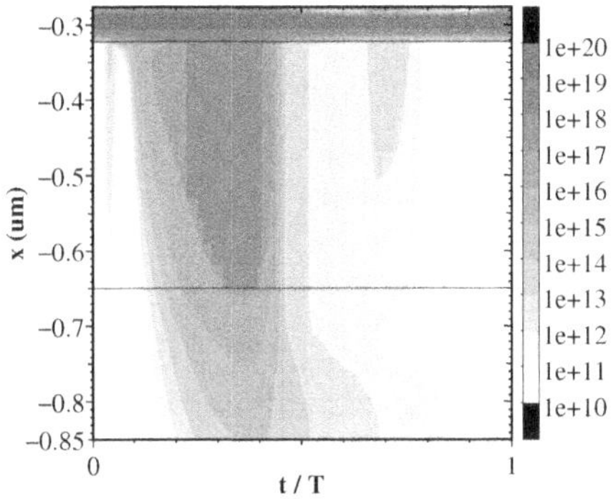

Figure15. p (cm^{-3}), $U_{in} = 6.4\,V$, $f = 1.9\,GHz$, "A"

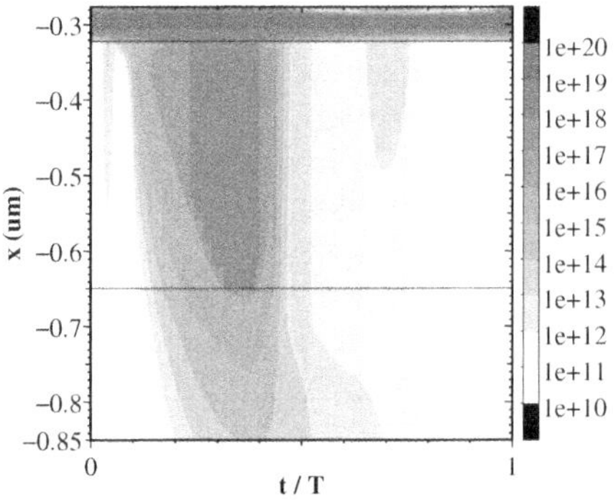

Figure16. p (cm^{-3}), $U_{in} = 6.4\,V$, $f = 1.9\,GHz$, "B"

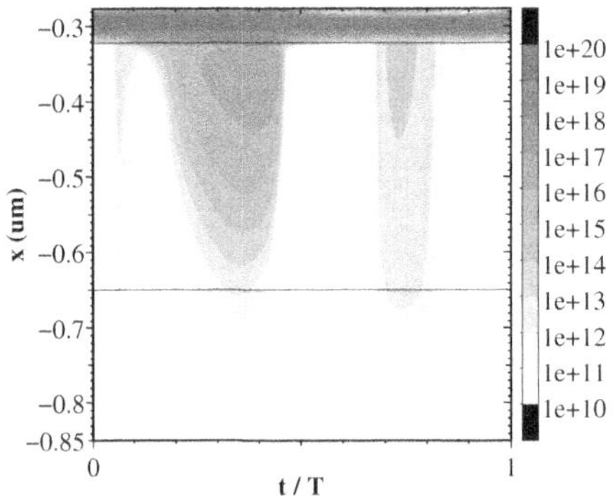

Figure17. p (cm^{-3}), $U_{in} = 6.4\,V$, $f = 3.8\,GHz$, "A"

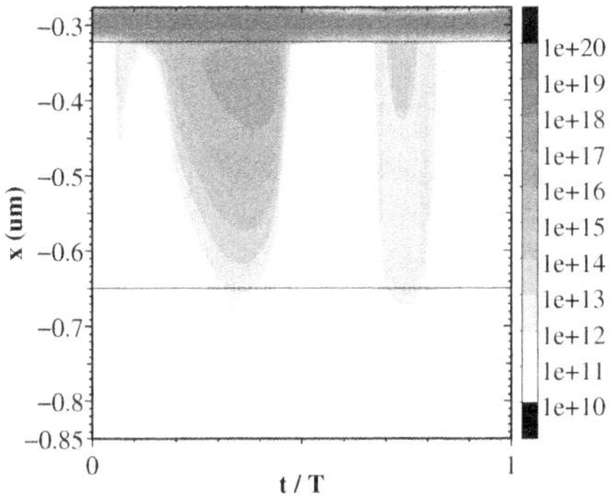

Figure18. p (cm^{-3}), $U_{in} = 6.4\,V$, $f = 3.8\,GHz$, "B"

Symmetrical Retarded Current Source Model for Transient Field Coupling on Interconnects

M. Kurten, U. Keller, W. John, K. Meerkötter, and H. Reichl

[1] FhG IZM Berlin/Paderborn, Fürstenallee 11, D-33102 Paderborn
[2] Universtiy of Paderborn, Fachgebiet Nachrichtentheorie, Warburgerstr. 100, 33098 Paderborn

Abstract In this article the model of retarded equivalent current sources for multiconductor transmission lines is extended to take incident field coupling of plane waves into account. The functions describing the transmission line effects and the effects of the incident field on to the transmission line are approximated in Laplace domain by Padé approximation and therafter analytically transformed back in time domain employing standard correspondences. The transient simulation is performed then in time domain utilizing a fast semianalytic recursive convolution approach which allows an accurate solution of the convolution integral.

1 Introduction

Due to the increase of electromagnetic pollution of the environment by electric and electronic applications on the one hand, and the tendency towards a decrease of operation levels from 5 down to 1.2 volts, the problem of susceptibility becomes more destinct. Accordingly it is helpful at early design stages to gain information concerning the noise levels the logic has to face. Typically, such noise is either coupled directly into the circuit or is collected by the interconnects and transmitted to the gates of the circuitry.
For estimation of the noise in the latter case simulation software has become quite popular in recent years. While field solvers provide more detailed solutions, the modelling as well as the computational expense is prohibitive for many practical problems. Therefore many attempts have been made to reduce the effort by mapping the problem onto simpler approaches. Here, transmission line theory appears to be the base of one of the most important procedures, mainly due to its widespread use within the signal integrity community.
In order to model signal integrity issues in time domain various numerical procedures may be applied. The most obvious is replacing the differential equations by finite difference formulations [4,5,6], where any mathematical extension may be incorporated with relative ease and elegance. However, one of the most traditional methods is based on the conception of equivalent source models for the signal transition [2,4,7,3]. The great advantage of this method is the mapping of the transmission line features onto the capibilities of a circuit simulator. For non–lossy structures this procedure is much

faster and more easily applied to multi–conductor transmission lines, in the contrary lossy structures are not satisfactorily modelled yet and subject of current research.
In the following a field coupling procedure based on the retarded equivalent current source model for transmission lines is presented allowing for plane wave being incident from virtually everywhere. The extend of the transmission line in the examples discussed within this contribution is still identified with the z–coordinate as it is no restriction.

2 Field Coupling within the Retarded Current Source Model

As indicated in fig. 1 the model of retared equivalent current sources maps the transmission properties of an assembly of parallel interconnects onto a $2N$–port [2]. In order to achieve a mathematical formulation the relations are first formulated in Laplace domain and thereafter transformed back into time domain. For the signal transmission the equations may be stated for the two port quantities $\boldsymbol{I}_\alpha$, $\boldsymbol{U}_\alpha$, $\alpha \in \{1,2\}$

$$\boldsymbol{I}_\alpha = \boldsymbol{I}_{T\alpha} + \boldsymbol{Y}_c \boldsymbol{U}_\alpha, \qquad \boldsymbol{I}_{T\alpha} = \boldsymbol{G}\left(\boldsymbol{I}_{T\alpha^*} - 2\boldsymbol{I}_{\alpha^*}\right),$$

with $\alpha^* = 3 - \alpha$, $\boldsymbol{I}_{T,\chi}$ the retarded equivalent current source, the characteristic admittance and the transfer function $\boldsymbol{G}$ given by

$$\boldsymbol{Y}_c = (\mathbf{R} + p\mathbf{L})^{-1} \sqrt{(\mathbf{R} + p\mathbf{L}) \cdot (\mathbf{G} + p\mathbf{C})} = \boldsymbol{Z}^{-1}\boldsymbol{\Gamma}, \tag{1}$$

and the current transfer function

$$\boldsymbol{G} = \boldsymbol{Y}_c \cdot \mathrm{e}^{-\boldsymbol{\Gamma}\Lambda} \cdot \boldsymbol{Y}_c^{-1}. \tag{2}$$

The corresponding time functions, using the convention that $[\cdot]_{i,j}$ refers to the entry of a matrix, are then

$$\boldsymbol{y}_c = \mathbf{m}_0\, \sigma(t) \left(\mathbf{1}\, \delta(t) + \left[\sum_{\nu=1}^{n} \mathrm{q}_\nu \cdot \mathrm{e}^{\mathrm{p}_\nu \cdot t} \right]_{i,j} \right), \tag{3}$$

$$\boldsymbol{g} = \sum_{\mu=1}^{m} \mathbf{m}_0\, \sigma(t - \mathrm{T}_\mu) \left(\mathbf{k}_\nu\, \delta(t - \mathrm{T}_\mu) + \left[\sum_{\nu=1}^{n} \mathrm{a}_{\nu,\mu} \cdot \mathrm{e}^{\mathrm{b}_\nu \cdot (t - \mathrm{T}_\mu)} \right]_{i,j} \right). \tag{4}$$

In much the same manner the effect of the coupling of an external field to the transmission line may be considered. In that case one has to employ an augmented set of equivalent sources

$$\tilde{\boldsymbol{I}}_{T\alpha} = \boldsymbol{G}\left(\boldsymbol{I}_{T\alpha^*} - 2\boldsymbol{I}_{\alpha^*}\right) + \boldsymbol{\mathcal{I}}_{T\alpha}, \tag{5}$$

where the effect of the external plane wave is accounted for by the the term $\boldsymbol{\mathcal{I}}_{T\alpha}$. The field contribution may be decomposed into three parts

$$\boldsymbol{\mathcal{I}}_{T\alpha} = (-1)^{\alpha^*} \left(\boldsymbol{\mathcal{I}}_{T\alpha,1} + \boldsymbol{\mathcal{I}}_{T\alpha,2} - \boldsymbol{\mathcal{I}}_{T\alpha,3} \right), \tag{6}$$

that represent the effect of the longitudinal component of the electric field along the interconnect and — asuming a negligible distance within the cross–sectional dimensions — the effect of the transversal electric field at both terminations:

$$\mathcal{I}_{T\alpha,1} = \boldsymbol{Y}_c \left(\mathrm{e}^{-\boldsymbol{\Gamma}\Lambda}\right)^{\alpha-1} \int_0^{\Lambda} \mathrm{e}^{(-1)^{\alpha}\boldsymbol{\Gamma}\zeta}\, \boldsymbol{E}_l \,\mathrm{d}\zeta, \tag{7}$$

$$\mathcal{I}_{T\alpha,2} = \boldsymbol{Y}_c \left(\mathrm{e}^{-\boldsymbol{\Gamma}\Lambda}\right)^{\alpha-1} \boldsymbol{V}_{t,0}, \tag{8}$$

$$\mathcal{I}_{T\alpha,3} = \boldsymbol{Y}_c \left(\mathrm{e}^{-\boldsymbol{\Gamma}\Lambda}\right)^{2-\alpha} \boldsymbol{V}_{t,0}\, \mathrm{e}^{-\gamma_z\Lambda}. \tag{9}$$

The concept of this procedure is represented in fig. 1, where the effect of the external field is described solely by the current sources at the terminations $\mathcal{I}_{\alpha}$ while the characteristic of the transmission line without field illumination is described by the internal current sources $\boldsymbol{I}_{T\alpha}$, and $\boldsymbol{Y}_c$. Moreover, since $\boldsymbol{Y}_c$ and $\boldsymbol{\Gamma}$ are known already from the signal transmission part, just the integral in (7) and the quantity $\boldsymbol{V}_{t,0}$ remains to be determined, assuming a plane wave

$$\boldsymbol{E}^i = \left[\mathrm{E}_x\ \mathrm{E}_y\ \mathrm{E}_z\right]^{\mathrm{T}}\ \mathrm{e}^{-(\gamma_x x+\gamma_y y+\gamma_z z)} \cdot T(p). \tag{10}$$

The function $T(p)$ represents the temporal behaviour of the incident field in Laplace domain. Since the z–coordinate correlates with the extent of the transmission line, the transversal part at the beginning $\boldsymbol{E}_{t,0}$ translates into the corresponding voltage term by integration from the reference conductor — assumed to be situated at 0 — to the ν–th signal line at distance d_{ν} in the cross–sectional plane

$$\boldsymbol{V}_{t,0} = \left[\int_0^{\mathrm{d}_{\nu}} \boldsymbol{E}_t^i \,\mathrm{d}l\right]_{\nu} = \left[\mathrm{E}_{\nu} \cdot \frac{e^{p\mathrm{T}_{t,\nu}} - e^{-p\mathrm{T}_{t,\nu}}}{p} \cdot T(p)\,\mathrm{e}^{-p\mathrm{T}_{t,\nu}}\right]_{\nu}, \tag{11}$$

where the notation $[\cdot]_{\nu}$ represents the ν–th entry of an algebraic vector corresponding to the ν-th signal line, and

$$\mathrm{E}_{\nu} = p \cdot \frac{\mathrm{E}_x \cdot \mathrm{x}_{\nu} + \mathrm{E}_y \cdot \mathrm{y}_{\nu}}{\gamma_x \cdot \mathrm{x}_{\nu} + \gamma_y \cdot \mathrm{y}_{\nu}}, \qquad \mathrm{T}_{t,\nu} = \frac{\gamma_{\mathrm{x}} \cdot \mathrm{x}_{\nu} + \gamma_{\mathrm{y}} \cdot \mathrm{y}_{\nu}}{2 \cdot p},$$

It should be remarked that the denominator in the definition of $\mathrm{T}_{t,\nu}$ compensates the dependence of the $\gamma_{\mathrm{x,y}}$ in p and hence $\mathrm{T}_{t,\nu}$ appears as constant. The expression (11) has to be transformed to time domain and it suffices to focus on the dependency in p, where the fraction in (11) corresponds to a rectangular window function, being 1 within the interval $[-\mathrm{T}_{t,\nu}, \mathrm{T}_{t,\nu}]$ and zero elsewhere, and the exponential causes a shift of the time function of the incident field $T(t)$. Using the trapezoidal rule for the evaluation of the convolution integral, one obtains $-\left[2\,\mathrm{T}_{t,\nu} \cdot T(t-\tau_{t,\nu})\right]_{\nu}$, and for (8)

$$i_{T\alpha,2} = -2\,\boldsymbol{y}_c * \left[\mathrm{T}_{t,\nu} \cdot T(t-\tau_{t,\nu})\right]_{\nu}, \qquad \tau_{t,\nu} = \frac{\gamma_z \mathrm{z}_{\nu}}{p} + \mathrm{T}_{t,\nu} + \mathrm{T}_0. \tag{12}$$

The delay $\tau_{t,\nu}$ is composed of the $\mathrm{T}_{t,\nu}$, a signal delay and a term that represents the "first contact" of the wave with the structure, which has to be introduced since the excitation is defined with regard to the origin. The $\boldsymbol{i}_{T\alpha,3}$ may easily be obtained from (12) by convolution with the negative current transfer function $-\boldsymbol{g}$.

In order to obtain an expression for the integral in (7) the z–component of the electric field incident on the ν–th signal line may be extracted as

$$\boldsymbol{E}_l = \boldsymbol{K}_0 \cdot \mathrm{e}^{-\gamma_z z}, \qquad \boldsymbol{K}_0 = \mathrm{E}_z \cdot T(p) \cdot \left[\mathrm{e}^{-(\gamma_x \mathrm{x}_\nu + \gamma_y \mathrm{y}_\nu)} - 1\right]_\nu . \quad (13)$$

Performing the integration (7) yields an analytical formulation for the longitudinal part as well

$$\mathcal{I}_{T\alpha,1} = \underbrace{-\boldsymbol{Y}_c \left(\mathrm{e}^{-\boldsymbol{\Gamma}\Lambda}\right)^{\alpha-1} \left(\mathrm{e}^{-(\boldsymbol{\Gamma} - (-1)^\alpha \gamma_z \mathbf{1})\Lambda} - \mathbf{1}\right) \boldsymbol{Y}_c^{-1}}_{(-1)^\alpha \left(\boldsymbol{G}\left(\mathrm{e}^{-\gamma_z \Lambda}\right)^{2-\alpha} - \mathbf{1}\left(\mathrm{e}^{-\gamma_z \Lambda}\right)^{\alpha-1}\right)} \cdot \underbrace{\boldsymbol{Y}_c \left(\boldsymbol{\Gamma} - (-1)^\alpha \gamma_z \mathbf{1}\right)^{-1}}_{\mathcal{Y}_{(-)^{\alpha *}}} \boldsymbol{K}_0 \quad (14)$$

According to the mapping on the current transfer function $\boldsymbol{G}$, only an explicit time domain representation of the terms $\mathcal{Y}_\pm$ and $\boldsymbol{K}_0$ is necessary. The first can be written as

$$\mathcal{Y}_{(-)^{\alpha *}} = \boldsymbol{Y}_c \left(\boldsymbol{\Gamma} - (-1)^\alpha \gamma_z \mathbf{1}\right)^{-1} = \boldsymbol{Z}^{-1} \cdot \left(\mathbf{1} - (-1)^\alpha \gamma_z \cdot \boldsymbol{\Gamma}^{-1}\right)^{-1} \quad (15)$$

which as well as $\boldsymbol{\Gamma}^{-1}$ are expanded separately into series. Due to convergence issues [7] the expension has to be performed at infinity, i.e. by replacing the p variable by $1/\mathcal{P}$. After a lengthy calculation the following expression is obtained

$$\mathcal{Y}_\pm = \mathcal{P} \cdot \mathbf{L}^{-1} \cdot \left(\mathbf{1} - (-1)^\alpha \gamma_z \mathcal{P} \left(\mathbf{CL}\right)^{-1/2}\right)^{-1} \left(\mathbf{1} + \sum_{\nu=1}^{M} \mathbf{c}_\nu \cdot \mathcal{P}^\nu\right) \quad (16)$$

with $\mathbf{c}_\nu$ representing the sum of the coefficients of terms of equal powers within the expansions. In order to achieve a time domain expression, (16) is transferred in a rational formulation via Padé approximation. Expanding this approximation into partial fractions and employing standard correspondences for the Laplace transform, a time domain formulation is achieved, that solely contains exponential time functions, necessary for recursive convolution.
Insertion of the time dependence of the field and transform back to time domain yields the relation corresponding to the elements of $\boldsymbol{K}_0$

$$2 \cdot \mathrm{E}_z \cdot \left(\mathrm{T}_{l,\nu}\, T'(t - \tau_{l,\nu})\right), \qquad \mathrm{T}_{l,\nu} = \frac{(\gamma_x \mathrm{x}_\nu + \gamma_y \mathrm{y}_\nu)}{2 \cdot p},$$

$$\tau_{l,\nu} = \frac{\gamma_z \mathrm{z}_\nu}{p} + \mathrm{T}_{l,\nu} + \mathrm{T}_0, \quad (17)$$

which is to be convolved with the time domain equivalent of (16) and the appropiately delayed current transfer function as indicated in (14) to obtain a time domain representation of (7)

$$\begin{aligned} \boldsymbol{i}_{T\alpha,1} &= (-1)^{\alpha} 2\,\mathrm{E}_z \Big(\boldsymbol{g} * \boldsymbol{y}_{(-)^{\alpha^*})} * [\mathrm{T}_{l,\nu} \cdot T'(t - \tau_{l\alpha})]_{\nu} \\ &\quad - \boldsymbol{y}_{(-)^{\alpha^*})} * [\mathrm{T}_{l,\nu} \cdot T'(t - \tau_{l\alpha^*})]_{\nu} \Big), \qquad (18) \\ \tau_{l\xi} &= \mathrm{T}_{l,\nu} + \mathrm{T}_0 + \left(\frac{\gamma_z \Lambda}{p} \right)^{\xi}, \qquad \xi \in [\alpha, \alpha^*]. \end{aligned}$$

Since the method propagates the quantities incident at one port to the oposite and constitutes an iteration procedure, it is obvious that the incident field being considered at both ports of the line is considered twice. This effect may be compensated formally by a substraction of the terms $\mathcal{I}_{T1,1} = \mathcal{Y}_{+}\boldsymbol{K}_0$ and $\mathcal{I}_{T2,1} = -\mathcal{Y}_{-}\boldsymbol{K}_0$ as well as the corresponding terms of (8) at the opposite ports, resp. This leads to the final formulation

$$\boldsymbol{i}_{\alpha} = \boldsymbol{y}_c * \boldsymbol{u}_{\alpha} + \boldsymbol{g} * \boldsymbol{i}_{T\alpha i} + \boldsymbol{i}_{M\alpha i}, \qquad (19)$$

$$\boldsymbol{i}_{T\alpha i} = \boldsymbol{g} * \left(\boldsymbol{i}_{T\alpha^*} - 2\,\boldsymbol{i}_{\alpha^*} - 2\,\boldsymbol{i}_{T\alpha^*,2} - \left(\boldsymbol{i}_{T,\alpha,1} - \boldsymbol{i}_{T\alpha^*,1} \right) \left(\mathrm{e}^{-\gamma_z \Lambda} \right)^{2-\alpha} \right),$$

$$\boldsymbol{i}_{M\alpha i} = \boldsymbol{i}_{T\alpha,2} + \boldsymbol{i}_{T\alpha,1} \left(\mathrm{e}^{-\gamma_z \Lambda} \right)^{\alpha-1}.$$

In that manner a symmetrical formulation is obtained allowing the electric wave being incident from any angle, and thereby, allowing for sucsessive treatment of arbitrarily oriented transmission line sections in complex networks.

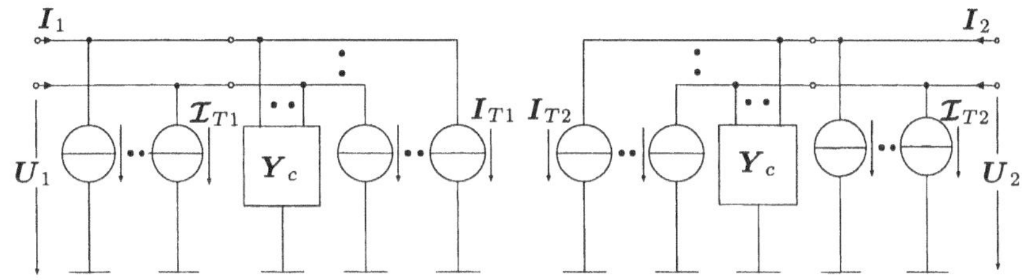

Figure1. Retarded equivalent current source model for signal transmission and noise coupling

3 Numerical Results

In order to validate the presented method an FDTD approach as presented in [4] is used. In the presented examples the excitation was a time slope function with unit magnitude and 1 ns transition time. The first example was a single transmission line as shown in fig. 2, where also the characteristic parameters may be found. Fig. 3 depicts the response of the single transmission line at both terminations, which are assumed to be passive, illuminated by a plane wave normal from above and the polarisation vector indicating in extent of the line. The bold dashed lines indicate the effect at the entrance, while the bold dotted lines the effect at the exit of the line as calculated by the presented method. The fine lines represent the voltages obtained via FDTD. The excellent correlation is obvious. The same configuration was then illuminated by a plane wave from the left propagating along the extend of the transmission line with the polarisation vector normal to the transmission line loop. The results computed with both methods at both termination again are shown in fig. 4. Also here the agreement is excellent. The most simple example for multiconductor transmission lines consists of a single pair as indicated by fig. 5. Again the characteristic parameters are found next to the sketch of the configuration. Figures 6 and 7 show the transmission lines to an arbitrary illumination. Also in this case an excellent match between the two approaches can be found. The employed semianalytic recursive convolution approach allows a very fast simulation with high accuracy in solving the convolution integral.

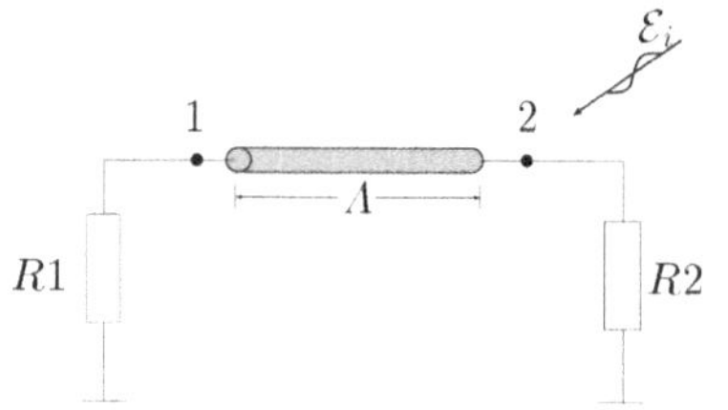

characteristic parameters
$L = 700\frac{\text{nH}}{\text{m}}$, $C = 40\frac{\text{nF}}{\text{m}}$,
$R = 10\frac{\Omega}{\text{m}}$, $G = 10\frac{\mu\text{S}}{\text{m}}$,
$R_1 = R_2 = 500\Omega$,
height $d = 0.05$ m,
length $\Lambda = 1$ m

Figure2. Transmission line excited by an electromagnetic plane wave

4 Summary

A method has been presented that allows the computation of the effect of the illumination of transmission lines by plane waves in a retarded equivalent current source setting. A symmetrical formulation has been presented that permitts arbitray angle of incidence, therby allowing for the treatment of arbitrarily oriented sections in sequel. It has been shown that the results obtained by the presented method are in excellent agreement with those from other methods.

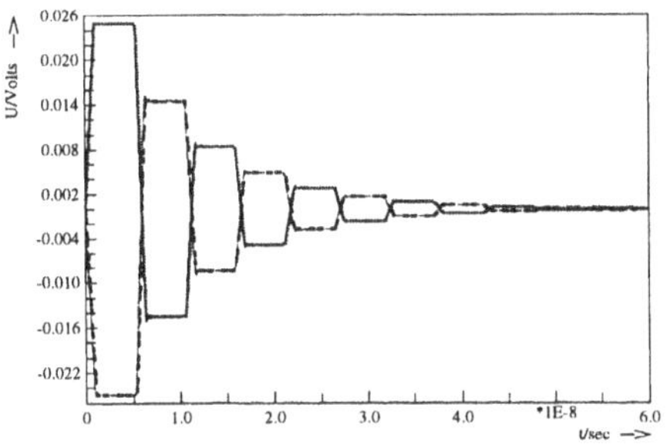

Figure3. Normal from above

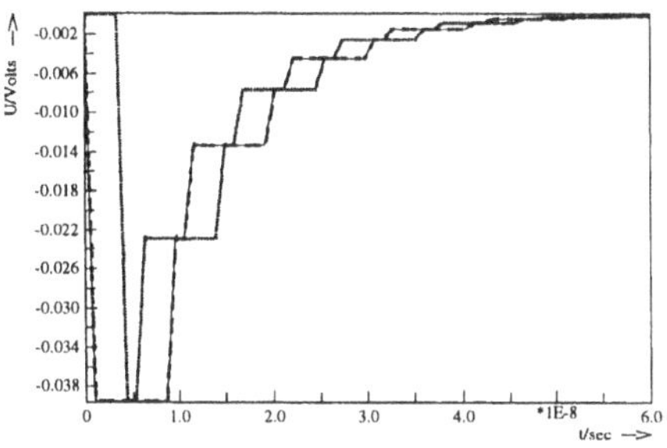

Figure4. Crazing angle from left

Transmission line response to plane wave excitation

FREACS ... bold, broken lines, FDTD ... small lines

References

1. Blakiewicz, G., Janke, W.: Semi Analytical Recursive Algorithms for Convolution Calculations. IEE Proceedings Circuits, Devices, Systems **142** (1995) 125 ff.
2. Griese, E., Ramme, M., Kurten, M.: Fast Simulation Method for a transient Analysis of Lossy Coupled Transmission Lines using a Semi Analytical Recursive Convolution Procedure. IEEE Symposium on Electromagnetic Compatibility (1997) 277 ff.
3. Tesche, F. M., Ianoz, M. V., Karlsson T.: EMC Analysis Methods and Computational Models. John Wiley and sons, (1996)
4. Paul, C.: Analysis of Multiconductor Transmission Lines. John Wiley and sons, (1994)
5. Paul, C.: A Spice Model for Multiconductor Transmission Lines Excited by an Incident Electromagnetic Field. IEEE Transactions on Electromagnetic Compatibility **36** (1994) 342 ff.
6. Paul, C., Orlandi, A.: FDTD Analysis of Lossy, Multiconductor Transmission Lines terminated in Arbitrary Loads IEEE Transactions on Electromagnetic Compatibility **38** (1996) 388 ff.
7. Shen, L., Kuh, E. S.: Transient Simulation of Lossy Interconnects Based on the Recursive Convolution Formulation IEEE Transactions on Circuits and Systems **39** (1992) 879 ff.

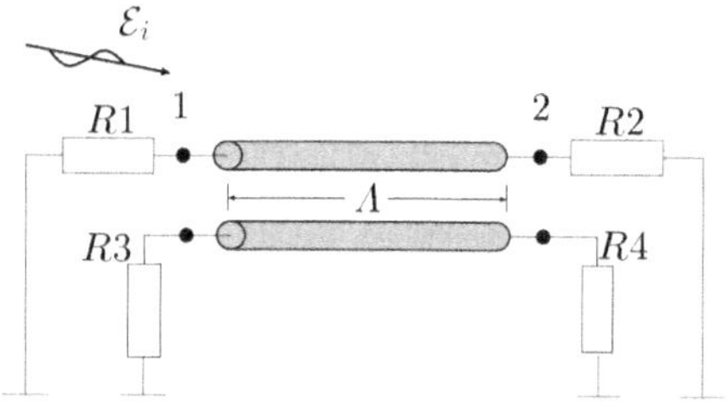

$$\mathbf{C} = \begin{bmatrix} 40.3439 & -20.1719 \\ -20.1719 & 29.491 \end{bmatrix} \frac{pF}{m},$$

$$\mathbf{L} = \begin{bmatrix} 1.1117 & 0.693901 \\ 0.693901 & 1.3878 \end{bmatrix} \frac{\mu H}{m},$$

$$\mathbf{R} = \begin{bmatrix} 1 & 0.6 \\ 0.6 & 1 \end{bmatrix} \frac{m\Omega}{m},$$

$$\mathbf{G} = \begin{bmatrix} 3 & -1 \\ -1 & 3 \end{bmatrix} \frac{\mu S}{m},$$

$R_\nu = 400\Omega, \qquad \nu = 1 \ldots 4,$
height $d = 0.15$ m, separation $s = 0.2$ m,
length $\Lambda = 1$ m

Figure5. Transmission line pair excited by an electromagnetic plane wave

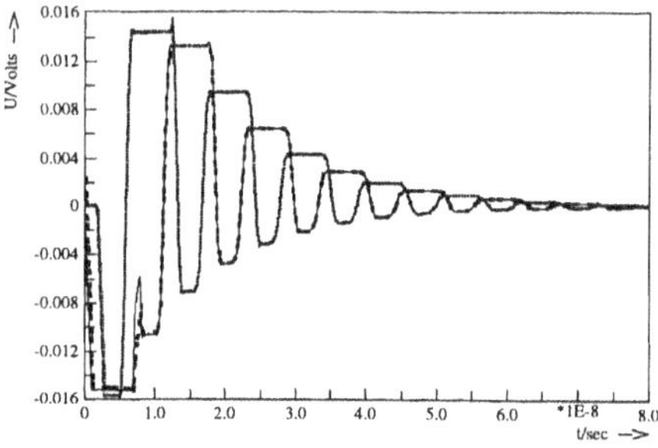

Figure6. Line 1

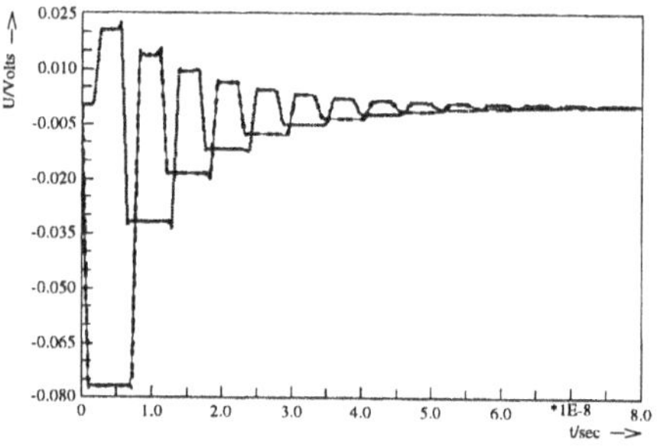

Figure7. Line 2

Response of the transmission line pair excited by a plane wave,
FREACS ... bold, broken lines, FDTD ... small lines

Iterative Solution of Systems of Linear Equations in Microwave Circuits Using a Block Quasi-Minimal Residual Algorithm

R. Schlundt[1], G. Hebermehl, F.-K. Hübner, W. Heinrich[2], and H. Zscheile

[1] Weierstrass Institute for Applied Analysis and Stochastics, Berlin, Germany
[2] Ferdinand-Braun-Institut für Höchstfrequenztechnik, Berlin, Germany

Abstract The electrical properties of monolithic microwave integrated circuits that are connected to transmission lines are described in terms of their scattering matrix using Maxwell's equations. Using a finite-volume method the corresponding three-dimensional boundary value problem of Maxwell's equations in the frequency domain can be solved by means of a two-step procedure. An eigenvalue problem for non-symmetric matrices yields the wave modes. The eigenfunctions determine the boundary values at the ports of the transmission lines for the calculation of the fields in the three-dimensional structure. The electromagnetic fields and the scattering matrix elements are achieved by the solution of large-scale systems of linear equations with indefinite complex symmetric coefficient matrices. In many situations, these matrix problems need to be solved repeatedly for different right-hand sides, but with the same coefficient matrix. The block quasi-minimal residual algorithm is a block Krylov subspace iterative method that incorporates deflation to delete linearly and almost linearly dependent vectors in the underlying block Krylov sequences.

1 Introduction

The design of monolithic microwave integrated circuits requires efficient CAD tools in order to avoid costly and time-consuming redesign cycles. The electromagnetic characteristics of microwave integrated circuits can be described by equivalent circuits in terms of voltages and currents. With growing frequencies the voltage and current definitions become ambiguous. Therefore, the description of the electromagnetic characteristics of microwave circuits by the scattering matrix is more appropiate. As an example, Fig. 1 illustrates the principle structure under investigation. In order to determine the scattering matrix, the circuit is inserted between transmission lines. The scattering matrix describes the structure in terms of wave modes on these lines, which can be computed from the electromagnetic field. A three-dimensional boundary value problem can be formulated using Maxwell's equations in frequency domain in order to compute the electromagnetic field. Using the fimite-volume method to the three-dimensional boundary value problem for the Maxwell's

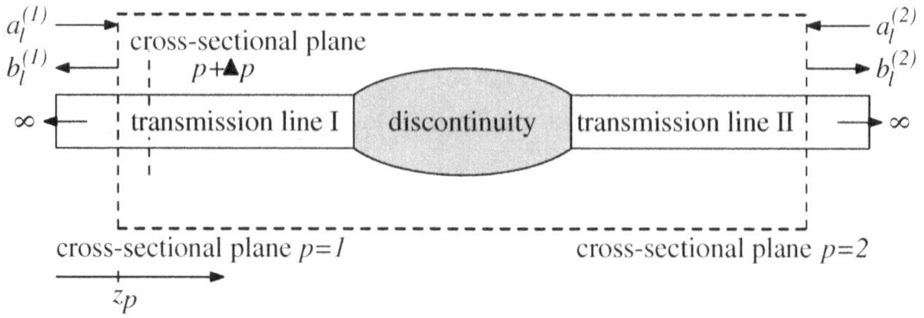

Figure1. Structure under investigation.

equations results in the so-called Finite-Difference method in the Frequency Domain (FDFD) (see [1], [2], and [3]). The field volume is subdivided into rectangular cells. The Maxwell's equations in integral formulation are approximated by the method of finite integration for each cell.

2 Scattering Matrix

The structure under investigation (see Fig. 1) consists of infinitly long transmission lines and a discontinuity. The transmission lines are assumed to be longitudinally homogeneous. The discontinuity may have an arbitrary structure. The scattering matrix S describes the energy exchange and phase relation between all outgoing modes $b_l^{(p)}$ and all incoming modes $a_l^{(p)}$ on all ports p on the transmission lines (see [6]):

$$b_{\rho,\nu} = \sum_{\sigma=1}^{m_s} S_{\rho,\sigma} a_{\sigma,\nu}, \quad S = \begin{pmatrix} S_{11} & S_{12} & \cdots & S_{1m_s} \\ S_{21} & S_{22} & \cdots & S_{2m_s} \\ \cdots & \cdots & \cdots & \cdots \\ S_{m_s1} & S_{m_s2} & \cdots & S_{m_sm_s} \end{pmatrix} = (S_{\rho,\sigma}), \tag{1}$$

$$\nu, \rho, \sigma = 1, \ldots, m_s, \; m_s = \sum_{p=1}^{\bar{p}} m^{(p)}, \; \rho = l + \sum_{q=1}^{p-1} m^{(q)}, \; l = 1, \ldots, m^{(p)}. \tag{2}$$

$\bar{p}$ is the number of ports. $m^{(p)}$ is the number of modes on the port p. The scattering matrix can be extracted from the orthogonal decomposition of the electric field at a pair of two neighboring cross-sectional planes p and $p + \Delta p$ (see Fig. 1) on each waveguide.

3 Boundary Value Problem

We use the integral form of the Maxwell's equations in the frequency domain:

$$\begin{aligned} \oint_{\partial\Omega} \boldsymbol{E} \cdot d\boldsymbol{s} &= -j\omega \int_\Omega \boldsymbol{B} \cdot d\boldsymbol{\Omega}, \; \oint_\Omega \boldsymbol{B} \cdot d\boldsymbol{\Omega} = 0 \\ \oint_{\partial\Omega} \frac{1}{[\mu]} \boldsymbol{B} \cdot d\boldsymbol{s} &= j\omega \int_\Omega [\epsilon] \boldsymbol{E} \cdot d\boldsymbol{\Omega}, \; \oint_\Omega [\epsilon] \boldsymbol{E} \cdot d\boldsymbol{\Omega} = 0 \end{aligned} \tag{3}$$

taking into account the constitutive relations

$$B = [\mu]H, \ D = [\epsilon]E, \ [\epsilon] = diag(\epsilon_x, \epsilon_y, \epsilon_z), \ [\mu] = diag(\mu_x, \mu_y, \mu_z). \quad (4)$$

The electric and magnetic field intensities $\boldsymbol{E}$ and $\boldsymbol{H}$ and the electric and magnetic flux densities $\boldsymbol{D}$ and $\boldsymbol{B}$ are complex functions of the spatial coordinates. The quantities $[\epsilon]$ and $[\mu]$ are diagonal tensors of dielectric permittivity and magnetic permeability, respectively. $\partial\Omega$ is the boundary of the open surface Ω.

At the port p the transverse electric field $\boldsymbol{E}_t(z_p)$ is given by superposing all transmission line modes $\boldsymbol{E}_{t,l}(z_p)$ with weighted mode-amplitude sums $w_l(z_p)$:

$$\boldsymbol{E}_t(z_p) = \sum_{l=1}^{m^{(p)}} w_l(z_p)\boldsymbol{E}_{t,l}(z_p). \quad (5)$$

The transverse electric mode fields $\boldsymbol{E}_{t,l}(z_p)$ are computed using an eigenvalue problem for transmission lines (see [7] and [8]). The boundary conditions at all other parts of the enclosure are $\boldsymbol{E}_{tang} = 0$ and $\boldsymbol{H}_{tang} = 0$, respectively.

4 Maxwell's Grid Equations

The integrals in (3) are applied to rectangular blocks using the following relations:

$$\oint_{\partial\Omega} \boldsymbol{f} \cdot d\boldsymbol{s} \rightarrow \sum_i (\pm f_i s_i) \quad \text{and} \quad \int_{\Omega} \boldsymbol{f} \cdot d\boldsymbol{\Omega} \rightarrow f_i \Omega_i, \quad (6)$$

where f_i is a center value associated with the ith cell. We use staggered grids (see [11]). Note that the discrete electric field is located at the center of an edge and the discrete magnetic flux density is normal to the center of the face. The electric field components form a primary grid and the magnetic flux density a dual grid. The discretized form of (3) results in an equation for each field component. Presenting each equation using matrices provides a compact form:

$$\begin{aligned} CD_s\boldsymbol{e} &= -j\omega D_A\boldsymbol{b}, \quad SD_A\boldsymbol{b} = 0, \\ \tilde{C}D_{\tilde{s}/\tilde{\mu}}\boldsymbol{b} &= j\omega\epsilon_0\mu_0 D_{\tilde{A}\tilde{\epsilon}}\boldsymbol{e}, \ \tilde{S}D_{\tilde{A}\tilde{\epsilon}}\boldsymbol{e} = 0. \end{aligned} \quad (7)$$

The diagonal matrices $D_{\tilde{s}/\tilde{\mu}}$, $D_{\tilde{A}\tilde{\epsilon}}$, D_s, and D_A represent all cell quantities. The so-called curl $(C, \tilde{C})$ and source matrices $(S, \tilde{S})$ describe the topology of the two grids with the following properties (see [10]):

$$SC = 0, \quad \tilde{S}\tilde{C} = 0, \quad C = \tilde{C}^T. \quad (8)$$

5 System of Linear Algebraic Equations

Using (8), eliminating the components of the magnetic flux density ($\boldsymbol{b}$) in (7), and multiplying by $D_s^{1/2}$ yields a symmetric form of linear algebraic equations:

$$\bar{A}_E(D_s^{1/2}e) = 0, \quad \bar{A}_E = D_s^{1/2}C^T D_{\tilde{s}/\tilde{\mu}} D_A^{-1} C D_s^{1/2} - k_0^2 D_{\tilde{A}\tilde{\epsilon}}, \tag{9}$$

where $k_0 = \omega\sqrt{\epsilon_0\mu_0}$. The gradient of the electric-field divergence

$$[\epsilon]\nabla([\epsilon]^{-2}\nabla \cdot [\epsilon]\boldsymbol{E}) = 0 \tag{10}$$

is equivalent to the matrix equation (symmetric form by multiplying by $D_s^{1/2}$)

$$\bar{B}_D(D_s^{1/2}e) = 0, \quad \bar{B}_D = D_s^{-1/2} D_{\tilde{A}\tilde{\epsilon}} \tilde{S}^T D_{\tilde{V}\tilde{\epsilon}\tilde{\epsilon}}^{-1} \tilde{S} D_{\tilde{A}\tilde{\epsilon}} D_s^{-1/2}. \tag{11}$$

The diagonal matrix $D_{\tilde{V}\tilde{\epsilon}\tilde{\epsilon}}$ is a volume matrix for the 8 partial volumes of the dual elementary cell.
Taking into account the boundary conditions the two equations (9) and (11) are transformed in

$$\bar{A}_E x = 0 \quad \longrightarrow \quad A_E x = b, \tag{12}$$

$$\bar{B}_D x = 0 \quad \longrightarrow \quad B_D x = 0, \tag{13}$$

where $x = D_s^{1/2}\boldsymbol{e}$.
The effect of the addition of the two equations (12) and (13) can be interpreted as preconditioning with the preconditioner $(I + B_D A_E^{-1})^{-1}$ for the system (12):

$$(I + B_D A_E^{-1}) A_E x = (I + B_D A_E^{-1}) b \quad \rightarrow \quad Ax = (A_E + B_D)x = b, \tag{14}$$

where A is symmetric, indefinite, and complex. Fig. 2 shows the typical structure of the system matrix A. Due to the structure of the matrices C, $\tilde{C}$, $\tilde{S}$,

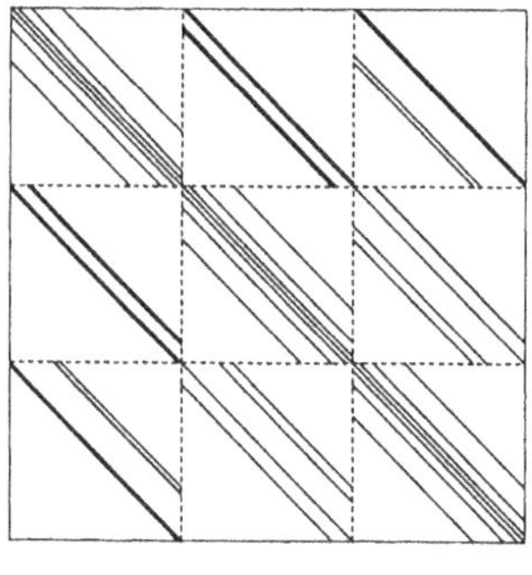

Figure2. Structure of the system matrix A.

and $\tilde{S}^T$ the diagonal entries of A are nonzero and sufficiently large compared with the off-diagonal entries.

5.1 Independent Set Ordering

A commonly used approach for solving large sparse linear systems (14) is to find sets of unknowns which are independent (see [9]). A set of such unknowns is called an independent set. Independent set orderings are permutations P_i to transform the matrix A_i with $A_0 = A$ (14) in the form

$$A_i \quad \longrightarrow \quad P_i A_i P_i^T = \begin{pmatrix} D_i & E_i^T \\ E_i & H_i \end{pmatrix}, \tag{15}$$

where D_i is a diagonal, E_i, and H_i are sparse matrices. The unknows of the independent set D_i are eliminated to get the next reduced matrix

$$A_{i+1} = H_i - E_i D_i^{-1} E_i^T. \tag{16}$$

We get a system of linear equations

$$P_i A_i P_i^T P_i x_i = P_i b_i \tag{17}$$

with $y_i = P_i x_i = (y_{i,1}, y_{i,2})^T$ and $c_i = P_i b_i = (c_{i,1}, c_{i,2})^T$ and have to solve the reduced system of linear equations

$$A_{i+1} x_{i+1} = b_{i+1}, \quad x_{i+1} = y_{i,2}, \quad b_{i+1} = c_{i,2} - E_i D_i^{-1} c_{i,1} \tag{18}$$

for $y_{i,2}$, and then we get

$$y_{i,1} = D_i^{-1}(c_{i,1} - E_i^T y_{i,2}). \tag{19}$$

Then we have to permute the solution vector y_i back to the vector x_i.

5.2 Preconditioning

Let D_A the diagonal matrix of A, where $A \equiv A_{i+1}$. Using $x \equiv x_{i+1}$ and $b \equiv b_{i+1}$ then we set $M_1 = M_2 = D_A^{1/2}$ and apply the Jacobi preconditioning

$$Ax = b \quad \longrightarrow \quad \hat{A}\hat{x} = D_A^{-1/2} A D_A^{-1/2} D_A^{1/2} x = D_A^{-1/2} b = \hat{b}. \tag{20}$$

The SSOR preconditioner matrices for the matrix $\hat{A}$ parametrized by ω are defined as

$$\hat{A} = L + I + L^T \quad \longrightarrow \quad \hat{M}_1 = (I + \omega L),\ \hat{M}_2 = \hat{M}_1^T,\ 0 < \omega < 2, \tag{21}$$

where L is the strictly lower triangular part of $\hat{A}$. The optimal value of the ω parameter reduces the number of iterations to a lower order (Table 3). The matrix vector products $r = (\hat{M}_1^{-1} \hat{A} \hat{M}_2^{-1}) v$ can be computed very efficiently by using Eisenstat's trick (see [4]) for any vector v:

$$\begin{aligned} r &= (\hat{M}_1^{-1} \hat{A} \hat{M}_2^{-1}) v \\ &= \tfrac{1}{\omega}[(I + \omega L^T)^{-1} v + (I + \omega L)^{-1}(v - (2 - \omega)(I + \omega L^T)^{-1} v)]. \end{aligned} \tag{22}$$

5.3 Block Quasi-Minimal Residual Algorithm

We first describe the general block Krylov subspace method. Then, we describe a J-symmetric variant of the block Krylov method, which includes the complex symmetric form as a special case.

The system of linear equations (14) is solved repeatedly with the same coefficient matrix A, but m_s different right-hand sides, i.e., we have linear systems

$$Ax^{(j)} = b^{(j)},\ j = 1, 2, \ldots, m = m_s \quad \longrightarrow \quad AX = B, \tag{23}$$

where $A \in \mathbb{C}^{n\times n}$, $B \in \mathbb{C}^{n\times m}$, and $X \in \mathbb{C}^{n\times m}$. An iterative scheme for solving (23) is said to be a block Krylov subspace method if it generates block iterates $X_l \in \mathbb{C}^{n\times m}, l = 1, 2, \ldots,$ with $x_l^{(j)} \in x_0^{(j)} + \mathcal{K}_l^{\text{dl}}(A, R),\ j = 1, 2, \ldots, m$. The block Krylov subspace $\mathcal{K}_l^{\text{dl}}(A, R)$ is spanned by the first l linearly independent vectors in the block Krylov sequence

$$R, AR, A^2R, \ldots, A^{k-1}R, \ldots,\ R = B - AX_0\ . \tag{24}$$

All linearly dependent or almost linearly dependent vectors on previous vectors are deleted by scanning the block Krylov sequence from left to right. We obtain the deflated block Krylov sequence

$$R^{(1)}, AR^{(2)}, A^2R^{(3)}, \ldots, A^{k-1}R^{(k)}, \ldots,\ R^{(0)} = R, \tag{25}$$

where $R^{(k)}$ is a submatrix of $R^{(k-1)}$. The Lanczos-type algorithm described in [5] generates a sequence of basis vectors $v_1, v_2, \ldots$ for the block Krylov subspaces $\mathcal{K}_j^{\text{dl}}(A, R)$ and a sequence of basis vectors $w_1, w_2, \ldots$ for the block Krylov subspaces $\mathcal{K}_j^{\text{dl}}(A^T, L)$ with $L \in \mathbb{C}^{n\times m}$. The vectors in the v and w sequences are constructed to be biorthogonal, i.e., $w_i^T v_k = \delta_{ik}$. Thus, all block iterates X_l can be represented in the form

$$X_l = X_0 + V^{(l)}Z, \quad V^{(l)} = [v_1, v_2, \ldots, v_l], \quad Z \in \mathbb{C}^{l\times m}. \tag{26}$$

The recursions for constructing the first ν basis vectors $v_1, v_2, \ldots, v_\nu$ can be summarized in matrix formulation as follows:

$$AV^{(l)} = V^{(\nu)}T^{(l)} + V_{\text{dl}}^{(l)}\ , \quad l = \nu - m_{\text{cr}} > 0, \tag{27}$$

where m_{cr} denotes the reduced size of the current block in the block Krylov sequence. The matrix $T^{(l)} \in \mathbb{C}^{\nu\times l}$ contains coefficients of the recursions in the Lanczos-type algorithm and the matrix $V_{\text{dl}}^{(l)} \in \mathbb{C}^{n\times l}$ consists of mostly zero column vectors and at most $m - m_{\text{cr}}$ nonzero column vectors corresponding to the deflated v vectors. The recurrences for the initial block of Lanczos vectors $v_1, v_2, \ldots, v_{m_1}$ can be stated as

$$V^{(m_1)}T^{(0)} + V_{\text{dl}}^{(0)} = R_0, \tag{28}$$

where $T^{(0)} \in \mathbb{C}^{m_1\times m}$ and $V_{\text{dl}}^{(0)} \in \mathbb{C}^{n\times m}$. After each deflation in the v sequence, the size of the current block $R^{(k)}$ in the corresponding deflated block Krylov sequence (25) decreases by one. Let $X_l^{\text{cr}} \in \mathbb{C}^{n\times m_{\text{cr}}}$ denote the current block iterate after each deflation, then we can rewrite (26) for X_l^{cr} as

$$X_l^{\text{cr}} = X_0^{\text{cr}} + V^{(l)}Z^{\text{cr}},\ Z^{\text{cr}} \in \mathbb{C}^{l\times m_{\text{cr}}},\ m_{\text{cr}} < m, \text{ and } AX^{\text{cr}} = B^{\text{cr}} \tag{29}$$

is the subset of linear systems in (23) that correspond to the current active block iterate X_l^{cr}. Using (27), (28), and (29), the residual block R_l^{cr} corresponding to X_l^{cr} satisfies:

$$\begin{aligned} R_l^{\mathrm{cr}} &= R_0^{\mathrm{cr}} - V^{(\nu)} T^{(l)} Z^{\mathrm{cr}} - V_{\mathrm{dl}}^{(l)} Z^{\mathrm{cr}} \\ &= V^{(\nu)} \left[\begin{pmatrix} T^{(0)} \\ 0 \end{pmatrix} - T^{(l)} Z^{\mathrm{cr}} \right] - V_{\mathrm{dl}}^{(l)} Z^{\mathrm{cr}}. \end{aligned} \tag{30}$$

We would like to choose the free parameter matrix Z^{cr} in (30) such that $\|R_l^{\mathrm{cr}}\|$ is minimal. In general, $V^{(\nu)}$ is not unitary and $V_{\mathrm{dl}}^{(l)}$ has some nonzero columns. Thus, we will minimize the Euclidean norm of the bracketed term in (30). Therefore, the block iterates X_l are characterized by a quasi-minimization of the residuals. For further implementation details of the block-QMR method see [5].

Symmetric Systems. The classical Lanczos process can be simplified when A is a J-symmetric matrix by choosing appropriate left starting vectors, i.e.,

$$A^T J = JA, \; L = JR \; \longrightarrow \; w_l = \eta_l J v_l, \; \eta_l \in \mathbb{C}, \eta_l \neq 0, \; l = 1, 2, \ldots . \tag{31}$$

We consider the solution of the complex symmetric block system (23), where A itself is complex symmetric, i.e., $A = A^T$. We want to use the matrix $M \in \mathbb{C}^{n \times n}$ with

$$M = M_1 M_2 = M_2^T M_1^T = M^T \tag{32}$$

as a preconditioner for the system (23). The two-sided preconditioned block system

$$\hat{A}\hat{X} = \hat{B}, \quad \hat{A} = M_1^{-1} A M_2^{-1}, \quad \hat{X} = M_2 X, \quad \hat{B} = M_1^{-1} B \tag{33}$$

is J-symmetric using $J = M_1^T M_2^{-1}$.

6 Numerical Results

The reduction of the iteration number is demonstrated calculating the systems of linear algebraic equations of a structure under investigation (two coupled rectangular waveguides with different diameters).
The order of the system of linear algebraic equations is 86 016. The number of stored nonzeros amounts to 593 141. We apply an independent set ordering to obtain the reduced matrix. In general, for a linear system arising in the design of monolithic microwave integrated circuits one reduction is more efficient than no, two, or more reductions. In addition the preprocessing cost for independent set ordering with one level of reduction is modest. The order of the reduced system of linear equations is 62 542, the number of stored nonzeros amounts to 1 017 395. The stopping criterion was a reduction of the norm of the residual for the preconditioned system by 10^{-8}, i.e., $\|r_l^{(j)}\| \leq 10^{-8} \cdot \|r_0^{(j)}\|$ for $j = 1, 2, \ldots, m_{\mathrm{cr}}$. In Table 1, we compare cumulative iteration counts required to individually solve each of the m linear systems using QMR with the number of block-QMR iterations required to solve all the m systems simultaneously.

Table1. Comparison of block-QMR with individual solution of each right-hand side using QMR with $\omega = 1.66$.

Number of RHS (m)	1	2	3	4	5	6	7	8	9	10
Cumulative QMR iterations	176	351	525	699	874	1052	1232	1412	1595	1776
Block-QMR iterations	176	261	318	381	441	511	610	671	693	709
Block-QMR iter./RHS	176	131	106	96	89	86	88	84	77	71

Table2. Comparison of execution times of block-QMR with individual solution of each right-hand side using QMR with $\omega = 1.66$.

Number of RHS (m)	1	2	3	4	5	6	7	8	9	10
Cumulative QMR iter. time (s)	98	196	293	389	487	587	687	787	889	990
Block-QMR iter. time (s)	109	172	219	278	331	396	491	557	608	630
Block-QMR iter. time/RHS (s)	109	86	73	70	66	66	70	70	68	63

Table3. Number of block-QMR iterations for different values of ω

ω	0.1	0.3	0.5	0.7	0.9	1.0	1.1	1.3	1.4	1.5	1.57	1.6	1.66	1.7	1.8	1.9
Iterations	733	600	485	391	313	279	249	201	186	182	178	179	176	177	193	250

Table 2 shows the corresponding cumulative execution times using QMR and the execution times of block-QMR.

Furthermore, Table 3 shows the number of block-QMR iterations for different values of ω for one right-hand side. The choice of ω affects the performance of preconditioner for the class of problems considered here.

Concluding Remarks. We have considered the iterative solution of systems of linear equations with multiple right-hand sides. We used a J-symmetric variant of the block-QMR method. The block-QMR method constructs basis vectors for block Krylov subspaces. The block iterates are characterized by a quasi-minimal residual property. The choice of ω affects the performance of the SSOR preconditioner. Furthermore, Eisenstat's trick reduces the time for the iteration algorithm.

References

1. Beilenhoff, K., Heinrich, W., Hartnagel, H. L.: Improved finite-difference formulation in frequency domain for three-dimensional scattering problems. IEEE Transactions on Microwave Theory and Techniques **40** (1992) 540–546
2. Christ, A., Hartnagel, H. L.: Three-dimensional finite-difference method for the analysis of microwave-device embedding. IEEE Transactions on Microwave Theory and Techniques **MTT-35** (1987) 688–696
3. Christ, A.: Streumatrixberechnung mit dreidimensionalen Finite-Differenzen für Mikrowellen-Chip-Verbindungen und deren CAD-Modelle. Fortschrittberichte VDI, Reihe 21: Elektrotechnik **31** (1988) 1–154
4. Eisenstat, S. C.: Efficient implementation of a class of preconditioned conjugate gradient methods. SIAM J. Sci. Statist. Comput. **2** (1981) 1–4
5. Freund, R. W., Malhotra, M.: A block-QMR algorithm for non-Hermitian linear systems with multiple right-hand sides. Linear Algebra and Its Applications **254** (1997) 119–157
6. Hebermehl, G., Schlundt, R., Zscheile, H., Heinrich, W.: Improved numerical solutions for the simulation of monolithic integrated circuits. Preprint Weierstraß-Institut für Angewandte Analysis und Stochastik im Forschungsverbund Berlin e.V. **236** (1996) 1–43
7. Hebermehl, G., Schlundt, R., Zscheile, H., Heinrich, W.: Eigen mode solver for microwave transmission lines. The International Journal for Computation and Mathematics in Electrical and Electronic Engineering **16** (1997) 108–122
8. Hebermehl, G., Schlundt, R., Zscheile, H., Heinrich, W.: Improved numerical methods for the simulation of microwave circuits. Surveys on Mathematics for Industry **9** (1999) 117–129
9. Saad, Y.: Iterative methods for sparse linear systems. PWS Publishing Company (1996)
10. Weiland, T.: On the unique numerical solution of Maxwellian eigenvalue problems in three dimensions. Particle Accelerators **17** (1985) 227–242
11. Yee, K. S.: Numerical solution of initial boundary value problems involving Maxwell's equations in isotropic media. IEEE Transactions on Antennas and Propagation **AP-14** (1966) 302–307

Part III

Coupled Problems

A Survey on Parameter Extraction Techniques for Coupling Electromagnetic Devices to Electric Circuits *

I. Munteanu[1 2] and D. Ioan[1]

[1] "Politehnica" University, Spl. Independenţei 313, Bucharest, Romania
[2] Darmstadt University of Technology, TEMF, Schlossgartenstr. 8, D-64289 Germany

Abstract The paper presents an overview of the state-of-the-art in the field of parameter extraction of electromagnetic devices. A classification of the most used system identification and reduction techniques is presented, together with algorithmic details.

1 Introduction

Electrical engineering is based on two fundamental theories: the electromagnetic field theory and the electrical circuits theory.

The first uses, as basic relations, the Maxwell equations. These are partial differential equations in which the independent variables are the three spatial coordinates and the time. By using the electromagnetic field theory, phenomena can be characterized locally, with maximum accuracy.

The second theory's basic relations are the Kirchhoff equations, to which the constitutive relations of the circuit elements (resistors, inductors, capacitors and sources) are added. The systems thus obtained consist of ordinary differential and algebraic equations in which the only independent variable is the time. Unlike the electromagnetic field theory, in which the physical (tridimensional) space has a metric structure, in the electrical circuits theory, space has only a topological structure. The characterization of the electromagnetic phenomena is done globally, the behavior of each twopole element being described by only two scalar quantities – the current and the voltage – which yields a system with a finite number of characteristic physical quantities.

Although the electrical circuits theory represents a coarse approximation of the field theory (obtained by adopting extremely simplifying hypotheses), it has a great practical importance due to its intrinsic simplicity, which allows the modeling of systems of extremely large complexity. Thus, it is inconceivable to analyze international power systems, a large automation installation or a complex electronic circuit (e.g. a VLSI circuit) using exclusively the field

* invited paper

theory, even on the most powerful computers worldwide. This is why, designers of electrical devices and systems prefer using the Kirchhoff equations instead of Maxwell's equations.

It is however obvious that the circuit theory in its classical form cannot describe important phenomena such as eddy currents, skin effects, signal propagation, charge diffusion, all of them being typical electromagnetic field effects.

In recent years, the impressive development of technologies lead the research to focus on extreme cases: devices of very large or of very small dimensions, functioning at very high frequencies, extreme powers, or extremely weak signals. To give just one example from the leading industry in electrical engineering, the microelectronic technology is heading towards a barrier which may lead to a major crisis if it cannot be overcome in due time: during the last 10 years, a progressive increase of miniaturization has taken place (the number of components on a chip doubles every 18 months!), accompanied by a quick increase of the working frequencies (values of 200-350 MHz are usual, and circuits working at 1 GHz are already available commercially).

The increase of the complexity leads to an increase of the total length of interconnections between components, which is already larger than 1 km on a current microprocessor. The delays of signals on these lines, as well as the crosstalk between lines limit the performances and make the 1 GHz–barrier impossible to cross if the modeling and design of these circuits is going to be made, as before, with tools based exclusively on the circuit theory. This is the reason why the EDA (Electronic Design Automation) community is extremely interested in modeling the field effects in VLSI interconnections.

The present paper offers a general view on the problem of circuit parameter extraction as a whole, as well as some algorithmic details for the most important stages in such an approach.

In Sect. 2, the problem formulation is presented. An insufficiently known, but vitally important concept, the Electromagnetic Circuit Element, is also shortly described. Without insisting on the stage of device analysis, the next section 3 focuses on the system identification and reduction stage, as this is the core part of the entire parameter extraction process. This section contains a review and an evaluation of the existing approaches, as well as an attempt of classification thereof. Several applications are presented in Sect. 4. Finally, in Sect. 5, a list of the still unsolved difficulties related to parameter extraction algorithms is presented and some general conclusions are drawn.

2 Problem Formulation

2.1 The Problem – Vaguely Formulated

In its simplest form, the problem formulation is the following: given an electromagnetic device whose interaction with the exterior (e.g. with an external electric or electronic circuit) is done by means of a finite number of terminals, find *a way* of dealing with the device not through field theory, but through circuit theory techniques.

This kind of approach would be useful if, for instance, the electromagnetic device must be interconnected, through its terminals, to an external electric or electronic circuit. In this case, either a coupled problem electromagnetic field–electric circuit must be solved – paying the price of high computational cost, or, if an equivalent circuit model for the device can be found, then the whole can be analyzed using circuit analysis software. In other words, an approximate circuit model for the electromagnetic device has to be found, which should:

- have a smaller complexity than that of the initial one;
- reproduce accurately the terminal–behavior of the initial device.

These two requirements are contradictory and any solution will usually be a compromise between them, e.g. a very small-complexity "equivalent" circuit will either exhibit a coarse approximate of the initial device's behavior, or a good approximation but valid only within certain limits (within a certain frequency range, for instance).

Therefore, we are looking not only for one model, but for a series of circuit models, having a behavior convergent to that of the electromagnetic device. The distance between two models can be defined as the difference between models' circuit functions. We suppose that each term in this series has a finite complexity order, but the limit can have an infinite complexity order.

At this point, there is the need for the introduction of a new concept: the *transfinite series of models.* Each model of such a series is an optimal one, having minimal distance to the electromagnetic device, for given finite order. The transfinite series of models are not unique, but depend on the considered norm in time or frequency domain. Transfinite series play an important role, since they represent the best possible approximations, for a given order.

The model of the device must be such that it can be coupled with the external circuit, **as if it was yet another circuit**, so that the result of the interconnection is a circuit analyzable e.g. with Spice. Therefore, it must satisfy the hypotheses of the electrical circuits theory.

In summary, its complicated behavior must be completely described by only two types of quantities, those characterizing the electrical circuits: currents through the terminals and potentials (voltages to the ground) of the terminals. One of the fundamental hypotheses of the electrical circuits theory is that the only interaction of an electrical circuit with its exterior is done through the terminals (an often-forgotten condition!).

Therefore, the question is: under which conditions does an electromagnetic device provided with terminals behave like a circuit? A theoretical concept is needed at the interface of the two theories.

2.2 The Missing Link: the Electromagnetic Circuit Element

The *Electromagnetic Circuit Element*, ECE for short, is a concept first proposed by the founder of the modern school of electrical engineering in Romania, Prof. Remus Răduleţ back in 1966 [37]. The cited article contains a consistent definition of the **multipolar non–filamentary electric circuit element**.

An ECE is a domain of space $\Omega \equiv \mathcal{D}_\Sigma$ (Fig. 1), whose boundary Σ is disjointly partitioned in the terminals $S_1, S_2, \ldots S_n$ and the complementary surface S_e. The defining conditions for the ECE only refer to the boundary of the domain, and not to its internal structure and read:

$$\boldsymbol{n} \cdot \operatorname{curl} \boldsymbol{E}(M,t) = 0 \quad (M \in \Sigma) \ ; \tag{1}$$

$$\boldsymbol{n} \cdot \operatorname{curl} \boldsymbol{H}(M,t) = 0 \quad (M \in S_e) \ ; \tag{2}$$

$$\boldsymbol{n} \times \boldsymbol{E}(M,t) = 0 \quad (M \in S_k, k = 1, 2, \cdots, n) \ . \tag{3}$$

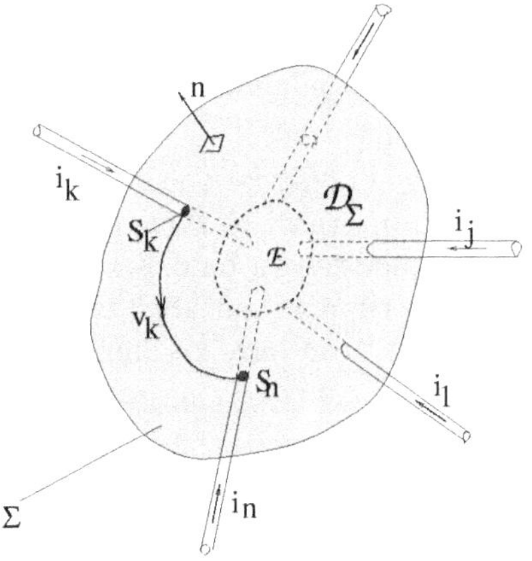

Figure1. Multipolar non-filamentary electrical circuit element

These boundary conditions mean: (1) forbids magnetic coupling between the interior and the exterior of the circuit element, (2) implies that (conduction or displacement) current can only flow through the terminals, while (3) imposes equipotentiality of the terminals.

An ECE model for multiply connected elements is proposed in [25], where the condition (1) is written in its global form:

$$\oint_\Gamma \boldsymbol{E} \cdot d\boldsymbol{r} = 0 \qquad \forall \Gamma \subset \Sigma \ , \tag{4}$$

which enforces the second Kirchhoff's relation and thus ensures that an interconnection of two circuit elements yield yet another circuit element without difficulties or restrictions of topological nature.

A further extension of the concept was proposed in [43], where an ECE with both electric and magnetic terminals is defined. Recently, another extension was introduced [33], the electromagnetic circuit element with propagation effects, on whose boundary patches exist through which waves can bidirectionally propagate towards/from the element's interior domain.

Uniqueness theorems for the solution of Maxwell's equations are demonstrated, for linear or nonlinear ECEs with Lipschitzian characteristics, with nonzero initial conditions, for ECE–type boundary conditions and with current- and/or voltage-excited terminals [37] [25] [33]. These uniqueness theorems show that an ECE with n–terminals can be considered as a system with $n-1$ inputs, currents i_k or potentials v_k:

$$i_k = \int_{\partial S_k} \boldsymbol{H} \cdot \mathrm{d}\boldsymbol{r}, \qquad v_k = \int_{C_k \subset \Sigma} \boldsymbol{E} \cdot \mathrm{d}\boldsymbol{r} \ ,$$

and $n-1$ outputs — those currents and potentials which are not input quantities. The input/output relation is, for given initial state, uniquely determined. Despite the finite number of inputs and outputs, the systems have an infinite dimension of the state space.

Both Kirchhoff's relations are automatically satisfied (because the terminal currents i_k include the displacement current and the voltages v_k are defined on curves belonging to the ECE boundary Σ) and, except for the radiating ECE, the power received by the ECE from the exterior circuit can be expressed similarly to the case of classical circuits, as the sum of current–potential products for all terminals. For instance, if the ECE only has electric terminals, the received power has the expression

$$p_\Sigma(t) = -\int_\Sigma (\boldsymbol{E} \times \boldsymbol{H}) \cdot d\boldsymbol{A} = \sum_{k=1}^{n-1} v_k i_k \ . \tag{5}$$

The concept proved to be very fruitful in establishing, through analytical approaches in the early years, and later also through numerical methods, "equivalent" schemes for various practical configurations. It was thus established that the linear ECE admits "equivalent" schemes of RL type in inductive (magnetic) quasistationary state, of RC type in capacitive (electric) quasistationary state, and of RLC type in the general case.

The ECE boundary conditions are, of course, satisfied only approximately in practice and the difference of the last two terms in (5) gives a measure of the approximation error. If this error is unacceptably high, one or more of the conditions (1) – (3) are not fulfilled and the domain $\mathcal{D}_\Sigma$ cannot be considered as a circuit element.

An English–language comprehensive review of the Electromagnetic Circuit Element bibliography can be found in [26].

Using the ECE concept, the parameter extraction problem can be formulated more precisely as:

Given a device occupying a domain in space Ω bounded by the closed surface $\partial\Omega$ on which n terminals are placed and ECE–type boundary con-

ditions are imposed, determine an electric circuit with the same number of terminals, which has approximately the same terminal–behavior as the initial device, within certain functioning limits and possibly for a certain type of excitation.

Additional constraints such as "the circuit should be as small as possible for a given approximation error", or "the error, measured in a certain norm should be as low as possible for a given circuit complexity", etc., could be added as design "wishes" to the above formulation.

3 State–of–the–Art in System Identification and Reduction, with Application to Electrical Engineering

3.1 The Three Steps in Circuit Parameter Extraction Approaches

Although many variations are possible, a parameter extraction approach is generally done in three stages.

The first stage consists in the **electromagnetic device's analysis**, by analytic, numerical or experimental techniques. The result of the first stage is the characterization, in some form, of the input and output quantities of interest – currents through terminals and terminal potentials. These results could consist of: "raw" values for the circuit quantities (voltages and currents), i.e. time (frequency) variation for all the voltages and for all the terminal currents, frequency dependence of characteristic parameters/matrices, such as $Z(j\omega)$, $Y(j\omega)$, or frequency dependence of the scattering matrix. [1]

Alternatively, the electromagnetic analysis could only provide a discrete system of equations which, once solved, would yield the values of the field quantities (e.g. $\boldsymbol{E}$ and $\boldsymbol{H}$) in the "nodes" of a spatial discretization mesh for the device under study.

The second stage implies the use of techniques which stem from systems and control theory, with the aim of determining a **system characterization** of the device, in the form of the transfer matrix (in a general sense, i.e. a matrix which maps inputs to outputs), the state space matrices, or even circuit–characteristic matrices such as the impedance or admittance matrices. This stage of the parameter extraction approach is often termed *system identification.*

In some cases (electrical circuits, spatially discretized electromagnetic devices), the system characterization is directly available, through the equations

[1] The type of input data (time- or frequency-domain) is crucial for the accuracy of the reduced model. For instance, it has been observed that much more accurate time-domain behaviour approximation can be obtained from time-domain than from frequency-domain input data.

characterizing the circuit or device, but the size of the system is very large. In this case, *order reduction techniques* are necessary in order to bring the system to a manageable size.

A final step is the utilization of techniques stemming from electric circuit theory for the **synthesis of the equivalent circuit**. The result of this stage is an equivalent scheme, usually using a circuit description language such as Spice.

Most of the known approaches for circuit synthesis date from the '50s and '60s and numerous books treating this subject exist, e.g. [22] sections IV and X, [30], [47]. The book [4] presents synthesis algorithms based on the state space matrices. This would eliminate the need to convert state space matrices into the transfer matrix, when using the state–space identification methods. For twopoles, the well–known Foster, Cauer, or Brune synthesis approaches are the most used. Their generalization to multipoles is not straightforward and most of the existing methods lead to circuits of quite complicated appearance and containing large numbers of ideal transformers [42].

It can be seen from this very general partition of a circuit parameter extraction algorithm that this very special field lies at the interface of three domains of science: electromagnetic devices analysis, system and control theory, and electric circuit synthesis. It has thus an interdisciplinary character, which makes it more difficult to deal with, but also very interesting.

3.2 Characterization of Linear Systems

As it is known, any well-formulated linear RLC circuit can be considered as a linear system. Linear systems with m inputs and l outputs are characterized by a system of equations of the form:

$$\begin{cases} \dot{x} = Ax + Bu \\ y = Cx + Du \end{cases} \tag{6}$$

where $x \in \mathbb{R}^n$ is the state vector, n is the dimension of the state space or system's order, $u \in \mathbb{R}^m$ is the vector of the inputs, $y \in \mathbb{R}^l$ is the vector of the outputs, and the system matrices $A \in \mathbb{R}^{n\times n}$, $B \in \mathbb{R}^{n\times m}$, $C \in \mathbb{R}^{l\times n}$, $D \in \mathbb{R}^{l\times m}$. The same type of equations is obtained for discretization of electromagnetic field linear problems [34] [8]. Alternatively, a circuit model can be obtained for the electromagnetic field device, by using the Partial Element Equivalent Circuit (PEEC) [24] [13].

With zero initial conditions, the Laplace transformation of the system (6) yields the transfer matrix

$$H(s) = C(sI - A)^{-1}B + D \ , \tag{7}$$

often used as an alternative way of characterizing the system's input–output behaviour.

From the point of view of input/output behaviour, the representation (6) is only unique up to a similarity transformation, i.e. the system $S' =$

(A', B', C', D) with $A' = T^{-1}AT$, $B' = T^{-1}B$, $C' = CT$ has the same transfer matrix.

A rational approximation of (7):

$$\widetilde{H}(s) = \frac{\sum_{k=0}^{M} a_k s^k}{\sum_{k=0}^{N} b_k s^k} \tag{8}$$

represents a *Padé approximation* if $\widetilde{H}$ matches the first $M + N + 1$ moments of the Taylor expansion of H around a complex frequency s_0.

3.3 Classification of Identification/Reduction Techniques

The variety of available approaches for system identification and reduction makes an attempt of classification a difficult task. From a practical point of view, the classification according to the type of input data available is probably the most useful, and that is what this section tries to present.

Input: State–Space Matrices. If the inputs are the (large) state–space matrices obtained by discretization of the field problem, then the problem is actually a reduction one.

Classical reduction techniques can be applied, such as the balanced truncation (truncated balanced realization — TBR), [32], the optimal Hankel norm reduction [19], or the improved frequency weighted optimal truncation [38]. An extension of the TBR technique for discrete systems was recently proposed [46], but was applied only to small-scale initial systems.

Recently, the Lanczos and Arnoldi methods [21] were proposed as reduction techniques for electrical circuits. Basically, the order reduction of the nth–order system (6) can be regarded as a change of variable for the state vector x (a subscript is added to emphasize the order): $x_n = V z_q$, $q \ll n$, with $V \in \mathbb{R}^{n \times q}$. The linear system thus becomes:

$$\begin{cases} \dot{z}_q = \widetilde{A} z_q + \widetilde{B} u \\ y = \widetilde{C} z_q + \widetilde{D} u \end{cases} \tag{9}$$

where the reduced system matrices are $\widetilde{A} = V^{-1}AV$, $\widetilde{B} = V^{-1}B$, $\widetilde{C} = CV$, $\widetilde{D} = D$ and, as it can be readily shown, the reduced transfer matrix is identical to the initial one: $\widetilde{H}(s) = H(s)$.

The various reduction algorithms use different types of transformation matrices V, such as to bring the reduced state matrix

$$\widetilde{A} = V^{-1}AV \tag{10}$$

to a desired form. Thus:

- The **Arnoldi** method constructs a unitary matrix $V^{-1} = V^T$, such that the reduced state matrix
$$\widetilde{A} = V^T AV$$

has an upper Hessenberg form; the method has been used e.g. in [40], [15]. A block version thereof (for multiple-input, multiple-output systems) which ensures passivity, called PRIMA, is proposed in [35].
- The **Symmetric Lanczos** algorithm uses a transformation matrix V with the same property as that used by Arnoldi, $V^{-1} = V^T$, such that the reduced state matrix is symmetric tridiagonal.
- The **Unsymmetric Lanczos** method, applicable to unsymmetric A matrices, constructs an orthogonal matrix V, with $V^{-1} = W^T \neq V^T$, such that the reduced state matrix:

$$\widetilde{A} = W^T A V$$

is tridiagonal. The promoters of the Lanczos algorithm in the electrical engineering community, Feldmann and Freund, first proposed the already consecrated term Padé via Lanczos (PVL) [17] and showed the connection between the Lanczos process, the Padé type approximations, and the moments of the system's impulse response.
- The **Split congruence transformation (SCT)** method / **Pole Analysis via Congruence Transformations (PACT)** [29] are two–step approaches, in which first a congruence transformation is applied to the matrix, in order to bring it to a (block) diagonal form, then the Lanczos algorithm is applied for order reduction[2].
- The **Truncated balanced realization (TBR)** [32] method uses a similarity transformation in the first stage, chosen such as to obtain an unreduced balanced system (A', B', C', D') (i.e. one for which $C'^T = B'$ holds). In the second stage the new state matrix A' is partitioned:

$$A' = \begin{bmatrix} A_1 & A_{12} \\ A_{21} & A_2 \end{bmatrix}$$

and $\widetilde{A} = A_1$ a matrix of size $q \times q$. Additionally, a so–called **compensation** can be applied, in order to preserve the d.c. behaviour:

$$\widetilde{A} = A_1 - A_{12} A_2^{-1} A_{21} \quad .$$

- The **Optimal Hankel norm reduction (OHNR)** due to Glover [19] first constructs a balanced realization as above, then truncates it according to the Hankel singular values of the system, by elimination of those rows and columns in A which correspond to "small" Hankel singular values.
A frequency–weighted version thereof is proposed in [38].

A reduction technique which does not fit in the scheme above (i.e. cannot be interpreted directly as a change of variable for the state vector) is SPACE [16]. The algorithm reduces large circuits by eliminating a part of the internal nodes according to an error criterion.

---

[2] A congruence transformation is a transformation of the form $A' = V^T A V$, with V a square matrix of the same dimension as A.

Input: Frequency–Domain Transfer Values. If the input data consist of frequency-domain "transfer" values (where "transfer" is meant in the general sense, as input-to-output relation), such as frequency-variation of the impedance, admittance or scattering-parameter matrices ($Z(j\omega)$, $Y(j\omega)$ or $S(j\omega)$, generically $H(j\omega)$), then the classical identification technique consists in imposing the interpolation conditions for the transfer matrix (8) at all the frequency points:

$$H(j\omega_k) = H_k \ , \quad k = 1, \ldots, N \tag{11}$$

then solving for the unknown coefficients a_k and b_k in (8).

Unfortunately, the resulting system of equations has a Vandermonde-type matrix,

$$\begin{bmatrix} 1 & j\omega_1 & (j\omega_1)^2 & \cdots \\ 1 & j\omega_2 & (j\omega_2)^2 & \cdots \\ \cdots & \cdots & \cdots & \cdots \end{bmatrix} \ .$$

which is known to be ill-conditioned. A solution to this difficulty is the use of a different set of basis functions in the numerator and denominator of (8) (the set $1, s, s^2, ...s^k, ...$ has almost linearly dependent functions, for big values of k, and this is what produces the ill-conditioning).

The most used approaches fall into the categories:

- The **Least squares (LSQ)** methods solve the system (11) by one of the least squares variants.
- The **Nonlinear optimization (Nl. Opt.)** techniques [48] regard the problem of determining the unknown coefficients as an optimization problem. The algorithms are complicated, and they suffer of ill-conditioning too, for the classical choice of the basis functions.
- The **Subspace-based State-Space System IDentification (4SID)** [44] method constructs a state-space model of type (6), based on the given input data.

Input: Moments of the System's Impulse Response. The impulse response of a linear system represents its time-domain characterization, just as the transfer matrix is its frequency-domain characterization. If the impulse response $h(t)$ is known, then the response $y(t)$ to an arbitrary excitation $u(t)$ can be determined by the convolution $y(t) = \int_{-\infty}^{\infty} h(\xi)u(t-\xi)d\xi$. The moments of the impulse response are defined as $m'_j = \int_{-\infty}^{\infty} \xi^j h(\xi)d\xi$.

It can be shown that the zero-order moment equals the area under $h(t)$, the first-order moment equals the same area multiplied by the distance of its "center of gravity" to the origin, etc. This similarity to the moments calculation in mechanics is at the origin of the name "moments".

On the other hand, the Laplace-domain transfer function of a linear system is equal to the output of the system when the inputs are of impulse type. A Taylor expansion around $s = 0$ of the transfer function $H(s)$ is

$$H(s) = H(0) + H'(0)s + H''(0)s^2/2! + \ldots = \sum_{j=0}^{\infty} m_j s^j \ , \tag{12}$$

The properties of the Laplace transform show the relationship of the Taylor expansion coefficients m_j of (12) and the moments m'_j of the impulse response (see e.g. [22], page 362). Namely, the coefficients m_j in (12) are given by $m_j = -C\ A^{-j-1}\ B$, and they are the same as the the moments m'_j of the impulse response.

The main identification/reduction approaches based on moment calculation are given below.

- The **Asymptotic Waveform Evaluation — AWE** [36] performs a Padé–type approximation based on moments of the Taylor expansion around a single frequency point. Since AWE recursion converges to the strongest eigenvector, low accuracy results for the rest of the spectrum.
- The **Complex Frequency Hopping — CFH** [7] and the **multipoint Padé approximation** [9] methods also determine a Padé–type approximation, but using several expansion points in order to obtain better approximation valid on a wider frequency range.
- The **Indirect 4SID** [11] method uses time–domain impulse response data and yields the discrete–time system matrices (the discretized form of (6)).

Since moments must be calculated in order to obtain the necessary input data, and since moment computation is an ill–conditioned problem, all these three approaches can fail to yield a valid approximation of the system.

Input: Time Domain Variation of Inputs/Outputs. When the input data resulting from the electromagnetic analysis consist of the values of input/output variables on a time–discretization grid, $u(t_k)$, $y(t_k)$, $k = 1, N$, the following methods can be used:

- **Least squares (LSQ)** method
- **Prediction error (PEM)** method [31]
- **Instrumental variable (IV)** method [41]
- The **4SID** method constructs the discrete–time equivalent of the state–space description (6), simultaneously with the determination of the system's order.

The first three methods yield a discrete–time transfer function, while the fourth constructs, as mentioned, discrete–time state–space matrices.

Since the 4SID method is less known, a short description will be provided below for the so-called *direct 4SID method.* As mentioned before, other versions, which use as inputs the system's discrete-impulse response, or frequency-domain values, are also available [45].

Consider the discrete-time analog of equations (6):

$$\begin{aligned} x_{k+1} &= Ax_k + Bu_k \ ; \\ y_k &= Cx_k + Du_k \ , \end{aligned} \tag{13}$$

where, for simplicity, the same notations A, B, C, D have been used for the discrete-time matrices (they are however different from the continuous-time matrices in (6)).

Writing the two equations for α timesteps ($k = 0, \ldots, \alpha - 1$), with $\alpha > n$, the following recurrence relations are obtained:

$$x_\alpha = A^\alpha x_0 + R_\alpha u^0 \ ; \quad y^0 = Q_\alpha x_0 + T_\alpha u^0 \ ,$$

where the controllability matrix in α steps, R_α (with transposed block columns), the observability matrix in α steps Q_α, and the causal block-Toeplitz matrix T_α, are given by

$$R_\alpha = \left[A^{\alpha-1}B \ \ldots \ AB \ B\right]$$

$$Q_\alpha = \begin{bmatrix} C \\ CA \\ \vdots \\ CA^{\alpha-1} \end{bmatrix} , \quad T_\alpha = \begin{bmatrix} D & 0 & \cdots & 0 \\ CB & D & \cdots & 0 \\ \vdots & \ddots & \ddots & \vdots \\ CA^{\alpha-2}B & \ldots & CB & D \end{bmatrix} . \tag{14}$$

The notations $u^k = [u_k, \ldots, u_{k+\alpha-1}]^T$ and $y^k = [y_k, \ldots, y_{k+\alpha-1}]^T$ denote the vectors of stacked inputs, respectively outputs, of length α and starting at timestep k ($u^k \in \mathbb{R}^{\alpha \cdot m}$, $y^k \in \mathbb{R}^{\alpha \cdot l}$).

Let us define the following input U, output Y and state X matrices: $U = [u^0 \ u^1 \ \cdots \ u^t]$, $Y = [y^0 \ y^1 \ \cdots \ y^t]$ and $X = [x_0 \ x_1 \ \cdots \ x_{t-1}]$. Then the following relation holds

$$Y = Q_\alpha X + T_\alpha U \ . \tag{15}$$

The 4SID algorithm determines the system matrices A, B, C, D (up to an unessential similarity transformation), as well as the system's order n. In order for the problem to have a solution, the matrix U must be epic (its rank equal to the number am of its rows), and the matrix Q_α is supposed to be monic (its rank equal to the number n of its columns). The 4SID algorithm, based on the expression (15), is shortly described below.

Input: $u(t_k), y(t_k)$, $k = 1, \ldots, N$.
Output: System order n, and matrices A, B, C, D.

Step 1: Determine an orthogonal projection matrix $U_\perp$ such that $U \, U_\perp = 0$ and multiply (15) to the right by $U_\perp$ to obtain $YU_\perp = Q_\alpha(XU_\perp)$.

Step 2: Perform singular value decomposition of $Z = YU_\perp$ to obtain an estimation for Q_α:

$$Z = V\Sigma W = [V_1 \ V_2] \begin{bmatrix} \Sigma_1 & 0 \\ 0 & 0 \end{bmatrix} \begin{bmatrix} W_1 \\ W_2 \end{bmatrix} \approx V_1 \Sigma_1 W_1^T \ ,$$

and take $Q_\alpha = V_1 \Sigma_1$. Order n is estimated as rank(Σ).

Step 3: Using (14), determine C by taking the first l lines of Q_α and determine A by solving (in a least squares sense) the matrix equation $\underline{Q_\alpha} A = Q_{\alpha-1}$, where $\underline{Q_\alpha}$ is the matrix obtained by deleting the first l rows of Q_α.

Step 4: Determine B and D, by using: A, C, the equation (15) and the properties of $U_{\perp}$. For more details on this step, the reader is referred to e.g. [11] [14].

3.4 Summary of Identification/Reduction Methods

The table below presents the main identification/reduction techniques currently used in conjunction with electrical engineering problems. The methods which result in a Padé approximation of the transfer function are marked with an asterisk.

Inputs → Results ↓	State–Space A, B, C, D	Frequency $Z(j\omega), Y(j\omega)$	Moments m_k	Time $u(t_k), y(t_k)$
$\tilde{A}, \tilde{B}, \tilde{C}, \tilde{D}$ or $\tilde{H}(s)$	Arnoldi Lanczos*			
$\tilde{H}(s)$		LSQ Nl. Opt.	AWE* CFH*	
$\tilde{A}, \tilde{B}, \tilde{C}, \tilde{D}$	PACT TBR OHNR SPACE	4SID	4SID	4SID
$\tilde{H}(z)$				LSQ PEM IV

3.5 Special Issues: Passivity, Accuracy, Nonlinearity

An important problem with the Krylov subspace algorithms is preservation of the initial system's stability and passivity.

Stability criteria for the Arnoldi–based model reduction are presented in 1996 by Elfadel & al. [15]. The paper [39] proposes a coordinate transformation for ensuring the preservation of stability. In conjunction with the Lanczos method, a technique which ensures stability and passivity is proposed in [5]: the partial Padé approximation (the corresponding algorithm is called PVLπ). In [3], passivity is ensured by adding to the reduced-order transfer function an appropriate function which compensates the minimum negative value of the transfer function's real part and does not affect the frequency response below the maximum frequency of interest. In [12], unstable poles are simply discarded and the residues of the remaining ones are then recomputed. In [18], the SyMPVL method is described, which performs a supplementary projection of the transfer matrix on the Krylov subspace, ensuring thus the passivity of RLC circuits.

A recently recognized problem with subspace-based methods is that they tend to capture a quite large number of nondominant poles with usually large real parts [3], [28]. To cure this, several two-stage approaches have been proposed. In [28], the first reduction stage with PRIMA is followed by a second one, based on truncated balanced realization (TBR) techniques, which eliminate the weakly controllable or observable modes from the PRIMA model. In [3], the initial model is reduced by a Krylov subspace technique, and further reduction is obtained by Block Complex Frequency Hopping. Reference [13] uses a first stage in which local (in frequency domain) approximations of the PEEC model with retardation are obtained, while in the second stage a block Arnoldi algorithm is used to reduce the obtained model.

Relatively few references deal with parameter extraction of nonlinear devices. A symbolic approach for generation of behavioral model of low-complexity circuits is presented in [6]. For large-scale circuits, [23] uses a Krylov subspace projection technique which preserves the time derivatives of the initial system.

4 Several Applications

Most of the applications of order-reduction and parameter extraction are in the field of design of integrated circuits.

An excellent reference book on VLSI interconnections modeling, analysis and simulation is [20]. A recent overview of the main problems arising in the modeling of interconnects is [10]. See also the special number of IEEE Microwave Theory and Techniques [2] on this subject.

Since other methods have been extensively presented in the scientific literature, the examples below illustrate the applicability of the less used 4SID-type methods to the identification and reduction of the parameters of electromagnetic devices.

4.1 FLUXSET Sensor

The first example is a sensor used in electromagnetic nondestructive evaluation, shown in Fig. 2. Its construction comprises the driving coil (1) with n_1 turns and the pick-up coil (2) with n_2 turns, winded around a very thin ferromagnetic ribbon core (3) of length l_c. The first coil is fed with a periodical, triangular–shaped AC driving current. The sensor is introduced in an external magnetic field to be measured.

Parameter extraction was necessary because of the extremely large ratio between the geometric dimensions (the length of the core is 500 times larger than its thickness, the insulation between coils is 1000 times smaller than the coil's length), and also to complicated effects (like hysteresis and displacement currents), which made the "brute force" approach based on the finite element method (FEM) inapplicable. Each electromagnetic effect present in the sensor (voltage induced in the coils, nonlinear magnetic characteristic of the ribbon core, eddy currents induced in the ribbon core, capacitive effects

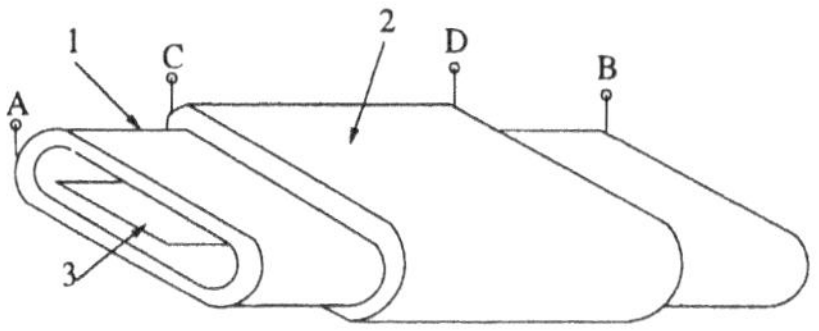

Figure2. FLUXSET sensor configuration

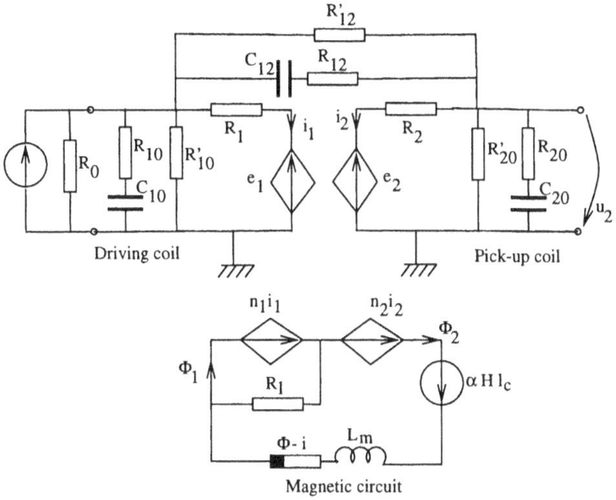

Figure3. The SPICE model of the FLUXSET sensor

in the insulation between coils, etc.) was analyzed independently in order to obtain characteristic parameters, then the subcircuits thus synthesized were integrated in a global circuit. The parameter extraction for two of the effects will be shortly described in what follows.

For the extraction of the **capacitive effects** equivalent circuit, an extended π scheme was proposed [27], which can be synthesized as an infinite RC-parallel circuit. The values of the resistances and capacitances were determined based on numerical simulations. This infinite circuit was subsequently reduced to a third-order one, using balanced realization technique.

For extracting the equivalent circuit for modeling the **eddy currents** in the core, the following technique was applied: the transient quasi-static problem with constant value for the input signal is solved, for $t > 0$, by a numerical method; using the time-domain 4SID technique [11], an equivalent system $S = (A, B, C, D)$ is generated; a reduced optimal Hankel approximation S' of the system S (with a desired order) is generated [19]; the state model S' is converted to the corresponding transfer model T; and finally the

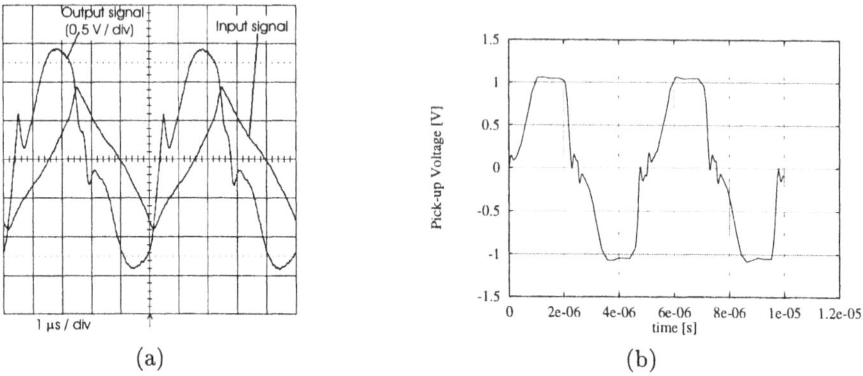

Figure4. Characteristics of the FLUXSET sensor: (a) Oscillogram of the input and output signals (b) Simulated output signal

classical Foster synthesis technique is used to synthesize the lumped circuit corresponding to the T model.

The final equivalent circuit of the whole FLUXSET sensor, including all the above-mentioned effects, is shown in Fig. 3. The simulated and measured output signals are depicted in Fig. 4. By including the capacitive effects in the model, qualitative agreement between measured and simulated results was obtained also at higher frequencies (otherwise, at low frequencies, the agreement is within 1% error, which is acceptable for the FLUXSET design).

4.2 Dielectric Filter

The dielectric filter (Fig. 5) – product of Siemens company was numerically analyzed using the Mafia program [1], in order to determine the frequency variation of the scattering parameters.

The filter can be considered as a linear system with $m = 2$ inputs and $l = 2$ outputs. Due to symmetry, $S_{11} = S_{22}$ and $S_{12} = S_{21}$, so that, for identification purposes, a model with only one output and two inputs was considered.

The frequency–domain 4SID algorithm was applied to the available input data, for the frequency band between 6.4 and 8.0 GHz.

Fig. 6 shows the initial and the identified variation of $|S_{12}|$, for the orders $n = 30$ of the equivalent system. The two plots are indistinguishable. It must be noted that models with lower order did not succeed to capture the whole behavior (two peaks are missing from the identified system's frequency response), unlike the model of order $n = 30$. However, accumulation of numerical errors may lead to unstable poles in the case of systems of higher order, which is not the case at low orders.

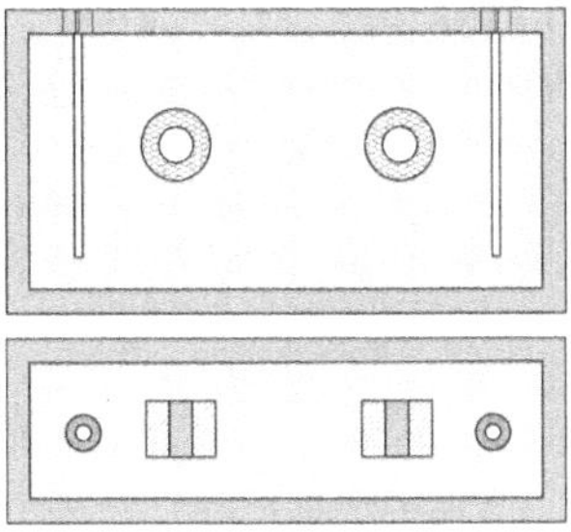

Figure5. Crossection of the dielectric filter (Siemens)

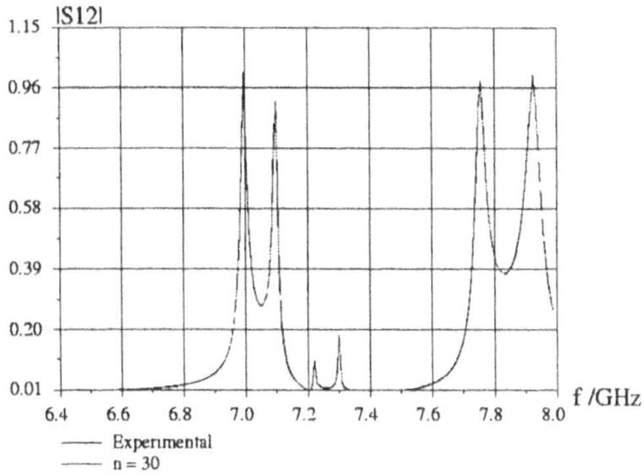

Figure6. Absolute value of S_{12}, initial and identified of order 30

5 Conclusions

Parameter extraction for general electromagnetic devices is still a challenge for researchers in the field. The classical, existing models for some classes of devices become either too imprecise in the new technological conditions (high frequency, low dimensions), or inappropriate for the complexity of real devices.

Among the difficult problems which have not found a definitive robust solution yet, are:

- Ensuring stability and passivity of the reduced models;
- Selection of the initial device's relevant modes to be preserved in the reduced model; in this respect, for some classes of the methods regularization techniques might need to be sought;
- Automatic optimal order determination of the reduced model, since neither the Krylov-subspace techniques, nor the SVD-based techniques such as 4SID don't succeed to solve this problem in a robust manner in the case of general devices; as mentioned before, this problem has been tackled by using two-step reduction algorithms, based on two different reduction techniques;
- Automatic selection of expansion points in multipoint matching methods;
- Dealing with the enormous complexity of modern devices such as integrated circuits or multichip modules (a comparison between different techniques to deal with parasitic extraction for a large, 1.7 million-transistor integrated circuit, is presented in [49]); finding appropriate partitioning techniques, in order to approach this problem in a "divide and conquer" fashion, is still an unsolved problem for the general case and the Electromagnetic Circuit Element concept may provide a solution to this problem;

– Circuit synthesis techniques for multi-input multi-output systems which could provide circuits meaningful to the designer; unfortunately, the synthesis techniques that can solve such a problem in a general manner generally yield hard-to-manage synthesized circuits.

While adapted programs can be thought of, tailored to the needs of a specific design and heavily based on the designer's experience, automatic parameter extraction will probably continue to be a field of intensive research.

Acknowledgments

I. M. acknowledges the support of the *Alexander von Humboldt Foundation* for the final part of this work, as well as of *Graduiertenkolleg "Physik und Technik von Beschleunigern"* which financed a part of the early research in 1998 and 1999.

References

1. Mafia Manual Version 4.00. CST, Darmstadt, Germany (1997)
2. Special issue on interconnects and packaging. Microwave Theory and Techniques, Vol. 45, No. 10, Part II (1997)
3. Achar, R., Gunupudi, P. K., Nakhla, M., Chiprout, E.: Passive interconnect reduction algorithm for distributed/measured networks. IEEE Trans. Circ. Syst. II **47** (2000) 287–301
4. Anderson, B. D. O., Vongpanitlerd, S.: Network Analysis and Synthesis. Prentice-Hall, Englewood Cliffs (1973)
5. Bai, Z., Feldmann, P., Freund, R. W.: Stable and passive reduced-order models based on partial Padé approximation via the Lanczos process. Numerical Analysis Manuscript 97–3–10 Bell Laboratories (1997). URL `http://cm.bell-labs.com/cm/cs/doc/97/3-10.ps.gz`
6. Borchers, C.: Symbolic behavioral model generation of nonlinear analog circuits. IEEE Trans. Circ. Syst. II **45** (1998) 1362–1371
7. Bracken, J. E., Sun, D.-K., Cendes, Z. J.: S–domain methods for simultaneous time and frequency characterization of electromagnetic devices. IEEE Trans. Microwave Theory Tech. **46** (1998) 1277–1290
8. Cangellaris, A. C., Zhao, L.: Rapid FDTD simulation without time stepping. IEEE Microwave Guided Wave Lett. **9** (1999) 4–6
9. Celik, M., Ocali, O., Tan, M. A.: Pole–zero computation in microwave circuits using multipoint Padé approximation. IEEE Trans. Computer-Aided Design Integrated Circ. and Sys. **42** (1995) 6–13
10. Chiprout, E.: Interconnect and substrate modeling and analysis: An overview. IEEE J. Solid-State Circ. **33** (1998) 1445–1452
11. Cho, Y., Xu, G., Kailath, T.: Fast identification of state-space models via exploitation of displacement structure. IEEE Trans. AC **26** (1994) 2004–2017

12. Choi, K. L., Swaminathan, M.: Development of model libraries for embedded passives using network synthesis. IEEE Trans. Circ. Syst. II **47** (2000) 249–260
13. Cullum, J., Ruehli, A., Zhang, T.: A method for reduced-order modelling and simulation of large interconnect circuits and its application to PEEC models with retardation. IEEE Trans. Circ. Syst. II **47** (2000) 261–273
14. De Moor, B., Van Overschee, P., Favoreel, W.: Numerical algorithms for subspace state space system identification – An overview. Internal Report 97-93 ESAT–SISTA, K. U. Leuven Leuven, Belgium (1995)
15. Elfadel, I. M., Silveira, L. M., White, J.: Stability criteria for Arnoldi–based model–order reduction. In Proc. of IEEE Conference on Acoustics, Speech and Signal Proc. ICASSP'96 volume 5 Atlanta, GA, USA (1996) 2642–2644
16. Elias, P. J. H., van der Meijs, N. P.: Extracting circuit models for large RC interconnections that are accurate up to a predefined signal frequency. In Proc. of 33rd Design Automation Conference Las Vegas, NV, USA (1996)
17. Feldmann, P., Freund, R. W.: Efficient linear circuit analysis by Padé approximation via the Lanczos process. IEEE Trans. Computer-Aided Design **14** (1995) 639–649
18. Freund, R. W.: Passive reduced-order models for interconnect simulation and their computation via Krylov-subspace algorithms. In Proc. of 36th Design Automation Conference DAC99 New Orleans, LA, USA (1999) 195–200
19. Glover, K.: All optimal Hankel-norm approximations of linear multivariable systems and their L^{∞}-error bounds. Int. J. Control **38** (1984) 1115–1193
20. Goel, A. K.: High-Speed VLSI Interconnections: Modeling, Analysis and Simulation. A Wiley-Interscience Publication, John Wiley & sons, Inc. (1994)
21. Golub, G. H., Loan, C. F. V.: Matrix computations. The Johns Hopkins University Press, Baltimore (1996) 3 edition
22. Guillemin, E. A.: Theory of Linear Physical Systems. John Wiley and Sons, Inc., New York, London (1963)
23. Gunupudi, P. K., Nakhla, M. S.: Model-reduction of nonlinear circuits using Krylov subspace techniques. In Proc. of 36th Design Automation Conference DAC99 New Orleans, LA, USA (1999) 13–16
24. Ho, C.-W., Ruehli, A. E., Brennan, P. A.: The modified nodal approach to network analysis. IEEE Trans. Circ. Syst. **CAS-22** (1975) 504–509
25. Hănţilă, F., Ioan, D.: Voltage–current relation of circuit elements with field effects. In 6th International IGTE Symposium Graz, Austria (1994) 41–46
26. Ioan, D., Munteanu, I.: Missing link rediscovered: The electromagnetic circuit element concept. JSAEM Studies in Applied Electromagnetics and Mechanics **8** (1999) 302–320
27. Ioan, D., Munteanu, I., Popeea, C.: Capacitive effects models for a magnetic field sensor. COMPEL, Int. J. Comput. Math. Electr. Electron. Eng. **18** (1999) 525–537
28. Kamon, M., Wang, F., White, J.: Generating nearly optimally compact models from Krylov-subspace based reduced order models. IEEE Trans. Circ. Syst. II **47** (2000) 239–248

29. Kerns, K. J., Yang, A. T.: Stable and efficient reduction of large, multiport RC networks by pole analysis via congruence transformation. In Proc. of 33rd Design Automation Conference Las Vegas, NV, USA (1996)
30. Kuh, E. S., Pederson, D. O.: Principles of Circuit Synthesis. McGraw–Hill Book Company, Inc., New York, Toronto, London (1959)
31. Ljung, L.: System Identification: Theory for the User. Prentice Hall Information and System Sciences Series (1999)
32. Moore, B. C.: Principal component analysis in linear systems controlability, observability and model reduction. IEEE Trans. AC **25** (1991) 17–32
33. Munteanu, I.: Two uniqueness theorems for electromagnetic field computation in domains with absorbing boundary conditions. Rev. Roum. Sci. Techn. Electrotechn. Energ. **42** (1997) 321–336
34. Munteanu, I., Wittig, T., Weiland, T., Ioan, D.: FIT/PVL circuit-parameter extraction for general electromagnetic devices. IEEE Trans. Magn. **36**
35. Odabasioglu, A., Celik, M., Pileggi, L. T.: PRIMA: Passive reduced-order interconnect macromodelling algorithm. In Int. Conf. on Computer-Aided Design San Jose, California (1997) 58–65
36. Pillage, L. T., Rohrer, R. A.: Asymptotic waveform evaluation for timing analysis. IEEE Trans. on CAD **9** (1990) 352–366
37. Răduleţ, R., Timotin, A., Ţugulea, A.: Introduction of transient parameters in the study of linear electric circuits with non-filamentary elements and supplementary losses (in Romanian language). St. cerc. energ. electr. **16** (1966) 857–929
38. Schefelhout, G., De Moor, B.: Frequency weighted H_2 and Hilbert–Schmidt–Hankel model reduction. In Proc. 33rd IEEE Conf. on Decision Control Lake Buena Vista, Florida, USA (1994) 3215–3216
39. Silveira, L. M., Kamon, M., Elfadel, I., White, J.: A coordinate–transformed Arnoldi algorithm for generating guaranteed stable reduced–order models of arbitrary RLC circuits. In Proc. International Conference on Computer Aided Design of IC San Jose, California, USA (1996)
40. Silveira, L. M., Kamon, M., White, J.: Efficient reduced-order modeling of frequency-dependent coupling inductances associated with 3–D interconnect structures. In Proc. 32nd Design Automation Conference San Francisco, California, USA (1995) 376–380
41. Soderstrom, T., Stoica, P.: Instrumental variable methods for systems identification. Springer-Verlag, New York (1983)
42. Tellegen, B. D. H.: Synthesis of $2n$-poles by networks containing the minimum number of elements. J. Math. Phys. **32** (1953) 1–18
43. Timotin, A.: The passive electromagnetic circuit element (in Romanian language). St. cerc. energ. electr. **21** (1971) 347–362
44. Van Overschee, P., De Moor, B.: Continuous–time frequency domain subspace system identification and stochastic realization. In Proc. 13th IFAC World Congress San Francisco, California (1996) 157–162
45. Subspace Identification for Linear Systems; Theory, Implementation, Applications. Kluwer Academic Publishers, Dordrecht (1996)

46. Wang, D., Zilouchian, A.: Model reduction of discrete linear systems via frequency-domain balanced structure. IEEE Trans. Circ. Syst. I **47** (2000) 830–837
47. Weinberg, L.: Network Analysis and Synthesis. McGraw–Hill Book Company, Inc., New York, Toronto, London (1962)
48. Wittig, T.: Implementierung eines Filtersyntheseverfahrens zur Weiterverarbeitung numerischer Simulationsergebnisse. Diplomarbeit D 179 Technische Universität Darmstadt, Fachbereich Theorie elektromagnetischer Felder (1998)
49. You, E., Varadadesikan, L., MacDonald, J., Xie, W.: A practical approach to parasitic extraction for design of multimillion-transistor integrated circuits. In Proc. of 37th Design Automation Conference DAC99 Los Angeles, CA, USA (2000) 69–74

A Mathematical Framework for the Finite Element Modelling of Electromechanical Problems

F. Henrotte and K. Hameyer *

K.U.Leuven, Dept. ESAT, Div. ELEN,
94, Kardinaal Mercierlaan, B-3001, Leuven, Belgium

Abstract Besides differential forms, special tensor fields are introduced and discussed for the representation of physical fields in continuous medium problems.

Introduction

When adressing coupled problems, experience shows that it is worthwhile to avoid notions that are specific to a particular branch of Physics (e.g. vectors in electromagnetism and tensors in elasticity) for the benefit of more abstract mathematical notions. These unified notions are less restrictive and more richly endowed with mathematical properties. They are thus liable to be implemented as properly designed objects with an *object-oriented* programming language in a *finite element* programme. On the way, one has been led to borrow notions from different mathematical disciplines. Conservation laws can be written in a metric-independent manner, thanks to *differential geometry* notions. Also, *group theory* and *functional analysis* are appropriate tools for tackling with the constitutive laws.

1 Differential geometry for Electromagnetism

Manifold: A *manifold* M of dimension n is a continuous set of points of which any neighborhood can be mapped by a differentiable 1-1 mapping onto a subset of $\mathbb{R}^n$. By assuming only the *existence* of the mapping, without requiring it to be explicitly specified, the manifold is endowed with a *differentiable structure* (it actually inherits that of $\mathbb{R}^n$) without being equipped with a preferred global co-ordinate system or even the notion of *distance.* These notions will be introduced at the right time in Sect. 3, leaving thus a substantial part of the theory foreign to any consideration of measure or distance. This will turn out to have great computational advantages.

---

* This text presents research results of the Belgian programme on Interuniversity Poles of Attraction initiated by the Belgian State, Prime Minister's Office, Science Policy Programming.

Vectors, covectors and tensors: The simplest object one can build on a manifold is a parametric *curve*, i.e. a differentiable rule $t \in]t_A, t_B[\subset \mathbb{R} \mapsto M$. The mathematical quantities that may rightly be called *vectors* are the tangent vectors to such curves. The set of all vectors that are tangent to curves going through a point $P \in M$ forms a *linear space* of dimension n. This result, which is true whatever the complexity of M itself, comes merely from the definition of a manifold and the fact that partial derivatives are linear operators in $\mathbb{R}^n$. This linear space is called *tangent space* and noted $T_P(M)$.

A *covector* a is defined as a real-valued linear operator on vectors:

$$a : v \in T_P(M) \mapsto a(v) \in \mathbb{R}. \tag{1}$$

The set of all covectors at point P also forms a linear space of dimension n that is called *cotangent space* and noted $T^*_P(M)$. Generalizing further by means of the tensor product $\otimes$, a *tensor* A is defined as a real-valued multilinear operator on ordered sets of vectors and covectors

$$A : v, \dots \in T_P(M),\ a, \dots \in T^*_P(M) \mapsto A(v, \dots; a, \dots) \in \mathbb{R}. \tag{2}$$

Tensors with p vector arguments and q covector arguments form a linear space of dimension n^{p+q}. In a co-ordinate system $\{x^i, i = 1, \dots, n\}$, any tensor admits an expansion

$$A = A^{k \dots l}_{i \dots j}\, dx^i \otimes \dots \otimes dx^j \otimes \frac{\partial}{\partial x^k} \otimes \dots \otimes \frac{\partial}{\partial x^l} \tag{3}$$

where the dx^k's (i.e. the gradients of the co-ordinates) form a convenient basis for covectors and the $\frac{\partial}{\partial x^k}$'s (i.e. the vectors tangent to co-ordinate lines) form a convenient basis for vectors. A *tensor field* is a rule that associates a tensor to each point of M.

Integration and p–forms: What is expected from a tensor field A for being an argument for a real-valued p–*integral*

$$I = \int_\Omega A \tag{4}$$

where Ω is a domain of dimension p? To answer this question, an infinitesimal piece $\Delta\Omega$ of a p–dimensional domain Ω is considered. It can be seen as being spanned with a set of p linearly independant infinitesimal vectors. It is also required that its orientation changes either by reversing one vector or swapping two of them. This amounts to saying that $\Delta\Omega$ must be proportional to the completely antisymmetrised tensor product of a set of p vectors. This is called a p–*vector*. The assertion is actually general: *antisymmetrisation* is the operation that selects the particular tensors that play a role for integration.

Now, the argument of a p–integral is expected to associate a real value (the local infinitesimal contribution to the integral) to each infinitesimal piece of Ω. It is thus defined as a real-valued linear operator on p–vectors or, equivalently, a tensor with p vector-arguments, that has been particularised in order that $A(..,v,..,w,..) = -A(..,w,..,v,..)$ holds for any pair of arguments. This is called a $p-$*covector*. In a 3D space, antisymmetrisation only allows non-zero $p-$covectors for $p = 0, 1, 2, 3$.

Fields of $p-$covectors are (by construction) arguments for $p-$integrals. They are called $p-$*form*. They respectively span linear spaces of dimension $C_p^n = \frac{n!}{(n-p)!p!}$ that are noted $F^p(M)$, $p = 0, 1, 2, 3$. They need to be handled by specific intrinsic antisymmetry-preserving operators. The antisymmetry-preserving tensor product is the *exterior product* $\wedge : F^p(M) \times F^q(M) \mapsto F^{p+q}(M)$ that verifies

$$\alpha \wedge \beta = (-1)^{pq} \beta \wedge \alpha. \tag{5}$$

The antisymmetry-preserving derivative is the *exterior derivative* $d : F^p(M) \mapsto F^{p+1}(M)$ that obeys *Leibniz rule* (6) and *Stokes theorem* (7)

$$d(\alpha \wedge \beta) = (d\alpha) \wedge \beta + (-1)^p \alpha \wedge d\beta \tag{6}$$

$$\int_\Omega d\alpha = \int_{\partial\Omega} \alpha \tag{7}$$

where $\partial\Omega$ is the *boundary* of $\Omega \subset M$. Another fundamental differential operator is the *Lie derivative* $\pounds_{\frac{d}{dt}}$ that allows us to compute time derivatives of integrals over moving and deforming domains. One has

$$\partial_t I(t) = \partial_t \int_\Omega A = \int_\Omega \pounds_{\frac{d}{dt}} A \tag{8}$$

where $\frac{d}{dt}$ stands for the bundle of trajectories (curves) generated by the movement of each point of Ω. Complete definitions of the notion presented in this section can be found in any Differential geometry treatise [?,2].

Conservation laws in Electromagnetism: *Electromagnetic fields* are paradigms for $p-$forms. If the magnetic field h and the electric field e are represented by 1–forms; the induction field b, the electric displacement d and the current density j by 2–forms; the charge density ρ^Q by a 3–form, one can write with the notions introduced so far the *Poynting relation* (9) and *charge conservation* (10) that are the *global conservation principles* electromagnetic fields must obey:

$$\int_\Omega (\partial_t b \wedge h + e \wedge \partial_t d) + \int_\Omega e \wedge j - \int_{\partial\Omega} e \wedge h = 0, \tag{9}$$

$$\partial_t \int_\Omega \rho^Q + \int_{\partial\Omega} j = 0. \tag{10}$$

As they must hold for all e, h and Ω, their famous *local* forms immediately follow from (7), (6) and (8):

$$\begin{cases} de + \partial_t b = 0 \\ dh - \partial_t d = j \\ dj + \pounds_{\frac{d}{dt}} \rho^Q = 0 \end{cases} \tag{11}$$

2 Continuous medium mechanics

Vector and covector valued p–forms: Whereas p–forms are the tensors that perfectly suit the electromagnetic fields, other kinds of tensors have to be selected in order to represent adequately the mechanical tensors (stress and strain).

A tensor α with $p+1$ vector arguments but that has been antisymmetrised only for its first p arguments is called a *covector valued p–form.* Similarly, a tensor β with p antisymmetrised vector arguments and one covector argument is called a *vector valued p–form.* Both have nC_n^p linearly independent components. They are noted

$$\alpha = \alpha_k \otimes dx^k \in F^{p\oplus 1}(M) \quad , \quad \beta = \beta^k \otimes \frac{\partial}{\partial x^k} \in F^p_{\oplus 1}(M) \tag{12}$$

where the α_k's and the β^k's are n p–forms.

As the exterior product $\wedge$ and the exterior derivative d are defined on p–forms only, they may operate on the α_k's and the β^k's but need to be generalised for tensors like α and β. One *defines*, $\forall \alpha \in F^{p\oplus 1}(M)$, $\forall \beta \in F^q_{\oplus 1}(M)$,

$$\alpha \wedge \beta = (\alpha_i \otimes dx^i) \wedge (\beta^j \otimes \frac{\partial}{\partial x^j}) = \alpha_i \wedge \beta^i, \tag{13}$$

$$d\alpha = (d\alpha_i) \otimes dx^i, \; d\beta = (d\beta^j) \otimes \frac{\partial}{\partial x^j}, \tag{14}$$

so that the Leibniz rule (6) still holds. [1]

Mechanical tensors: We are now ready to present the tensor structure of the mechanical tensors. Let $\frac{d}{dt} \in F^0_{\oplus 1}(M)$ be a *velocity field* on Ω, i.e. the field of the vectors tangent to the bundle of trajectories generated by the movement of the points of Ω. As a force f is a covector, a *volume density of force* ρ^f is a volume density of covector, i.e. $\rho^f \in F^{3\oplus 1}(M)$, of which the developed *mechanical power density* is, by (13), $\rho^f \wedge \frac{d}{dt} \in F^3(M)$. Similarly, a *stress tensor* T is a surface density of force, i.e. $T \in F^{2\oplus 1}(M)$, developing $T \wedge \frac{d}{dt} \in F^2(M)$. The (linear) *momentum density* $\rho^p \in F^{3\oplus 1}(M)$ has the same tensor structure as ρ^f and the density of *kinetic energy* is $\rho^p \wedge \frac{d}{dt} \in F^3(M)$.

The *global conservation law for momentum* is

$$\partial_t \int_\Omega \rho^p \wedge \frac{d}{dt} + \int_\Omega T \wedge D = \int_\Omega \rho^f \wedge \frac{d}{dt} + \int_{\partial\Omega} T \wedge \frac{d}{dt}. \tag{15}$$

[1] It may rightly be objected that (13) and (14) assume the implicit definition of a particular connection on the manifold which, moreover, relies on the coordinates. This lack of generality is certainly a theoretical shortcoming of this approach but it allows considerable simplifications in the algebraic developpements and for the implementation. It also makes the parallelism between the electromagnetic and mechanic systems more clear. As, from a theoretical point of view, a connection must be defined in any case and engineers are not used to such a notion, this implied introduction of the most customary connection has a virtue of simplicity.

wherein the mechanical power density developed by the stresses has been noted $T \wedge D \in F^3(M)$ with $D \in F^1_{\oplus 1}(M)$ the energetic dual of T. As (15) must hold for all T, $\frac{d}{dt}$, Ω, it has the *local* form

$$\begin{cases} \pounds_{\frac{d}{dt}} \rho^p = dT + \rho^f \\ D = d(\frac{d}{dt}) \end{cases} \tag{16}$$

where the first equation is *Newton's law* and the second one determines D as being the *gradient of the velocity field.* The striking similarity of (9,15) for the global equations and (11,16) for the local ones was already pointed out by Tonti in [3].

Electromagnetic forces : A first advantage derived from the definition of theses new kind of tensors is that they allow a much easier expression of the **electromagnetic forces** f_{EM}. Indeed, thermodynamical considerations allow to define them via the **virtual work principle** [4] by

$$\int_\Omega \rho^{f_{EM}} \wedge \frac{d}{dt} = \int_\Omega \pounds_{\frac{d}{dt}} \rho^{\Psi_{EM}} \text{ with } \pounds_{\frac{d}{dt}} b = 0 \tag{17}$$

where $\frac{d}{dt}$ is a virtual velocity field and $\rho^{\Psi_{EM}}$ is the density of electromagnetic energy, which is a function of the induction field b.

3 Constitutive Laws

The evolution of a physical system is not fully determined by conservation laws. *Constitutive laws* are also necessary to express the behavior of the matter involved. In a finite element programme, especially when multiphysic interactions are considered (e.g. electromechanical), the constitutive laws are advantageously represented by *energy functionals* in order to track as closely as possible the thermodynamics of the interaction. An energy functionals is a rule that associates a certain amount of energy to a field. This requires therefore to attribute a certain idea of *intensity* to the field. As, for a given flux through a surface, the intensity of the field depends on the size of the surface, defining an intensity amounts to the definition of a *metric* on M, i.e. the notion of distance. Practically, the metric g is a symmetric tensor with two vector-arguments, i.e. $g(v, w) = g(w, v)$, which allows to define a *norm* $|x|$ for the tensors. Together with a *volume form* $\pi \in F^3(M)$, it also allows to define the *Hodge operator*,

$$* : F^p(M) \mapsto F^{n-p}(M) \tag{18}$$

$$*\alpha = (*\alpha_k) \otimes dx^k, *\beta = (*\beta^k) \otimes \frac{\partial}{\partial x^k}. \tag{19}$$

Discussion: This paper stays at a modest level with regard to constitutive laws, i.e. linear elasticity, infinitesimal displacements. It is only shown that the Hooke law becomes a scalar one, and no more a tensorial one, in the group structure of the tensor sets we have defined.

The tensors T and D appearing in (16) are not the classical σ and ε tensors. This is shown by analysing the group structure of the 9-dimensional linear spaces they span, which are

$$D \in F^1_{\oplus 1}(M) \overset{g}{\longleftrightarrow} D^\flat \in F^{1\oplus 1}(M) \overset{*}{\longleftrightarrow} T \in F^{2\oplus 1}(M) \qquad (20)$$

where the intermediary space is $D^\flat = D^k_i g_{kj} dx^i \otimes dx^j \in F^{1\oplus 1}(M)$. By a decomposition of these three spaces into irreducible subspaces, one can split each of them into a subspace of dimension 3 with *rotation* modes (indexed R) and a subspace of dimension 6 with *deformation* modes (indexed D). In the latter, a subspace of dimension 5 with *shear* modes (indexed S) and a subspace of dimension 1 with *compression* modes (indexed C) may be further distinguished. With $tr_g T = T_{ijk} g^{jk} dx^i \in F^1(M)$, this can be summarised as follows:

		$D^\flat \in F^{1\oplus 1}(M)$	$T \in F^{2\oplus 1}(M)$
R		$D^\flat$ is antisym.	$tr_g T \neq 0$
D	S	$D^\flat$ is sym.	$tr_g T = 0$
	C	$D^\flat \propto g$	$T \propto \pi$

(21)

One can see that T and D have an antisymmetric part that σ and ε have not. This relaxation is the condition *sine qua non* to have the linear momentum conservation law (15) expressed in a metric-independent form.

The group decomposition (21) allows to state the constitutive law of an elastic medium in a very simple way. The physical idea of *deformation* is related with $D^\flat_D = \frac{1}{2}\pounds_{\frac{d}{dt}} g$ [5] where D_R plays no role because infinitesimal rotations do not participate in the stored elastic energy. One has now for the *Hooke law* of a linear isotropic material

$$T_C = 3K \, * D^\flat_C, \quad T_S = 2G \, * D^\flat_S, \quad T_R = 0, \qquad (22)$$

where K is the *bulk modulus* and G the *shear modulus*. The third relation is equivalent to the classical $\sigma_{ij} = \sigma_{ji}$ in orthonormal co-ordinates. It restores then, at the constitutive law level, the due symmetry of the stress tensor but it also makes the constitutive law *injective*, i.e. the knowledge of T does not fully determine D. Such constitutive laws can be properly handled by *Convex analysis* [6] which associates to the elastic energy functional directly derived from (22), i.e.

$$\Psi(D) = \frac{1}{2}\int_\Omega (3K\,|D_C|^2 + 2G\,|D_S|^2)\pi, \qquad (23)$$

its *Legendre transform* where D_R plays the role of a Lagrange multiplier for the $T_R = 0$ equation, i.e.

$$\Phi(T, D_R) = \frac{1}{2}\int_\Omega (\frac{|T_C|^2}{3K} + \frac{|T_S|^2}{2G})\pi - \sup_{D_R}\int_\Omega T_R \wedge D_R. \qquad (24)$$

Conclusion

A finite element formulation of a continuous medium problem relies on a *minimum principle* which itself relies on a *function space* F and a *functional* Λ defined on F. Provided the fields in presence are represented by $p-$forms or by (co)vector-valued $p-$forms, they can be interpolated with *Whitney forms* (special shape functions that ensure an interpolation that are always conforming and that preserve the right continuity properties at the discrete level [7,8]). Moreover, global and local conservation laws in electromagnetism and elasticity can be expressed as metric-independent differential relations in a strikingly similar manner. These are the main achievements of this paper. There are practical implications. A great part of the desirable properties of the Whitney forms that are exploited in electromagnetic computations can be translated straightfully into equivalent techniques for elasticity computations. This helps a lot for the implementation of an electromechanical finite element programme and to the understanding of the difficult problem of electromagnetic forces.

References

1. B. Schutz: Geometrical methods of mathematical physics, Cambridge University Press 1980
2. Y. Choquet-Bruhat and al, Analysis, manifolds and physics, Elsevier Science North Holland 1982
3. E. Tonti, On the mathematical structure of a large class of physical theories, Rend. Acc. Lincei (Vol.52 1974 pages:48-56)
4. A. Bossavit, On local computation of the electromagnetic force field in deformable bodies, nternational Journal of Applied Electromagnetics in Materials (Vol.2 1992 pages:333-343)
5. J.E. Marsden and T.J.R. Hughes, Mathematical foundations of elasticity, Dover 1983
6. R. T. Rockafellar, Convex analysis, Princeton University Press 1972
7. A. Bossavit, Whitney forms : a class of finite elements for 3D computations in electromagnetism, IEE Proceedings (Vol.135 A, No.8 1988 pages: 493-500)
8. J.C. Nedelec, Notions sur les techniques d'lments finis. Mathmatiques et applications, Ellipses 1991

Dual Magnetodynamic Finite Element Formulations with Natural Definitions of Global Quantities for Electric Circuit Coupling

P. Dular

Dept. of Applied Electricity, Montefiore Institute B28, University of Lige, B-4000 Lige, Belgium

1 Introduction

Both general families of magnetodynamic formulations, i.e. $\boldsymbol{h}$- and $\boldsymbol{b}$ -conform formulations, are studied in the frame of the finite element method, in 2D and 3D. These dual formulations are respectively weak forms of the Faraday and Ampere equations, and respectively use unknowns directly associated with the magnetic field $\boldsymbol{h}$ (e.g. magnetic field - magnetic scalar potential h-formulations) and with the magnetic flux density b (e.g. magnetic vector potential $\boldsymbol{a}$-formulations).
When such formulations are applied to systems coupled with electric circuits, not only local quantities, characterizing the unknown field, are involved, but also electric global quantities, i.e. currents and voltages. The conducting regions carrying these global quantities can be of massive or stranded types, each type necessitating a formulation dependent treatment. The mathematical and numerical tools for naturally coupling local and global quantities are studied for all these variants, using edge and nodal coupled finite elements and efficient forms of source magnetic and electric fields. The results are circuit relations characterizing the conducting regions, i.c. rclations relating currents and voltages.

2 Magnetodynamic Problem and Associated Constraints

A bounded domain Ω of the two- or three-dimensional Euclidean space is considered. Its boundary ($\partial\Omega$ is denoted Γ. The differential equations and material constitutive relations of the magnetodynamic, or eddy current, problem in Ω are

$$\mathbf{curl}\,\boldsymbol{h} = \boldsymbol{j}, \mathbf{curl}\,\boldsymbol{e} = -\partial_t \boldsymbol{b}, \mathbf{div}\,\boldsymbol{b} = 0 \tag{1}$$

$$\boldsymbol{b} = \mu\boldsymbol{h}, \boldsymbol{j} = \sigma\boldsymbol{e} \tag{2}$$

where $\boldsymbol{h}$ is the magnetic field, $\boldsymbol{b}$ is the magnetic flux density, e is the electric field, $\boldsymbol{j}$ is the electric current density, including source currents $\boldsymbol{j}_s$

and eddy currents, μ is the magnetic permeability and σ is the electric conductivity. Equations (1(and (1) are respectively the Ampere and Faraday equations. Conservation equation (1) carries a part of (1) and is usually associated with this latter. It gives the information needed for static problems as well as for nonconducting regions (in which $\sigma = 0$; field e is therefore not uniquely defined by the considered subset (1) of the Maxwell equations). Relations (2) are respectively the considered magnetic and Ohm laws.
The eddy current conducting part of Ω is denoted Ω_c and the complementary nonconducting one Ω_c^C, with $\Omega = \Omega_c \cup \Omega_c^C$. Massive inductors belong to Ω_c and stranded inductors are regions of Ω_s, with $\Omega_s \subset \Omega_c^C$. Such a distinction rests on the fact that unknown eddy currents appear in massive inductors while fixed current distributions $\boldsymbol{j_s}$ have to be given in stranded ones.
Boundary conditions on fields involved in (1) are defined on complementary - generally nonconnected - parts Γ_h and Γ_e of Γ, i.e.

$$\boldsymbol{n} \times \boldsymbol{h}|_{\Gamma_h} = 0, \qquad \boldsymbol{n} \times \boldsymbol{e}|_{\Gamma_e} = 0, \qquad \boldsymbol{n} \cdot \boldsymbol{b}|_{\Gamma_e} = 0 \tag{3}$$

where $\boldsymbol{n}$ is the unit normal vector on Γ exterior to Ω. Here again, (3) carries a part of (3), through (1). These homogeneous conditions are commonly encountered in electromagnetic problems for either physical (conditions at infinity or associated with idealized materials) or symmetry reasons (for normal or tangential fields). Their extension to nonhomogeneous ones can be done without difficulty.
In addition to local conditions (3), global conditions on voltages or currents in inductors are considered, through functionals on local fields (circulations and fluxes). An idealized inductor Ω_c is shown in Fig. (1) where a source of electromotive force is located between two sections, these being two electrodes or terminals very close to each other. The studied domain Ω contains all the regions except the very thin one $\Omega_{emf,i}$ separating the electrodes, called electromotive force region. Such a source is characterized by a voltage V_i and a current I_i flowing through one of its terminal surface $\Gamma_{j,i}$, considered as a cross-section of the inductor; the set of all such inductor cross-sections is noted γ_j. Surfaces $\Gamma_{j,i}$ can also be parts of the exterior boundary Γ_e of the studied domain either in presence of symmetry conditions (with perpendicular currents) or as exterior terminals for external connections.
For massive inductors, the electric field $\boldsymbol{e}$ in $\Omega_{emf,i}$ can be considered as being known, as a conservative electric field, and its circulation along any path γ_i from one terminal to the other in $\Omega_{emf,i}$ is actually the applied voltage V_i. One thus has

$$\int_{\gamma_i} \boldsymbol{e} \cdot d\boldsymbol{l} = V_i, \qquad \int_{\Gamma_{j,i}} \boldsymbol{n} \cdot \boldsymbol{j} ds = I_i \tag{4}$$

where Γ_i is any path in $\Omega_{emf,i}$ connecting its two terminals; the convention used for both orientations of γ_i and $\boldsymbol{n}$ gives the same direction to I_i and V_i. For stranded inductors (included in Ω_s), condition (4) holds while (4) has to

be expressed as the sum of the circulations of $\boldsymbol{e}$ for all the wires.
Moreover, for both kinds of inductors, one must satisfy local conditions [12]

$$\boldsymbol{n} \times \boldsymbol{e}|_{\Gamma_j} = 0, \qquad \boldsymbol{n} \cdot \boldsymbol{j}|_{\partial\Omega_c} = 0, \qquad \boldsymbol{n} \cdot \mathbf{curl}\, \boldsymbol{e}|_{\Gamma} = 0 \tag{5}$$

Conditions (5) respectively ensure that each terminal is equipotential, the total current flows through terminals only and there is no magnetic coupling between Ω and its exterior. In order to satisfy (5) with closed inductors (such as the idealized one in Fig. 1), an assumption must be made about the connection of the external circuit: this circuit is considered to be located outside Ω far enough from the main structure and is connected to it through twisted pairs.

3 Magnetic Field Weak Formulation

3.1 h-Formulation with Reaction and Source Magnetic Fields

The general expression of the magnetic field $\boldsymbol{h} \in F_h(\Omega)$ in an $\boldsymbol{h}$-formulation is

$$\boldsymbol{h} = \boldsymbol{h}_s + \boldsymbol{h}_r - \mathbf{grad}\, \phi, \qquad \text{with} \tag{6}$$

$$\mathbf{curl}\, \boldsymbol{h}_s = \boldsymbol{j}_s \tag{7}$$

involving complementary fields $\boldsymbol{h}_s \in F_{h_s}(\Omega)$, $\boldsymbol{h}_r$ and $\phi(\boldsymbol{h}_r - \mathbf{grad}\, \phi \in F_{h\phi}(\Omega))$, with

$$F_h(\Omega) = \left\{\boldsymbol{h} \in \boldsymbol{L}^2(\Omega); \mathbf{curl}\, \boldsymbol{h} \in \boldsymbol{L}^2(\Omega), \boldsymbol{n} \times \boldsymbol{h}|_{\Gamma_h} = 0\right\}, \tag{8}$$

$$F_{h_s}(\Omega) = \left\{\boldsymbol{h}_s \in F_h(\Omega); \mathbf{curl}\, \boldsymbol{h}_s = 0 \quad \text{in} \quad \Omega_s^C\right\} \quad , \text{with} \quad \Omega_s^C = \Omega / \Omega_s \tag{9}$$

$$F_{h\phi}(\Omega) = \left\{\boldsymbol{h} \in F_h(\Omega); \mathbf{curl}\, \boldsymbol{h} = 0 \quad \text{in} \quad \Omega_c^C\right\} \tag{10}$$

Field $\boldsymbol{h}_s$ is a source magnetic field associated with the imposed current density $\boldsymbol{j}_s$ in stranded inductors Ω_s through (7). Field $\boldsymbol{h}_r$ is the associated reaction field in conducting regions Ω_c while ϕ is the associated reaction magnetic scalar potential in nonconducting regions Ω_c^C. Potential Φ can be multivalued when Ω_c^C is multiply connected, in which case cuts must be defined to make this domain simply connected [4].
The $\boldsymbol{h} - \phi$ magnetodynamic formulation is an example of h-formulation [2], [8]. It is obtained from the weak form of the Faraday equation (1), with (2), i.e.

$$\partial_t(\mu \boldsymbol{h}, \boldsymbol{h}')_\Omega + (\sigma^{-1}\, \mathbf{curl}\, \boldsymbol{h}, \mathbf{curl}\, \boldsymbol{h}')_{\Omega_c} + (\sigma^{-1} \boldsymbol{j}_s, \mathbf{curl}\, \boldsymbol{h}')_{\Omega_s} + \\ < \boldsymbol{n} \times \boldsymbol{e}_s, \boldsymbol{h}' >_{\Gamma_e} = 0, \qquad \forall \boldsymbol{h}' \in F_{h\phi}(\Omega) \tag{11}$$

where $(\cdot,\cdot)_\Omega$ and $< \cdot,\cdot >_\Gamma$ respectively denote a volume integral in Ω and a surface integral on Γ of products of their vector field arguments. Surface electric field $\boldsymbol{n} \times \boldsymbol{e}_s$ is a natural boundary condition on boundaries Γ_e of two kinds. Either, it can be a locally specified field, e.g. (3), or a field for which only associated global quantities are known (circulations of e_s), i.e. voltages through (4) [8].

3.2 Currents as Essential Global Quantities

Current driven massive inductors can be directly considered through the form of the reaction field in (7). Indeed, the circulation of $\boldsymbol{h}_r - \mathbf{grad}\,\phi$ along any closed path around each inductor i is equal to the current I_i flowing in this inductor.
For stranded inductors, the constraint of uniformity of the current density can be automatically satisfied through source magnetic fields $\boldsymbol{h}_{s,i}$, each one being associated with an inductor $i \in \Omega_s$ carrying a unit current, i.e. satisfying $\mathbf{curl}\,\boldsymbol{h}_{s,i} = \boldsymbol{j}_{s,i}$ (7) where $\boldsymbol{j}_{s,i}$ is the known equivalent current density of this unit current flowing in the N_i turns of the inductor. Then, the total source magnetic field $\boldsymbol{h}_s$ can be expressed as

$$\boldsymbol{h}_s = \Sigma_{i \in \Omega_s} I_{s,i} \quad , \quad \boldsymbol{h}_s \in F_{h_s}(\Omega) \tag{12}$$

where $I_{s,i}$ is the current flowing in stranded inductor i, thus associated with a global basis function being the unit source magnetic field $\boldsymbol{h}_{s,i}$. The actual source current density in inductor i is $\boldsymbol{j}_s = I_{s,i}\boldsymbol{j}_{s,i}$. Consequently, source magnetic fields $\boldsymbol{h}_s$ (12) in formulation (11), through expression (7) of the total field $\boldsymbol{h}$, enable to directly take current driven stranded inductors into account.

3.3 Weak Circuit Relations and Voltages as Natural Global Quantities

Voltage driven inductors need circuit relations to be considered. For stranded inductors, such relations are relative to the Faraday equation applied to the coil windings relating the current and the voltage for each inductor i, i.e. $V_i = R_i I_i + \partial_t \Phi_i$, where R_i is the coil resistance and Ψ_i is the magnetic flux linked to the coil.
The terms of the circuit relation can be expressed in various ways (all similar at the continuous level but different at the discrete level) from the quantities involved in formulation (11). Nevertheless, this formulation already contains the information concerning the Faraday law because it is precisely a weak form of this law for the whole studied domain, and thus also for the stranded inductors. In other words, the total magnetic flux linked to each coil is already taken into account in a weak way through (11) and should therefore be considered with exactly the same weak sense (the same approximation at the discrete level, not defining another nonconserved discrete flux) in the circuit relation, sense that will be called natural.
For massive inductors, a more general relation has to be used and is first developed. There exists for each inductor one basis function of $\boldsymbol{h}$ (called current basis function and denoted c_i) associated with its current with the property of having a unit circulation along any closed path around the inductor.
At the continuous level, it is known that function $\boldsymbol{c}_i$ can be expressed as the gradient of a scalar potential, i.e. [8]

$$c_i = -\,\mathbf{grad}\, q_i \text{ in } \Omega_c^C \text{ with } \boldsymbol{n} \times \mathbf{grad}\, q_i \text{ on } \partial\Omega_c \tag{13}$$

this potential qi being defined in domain Ω_c^C made simply connected by cut C_i (1). Its value can be fixed to 1 on one side of the associated cut, called side '+', and to 0 on the other side, called side '-'. The continuous transition of q_i between both these values can be performed in a transition layer in Ω_c^C adjacent to side '+', which reduces the support of q_i and $\boldsymbol{c}_i$.
The surface integral in (11) can be developed for $\boldsymbol{h}'$ equal to c_i. The only contributing part of Γ_e is then the lateral boundary of $\Omega_{emf,i}$ made simply connected. The other parts of the boundary of $\Omega_{emf,i}$, i.e. its terminals, give no contribution because $\boldsymbol{n} \times \boldsymbol{e} = \boldsymbol{n} \times \boldsymbol{e}_s = 0$ on them. Thus, using (13) and the Stokes formula, one has

$$< \boldsymbol{n} \times \boldsymbol{e}_s, \boldsymbol{c}_i >_{\Gamma_e} = \oint_{\partial\Gamma_e} q_i \boldsymbol{e}_s \cdot dl = \oint_{\gamma_i} \boldsymbol{e}_s \cdot dl = V_i \tag{14}$$

because only the part γ_i of the oriented contour $\partial\Gamma_e$ in contact with ∂C_i^+ gives a nonzero contribution (1). Consequently, for test function $\boldsymbol{h}' = \boldsymbol{c}_i$, (11) becomes

$$\partial_t(\mu\boldsymbol{h}, \boldsymbol{c}_i)_\Omega + (\sigma^{-1}\,\mathbf{curl}\,\boldsymbol{h}, \mathbf{curl}\,\boldsymbol{c}_i)_{\Omega_c} = -V_i \tag{15}$$

which is the natural weak circuit relation for massive inductor i, in which current I_i is strongly defined through a circulation constraint on $\boldsymbol{h}$.
Similarly, for stranded inductors, the basis functions of $\boldsymbol{h}_s$ in (12), i.e. the unit source magnetic fields $\boldsymbol{h}_{s,i}$, when used as test functions h', lead to circuit relation

$$\partial_t(\mu\boldsymbol{h}, \boldsymbol{h}_{s,i})_\Omega + I_{s,i}(\sigma^{-1}\boldsymbol{j}_{s,i}\,\mathbf{curl}\,\boldsymbol{h}_{s,i})_{\Omega_{s,i}} = -V_i \tag{16}$$

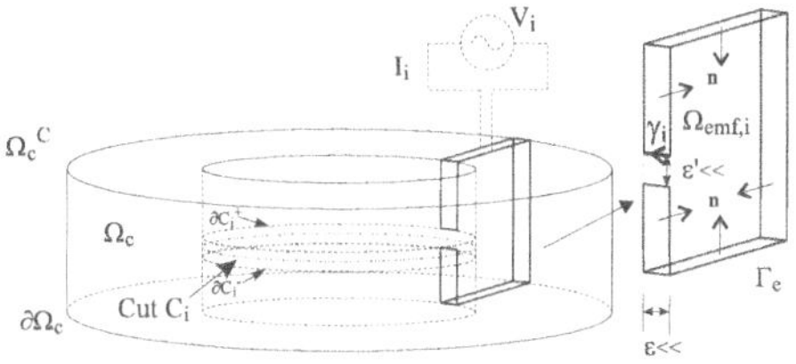

Figure1. Inductor with a source of electromotive force $\Omega_{emf,i}$ and its cut C_i; ∂C_i^+ and ∂C_i^- are the traces of both sides of C_i on $\partial\Omega_c$

3.4 Discretization of the Magnetic Field

At the discrete level, the use of edge finite elements [3], [5], [13] either in 2D or 3D to interpolate a curl-conform field, such as the magnetic field $\boldsymbol{h}$, first gives facilities in defining currents. Indeed, the circulation of such a field along a closed path, being also the flux of its curl and thus the current, is directly obtained from coefficients of the interpolation, i.e. those associated with the edges of the path.
The magnetic field $\boldsymbol{h}$ is thus interpolated with edge finite elements, generating the function space $S^1(\Omega)$ defined on a mesh of Ω, i.e.

$$\boldsymbol{h} = \Sigma_{e \in E} \boldsymbol{h}_e \boldsymbol{s}_e \quad , \quad \boldsymbol{h} \in S^1(\Omega) \tag{17}$$

where E is the set of edges of Ω, $\boldsymbol{s}_e$ is the edge basis function associated with edge $\boldsymbol{e}$ and $\boldsymbol{h}_e$ is the circulation of $\boldsymbol{h}$ along edge $\boldsymbol{e}$. Geometrical elements of the mesh can be triangles and quadrangles in 2D, or tetrahedra, hexahedra and prisms in 3D [5].
Then, this characterization (17) can be transformed to give explicitly the basis functions of the considered discrete space for $F_{h\Phi}(\Omega)$ with the essential constraints (4) and (7). The $\boldsymbol{h} - \Phi$ coupling, by (7), with Ω_c^C multiply connected, leads to

$$\boldsymbol{h} = \boldsymbol{h}_s + \Sigma_{k \in E_c} \boldsymbol{h}_k \boldsymbol{s}_k + \Sigma_{n \in N_c^C} \Phi_n \boldsymbol{v}_n + \Sigma_{i \in C} I_i \boldsymbol{c}_i \tag{18}$$

where E_c is the set of inner edges of Ω_c, N_c^C is the set of nodes inside Ω_c^C and on its boundary $\partial\Omega_c^C$, and C is a set of cuts making Ω_c^C simply connected [4], [6]. Indeed, the circulation h_e along an edge in Ω_c^C is actually the difference of the scalar potentials at both ends of the edge, which leads to define new basis functions v_n associated with the nodes of Ω_c^C and the nodal values Φ_n of Φ. Actually, potential Φ in Ω_c^C is decomposed in continuous and discontinuous parts, of which the gradients are respectively given by the two last sums in (18). Coefficients I_i represent circulations of $\boldsymbol{h}$ along well defined paths (equal to the fluxes of their curl and thus to the currents through associated surfaces) and functions $\boldsymbol{c}_i$ are vector basis functions associated with cuts C_i. Note that such a characterization enables function $\boldsymbol{v}_n$, and thus the associated scalar potential, to be fully continuous in a multiply connected domain, the discontinuity being taken into account by $\boldsymbol{c}_i$.
Source magnetic fields can benefit from the same approximation as the one in (18). The support of each source field can then be limited to the associated inductor and a transition layer associated with a cut, i.e. the associated source scalar potential out of the inductor is freely fixed to zero [6]. The co-tree gauge condition is used, i.e. the circulation of the source field is fixed to zero along all the edges of a tree built in the inductor. This limited support leads to an elegant way to strongly reduce the cancellation error in magnetic materials [15]. Indeed, this implicitly defines a total magnetic scalar potential in a formulation using a reduced magnetic scalar potential Φ, in addition to have

a perfect complementarity between source and reaction fields finite element spaces.

4 Magnetic Vector Potential Weak Formulation

4.1 a-Formulation with Source Electric Scalar Potential

The general expression of the electric field $\boldsymbol{e}$ via a magnetic vector potential $\boldsymbol{a} \in F_a(\Omega)$ involves the gradient of an electric scalar potential $v \in F_v(\Omega_c)$, i.e.

$$\boldsymbol{e} = -\delta_t \boldsymbol{a} - \mathbf{grad}\, v \text{ in } \Omega_c \text{ with} \tag{19}$$

$$\boldsymbol{b} = \mathbf{curl}\, \boldsymbol{a} \text{ in } \Omega \tag{20}$$

so that the Faraday equation (1) is satisfied, with

$$F_a(\Omega) = \{\boldsymbol{a} \in \boldsymbol{L}^2(\Omega); \mathbf{curl}\, \boldsymbol{a} \in \boldsymbol{L}^2(\Omega), \boldsymbol{n} \times \boldsymbol{a}|_{\Gamma_e} = 0\} \tag{21}$$

$$F_V(\Omega_c) = \{v \in L^2(\Omega_c); \mathbf{grad}\, v \in \boldsymbol{L}^2(\Omega_c), [v]_{\Gamma_{j,i}} = V_i\} \tag{22}$$

$[\cdot]_\Gamma$ stands for a discontinuity through a boundary Γ.

Because $\boldsymbol{e}$ is not uniquely defined in Ω_c^C by the considered magnetodynamic equations, there is no need to define it through (19) in Ω_c^C. Equation (20) is the only one to be considered there, giving a the nature of a vector potential. If potential v is fixed a priori, $\boldsymbol{a}$ will not have this nature in Ω_c^C and will be rather a kind of reaction electric field by (19), with no need to be gauged. With these potentials, the $\boldsymbol{a} - v$ magnetodynamic formulation is obtained from the weak form of the Ampere equation 1, with (2), i.e. [2]

$$(\mu^{-1}\, \mathbf{curl}\, \boldsymbol{a}, \mathbf{curl}\, \boldsymbol{a}')_\Omega + (\sigma \delta_t \boldsymbol{a}, \boldsymbol{a}')_{\Omega_c} + (\sigma\, \mathbf{grad}\, v, \boldsymbol{a}')_{\Omega_c} \tag{23}$$

$$-(\boldsymbol{j}_s, \boldsymbol{a}')_{\Omega_s} = 0, \qquad \forall \boldsymbol{a}' \in F_a(\Omega) \tag{24}$$

Formulation (24) implies, with test function $\boldsymbol{a}' = \mathbf{grad}\, v'$, that

$$(\sigma \delta_t \boldsymbol{a}, \mathbf{grad}\, v')_{\Omega_c} + (\sigma\, \mathbf{grad}\, v\, \mathbf{grad}\, v')_{\Omega_c} = < \boldsymbol{n} \cdot \boldsymbol{j}, v' >_{\Gamma_j}, \forall v' \in F_v(\Omega_c) \tag{25}$$

where Γ_j is the part of the boundary of Ω_c crossed by currents. Formulation (25) is actually also the weak form of $\mathbf{div}\, j = 0$ (coming from the Ampere equation) in Ω_c.

4.2 Voltages or Currents as Essential Global Quantities

Certain kinds of source fields can be defined and used in $\boldsymbol{a}$-formulations to lead to circuit relations.
Voltage driven massive inductors can be considered through a unit source electric scalar potential $v_{s,i}$ associated with a unit voltage for each inductor i

($v_{s,i}$ is equal to 1 on one electrode, to 0 on the other and varies continuously in $\Omega_{c,i}$, which gives the electric field needed in region $\Omega_{emf,i}$), leading to

$$v = \Sigma_{i \in \Gamma_j} V_i v_{s,i} \tag{26}$$

The voltage associated with a massive inductor then appears explicitly in formulation (24) as a global quantity defined in a strong sense.
The treatment of stranded inductors is somewhat different. What is actually given for a stranded inductor is a current density distribution and no a priori information about one kind of unit voltage can be known. This points out a loss of a part of the duality between $\boldsymbol{h}-$ and $\boldsymbol{a}-$ formulation: the current will be rather a strong quantity for stranded inductors, through source current density $\boldsymbol{j}_s$ in (24).

4.3 Weak Circuit Relations and Natural Global Quantities

With the a-v formulation, the total current flowing in a massive conductor can only be expressed in a weak sense, because it comes from the Ampere equation which is itself expressed in a weak form.
The current I_i through section $\Gamma_{j,i}$ of a massive inductor can be obtained from (25) with v' equal to the previously defined source scalar potential $v_{s,i}$. This gives

$$< \boldsymbol{n} \cdot \boldsymbol{j}, v_{s,i} >_{\Gamma_{j,i}} = I_i = (\sigma \delta_t \boldsymbol{a}, \mathbf{grad}\, v_{s,i})_{\Omega_c} + V_i (\sigma\, \mathbf{grad}\, v_{s,i}, \mathbf{grad}\, v_{s,i})_{\Omega_c} \tag{27}$$

which is the natural weak circuit relation associated with massive inductor i.
For a stranded inductor, the circuit relation cannot be issued from (27). Indeed, the gradient of the source scalar potential $v_{s,i}$ should then be the electric field in the stranded inductor i associated with a unit voltage. It is however generally impossible to determine a priori such a field because the distribution of the voltage in the wires is precisely a part of the solution of the magnetodynamic problem.
One has rather to resort to the circuit relation obtained from the $\boldsymbol{h}$-formulation, i.e. (16), and to express it in term of the magnetic vector potential [10], i.e.

$$\delta_t (\boldsymbol{a}, \boldsymbol{j}_{s,i})_{\Omega_{s,i}} + I_{s,i} (\sigma^{-1} \boldsymbol{j}_{s,i}, \boldsymbol{j}_{s,i})_{\Omega_{s,i}} = -V_i \tag{28}$$

which is the circuit relation associated with stranded inductor i in the $\boldsymbol{a}$-formulation. Current densities $\boldsymbol{j}_{s,i}$ are then source fields giving the circuit relations.
Different expressions of the unit current density $\boldsymbol{j}_s$ can be considered, which will lead to different approximations of winding distributions. The most commonly used form for $\boldsymbol{j}_s$ is $\boldsymbol{j}_s = \boldsymbol{t} N_i / S_i$, where t is a unit vector tangent to the coil direction and S_i is the surface area of the inductor. The distribution

of $\boldsymbol{j}_s$ can also be considered as the curl of an electric vector potential (having the same definition as the source magnetic field and used to enable a good convergence of a nongauged resolution [14]) or as the gradient of a source scalar potential [10].

4.4 Discretization of the Magnetic and Electric Potentials

At the discrete level, formulation (24) includes (25) when the gradient of v' is included in the space of $\boldsymbol{a}'$, which is the case when edge and nodal finite elements are used respectively for $\boldsymbol{a}'$ and v', and thus for a and v [3], [5], [13]. $F_a(\Omega)$ is thus chosen here as an edge finite element function space $S^1(\Omega)$ built on a mesh of Ω and associated with a gauge condition in Ω_c^C [1], and $F_v(\Omega_c)$ as the associated nodal finite element space $S^0(\Omega_c)$, with the relation $\text{grad}S^0(\Omega_c) \subset S^1(\Omega)$ [3], [5], [13]. Note that nodal finite elements for a would lead to the loss of important properties: a penalty term would have to be added in (24) and the exact complementarity between $\delta_t \boldsymbol{a}$ and $\mathbf{grad}\, v$ in (19) would not be satisfied anymore.
Source potential $v_{s,i}$ is then discretized as the sum of the nodal basis functions s_n of all the nodes located on the cross-section $\Gamma_{j,i}$, with a support limited to a transition layer containing all the elements adjacent to one side of the cross section (Fig. 2). Potential $v_{s,i}$ is indeed equal to 1, the unit voltage, on $\Gamma_{j,i}$ [7], [9]. Another form of $v_{s,i}$ can be given by the solution of the electrokinetic problem in Ω_c, which nevertheless presents the disadvantage of necessitating a pre-computation.

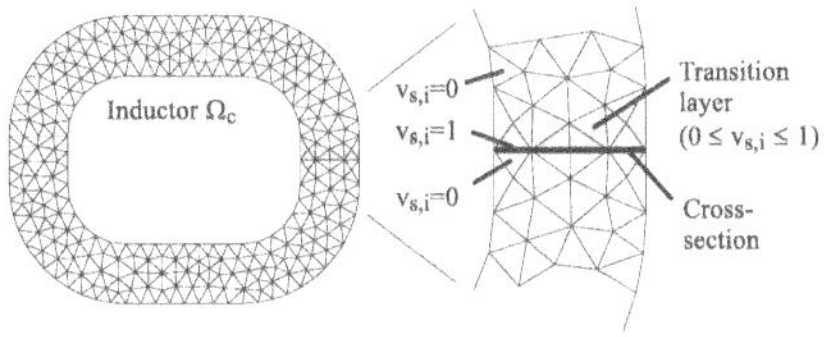

Figure2. Cross-section and associated transition layer in an inductor

So, for massive inductors, using (27), the current is obtained rather from a volume integration in a transition layer located on one side of the cross-section (Fig. 2; because the support of $v_{s,i}$ is reduced to this layer in Ω_c) than from a numerical surface integration of $\boldsymbol{n} \cdot \boldsymbol{j} = \boldsymbol{n} \cdot \sigma \boldsymbol{e}$ on this section. Such an explicit surface integration would be affected by the choice of the integration surface and there would be generally no reason for the so-computed current to be equal to the current given by the volume integral in the transition layer, even if the surface is the actual cross-section. These methods to compute the current are totally different in 3D, although they are identical when applied in 2D. The proposed method then appears to be a generalization from 2D to 3D.

For stranded inductors, in 2D, all the defined forms of $\boldsymbol{j}_s$ are equivalent. Indeed, j_s is a constant on the studied section of the inductor, constant which is exactly taken into account whatever the method is. This is nevertheless not the case in 3D where the direction of the winding is not constant.

5 Conclusions

Methods for defining currents and voltages in finite element magnetodynamic $\boldsymbol{h}-$ and $\boldsymbol{a}-$formulations have been developed for both massive and stranded inductors, in 2D and 3D. It has been shown that the use of edge and nodal coupled finite elements and adapted source fields enables to define currents and voltages in a natural and efficient way. All the advantages of the methods appear when local and global quantities have to be coupled, either within a finite element problem or through external lumped circuits. Computational tools adapted to these methods have been implemented in the GetDP software [11].

References

1. R. Albanese, G. Rubinacci, "Magnetostatic field computation in terms of two-component vector potentials", International Journal for Numerical Methods in Engineering, Vol. 29, pp. 515-532, 1990.
2. A. Bossavit, "Two dual formulations of the 3D eddy currents problem", COMPEL, No. 4, pp. 103-116, 1984.
3. A. Bossavit, "A rationale for edge-elements in 3-D fields computations", IEEE Trans. Magn., Vol. 24, No. 1, pp. 74-79, 1988.
4. A. Bossavit, "Magnetostatic problems in multiply connected regions: some properties of the curl operator", IEE Proceedings, Vol. 135, Pt. A, No. 3, pp. 179-187, 1988.
5. P. Dular, J.Y. Hody, A. Nicolet, A. Genon, W. Legros, "Mixed finite elements associated with a collection of tetrahedra, hexahedra and prisms", IEEE Trans. Magn., Vol. 30, No. 5, pp. 2980-2983, 1994.
6. P. Dular, F. Henrotte, A. Genon, W. Legros, "A generalized source magnetic field calculation method for inductors of any shape", IEEE Trans. Magn., Vol. 33, No. 2, pp. 1398-1401, 1997.
7. P. Dular, W. Legros, A. Nicolet, "Coupling of local and global quantities in various finite element formulations and its application to electrostatics, magnetostatics and magnetodynamics", IEEE Trans. Magn., Vol. 34, No. 5, pp. 3078-3081, 1998.
8. P. Dular, C. Geuzaine, W. Legros, "A natural method for coupling magnetodynamic h-formulations and circuit equations", IEEE Trans. Magn., Vol. 35, No. 3, pp. 1626-1629, 1999.

9. P. Dular, F. Henrotte, W. Legros, "A general and natural method to define circuit relations associated with magnetic vector potential formulations", IEEE Trans. Magn., Vol. 35, No. 3, pp. 1630-1633, 1999.
10. P. Dular, N. Sadowski, J.P.A. Bastos, W. Legros, "Dual complete procedures to take stranded inductors into account in magnetic vector potential formulations", IEEE Trans. Magn., Vol. 36, No. 4, pp. 1600-1605, 2000.
11. P. Dular, C. Geuzaine, GetDP manual: the documentation for GetDP, a general environment for the treatment of discrete problems, http://www.montefiore.ulg.ac.be/services/elap/elm.html, 2000.
12. D. Ioan, I. Munteanu, C.-G. Constantin, "The best approximation of the field effects in electric circuit coupled problems", IEEE Trans. Magn., Vol. 34, No. 5, pp. 3210-3213, 1998.
13. J.C. Ndlec, "Mixed finite elements in R3", Numerische Mathematik, Vol. 35, pp. 315-341, 1980.
14. Z. Ren, "Influence of the R.H.S. on the convergence behaviour of the curl-curl equation", IEEE Trans. Magn., Vol. 34, No. 5, pp. 2509-2512, 1998.
15. J.P. Webb, B. Forghani, "A single scalar potential method for 3D magnetostatics using edge elements", IEEE Trans. Magn., Vol. 25, No. 5, pp. 4126-4128, 1989.

Induction Heating of Thin Slabs in Nonmagnetic Media

P. Šolín[1], I. Doležel[2], M. Škopek, and B. Ulrych[3]

[1] Johannes Kepler University, Linz, Austria
[2] Institute of Electrical Engineering AS CR, Prague, Czech Republic
[3] Electrotechnical Faculty of the UWB, Pilsen, Czech Republic

Abstract The paper deals with the mathematical and computer modelling of the induction heating of thin non-ferromagnetic slabs in transverse electromagnetic fields. The task is formulated as a non-stationary quasi-coupled problem, with respecting the temperature dependencies of all important material parameters. Distribution of the eddy currents and Joule losses in the slab is solved by integral equations while the temperature field by means of non-stationary balance equations. The theoretical analysis is supplemented with an illustrative example.

1 Introduction

Mathematical modelling of the induction heating belongs to relatively well mapped disciplines. The basic model consists of two second-order (generally non-linear) partial differential equations of the elliptical and/or parabolic types, whose solutions yield distribution of the electromagnetic field, eddy currents, corresponding Joule losses and consequent temperature rise of the heated body. Sometimes, however, various difficulties have to be overcome in order to obtain correct results. We can mention, for example, the temperature dependent parameters of the materials involved, specific arrangements of the heaters etc. Nevertheless, in many geometries the field equations supplemented with correct boundary conditions may be solved by the existing FEM-based professional programs (FLUX, ANSYS, MARC etc.), and the results relatively well correspond with the physical reality.

In case of thin metal slabs or strips the basic complication consists in the geometrical incommensurability of particular subregions in the investigated area (heated slab versus 3D inductor and/or practically unbounded air). Using methods such as the FEM can lead to serious problems associated with generation of the discretisation grid that may result in unacceptable errors occurring particularly at the electromagnetic field calculation.

The paper offers an alternative algorithm for direct determination of the local Joule losses (representing the input data for the consequent thermal calculations) based on solution of an integral equation for the eddy current density in the slab. Detailed knowledge of the 3D electromagnetic field is, therefore, unnecessary and the above problems are avoided. The temperature field distribution is then solved by means of the non-stationary balance equations with a special procedure for respecting the temperature dependence of the material properties.

2 Description of the Technical Problem

A thin non-ferromagnetic slab Ω_1 of sizes a, b and h is heated by an inductor consisting of two coils Ω_2 and Ω_2' of any shape placed symmetrically with respect to the slab (Fig. 1). The coils carry identical harmonic currents I_{ext} and I_{ext}' with angular frequency ω. Thickness h of the slab is very small, so that the field in it is practically transversal and independent of co-ordinate y. As the inductor contains no ferromagnetic parts, all electromagnetic quantities may be expressed in terms of their phasors.

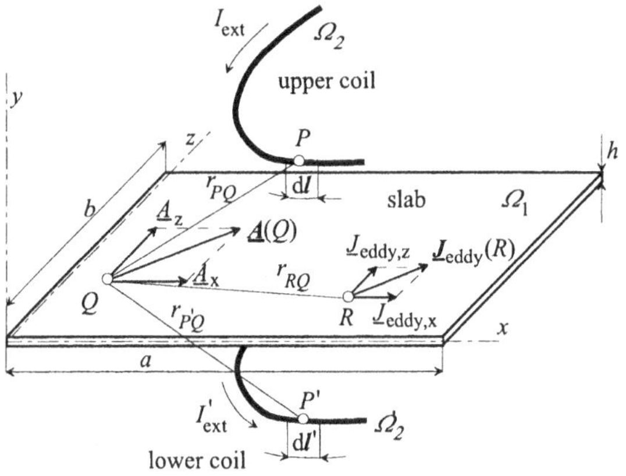

Figure1. The investigated arrangement

3 Computation of the Electromagnetic Quantities

Let Q be a point within the slab placed in plane $y = 0$. Phasor $\underline{\mathbf{A}}$ of the vector potential at this point [1] is given by superposition of three components excited by the field currents I_{ext} and I_{ext}' and the eddy currents in the slab

$$\underline{\mathbf{A}}(Q) = \underline{\mathbf{A}}(PQ) + \underline{\mathbf{A}}(P'Q) + \underline{\mathbf{A}}(RQ) =$$

$$= \frac{\mu_0}{4\pi}\left(\underline{I}_{\text{ext}}\int_{\Omega_2}\frac{\mathrm{d}\boldsymbol{l}(P)}{r_{PQ}} + \underline{I}_{\text{ext}}'\int_{\Omega_2'}\frac{\mathrm{d}\boldsymbol{l}'(P')}{r_{P'Q}} + \iint_{\Omega_1}\frac{\underline{\boldsymbol{J}}_{\text{eddy}}(R)}{r_{RQ}}h\cdot\mathrm{d}S\right) \tag{1}$$

Here, $\mathrm{d}\boldsymbol{l}$ and $\mathrm{d}\boldsymbol{l}'$ are the vectors denoting the elementary lengths of conductors of the field coils and S the area of the slab. All remaining quantities follow from Fig. 1.

Phasors $\underline{\boldsymbol{A}}(PQ)$ and $\underline{\boldsymbol{A}}(P'Q)$ have generally three components corresponding to the shape of the coils. In our case these components in direction x are equal, in direction z as well, and components in direction y eliminate one another. Equation (1) may be now rewritten as follows

$$\underline{\boldsymbol{A}}(Q) = \frac{\mu_0}{4\pi}\left(2\underline{I}_{\text{ext}}\int_{\Omega_2}\frac{\mathrm{d}\boldsymbol{l}(P)}{r_{PQ}} + h\int_{\Omega_1}\frac{\underline{\boldsymbol{J}}_{\text{eddy}}(R)}{r_{RQ}}\mathrm{d}S\right). \quad (2)$$

For next considerations it is necessary to express vector potential $\boldsymbol{A}(Q)$ in the slab in terms of the eddy current density $\boldsymbol{J}_{\text{eddy}}(Q)$. Starting from the second Maxwell equation

$$\text{rot}\boldsymbol{E} = -\frac{\partial \boldsymbol{B}}{\partial t} = -\frac{\partial \text{rot}\boldsymbol{A}}{\partial t} \quad (3)$$

and interchanging the order of the operators we get

$$\boldsymbol{E} = -\frac{\partial \boldsymbol{A}}{\partial t} - \text{grad}\varphi, \quad (4)$$

where φ denotes the scalar potential. Applying this equation to the slab that is not connected to any external source of voltage ($\varphi = 0$) and rewriting it in terms of the corresponding phasor quantities we obtain

$$\underline{\boldsymbol{E}} = -\mathrm{j}\cdot\omega\underline{\boldsymbol{A}} \Rightarrow \underline{\boldsymbol{J}}_{\text{eddy}} = -\mathrm{j}\cdot\omega\gamma\underline{\boldsymbol{A}}, \quad (5)$$

where γ denotes the electrical conductivity of the slab. Substitution of (5) (for point Q) into (2) provides the basic integral equation for $\underline{\boldsymbol{J}}_{\text{eddy}}$

$$\mathrm{j}\cdot\underline{\boldsymbol{J}}_{\text{eddy}}(Q) - \kappa_1\cdot\iint_{\Omega_1}\frac{\underline{\boldsymbol{J}}_{\text{eddy}}(R)}{r_{RQ}}\mathrm{d}S = \kappa_2\cdot\underline{I}_{\text{ext}}\int_{\Omega_2}\frac{\mathrm{d}\boldsymbol{l}(P)}{r_{PQ}}, \quad (6)$$

where

$$\kappa_1 = \frac{\omega\gamma\mu_0 h}{4\pi}, \qquad \kappa_2 = \frac{\omega\gamma\mu_0}{2\pi}.$$

This phasor equation may easily be subdivided into two equations (for the components in directions x and z) of the complex character whose solvability is discussed in the appendix.

The continuous model is then discretised [2]. We used a collocation scheme with by part constant approximation of the solution on quadrilateral elements. The discretised integrals were calculated accurately including 2D integration of function $1/r$ in the neighbourhood of singular points. In this manner we obtained a system of linear equations for the eddy current densities in particular elements. The specific average Joule losses w_{Ja} in an element of the slab are then given by formula

$$w_{\text{Ja}} = \frac{\underline{\boldsymbol{J}}_{\text{eddy}}\cdot\underline{\boldsymbol{J}}^*_{\text{eddy}}}{\gamma}, \quad (7)$$

where $\underline{\boldsymbol{J}}^*_{\text{eddy}}$ is the complex conjugate to $\underline{\boldsymbol{J}}_{\text{eddy}}$.

4 Thermal Calculation

The non-stationary distribution of the temperature in the slab is generally described (for example [3]) by equation

$$\mathrm{div}(\lambda \,\mathrm{grad}\, T) = \rho c \cdot \frac{\partial T}{\partial t} - w_{\mathrm{Ja}}, \tag{8}$$

where λ denotes the thermal conductivity, ρ the specific mass of the heated material, c its specific heat and w_{Ja} the specific Joule losses given by (7). The boundary condition along the whole surface of the slab reads (radiation is not considered)

$$-\lambda \cdot \frac{\partial T}{\partial n} = \alpha \cdot (T - T_{\mathrm{ext}}),$$

where α denotes the coefficient of the convective heat transfer, T_{ext} the temperature of the surrounding medium (air) and n direction of the outward normal. Solution to (8) is performed by means of the discretised balance equations. Such an equation for any internal element of the grid reads

$$\frac{\partial}{\partial x}\left(\lambda \frac{\partial T}{\partial x}\right) + \frac{\partial}{\partial z}\left(\lambda \frac{\partial T}{\partial z}\right) = \rho c \cdot \frac{\partial T}{\partial t} - w_{\mathrm{Ja}} + \frac{2\alpha}{h} \cdot (T - T_{\mathrm{ext}}). \tag{9}$$

Similar formulas can be derived even for the edge and corner elements. Numerical calculation of the temperature field was realised by combination of the FDM in space and explicit Euler method in time. Its stability was secured by satisfying the conditions for stability of the explicit time integration of the parabolic equations. At present, this procedure is being improved by implementation of more efficient algorithms.

5 Algorithm of Solution

Numerical processing of the task consists of several steps specified as follows:

1. Selection of the initial distribution of temperature T_0 of the slab.
2. Determination of the corresponding distributions of its electrical and thermal conductivities γ and λ. Computation of distribution of the eddy currents and specific Joule losses in the slab.
3. The non-stationary temperature equation is solved numerically with the Joule losses that are considered constant unless the maximum temperature change ΔT in the slab exceeds some prescribed parameter D_TEMP or the time interval exceeds another parameter D_TIME. Parameters ρ and c vary with the temperature.
4. Return to point 2. The cycle described by points 2.–4. is repeated until the temperature change within interval D_TIME in no element exceeds prescribed tolerance TEMP_TOLL. At this moment the process stops.

The level of coupling is influenced by selection of parameter D_TEMP (quasi-coupling - D_TEMP> 0, hard-coupling - D_TEMP→ 0). As the computation of the integral equations takes much more time than calculation of the temperature evolution, the choice of parameter D_TEMP depends on how this system is large.

We tested a number of bi-conjugate gradient methods with the ILU preconditioning [4]. Their behaviour was very similar. The most important factor for their convergence is not the temperature field, but analytically calculated coefficients obtained from integration of function $1/r$.

The above algorithm represents the basis of a special user program package developed by the authors and written in C++. Its latest version allows computation of arrangements with quite arbitrary inductors (their shapes are parametrised) that can also move.

6 Illustrative Example

The suggested methodology has been used for solution of a task depicted in Fig. 2. A copper slab of sizes 0.3×0.6 m was heated by two helicoidal inductors (number of turns $N = 10$) of the given geometry.

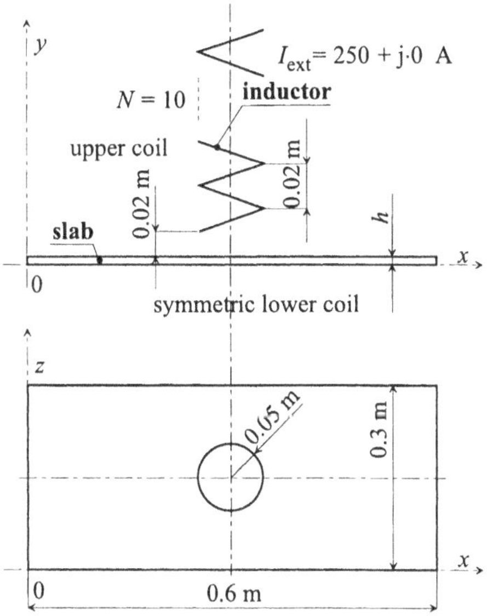

Figure2. The investigated arrangement

The dependencies of parameters γ and λ on temperature T [5] are shown in Fig. 3. The temperature dependence of product ρc is considered linear: $\rho c = 3.6312 \cdot 10^6 + 934.5 \cdot (T - 20)$ J/deg m^3. Parameter $\alpha = 25$ W/m^2 deg and $T_{ext} = 20$ °C. The discretisation grid for both electromagnetic and thermal calculations had $30 \times 60 = 1800$ square elements by 1 cm^2 (7200 component equations). Computations were carried out on SGI Origin 2000 (in this version without parallelization that is only developed) and took several hours (D_TEMP being 5 °C).

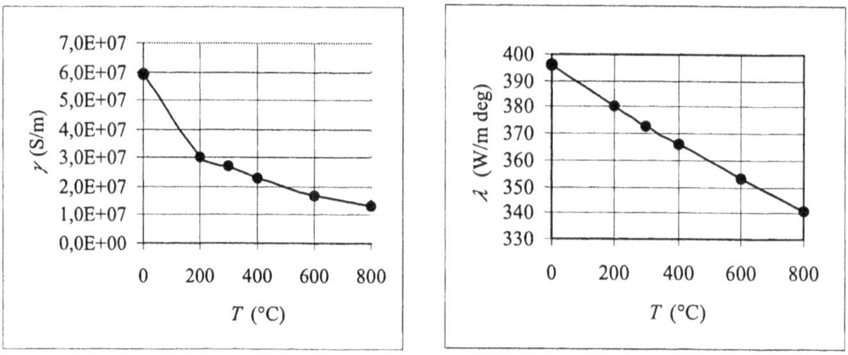

Figure3. Temperature dependencies of γ and λ for copper

Several results for parameters $f = 1000\,\text{Hz}$ and $h = 0.0005\,\text{m}$ are shown in Figs. 4–6. Fig. 4 depicts the distribution of the temperature (at various time levels) in the slab along line $z = 0.05\,\text{m}$ and Fig. 5 the same distribution along line $z = 0.15\,\text{m}$. Fig. 6 shows the time evolution of temperature T at the "hottest" point of the slab (co-ordinates 0.25 m and $z = 0.15\,\text{m}$).

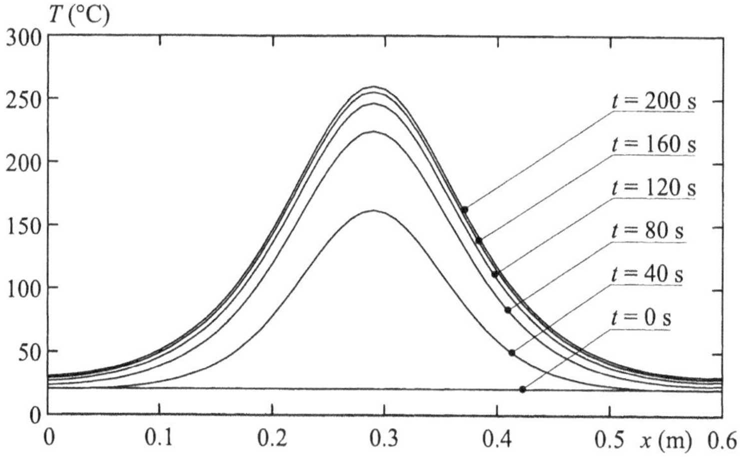

Figure4. Distribution of the temperature along line $z = 0.05\,\text{m}$

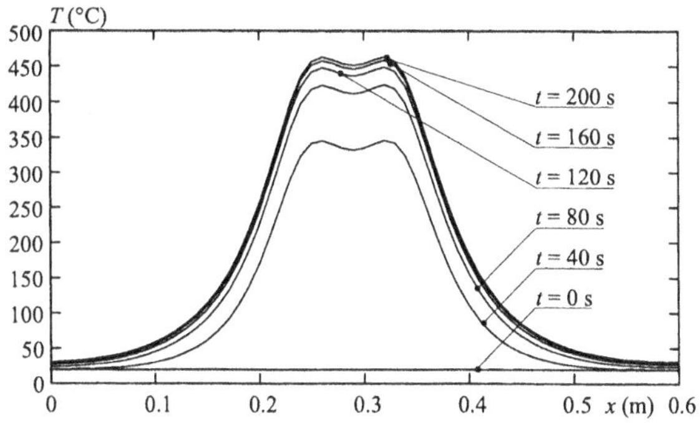

Figure5. Distribution of the temperature along line $z = 0.15\,\mathrm{m}$

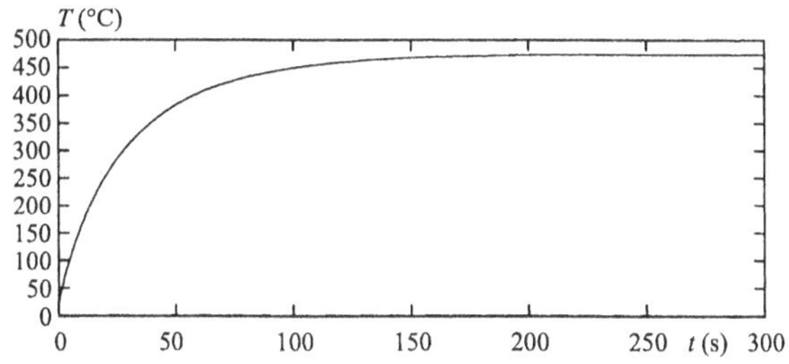

Figure6. Evolution of the temperature at point $x = 0.25\,\mathrm{m}$, $z = 0.15\,\mathrm{m}$

Appendix: Solvability of Integral Equation (6)

Let us analyse equation (6) for the x-components of particular phasors. Its reformulation into the operator form yields

$$\boldsymbol{L}_{\mathrm{x}} - \boldsymbol{K}\boldsymbol{L}_{\mathrm{x}} = \boldsymbol{F}_{\mathrm{x}}, \tag{10}$$

where

$$\boldsymbol{L}_{\mathrm{x}}(\mathbf{v}) = (-\mathrm{Im}\{\underline{J}_{\mathrm{eddy,x}}\}, \mathrm{Re}\{\underline{J}_{\mathrm{eddy,x}}\})^T(\mathbf{v}).$$

Operator

$$(\boldsymbol{K}\boldsymbol{L}_{\mathrm{x}})(\mathbf{v}) = \iint_S \boldsymbol{k}(\mathbf{v}, \mathbf{w})\boldsymbol{L}_{\mathrm{x}}(\mathbf{w})\mathrm{d}\mathbf{w} \tag{11}$$

is given by kernel function

$$\boldsymbol{k}(\mathbf{v}, \mathbf{w}) = \frac{\kappa_1}{|\mathbf{v} - \mathbf{w}|} \cdot \boldsymbol{M}, \quad \boldsymbol{M} = \begin{pmatrix} 0 & 1 \\ -1 & 0 \end{pmatrix},$$

and the right-hand side function has the form

$$\boldsymbol{F}_{\mathrm{x}}(\mathbf{v}) = \kappa_2 \int_{\Omega_2} \frac{\mathrm{d}x}{|\mathbf{u} - \mathbf{v}|} \cdot (\mathrm{Re}\{\underline{I}_{\mathrm{ext}}\}, \mathrm{Im}\{\underline{I}_{\mathrm{ext}}\})^T.$$

Here $\mathbf{v}$, $\mathbf{w} \in S$, $\mathbf{u} \in \Omega_2$.

As is usual for this class of problems, the Fredholm alternative will be used. The first step is to verify the compactness of operator $\boldsymbol{K}$ (or, in our case, to find suitable function spaces where operator $\boldsymbol{K}$ is compact). Function $\boldsymbol{k}(\mathbf{v},\mathbf{w})$: $S \times S \setminus \{(\mathbf{w},\mathbf{w});\ \mathbf{w} \in S\} \rightarrow \mathbb{R}^2 \times \mathbb{R}^2$ can be written as

$$\boldsymbol{k}(\mathbf{v}, \mathbf{w}) = \frac{\kappa_1}{|\mathbf{v} - \mathbf{w}|^{\alpha}} \cdot \boldsymbol{M},$$

with $\alpha = 1$. As

$$||\boldsymbol{k}(\mathbf{v}, \mathbf{w})|| \leq m|\mathbf{v} - \mathbf{w}|^{\alpha - D}$$

where $m = \sqrt{2}\kappa_1$ and $D = 2$ is the space dimension, we can say that $\boldsymbol{k}(\mathbf{v},\mathbf{w})$ is weakly singular ([6], Def. 2.12). Therefore, operator $\boldsymbol{K}$ is defined as $\boldsymbol{K}$: $C(S) \rightarrow C(S)$ and compact ([6], Th. 2.13).

As operator $\boldsymbol{K}$ is not defined on $L^2(S)$, we cannot directly use the duality properties of this space to satisfy the requirements of the Fredholm alternative. Nevertheless, as $C(S)$ is continuously embedded to $L^2(S)$ (i.e. identity $\boldsymbol{I}$:$C(S) \rightarrow L^2(S)$ is continuous and $\boldsymbol{I}(C(S))$ is dense in $L^2(S)$ - [6], Ex. 2.23c, Th. 2.26), we can define a dual system $(C(S), C(S))$ with

$$(\psi_1, \psi_2) = \iint_S \psi_1(\mathbf{v})\psi_2(\mathbf{v})\mathrm{d}\mathbf{v}, \qquad \psi_1, \psi_2 \in C(S).$$

Then, as $\boldsymbol{k}(\mathbf{v},\mathbf{w}) = \boldsymbol{k}(\mathbf{w},\mathbf{v})$ for all $\mathbf{v}$, $\mathbf{w} \in S$, $\mathbf{w} \neq \mathbf{v}$, we can say that operator $\boldsymbol{K}$ is self-adjoint in this dual system.

Finally, using the fact that homogeneous equation

$$\boldsymbol{L}_{\mathrm{x}} - \boldsymbol{K}\boldsymbol{L}_{\mathrm{x}} = 0$$

has only trivial solution, we may conclude that for each $\boldsymbol{F}_{\mathrm{x}} \in C(S)$ equation (10) has a unique solution $\boldsymbol{L}_{\mathrm{x}} \in C(S)$. An analogous conclusion holds, of course, for equation (10) written for $\boldsymbol{L}_{\mathrm{z}}$.

References

1. Stratton, J.: *Electromagnetic Theory* Mc Graw Hill Book Co., NY, 1941.
2. Šolín, P., Doležel, I., Škopek, M., Ulrych, B.: *Stationary Temperature Field in a Non-Magnetic Thin Plate Heated by Transversal Electromagnetic Field.* Acta Technica CSAV **45** (2000) 105–128.
3. Schneider, P. J.: *Conduction Heat Transfer.* Addison-Wesley, Pub. Comp., Cambridge, 1955.
4. Thomee, V.: *Galerkin Finite Element Methods for Parabolic Problems.* Berlin, Springer, 1997.
5. Davies, P. G.: *Conduction and Induction Heating.* Peter Pelegrinus Ltd., London, 1990.
6. Engl, H. W.: *Integralgleichungen.* Springer-Verlag, Wien (1997), in German.

Krylov Subspace Methods for Harmonic Balanced Finite Element Methods

H. De Gersem[1*], Stefan Vandewalle[2], and Kay Hameyer[1]

[1] Katholieke Universiteit Leuven, Dep. EE (ESAT) / Div. ELEN
Kardinaal Mercierlaan 94, B-3001 Leuven, Belgium
[2] Katholieke Universiteit Leuven, Dep. Computer Science
Celestijnenlaan 200A, B-3001 Leuven, Belgium

Abstract A Conjugate Gradient solver is developed for harmonic balanced finite element systems. This approach avoids the construction of real equivalent systems which suffer from worse spectral conditions for Krylov subspace solvers. The application to a transformer with ferromagnetic material shows the better convergence of the Conjugate Gradient solver.

1 Introduction

Alternating current electrical energy transducers in steady-state regime are commonly simulated by linearised time-harmonic finite element (FE) models. As phenomena at higher harmonics due to power electronic supply or ferromagnetic saturation have an increasing technical importance, transient FE or harmonic balanced FE (HBFEM) simulation is becoming more and more important. In the applications considered here, periodic steady-state operations are simulated. Since only a few, a priori known frequencies are technically relevant, multi-harmonic simulation is favoured over time stepping.

The use of HBFEM is not a very common practice, mainly because of the expensive solution of the corresponding matrix system. Previously published research and existing implementations rely upon a weakly coupled time-harmonic scheme [1] or a real-valued splitting of the system matrix [2]. Here, a different, strongly coupled approach without splitting is presented. The emphasis is on the development of appropriate Krylov subspace solvers for the particular systems of equations arising from this approach.

2 Finite Discrete Spectra

Consider a periodic signal $a(t)$ and its corresponding discrete spectrum $\underline{a}(\omega)$ with ω the pulsation. The period is t_1 and hence, the lowest participating

* The authors are grateful to the Belgian "Fonds voor Wetenschappelijk Onderzoek Vlaanderen" (G.0427.98) and the Belgian Ministry of Scientific Research (IUAP No. P4/20) for the financial support of this work.

frequency, besides the DC-component, is $f_1 = 1/t_1$. All other participating frequencies are multiples of f_1, i.e. $f_k = kf_1, k \in \mathbb{N}$. The corresponding pulsations are $\omega_k = 2\pi f_k$. The *harmonic component* of *harmonic order* k,

$$\underline{a}|_k = \underline{a}(\omega_k) = \frac{2}{T}\int_0^T a(t)e^{\jmath\omega_k t}dt, \tag{1}$$

represents the magnitude of the discrete spectrum at pulsation ω_k (Fig. 1). *Even* and *odd harmonics* are distinguished by their even and odd orders respectively.

For particular applications, only a finite set of harmonics is present or technically relevant. The set is called a *harmonic pattern* and is denoted by $P = (p_1, p_2, \cdots)$ with $p_1, p_2, \cdots$ positive numbers indicating the orders of the considered harmonics. The *finite discrete spectrum* is denoted by $\underline{a} = (\underline{a}|_{p_1}, \underline{a}|_{p_2}, \cdots)$. This notation only gathers the complex coefficients corresponding to positive orders. Because $\underline{a}|_{-p} = \overline{\underline{a}|_p}$, only those have to be computed. All spectra $\underline{a}, \underline{b}, \cdots$ based on the same harmonic pattern P form a Hilbert space G_P, similar to $\mathbb{C}^{\#P}$ endowed with the inner product

$$(\underline{a}, \underline{b})_{G_P} = \sum_{p \in \pm P} \overline{\underline{a}}|_p \underline{b}|_p, \tag{2}$$

corresponding to the DC component of the convolution of $\underline{a}$ and $\underline{b}$, and the norm

$$\|\underline{a}\|_{G_P} = \sqrt{(\underline{a}, \underline{a})_{G_P}}. \tag{3}$$

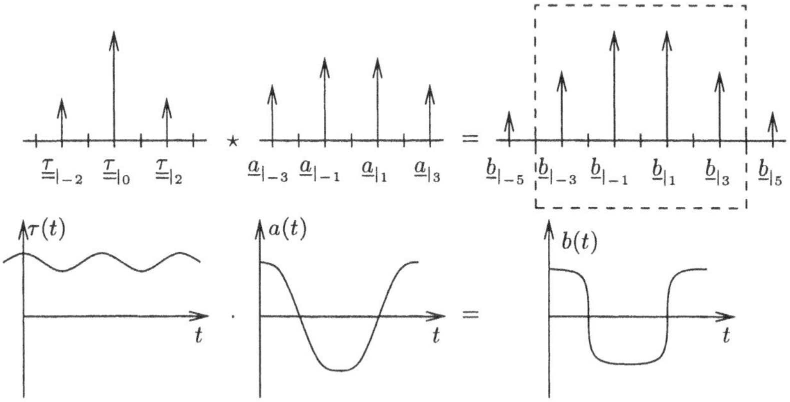

Figure1. Truncated convolution $\underline{\underline{T}} \star \underline{a} = \underline{b}$.

3 Ferromagnetic Saturation

In the design of electrical energy transducers, ferromagnetic materials are applied because of their high permeabilities. The unwanted influence of saturation is kept acceptable by operating these materials at modest levels of saturation. As a consequence, the magnitudes of the field harmonics decrease considerably when the frequency increases. The discretisation of the frequency domain is straightforward:

- the periodicity of the excitation and the assumption that all materials are in steady-state behaviour, allows to consider only discrete spectra;
- technical considerations guide the selection of the harmonics to which the simulation is restricted.

The determination of the harmonic pattern can be done a priori and adaptively refined during the simulation. All excitating signals only contain odd harmonics. At higher amplitudes, saturation occurs and an infinite set of even harmonics is introduced in the pattern of the permeability. These excite additional harmonics in the magnetic field. In all examples here, the field quantities have the odd pattern $P = (1, 3, \cdots, 2n_h - 1)$ whereas the materials have the even pattern $Q = (0, 2, \cdots, 2n_h - 2)$, each of them considering n_h harmonic components. Moreover, as the material is energy conserving, its harmonic components are in phase and can be represented by real values. Spectra of field quantities are denoted by underlined, small roman characters whereas material spectra are represented by Greek symbols with double underlining.

4 Truncated Convolution

The convolution of a Q-patterned material coefficient with a P-patterned field quantity yields a field quantity with additional harmonic components. Successive convolutions increase the set of harmonics present in the field quantities. Hence, in practice, a truncated convolution is applied (Fig. 1):

$$\begin{array}{lll} \star : G_Q \times G_P \to G_P : \underline{\underline{\tau}} \star \underline{a} = (\underline{b}|_k) & & \\ \textit{with } \underline{b}|_k = \sum_{p \in \pm P; q \in \pm Q; p+q=k} \underline{\underline{\tau}}|_q \underline{u}|_p & \text{if} \quad k \in P; & \\ \underline{b}|_k = 0 & \text{if} \quad k \notin P. & \end{array} \tag{4}$$

Due to energy considerations, the truncated convolution of a material coefficient is a hermitian and positive definite operator with respect to $(\cdot,\cdot)_{G_P}$, i.e. $\forall \underline{a}, \underline{b} \in G_P$:

$$\left(\underline{\underline{\tau}} \star \underline{a}, \underline{b}\right)_{G_P} = \left(\underline{a}, \underline{\underline{\tau}} \star \underline{b}\right)_{G_P} ; \tag{5}$$

$$\left(\underline{\underline{\tau}} \star \underline{a}, \underline{a}\right)_{G_P} \geq 0; \tag{6}$$

$$\left(\underline{\underline{\tau}} \star \underline{a}, \underline{a}\right)_{G_P} = 0 \Leftrightarrow \underline{a} = 0. \tag{7}$$

Consider e.g. $P = (1,3)$ and $Q = (0,2)$. Several matrix representations of the truncated convolution $\underline{\underline{\tau}} \star \underline{a} = \underline{b}$ are:

$$1. \quad \begin{bmatrix} \tau_0 + \tau_2 \mathrm{Cg} & \tau_2 \\ \tau_2 & \tau_0 \end{bmatrix} \begin{bmatrix} \underline{a}_{|1} \\ \underline{a}_{|3} \end{bmatrix} = \begin{bmatrix} \underline{b}_{|1} \\ \underline{b}_{|3} \end{bmatrix}; \tag{8}$$

$$2. \quad \begin{bmatrix} \tau_0 & \tau_2 & & \\ \tau_2 & \tau_0 & \tau_2 & \\ & \tau_2 & \tau_0 & \tau_2 \\ & & \tau_2 & \tau_0 \end{bmatrix} \begin{bmatrix} \underline{a}_{|3} \\ \underline{a}_{|1} \\ \underline{a}_{|-1} \\ \underline{a}_{|-3} \end{bmatrix} = \begin{bmatrix} \underline{b}_{|3} \\ \underline{b}_{|1} \\ \underline{b}_{|-1} \\ \underline{b}_{|-3} \end{bmatrix}; \tag{9}$$

$$3. \quad \begin{bmatrix} \tau_0 & \tau_2 & & \\ \tau_2 & \tau_0 + \tau_2 & & \\ & & \tau_0 - \tau_2 & \tau_2 \\ & & \tau_2 & \tau_0 \end{bmatrix} \begin{bmatrix} \mathrm{Re}\ \underline{a}_{|3} \\ \mathrm{Re}\ \underline{a}_{|1} \\ \mathrm{Im}\ \underline{a}_{|1} \\ \mathrm{Im}\ \underline{a}_{|3} \end{bmatrix} = \begin{bmatrix} \mathrm{Re}\ \underline{b}_{|3} \\ \mathrm{Re}\ \underline{b}_{|1} \\ \mathrm{Im}\ \underline{b}_{|1} \\ \mathrm{Im}\ \underline{b}_{|3} \end{bmatrix} \tag{10}$$

where Cg, Re and Im are operators selecting the conjugate, real or imaginary parts of their arguments. Considering appropriate isomorphisms inbetween G_P and the vector spaces of the real equivalents, it is straightforward to prove (5), (6) and (7) using matrix calculus. The third notation is implemented in [2]. The second one is inefficient because of the combination of the double length and the complex type of the solution vectors. All three notations store superfluous copies of the material harmonics. In the approach presented here, the original formulation $\underline{\underline{\tau}} \star \underline{a} = \underline{b}$ is maintained. $\underline{a}$, $\underline{b}$ and $\underline{\underline{\tau}}$ are represented by appropriate software objects and hence avoiding splitting and additional copies. Notice that, in general, there is no $\underline{\underline{\tau}}^{-1}$ in G_Q such that

$$\underline{\underline{\tau}} \star \underline{a} = \underline{b} \Leftrightarrow \underline{\underline{\tau}}^{-1} \star \underline{b} = \underline{a}. \tag{11}$$

5 Harmonic Balanced Finite Element Method

The governing partial differential equation (PDE) is

$$\nabla \times (\nu \nabla \times \mathbf{A}) = \mathbf{J} \tag{12}$$

with $\mathbf{A}$ the magnetic vector potential, ν the reluctivity and $\mathbf{J}$ the current density. The PDE, transferred to the frequency domain, is

$$\nabla \times \left(\underline{\underline{\nu}} \star \nabla \times \underline{\mathbf{A}}\right) = \underline{\mathbf{J}}. \tag{13}$$

The time scale t is transformed into the pulsation scale ω. Even without parabolic term, this PDE captures the time-varying effects due to the alternating excitation and the reluctivity being non-linear and hence excitating higher harmonic components [3]. The geometry and excitation $\underline{\mathbf{J}} = (0, 0, \underline{J}_z)$ of the models, developed here, allow for a 2D nodal FE discretisation of the z-component of the magnetic vector potential $\underline{A}_z$ by n_f linear triangular FEs N_i based upon a triangulation of the computational domain Ω. The resulting equation is

$$\mathcal{K}\underline{x} = \underline{f}, \tag{14}$$

expressing the relations

$$\sum_{j=1}^{n_f} \underline{\underline{k}}_{ij} \star \underline{x}_j = \underline{f}_i, \qquad i = 1, \cdots, n_f, \tag{15}$$

with

$$\underline{\underline{k}}_{ij} = \int_{\Omega} \underline{\underline{\nu}} \nabla N_i \cdot \nabla N_j d\Omega; \tag{16}$$

$$\underline{f}_i = \int_{\Omega} \underline{J}_z N_i d\Omega; \tag{17}$$

$$\underline{x}_j = \underline{A}_{zj}. \tag{18}$$

$\underline{x}$ and $\underline{f}$ are elements of $G_P^{n_f \times 1}$. $\mathcal{K}$ is an operator mapping $G_P^{n_f \times 1}$ into $G_P^{n_f \times 1}$. In practice, $\mathcal{K}$ is represented by an $n_f \times n_f$ sparse matrix of coefficients $\underline{\underline{k}}_{ij} \in G_Q$. The definitions of the inner product and norm for G_P extend in the classical way to $G_P^{n_f \times 1}$:

$$(\underline{u}, \underline{v})_{G_P^{n_f \times 1}} = \sum_{i=1}^{n_f} (\underline{u}_i, \underline{v}_i)_{G_P}; \tag{19}$$

$$\|\underline{u}\|^2_{G_P^{n_f \times 1}} = \sum_{i=1}^{n_f} \|\underline{u}_i\|^2_{G_P}. \tag{20}$$

6 Conjugate Gradients for HBFEM

The operator $\mathcal{K}$ is hermitian and positive definite with respect to (19). This is easily proved using (5), (6) and (7) together with the properties of the discretised diffusion operator. The method of choice for solving such a system is Conjugate Gradients (CG) [4]. The symmetry enables the application of the Lanczos algorithm to construct the Krylov subspace

$$\mathrm{K}_m (\underline{r}_0, \mathcal{K}) = \mathrm{span} \left\{ \underline{r}_0, \mathcal{K}\underline{r}_0, \cdots, \mathcal{K}^{m-1}\underline{r}_0 \right\} \tag{21}$$

by a three-terms recurrence relation

$$\mathcal{K}\underline{V}_m = \underline{V}_{m+1}\widetilde{T}. \tag{22}$$

$\underline{V}_m \in G_P^{n_f \times m}$ collects the base vectors of $\mathrm{K}_m (\underline{r}_0, \mathcal{K})$. $\widetilde{T} \in \mathbb{C}^{(m+1)\times m}$ is a tridiagonal matrix of recurrence coefficients. Its type is complex because the inner product (2) maps two finite discrete spectra onto $\mathbb{C}$. The positive definite $\mathcal{K}$ defines the norm

$$\|\underline{v}\|_{\mathcal{K}} = (\mathcal{K}\underline{v}, \underline{v})_{G_P^{n_f \times 1}} \tag{23}$$

with respect to which a new CG-approximant for $\underline{x}$ is computed. The classical CG algorithm is endowed with the operator $\mathcal{K}$ and the inner product (19).

The convergence of Krylov subspace solvers is poor unless preconditioning is applied to enhance the spectral properties of the system [4]. A considerable improvement is achieved by diagonal scaling. Explicit scaling is only possible with scalars as, in general, $\underline{\underline{k}}_{ii}^{-1}$, and hence $\underline{\underline{k}}_{ii}^{-1} \star \underline{\underline{k}}_{ij} \star \underline{\underline{k}}_{jj}^{-1}$ do not exist within G_Q. Implicit scaling, i.e. solving $\underline{\underline{k}}_{ii} \star \underline{z}_i = \underline{r}_i, \quad i = 1, \cdots, n_f$ within the CG process, is efficiently performed by a modified Levinson algorithm [5]. This techniques are extended to a Symmetric Successive Overrelaxation (SSOR) preconditioner as well.

7 Implementation

Preconditioned CG involves 1 preconditioning step, 1 matrix-vector convolution, a few vector inner products and a few vector updates. These operations are considerably more expensive when compared to static, time-harmonic or transient simulation because of the convolutions. An efficient implementation is not straightforward: convolutions of relatively small spectra do not make optimal use of the available cache memory. Computer architectures are more adapted to vector products than convolutions. The implementation of the matrix-vector convolution as such, results in bad performing code. The implementation as a convolution of matrix-vector products, is 10 times faster (Table 1). The same is a fortiori true for the preconditioning step. The implementation of e.g. SSOR in such a way, is however much more cumbersome. Notice that reversing the order of processing only affects the implementation and does not change the system that is solved.

8 Real Equivalent Systems

The matrix elements, split up in their real equivalents (10), form a real-valued system of dimension $2n_f n_h$. The spectrum is real and positive but counts twice as much eigenvalues when compared to the original system. Moreover, as it includes, and hence embraces, the original spectrum, it is less favourable for Krylov subspace solvers. A worse convergence of CG is observed.

9 Application

A single-phase transformer, featuring a ferromagnetic, laminated iron core, is excited by pure sinusoidal currents at the frequency f_1. Hysteresis effects are neglected. The FE simulation of the magnetic field applies adaptive mesh refinement to achieve a prescribed accuracy. The a-posteriori error estimator applies the stored magnetic energy and the curvature of the magnetic flux lines per element as indicators for the refinement. The non-linearity of (14) is resolved by a Picard iteration. Inbetween the successive steps k, adaptive underrelaxation is indispensable to ensure global convergence. The new iterate is

$$\underline{x}^{(k+1)} = \gamma\underline{\tilde{x}} + (1-\gamma)\underline{x}^{(k)} \tag{24}$$

where $\underline{\tilde{x}}$ is the solution of $\mathcal{K}^{(k)}\underline{\tilde{x}} = \underline{f}^{(k)}$ and the underrelaxation factor γ is chosen from $1, 0.5, 0.25, 0.125, 0.625$ minimising

$$\|\underline{f}^{(k+1)} - \mathcal{K}^{(k+1)}\underline{x}^{(k+1)}\|_{G_P^{n_f \times 1}}. \tag{25}$$

The updated spectrum $\underline{\underline{\nu}}^{(k+1)}$ follows from the application of the material characteristic in the time domain.

The solution is plotted for each harmonic component of the magnetic flux density separately (Fig. 3). Significant saturation and hence induction of higher harmonic fields, occur at the corners of the laminations. The convergences of CG applied to the original system and to its real equivalent, corresponding to the second refinement step and the third Picard step, are compared in Table 1 and Fig. 2. The mesh has 2201 nodes. 8 harmonic components are considered. CG applied to the system scaled by a scalar is denoted by CG*. The better spectral conditions of the original system is reflected in the smaller numbers of iteration steps and in the fact that CG still reaches convergence where CG applied to the real equivalent stagnates. The considerably more expensive matrix-vector multiplication and SSOR preconditioning in the case of CG for HBFEM causes these advantages not always to result in smaller computation times.

Table1. Timings (s) and iterations counts (between brackets) for the straightforward (str) and enhanced (enh) implementations of (SSOR)CG(*) applied to the HBFEM system of equations and its real equivalent system (real eq).

solver	HBFEM(str)	HBFEM(enh)	real eq
CG	318.73 (4841)	32.55 (4841)	-
CG*	58.12 (841)	5.45 (839)	-
SSORCG*	100.85 (291)	10.90 (287)	6.45 (327)

10 Conclusions

The system corresponding to a harmonic balanced finite element discretisation of a diffusion equation is positive definite and symmetric with respect to the proper inner product of the vector space of finite discrete spectra. Hence, the Conjugate Gradient method is applicable. This approach outperforms the solution of real equivalent systems because of their worse spectra. Attention has to be paid to an efficient implementation of the matrix-vector convolution and the preconditioning. The application to a single-phase transformer with ferromagnetic materials illustrates the benefits of this approach.

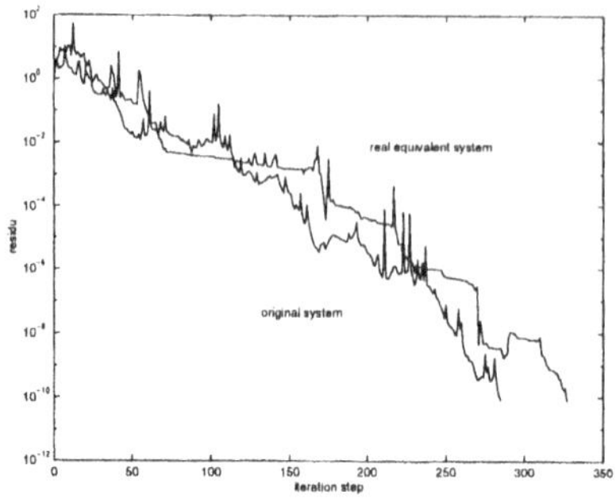

Figure2. Comparison of the convergence of SSORCG applied to the original and the real equivalent systems.

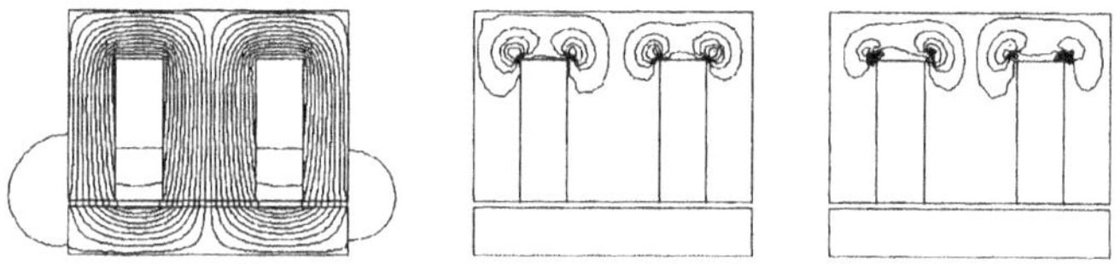

Figure3. 1st, 3rd and 5th harmonic of the magnetic flux density in a single-phase transformer.

References

1. J. Driesen and G. Deliége and Van Craenenbroeck, T. and K. Hameyer, Implementation of the harmonic balance FEM method for large-scale saturated electromagnetic devices, Software for Electrical Engineering and Analysis and Design IV, WIT Press 1999 pages:75-84
2. S. Yamada and K. Bessho and J. Lu, Harmonic balance finite element method applied to nonlinear AC magnetic analysis, IEEEmagn 1989 Vol.25 No.4 pages:2971-2973
3. L. Vandevelde and J. Gyselinck, J. Melkebeek, Steady-state finite element analysis in the frequency domain of inverter-fed squirrel cage induction motors, Proceedings of the Symposium on Power Electronics, Electrical Drives, Advanced Electrical Motors (SPEEDAM94), Proceedings of SPEEDAM94 1994 pages:29-34
4. Y. Saad, Iterative Methods for Sparse Linear Systems, PWS Publishing Company Boston 1996
5. G.H. Golub and Van Loan, C.F., Matrix Computations, The John Hopkins University Press Baltimore 1989

11 Reserve

The governing partial differential equation (PDE) is

$$\nabla \times (\nu \nabla \times \mathbf{A}) + \sigma \frac{\partial \mathbf{A}}{\partial t} = -\sigma \nabla \phi \tag{26}$$

with $\mathbf{A}$ the magnetic vector potential, ϕ the electric voltage, ν the reluctivity and σ the conductivity. The PDE, transferred to the frequency domain, is

$$\nabla \times \left(\underline{\underline{\nu}} \star \nabla \times \underline{\mathbf{A}}\right) + \underline{\underline{\sigma}} \star \jmath\omega \underline{\mathbf{A}} = -\underline{\underline{\sigma}} \star \nabla \underline{\phi}. \tag{27}$$

The time scale t is transformed into the pulsation scale ω. The geometry and excitation $\underline{\mathbf{J}} = (0, 0, \underline{J}_z)$ of the models, developed here, allow for a 2D nodal FE discretisation in term of n_f linear triangular FEs N_i for the z-component of the magnetic vector potential $\underline{A}_z$. The resulting equation is

$$\underline{\mathcal{K}\S} = \underline{\{} \tag{28}$$

and expresses the relations

$$\sum_{j=1}^{n_f} \underline{\underline{k}}_{ij} \star \underline{x}_j, i = 1, \cdots, n_f \tag{29}$$

with

$$\underline{\underline{k}}_{ij} = \int_{\Omega} \underline{\underline{\nu}} \nabla N_i \cdot \nabla N_j d\Omega; \tag{30}$$

$$\underline{f}_i = \int_{\Omega} \underline{J}_z N_i d\Omega; \tag{31}$$

$$\underline{x}_j = \underline{A}_{zj}. \tag{32}$$

$\underline{x}$ and $\underline{f}$ are elements of $G_P^{n_f \times 1}$. $\mathcal{K}$ is an operator mapping $G_P^{n_f \times 1}$ onto $G_P^{n_f \times 1}$. In practice, $\mathcal{K}$ is represented by an $n_f \times n_f$ sparse matrix of coefficients $\underline{\underline{k}}_{ij} \in G_Q$. The definitions of the inner product and norm for G_P extend in the classical way to $G_P^{n_f \times 1}$:

$$(\underline{u}, \underline{v})_{G_P^{n_f \times 1}} = \sum_{i=1}^{n_f} (\underline{u}_i, \underline{v}_i)_{G_P}; \tag{33}$$

$$||\underline{u}||_{G_P^{n_f \times 1}} = \sum_{i=1}^{n_f} ||\underline{u}_i||_{G_P}. \tag{34}$$

The time scale t is transformed into the pulsation scale ω. The geometry and excitation of the models, developed here, allow for a 2D nodal FE discretisation in term of n_f linear triangular FEs for the z-component of the magnetic vector potential. The resulting equation is

$$(\mathcal{K} + \mathcal{L})\, \underline{x} = \underline{f} \tag{35}$$

with

$$\mathcal{K}_{ij} = \int_{\Omega} \underline{\underline{\nu}} \nabla N_i \cdot \nabla N_j d\Omega; \tag{36}$$

$$\mathcal{L}_{ij} = \int_{\Omega} \underline{\underline{\sigma}} \star \jmath\omega N_i N_j d\Omega; \tag{37}$$

$$\underline{f}_i = - \int_{\Omega} \underline{\underline{\sigma}} \star \nabla\underline{\phi} N_i d\Omega; \tag{38}$$

$$\underline{x}_j = \underline{A}_{zj}. \tag{39}$$

$\underline{x}$ and $\underline{f}$ are elements of $G_P^{n_f \times 1}$. $\mathcal{K}$ and $\mathcal{L}$ are operators mapping $G_P^{n_f \times 1}$ upon $G_P^{n_f \times 1}$. In practice, $\mathcal{K}$ and $\mathcal{L}$ are represented by $n_f \times n_f$ sparse matrices of coefficients $\underline{\underline{k}}_{ij}, \underline{\underline{l}}_{ij} \in G_Q$. It is, however, more convenient to keep the operator notation in the following. The definitions of the inner product and norm for G_P extend in the classical way to $G_P^{n_f \times 1}$:

$$(\underline{u}, \underline{v})_{G_P^{n_f \times 1}} = \sum_{i=1}^{n_f} (\underline{u}_i, \underline{v}_i)_{G_P}; \tag{40}$$

$$\|\underline{u}\|_{G_P^{n_f \times 1}} = \sum_{i=1}^{n_f} \|\underline{u}_i\|_{G_P}. \tag{41}$$

Consider e.g. as a benchmark model, a single-phase transformer (Fig. 3) with a ferromagnetic and laminated iron core, operating at zero load, fed by a pure sinusoidal current source at frequency f_1 at the primary side.

The "Modified Trapezoidal Rule" (θ-Method) for the Integration of DAEs Modelling an Electro-Mechanical Drive Including External Circuitry and Magnetic Field-Distribution

M. Klocke

University of Dortmund, Institute of Electrical Machines, Drives and Power Electronics, Emil-Figge-Str. 70, 44221 Dortmund, Germany
E-mail: klocke@mal.e-technik.uni-dortmund.de

Abstract A simulation of an electro-mechanical drive including simultaneously the torsional behaviour of a mechanical load, a power electronic supply for soft-starting and a numerical transient field calculation requires the solution of a system of differential and algebraic equations. The Crank-Nicholson method for PDEs of parabolic type implies an application of the trapezoidal rule of integration with respect to time. This method of integration applied to a more general system of differential and algebraic equations can cause stability problems by numerically conditioned hazardous oscillations. These disturbances are suppressed by a modification of the trapezoidal rule sometimes referred to as θ-method with weighting-factors θ and $1 - \theta$ for the integration. However, the damping of natural oscillations of the system sometimes increases unacceptably. Best results concerning the damping are achieved, if θ remains at 0.5. Numerically critical eigenvalues of the system and their stimulation can be avoided by taking into account additional properties of the system in most practical cases.

1 The θ-Method

In a simulation program now called FELMEC developed at the Institute of Electrical Machines, Drives and Power Electronics at the University of Dortmund and applied e. g. in [6] spatially discretized field equations, network equations and equations describing the mechanical behaviour form a system of differential and algebraic equations for electro-mechanical drives. The θ-method [1,2] is illustrated for differential equations occurring in the system written in a most general way as

$$\sum_j a_j x_j(t) + \sum_j b_j \dot{x}_j(t) = r(t) \tag{1}$$

with a given right hand side $r(t)$ and unknown functions of time $x(t)$, which may represent e. g. vectorpotentials of nodes in the finite difference scheme or branch voltages and currents in the electrical network. The modified trapezoidal rule or θ-method transforms such an equation into a difference equation in time by integrating it over the interval $[t, t+h]$. The integrals of quantities not appearing as derivatives are approximated by linear combinations of the boundary values of the integrand with weighting factors θ and $1-\theta$, resp.:

$$\sum_j a_j \left[(1-\theta)\, hx_j|_{t+h} + \theta hx_j|_t\right] + \sum_j b_j \left(x_j|_{t+h} - x_j|_t\right) = \\ = (1-\theta)\, hr|_{t+h} + \theta hr|_t \ . \qquad (2)$$

If all coefficients b_j in (1) are zero, an algebraic equation without the need of discretization by the θ-method is obtained. However, if nevertheless applied, no difference occurs as long as consistent initial values are provided.

The so far unknown values $x_j|_{t+h}$ are put onto the left hand side. The right hand side of the equation is formed by the already known values at instant of time t and the right hand side of (2).

In literature the roles of the weighting factors θ and $1-\theta$ sometimes appear exchanged, e. g. in [3] or in [4], where the parameter is denoted α. For circuit simulation, in [5] a "floating order" method of integration with an order r, $1 \leq r \leq 2$ is proposed, where r is chosen automatically according to the behaviour of sources and the solution. $r = 1$ herein is equivalent to $\theta = 0$ and $r = 2$ to $\theta = 0.5$.

2 Some Physical Examples for Differential Equations Occurring in the System

2.1 Electrical Supply Network

The voltage u_j on an ohmic-inductive branch j contains resistive and self-inductive parts, $R_j i_j$ and $L_j \frac{d}{dt} i_j$, resp. Additionally mutually inductive voltage components $M_{j,k} \frac{d}{dt} i_k$ can occur. Applying (2) to the resulting voltage equation one obtains the following first order difference equation:

$$(1-\theta)\, hu_j|_{t+h} + \theta hu_j|_t = R_j \left[(1-\theta)\, hi_j|_{t+h} + \theta hi_j|_t\right] + L_j \left(i_j|_{t+h} - i_j|_t\right) + \\ + \sum_{k=1, k\neq j}^{N} M_{j,k} \left(i_k|_{t+h} - i_k|_t\right) \ . \qquad (3)$$

Ohmic-capacitive branches and resistors connected in parallel to capacitors or inductors can be treated similarly.

Each first order differential equation is transformed into a first order difference equation. Algebraic equations as given by merely ohmic branches and the topological descriptions $(C) \cdot \boldsymbol{i} = (0)$ and $(B) \cdot \boldsymbol{u} = (0)$ with cut set matrix (C), vector of branch currents $\boldsymbol{i}$, mesh matrix (B) and vector of branch voltages $\boldsymbol{u}$ remain untransformed.

Nonlinearities from semiconductor-valves are assumed to be only static. Their voltage-current characteristics are nonlinear algebraic equations taken into account iteratively by successive linearization.

2.2 Finite Difference Equations

Regions of a given cross-section of an electrical machine to be described by Poisson's equation after spatial discretization only contribute algebraic relations between the magnetic vectorpotentials A_{m} of nodes in immediate neighbourhood and exciting currents i_ρ of coil groups ρ. However, the exciting currents as stator- and rotor-winding-currents are unknown themselves and also appear in the voltage equations for the windings of the machine:

$$u_\rho(t) = R_\rho i_\rho(t) + \dot{\Psi}_\rho(t) = R_\rho i_\rho(t) + l_{\mathrm{e}} \sum_j F_{\mathrm{N},j} \dot{A}_{\mathrm{m},j}(t) \ . \tag{4}$$

The flux-linkage Ψ_ρ of coil group ρ herein is expressed by the vectorpotentials of the nodes in the cross-section of the winding and so are their derivatives. They are weightened by winding density factors F_{N} and the iron core lenght l_{e} of the machine. Voltage contributions from leakage inductances also have to be taken into account in (4), but are not mentioned here.

Further differential equations may be contributed by electrically conductive regions like short-circuited rotor-bars or unlaminated iron cores. The exciting current to a node under consideration is then not only determined by an external voltage-equation but also directly depends on the induced electric field strength $E_{\mathrm{i}} = -\dot{A}_{\mathrm{m},0}(t)$ in that very node. Additionally taking into account a gradient field strength E_{a} arising from external voltage drop in the region under consideration one obtains:

$$-\sum_{i=1}^{4} \alpha_i A_i(t) + \left(\sum_{i=1}^{4} \alpha_i \right) A_{\mathrm{m},0}(t) - F_{\gamma,0} \left(E_{\mathrm{a}}(t) - \dot{A}_{\mathrm{m},0}(t) \right) = 0 \ . \tag{5}$$

This is the spatially discretized form of a parabolic PDE, the eddy-current equation, but it is also coupled to network equations via E_{a}. The discretization in time of (4) and (5) is carried out according to (1) and (2), resp.

2.3 Mechanical System

The mechanical load is treated according to the analogy of current i and torque M or voltage u and angular velocity ω. So an equivalent network of the whole mechanical system can be formed.

A rotating mass is accelerated by a torque $J\dot{\omega}$, which equals the total torque acting M diminished by friction $d \cdot \omega$. The frictional torque is assumed to be proportional to the angular velocity ω. Discretization in time according to the θ-method yields:

$$\begin{aligned} (1-\theta)\, hM|_{t+h} + \theta hM|_t &= d \cdot \left[(1-\theta)\, h\omega|_{t+h} + \theta h\omega|_t \right] + \\ &\quad + J \left(\omega|_{t+h} - \omega|_t \right) \ . \end{aligned} \tag{6}$$

The mechanical system of equations is also transformed into a system of difference- and algebraic equations using the θ-method, although this system is rather weakly coupled to the electric and electro-magnetic part. This coupling is only given by the calculation of the electro-magnetic moment acting on the rotor. This is done by evaluating Maxwell's stress tensor in the air gap of the machine *after* having determined the field distribution for a given instant of time. Vice versa the new position of the rotor is obtained by integrating its angular velocity resulting only from the mechanical analysis.

Due to the weak coupling different integration methods like explicit Runge-Kutta-methods seem to be applicable. Nevertheless the θ-method has been chosen for the sake of stability: If nonlinear elasticity is given in an extreme case e. g. by tooth clearance of couplings, the eigenvalues of the system can vary in a wide range, which requires an absolutely stable method.

3 Theoretical Investigation of Effects of the θ-Method

Considerations are simplified in the following by assuming that a system of explicit first order ordinary differential equations can be extracted from a given system of differential and algebraic equations by means of linear transformations and defining appropriate quantities as a base for a state-space. The natural solutions of this system are given by the eigenvalues $\underline{\lambda}$ and eigenvectors $\underline{v}$ of the system matrix (A) as follows. Here it is assumed that each of the eigenvalues is different from any other one. In general $\underline{\lambda}$ and $\underline{v}$ have to be considered complex and therefore are underlined like all complex quantities in the following.

$$\underline{x}(t) = \underline{v} \cdot e^{\underline{\lambda} t} \ . \tag{7}$$

The modified trapezoidal rule transforms the system into a system of difference equations, which can explicitly be written as:

$$\underline{x}|_{t+h} = \left[(E) - (1-\theta)\, h\, (A)\right]^{-1} \cdot \left[(E) + \theta h\, (A)\right] \underline{x}|_t \ . \tag{8}$$

The solutions of this system are sequences of complex numbers for the quantities in $\underline{x}$ at instants of time $k \cdot h$ according to:

$$\underline{x}|_{t=kh} = \underline{v} \cdot \underline{\nu}^k = \underline{v} \cdot \underline{\nu}^{\frac{t}{h}} \ . \tag{9}$$

Here, $\underline{v}$ is a complex eigenvector of the matrix (A), $\underline{\nu}$ is the corresponding eigenvalue of the transformed system matrix. The following relation between the eigenvalues $\underline{\nu}$ of the time-discrete system and their continuous originals $\underline{\lambda}$ results from the transformation:

$$\underline{\nu} = \frac{1 + \theta h \cdot \underline{\lambda}}{1 - (1-\theta)\, h \cdot \underline{\lambda}} = -\frac{\theta}{1-\theta} + \frac{1}{1-\theta} \cdot \frac{1}{1 - (1-\theta)\, h \cdot \underline{\lambda}} \ . \tag{10}$$

For an exact agreement of the time-discrete solutions with the solutions for the continuous system $\underline{\nu}$ should be $e^{h\underline{\lambda}}$ as seen from (7) and (9). The transformation (10) can only be an approximation. Fig. 1a) shows how the complex $\underline{\lambda}$-plane is mapped onto the $\underline{\nu}$-plane by $e^{h\underline{\lambda}}$ or $\underline{\lambda} = \ln(\underline{\nu}/h)$. The parallels of constant damping $\mathrm{Re}\{\underline{\lambda}\}$ are mapped onto concentric circles. $|\underline{\nu}| = 1$ corresponds to $\mathrm{Re}\{\underline{\lambda}\} = 0$. Eigenvalues on these boundaries describe undamped oscillations. The diagram for θ=1/2, i. e. the trapezoidal rule, in

Fig. 1b) shows that the boundary of stability lies on the circle $|\underline{\nu}| = 1$. So undamped oscillations of the continuous system are numerically determined as undamped, too. If for a merely real and stable λ the expression $|h\lambda|$ is very large, an erroneous numerical oscillation in the solution can occur becoming nearly $\{-1\}^{t/h}$, an undamped alternating sequence. Hence, despite the stable mapping of all stable eigenvalues of the continuous system stability problems can occur for so-called stiff systems. This problem can be sharpened in the case of nonlinearities taken into account and by rounding errors with the result of even self-exciting or negatively damped numerical oscillations, see e.g. [4]. A similar phenomenon concerning the gradient electric field in massive conductors affected by eddy-currents is referred to as the stability paradox in [3]. How this is related to the loss of stability of the trapezoidal rule in the case of higher index DAEs, would be a worthy topic of further analysis.

The problem can be overcome by diminishing θ. The diagrams for θ=1/3 in Fig. 1c) and θ=0 in Fig. 1d) show that an undamped numerical oscillation even for real eigenvalues $\lambda = -\infty$ cannot occur. $\theta = 0$ yields the backward Euler-formula. However, with $\theta < 0.5$ erroneous damping occurs for natural oscillations of the continuous system, especially for those with only low or negligible damping. The deformation of the light gray area in the figures with θ decreasing illustrates the dependence of the damping error on θ and ω.

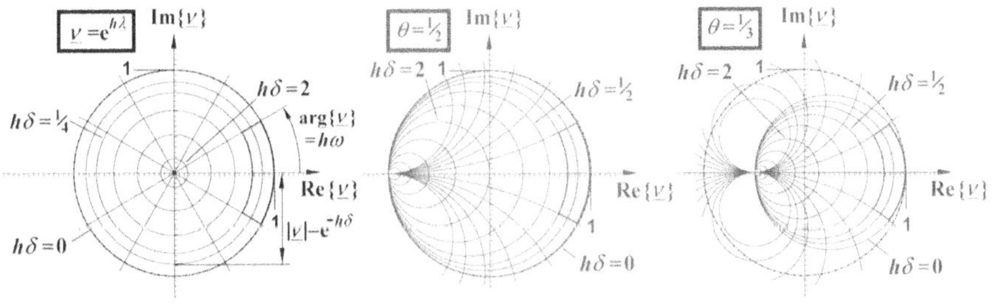

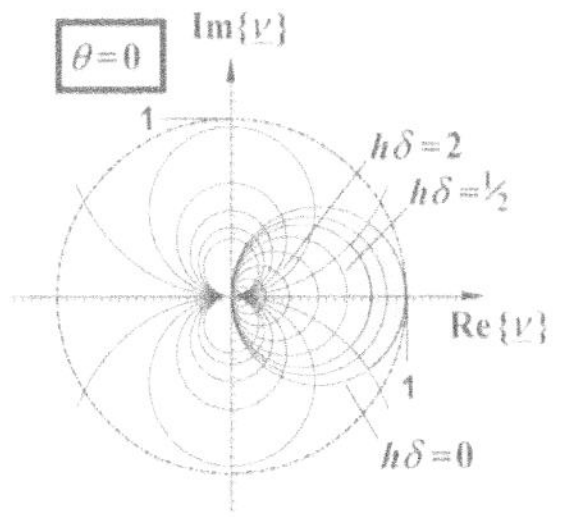

Figure1. Mappings of the eigenvalues $\underline{\lambda}$ into the $\underline{\nu}$-plane according to:
a) $\underline{\nu} = e^{h\underline{\lambda}}$, b) $\underline{\nu} = \frac{2+h\underline{\lambda}}{2-h\underline{\lambda}}$, c) $\underline{\nu} = \frac{3+h\underline{\lambda}}{3-2h\underline{\lambda}}$, d) $\underline{\nu} = \frac{1}{1-h\underline{\lambda}}$. In the diagram a) $\omega = \text{Im}\{\underline{\lambda}\}$ denotes the angular eigenfrequency. $\delta = \text{Re}\{\underline{\lambda}\}$ is the damping constant of the eigenfunction belonging to $\underline{\lambda}$, i. e. for a pair of conjugate-complex $\underline{\lambda}$'s the reciprocal time-constant of the enveloping e-function.

4 Actual Examples for Effects of the θ-Method

Fig. 2 shows a simple torsional oscillator consisting of two masses and one elastic shaft with damping. Oscillations are excited by a given time function of the electrical moment exerted on one mass representing the rotor of an induction motor. In this simple test-case the electrical moment is not a result of a coupled electro-mechanical analysis, but prescribed by a function of the rotor position angle. It is assumed to contain a large oscillating component with a constant amplitude of 24 Nm, which is 2.4 times the average value. Its frequency depends linearly on the speed with a rate of 100 Hz per 1500/7 rpm and becomes zero at a parasitic synchronous speed of 1500/7 rpm. Since the average value of the torque accelerates the drive, the exciting frequency varies and at a certain time the torsional oscillator becomes resonant. The natural oscillation once stimulated decreases very slowly due to only low damping.

For this simple arrangement a system of four explicit differential equations is easily set up and solved with high accuracy using the Runge-Kutta-method with an adequately small time-step [6]. The solution in Fig. 3 obtained by using the θ-method with $\theta = 0.5$ is nearly equal to that solution. However, already the solution in Fig. 4 obtained with $\theta = 1/3$ is unacceptable, since damping of the natural oscillation is obviously calculated far too large, according qualitatively to the predictions derived from the diagrams of eigenvalues.

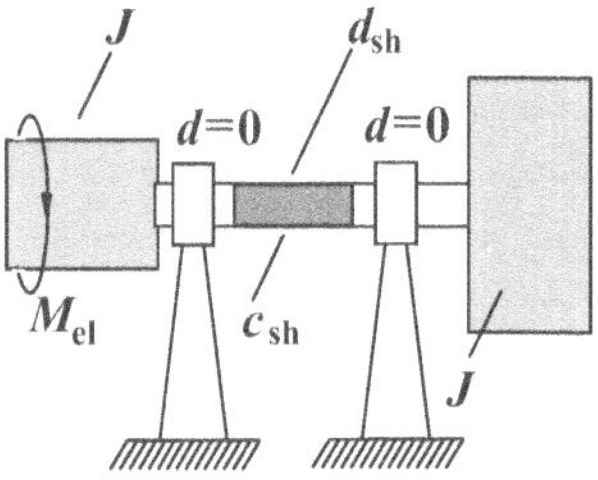

Figure2. Simple torsional oscillator excited by oscillating components of the electrical moment M_{el}. $J = 0.09\,\text{kgm}^2$, $c_{sh} = 32.55\,\text{kNm/rad}$, $d_{sh} = 0.015\,\text{Nms/rad}$. Natural frequency $f = 135.4\,\text{Hz}$, damping constant $\delta = 1/6\,\text{s}^{-1}$.

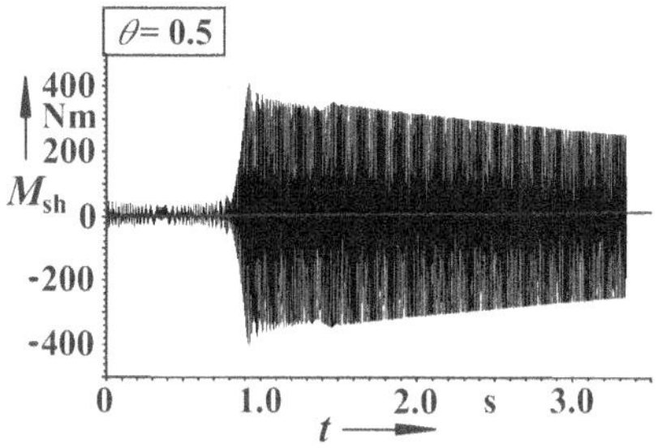

Figure3. Torque in the measuring shaft calculated with $\theta = 0.5$, time step $h = 66.6\,\mu\text{s}$.

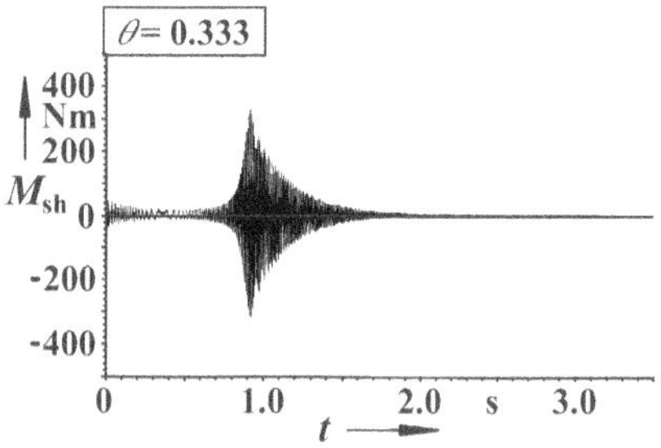

Figure4. Torque in the measuring shaft calculated with $\theta = 0.333$, time step $h = 66.6\,\mu\text{s}$.

The electric and electro-magnetic system consisting of a three phase ac power controller and the induction motor (Fig. 5) is discussed in the following. A typical voltage shape at the motor-terminals obtained by measurement is shown in Fig. 6. The voltage is nearly the grid-voltage when both pairs of thyristors connecting the motor-terminals with the power grid are conductive. The gaps in the shape mark the turn-off states of at least one of these thyristor-pairs. The voltage-shape obtained by simulation with $\theta = 1/2$ in Fig. 7 shows undamped oscillations in these gaps. The non-conductive state of the thyristors is simulated by a very high resistance. These resistances in conjunction with the inductances of the machine cause large negative eigenvalues. As a result the numerical solution contains voltage oscillations of period $2h$. If θ is decreased below 0.5 these oscillations are successfully suppressed, see e. g. [6].

Another method for reducing the undesirable erroneous alternating sequences in the solution is to take into account additional elements. In the example presented here these are the capacitors and the resistors connected in parallel to the thyristor pairs via four rectifying diodes for protecting the thyristors against over-voltage and steep increases of the voltage-drop between anode and cathode. The eigenvalue configuration is uncritical due to the capacitance in series to the inductance, when the rectifier is conducting. This is the case, while the thyristors are turned off. After turn off of the thyristor-branch the rectifier becomes non-conductive. But then the critical eigenvalues of the system are hardly excited because voltage only changes relatively slow now. Hence, even if $\theta = 1/2$ is chosen, numerical oscillations in the calculated voltage shape almost vanish as shown in Fig. 8.

Although the number of network equations is increased from 6 to 36 due to the number of branch elements being six times higher, computational effort in the presence of 4806 field equations only rises negligibly: The average cpu-time needed for one solution of the system of equations on an IBM RS6000/990 is about 0.94 s ... 0.97 s in both cases. The total cpu-time is rather influenced by accuracy requirements and the resulting number of iteration steps. In the calculation for Fig. 8 the average number of iterations per time-step is 10.4 due to iron saturation and semiconductor nonlinearities

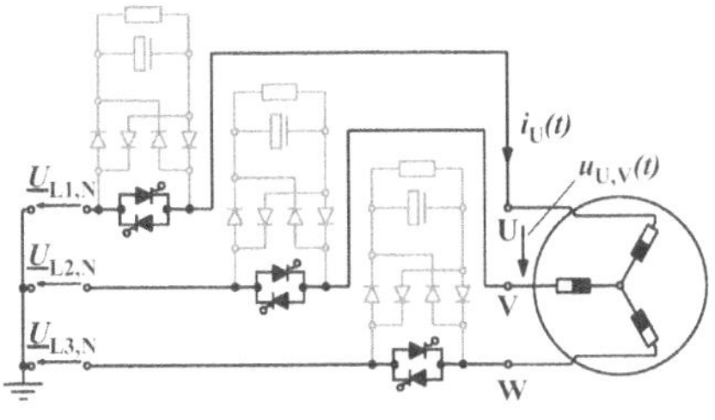

Figure5. Scheme of a three-phase ac power-controller with over-voltage-protection consisting of R and C in parallel connected to the thyristors via rectifier diodes.

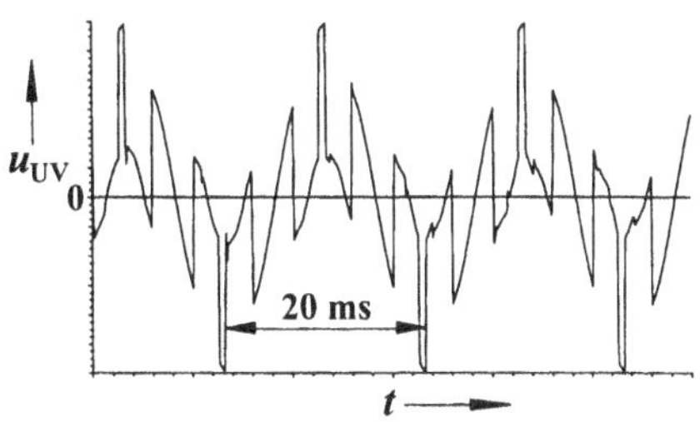

Figure6. Shape of voltage u_{UV} at the motor terminals from measurement (including over-voltage-protection).

approximated with higher accuracy (e. g. $\epsilon_{sat} = 2 \cdot 10^{-5}$ instead of $2 \cdot 10^{-4}$), whereas only 8.2 iterations per time-step are needed in Fig. 7.

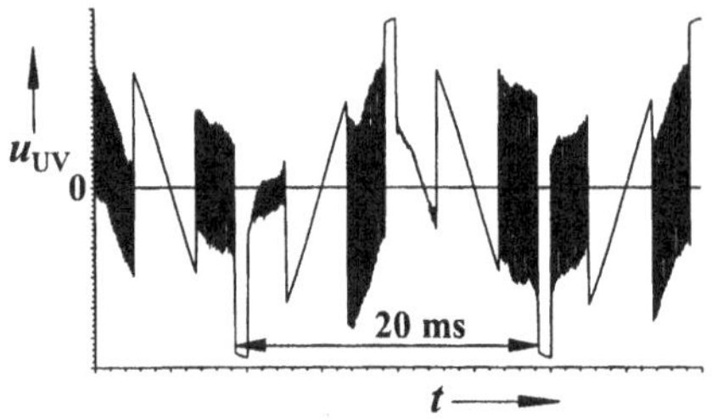

Figure7. Shape of voltage u_{UV} at the motor terminals calculated with $\theta = 0.5$, over-voltage-protection neglected.

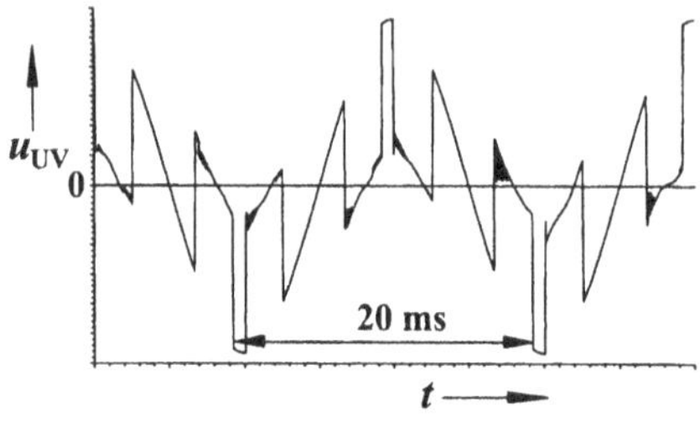

Figure8. Shape of voltage u_{UV} at the motor terminals calculated with $\theta = 0.5$, over-voltage-protection taken into account.

5 Concluding Remarks

Concerning the numerical simulation of electro-mechanical drives a conflict between the suppression of undesirable numerical oscillations caused by large negative eigenvalues in the system of differential and algebraic equations and a sufficiently accurate determination of the damping of natural oscillations arises, if the θ-method usually applied to parabolic PDEs [1,2] is uncritically transferred to arbitrary DAEs. However, problems conditioned by large negative eigenvalues (stiffness) can be reduced by additionally modelling properties previously neglected, a procedure quite similar to the regularization of DAEs [7]. θ may then be set very close to 0.5 or even remain 0.5, which usually yields most accurate results in the presence of nearly undamped natural oscillations.

References

1. Krawczyk A., Tegopoulos J. A.: Numerical Modelling of Eddy Currents. Clarendon Press, Oxford, 1993.
2. Krawczyk A.: The application of FEM to transient problems of electromagnetic engineering. Rozprawy Elektrotechniczne 29, (2), pp. 331-344, 1983.
3. Tsukerman, I.: A Stability Paradox for Time-Stepping Schemes in Coupled Field-Circuit Problems. IEEE Transactions on Magnetics, Vol. 31, No. 3, pp. 1857-1860, May 1995.
4. Belmans, R., Hameyer K. and Mertens, R.: Combined Time-Harmonic-Transient Approach to Calculate the Steady-State Behaviour of Induction Machines. IEEE Transactions on Magnetics, 1999.

5. Spiro, H.: Simulation integrierter Schaltungen, 2. Auflage. Oldenbourg Verlag München, Wien, 1990.
6. Klocke, M.: Zur Berechnung dynamischer Vorgänge bei von einem Drehstromsteller gespeisten Antrieben mit Asynchronmaschinen und mehreren gekoppelten Massen mittels Finite-Differenzen-Zeitschrittrechnung. Dissertation, University of Dortmund, 1999.
7. Feldmann, U., Günther, M.: Some Remarks on Regularization of Circuit Equations. Conference Proceedings of X. ISTET, 1999.

Analysis of Coupled Electromagnetic and Thermal Fields in the Instationary Case*

O. Michelsson, G. Scheinert, and F.H. Uhlmann[1]

Technical University of Ilmenau, Dept. of Fundamentals and Theory of Electrical Engineering, P.P.-Box 100 565, D-98684 Ilmenau, Germany
uhlmann@e-technik.tu-ilmenau.de

Abstract The handling of mission-critical devices, e.g. electro-mechanical actuators, requires an exact calculation of the electromagnetic field. For this task, thermal influences on the electrical behavior cannot be neglected. Furthermore stationary solutions will no longer be sufficient for the coupled problem. Therefore, we combine Maxwell's equations, excluding displacement currents, together with the instationary heat transfer equations. This combination is input to the existing finite difference based program, PROFI. The stationary solutions of this coupled problem provide initial conditions for the instationary problem only. For the analysis of instationary processes, both fields are solved simultaneously in the time domain. To ensure stability and physical reliability, the necessary criteria are fulfilled by means of automatically adapted coefficients for the implicit solution scheme. The accuracy of the method will be shown by several examples.

1 Introduction

Thermal influences on the electrical behavior of highly loaded electromagnetic devices is of much interest. In most cases stationary solutions are sufficient, but sometimes transient processes must be taken into account. This is especially true with magneto-mechanical actuators; where the dynamic behavior over a wide temperature range is of interest. Also, in micro-mechanical systems, these investigations become essential because coupling between electrical and thermal parameters can not be omitted. Since the simulation of coupling between transient electromagnetic and thermal fields is available only in very expensive simulators, we decided to extend the existing CAE-program, PROFI[3] to handle the required thermal capabilities.

The electromagnetic behavior is simulated using the magnetic vector potential $\boldsymbol{A}$. It is a solution to the following partial differential equation (PDE):

$$\nabla \times \left(\frac{1}{\mu(T)} \nabla \times \boldsymbol{A} \right) + \kappa(T) \frac{\partial \boldsymbol{A}}{\partial t} - \kappa(T) \cdot \boldsymbol{v} \times \nabla \times \boldsymbol{A} \tag{1}$$

$$= \boldsymbol{J}_s(t,T) + \kappa(T) \cdot \boldsymbol{E}(t) + \nabla \times \boldsymbol{M} \ , \tag{2}$$

* This project was supported by the BMBF under project-number 16 SV 680/5.

in the domain Ω_e, whereby in (1) the spatial distribution $x \in \Omega_e$ of all parameters is omitted for the sake of simplicity. Here $\boldsymbol{J}_s$ is an impressed current-source, $\boldsymbol{E}$ the electrical field strength and $\boldsymbol{M}$ is a given magnetization. The material parameters, κ as electrical conductivity and μ as permeability, are used with its thermal characteristics. The velocity $\boldsymbol{v}$ of a mechanical part is also taken into account. Restricted by a two-dimensional investigation, $\boldsymbol{A}$ may be described sufficiently by its scalar component perpendicular to the used two-dimensional plane.

The basis for thermal analysis is the Fourier's law in the domain Ω_T

$$\rho(T)c_p(T)\frac{\partial T}{\partial t} - \nabla\big(\lambda(T)\nabla T\big) = w(T,t) \ , \tag{3}$$

where ρ is the density, c_p the specific heat, T the temperature, t the time and λ is the thermal conductivity. Here, the spatial distribution of all parameters by $x \in \Omega_T$ was omitted. For a realistic thermal analysis, temperature dependencies as well as the time dependency of heat sources $w(T,t)$ must be taken into account. These heat sources are beside given direct heat-sources almost exclusively joule heat losses inside electrically conductive media. So both equations are strongly coupled in the stationary case and due to much different time-constants weakly coupled in the non-stationary case.

For the solution of (1) boundary conditions must be provided on $\Gamma_e = \partial\Omega_e$ for $A(x)\,, x \in \Gamma_e$. Also for the solution of (3) boundary conditions on $\Gamma_T = \partial\Omega_T$ are necessary for $T(x)\,, x \in \Gamma_T$. It is emphasized here, that in our investigations $\Omega_T \subseteq \Omega_e$ and therefore Γ_T may be different from Γ_e. In regions $\Omega_e \setminus \Omega_T$, where the temperature is not calculated, user provided constant values are assumed.

In addition to Dirichlet's or Neumann's boundary conditions, in thermal analysis the so called Newton's boundary condition

$$-\lambda(T)\frac{\partial T}{\partial n} = \alpha_t(T)\cdot(T - T_{\text{fluid}}) \quad \text{on} \quad \Gamma \in \Gamma_T$$

may be defined. Therewith, heat diffusion into a surrounding fluid with temperature T_{fluid} may be taken into account. The parameter α_t, the thermal diffusivity, includes convective losses as well as radiation losses according to the Stefan-Boltzmann equation and is therefore strongly temperature dependent. The radiation is described in terms of a black radiator multiplied by a radiation coefficient $\epsilon \in [0,1]$.

2 Instationary solution of the coupled magnetic and thermal field

2.1 Implementation in the CAE-system Profi

The CAE-system PROFI, which is a set of two- and three-dimensional field computation modules is based on a finite integration method. It is able to work with non-linear material parameters, which makes it suitable for the field coupling. To study the coupling of thermal and electromagnetic fields, the two-dimensional field calculation module PROFI 2DTRANSIENT has been used. It dealt so far with electromagnetic fields and mechanical transitions only, and has been extended by a thermal field solver.

For the solution of equations (1) and (3) the domain Ω_e is divided along the coordinate axes into finite volume elements. Herein piecewise constant material parameters are assumed so that the given PDEs may be integrated by parts using Gauss or Stokes theorem, respectively, and then approximated by finite differences. For a single volume element ΔV, bounded by its surface ΔA_o, equation (3) for instance may then be noted as

$$\int_{\Delta V} \rho(T) c_p(T) \frac{\partial T}{\partial t} \, dV = \oint_{\Delta A_o} \lambda(T) \nabla T \, dA_o + \int_{\Delta V} w(t, T) \, dV \qquad (4)$$

which represents the implemented finite difference approximation.

To ensure as much compatibility to the original code as possible, the thermal field solver has been added as an optional extension with slightly small changes in the existing routines only. Its user interface follows the rules of the PROFI environment.

The whole process for the solution of the coupled PDE is shown schematically in Fig.1. As can be seen here, the algorithm consists of two parts; first a solution of the steady-state problem for the determination of the initial conditions and second the calculation of the time-dependent field solutions. These two parts are explained in more detail below.

2.2 Stationary Solution

The initial conditions, $\boldsymbol{A}(x, t = t_0)$ and $T(x, t = t_0)$, for the instationary case are given by the stationary solution for the initial source values. As with all steady-state solutions, time derivatives are set to zero. The PDE for the magnetic vector potential $\boldsymbol{A}$ is then given as

$$\oint_{\Delta l} \left(\frac{1}{\mu(T)} \nabla \times \boldsymbol{A} + \kappa(T) \cdot \boldsymbol{v}_0 \times \boldsymbol{A} \right) dl$$
$$= \int_{\Delta A_q} \left(\boldsymbol{J}_s(t_0, T) + \kappa(T) \cdot \boldsymbol{E}_0(t_0) \right) d\boldsymbol{A}_q + \oint_{\Delta l} \boldsymbol{M} \, dl \ , \qquad (5)$$

where ΔA_q is the cross section of a volume element ΔV in the used two-dimensional plane and Δl is its surrounding curvature. The thermal PDE is simply given by

$$\oint_{\Delta A_o} \lambda(T) \nabla T \, dA_o + \int_{\Delta V} w(t_0, T) \, dV = 0 \ . \qquad (6)$$

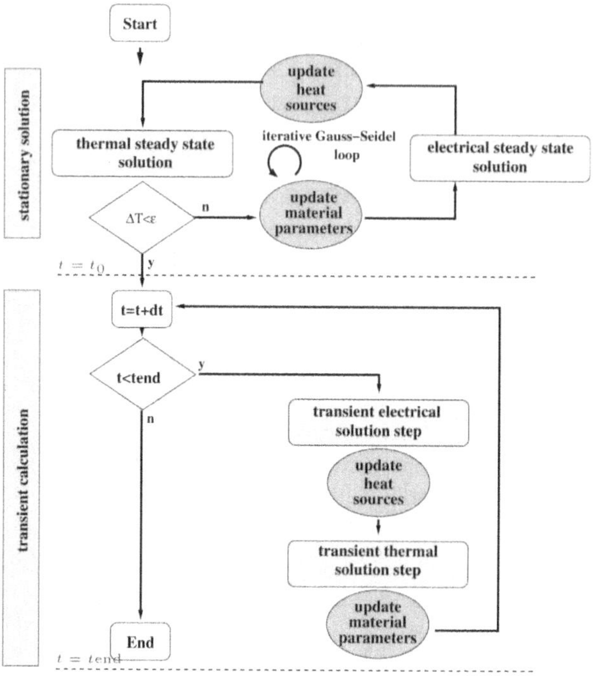

Figure1. Principle of the coupled transient electromagnetic thermal field calculation

Due to the strong mutual coupling between (5) and (6) by the material parameters and joule losses, both equations have to be solved simultaneously. The equation systems, resulting from the finite difference approximation, are solved iteratively in a block-wise Gauss-Seidel-like scheme. This iterative process is repeated until a user-defined temperature accuracy or iteration-number has been achieved. For nearly all physical systems where with increasing temperatures the electrical conductivity decreases (above a critical temperature at least) the system fulfills the Banach fixed-point criterion of a contractive transformation. Perhaps, for strong nonlinear materials or if radiation losses are taken into account, the convergence may be very slow. Therefore, a linearization of the temperature-dependence of the heat sources can be used to speed-up this iterative process. The right hand side c of a single row of the thermal equation system can be linearized in iteration step $i+1$ from the known values of two former iterations as

$$c_{i+1}(T) = c_i(T_i) + \frac{c_i(T_i) - c_{i-1}(T_{i-1})}{T_i - T_{i-1}} T \tag{7}$$

The far better convergence is demonstrated in Fig.2 for the steady-state solution of a self-heated relay coil, which later serves as an example for the instationary analysis.

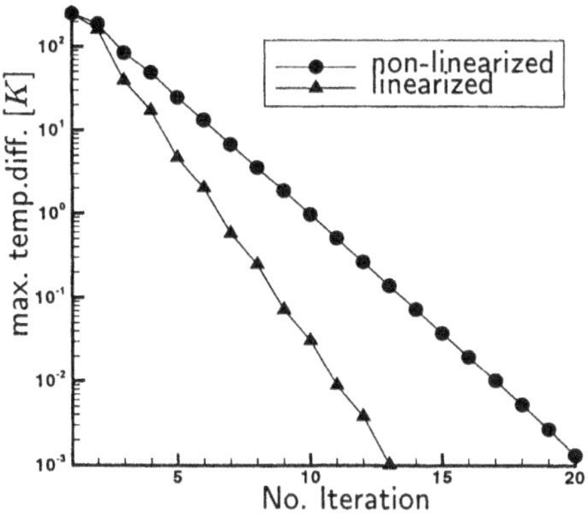

Figure2. Number of Iterations for steady state analysis of a relay coil with and without linearization of the temperature dependency of heat sources

For the solution of the sparse system matrices we use PROFI implemented SOR-solvers also CG- as well as GMRES-based solvers.

2.3 Instationary analysis

With knowledge of the initial conditions of $\boldsymbol{A}(t = t_0)$ and $T(t = t_0)$ the coupled instationary PDEs can be solved. Because the electromagnetic and mechanical module in PROFI have not been changed, we will focus on the thermal field solution here.

Therefore, (4) will be integrated at point p with respect to time t

$$\int_{\Delta V} \rho c_p \, dV (T_p^1 - T_p^0) = \int_t^{t+\Delta t} \left[\oint_{\Delta A_o} \lambda \nabla T \, d\boldsymbol{A}_o + \int_{\Delta V} w(t) \, dV \right] dt \qquad (8)$$

The remaining time integral may then be approximated according to [2] as

$$\int_t^{t+\Delta t} T_p \, dt \approx [f \, T_p^1 + (1-f) T_p^0] \, \Delta t \quad , \qquad (9)$$

where $f \in [0, 1]$ is a weighting factor. Special cases of f are: $f = 0 \rightarrow$ explicit scheme, $f = 1 \rightarrow$ fully implicit scheme. All others are implicit schemes, whereas $f = 0.5$ is the special case with linear approximation between the temperatures T_p at both time-steps, called Crank-Nicholson scheme. Applied to the finite difference approximation of (8) the approximation (9) leads to

$$a_p T_p^1 = \sum_k a_k \left[f T_k^1 + (1-f) T_k^0 \right] + a_{T0} T_p^0 \quad k = 1..4 \qquad (10)$$

where k marks the four neighboring grid points and

$$a_k = \frac{\lambda_k}{\Delta x_k}, \quad k = 1..4 \qquad a_p^0 = \frac{\rho c_p \Delta V}{\Delta t}$$

$$a_p = f \cdot (\sum_{k=1}^{4} a_k) + a_p^0 \qquad a_{T0} = [a_p^0 - (1-f)\sum_{k=1}^{4} a_k] .$$

The heat sources were omitted here because their time dependence is mostly affected by the electromagnetic system and therefore will be treated independent of the homogeneous thermal system.

To ensure physically realistic solutions, all coefficients a_i must be positive. Factor a_{T0} does not fulfill this requirement in all cases depending on the weighting factor f, the time- and spatial discretization and the material parameters. In the Crank-Nicholson scheme e.g., this can lead to oscillations although the algorithm is unconditionally stable in the mathematical sense. As a practical solution we may adapt the time-discretization or the weighting factor. As in most situations, the time-dependence of the thermal system is less complicated than the electromagnetic, therefore we decided to adapt the weighting factor f for the thermal system and leave the time-adaption to the electromagnetic solver. Hence, $f \in [0.5, 1.0]$ is estimated for each volume element ΔV to fulfill $a_{T0} > 0$. This is done in every time-step as long as thermally nonlinear materials are involved.

3 Numerical examples

In order to verify the implemented algorithm, we investigated the self-heating process of a relay coil. As an indicator of the coil temperature the time-dependence of the current was measured and calculated when a stabilized constant voltage is switched on. Beside these measurements, the surface temperature was recorded by an infrared camera system TVS 100 (Nippon Avionics Co., Ltd.). In Fig.3 an image of the investigated device and its infrared image are shown.

The material parameters we used are: radiation constant $\epsilon = 0.95$; core-material steel (1% C); the coil support polyamid-like and the winding material is E-CuF20. The winding isolation has been approximated by: $\lambda = 0.2\,W/(m^2K)$, $\rho = 800\,kg/m^3$ and $c_p = 1\,kJ/(kgK)$. The convection coefficient α has been estimated from [4] as a characteristic of a cylindrical structure. The importance of a good approximation of the inner coil-structure is shown in the calculated transient process. As shown in Fig.4, accurate results have been achieved only for a coil model consisting of air-surrounded insulated copper blocks, which represents the real coil structure best.

As a second example, the self-heating process in a hot-strip sensor according to [1] was investigated. As excitation a voltage pulse (12V) was imposed.

Special care must be taken to discretize this device appropriately. This is due to the high aspect ratio between the Pt-strip and sensor width. Al-

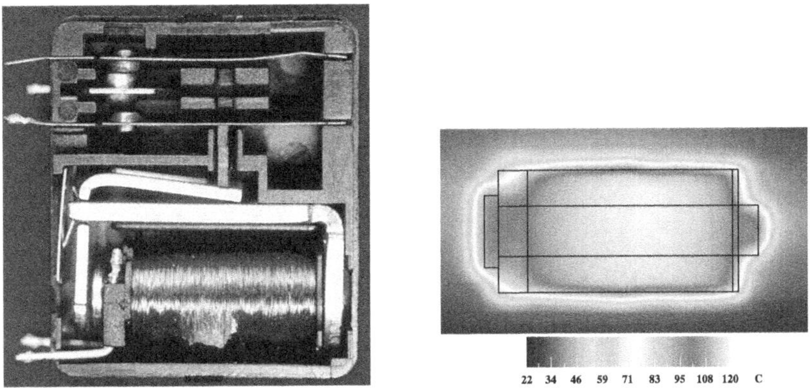

Figure3. Image and infrared image of the relay coil

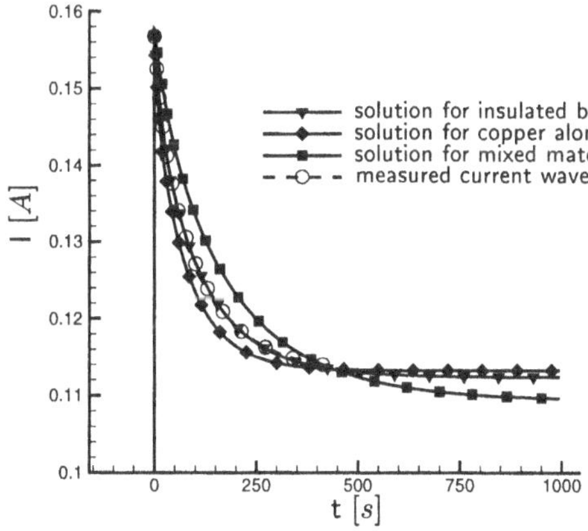

Figure4. Current waveform caused by self-heating for different material arrangements in the winding

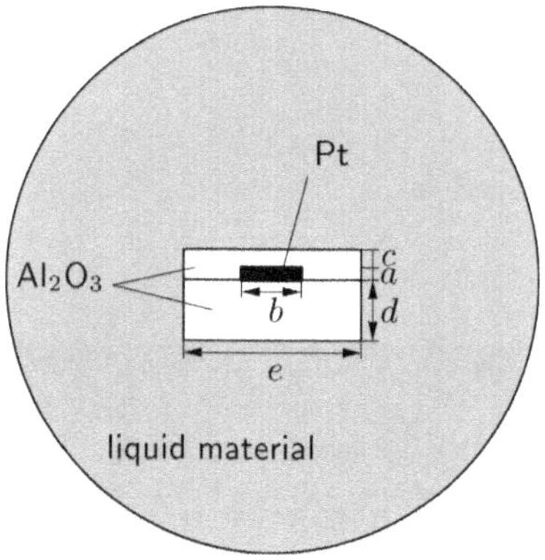

Figure5. Setup of the hot-strip sensor

geometrical parameter:

a: $5\mu m$ b: $100\mu m$

c: $5\mu m$ d: $400\mu m$

e: 58mm

material parameter:

material	λ $[W/Km]$	c_p $[J/kgK]$	ρ $[kg/m^3]$
substrate	32.3	754	4e3
coating	25.5	754	4e3
Pt-strip	71	133	21.4e3
water	0.598	4183	999.8

though the aspect ratio in certain elements is higher than 1:100, the solution converged stable. The influence of the weighting factor for the stability and accuracy of the thermal solution is shown in Fig.6. Obviously, the Crank-Nicholson scheme lead to an oscillating result, whereas for higher weighting factors this can be avoided. The adaptive procedure preserves the user from any sophisticated parameter estimation.

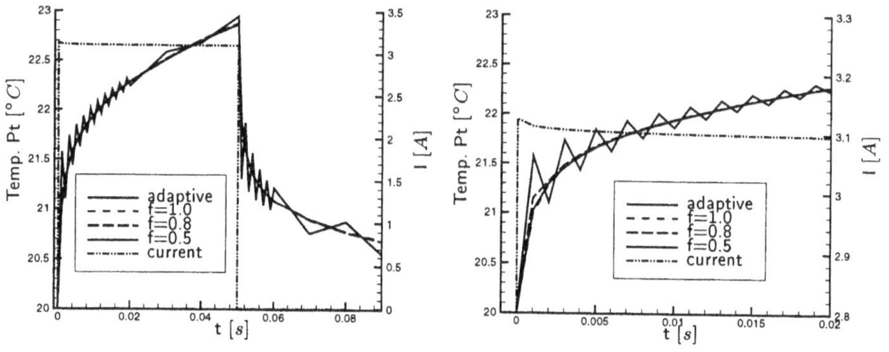

(a) whole impulse (b) Zoom near the starting point

Figure6. Temperature and current in the Pt-strip due to an imposed voltage pulse

4 Conclusion

The instationary analysis of coupled electromagnetic and thermal fields has been implemented for two-dimensional problems. The stability and accuracy of the algorithm was shown by two examples. Further issues will be higher order time-integration methods combined with possible error-controlled time discretization.

References

1. Assael, M.J. and Dix, M. and Drummond,I. and Karagiannidis, M.J. and Lourenço, M.J. and Nieto de Castro, C. and Papadaki, M. and Ramires, M.L. and van den Berg, H. and Wakeham, W.A., Toward Standard Reference Values for the Thermal Conductivity of High-Temperature Melts, Int. J. of Thermophysics 1997 Vol. 18 No.2 pages:439-446
2. Patankar, S.V., Numerical Heat Transfer and Fluid Flow, Hemisphere Publishing Corporation 1980, Series in computational methods in mechanics and thermal sciences
3. PROFI 2D transient, Version 2.0, PROFI Engineering Systems GmbH, PROFI ENGINEERING SYSTEMS GmbH Darmstadt 1997
4. Verein Deutscher Ingenieure, VDI-Wärmeatlas: Berechnungsblätter für den Wärmeübergang, Berlin, Heidelberg Springer-Verlag 1998,

On the Evaluation of Transients in Conducting Ferromagnetic Cores*

R. Van Keer[1], L.R. Dupre[2], J.A.A. Melkebeek[2],
Y.I. Moroz[3], S.E. Zirka*[3]

[1] Ghent Univ., Dept. Math. Anal., Galglaan 2, B-9000 Gent, Belgium
[2] Ghent Univ., Dept. Elect. Power Engin., St. Pietersnieuwstr. 41, B-9000 Gent, Belgium
[3] Dnepropetrovsk State University, Naukova 13, Dnepropetrovsk 49050, Ukraine

Abstract We deal with the numerical approximation of a parabolic boundary value problem (BVP) with Neumann boundary conditions arising from the evaluation of transients in conducting ferromagnetic cores. This BVP is derived from the proper Maxwell equations. A nonstandard finite-difference (FD) technique is compared with a standard finite element (FE) formulation. For nonlinear problems, characterized by abrupt changes in the magnetic permeability, being the material coefficient in the BVP, the 2nd-order FD-scheme is recommended.

1 Introduction

In modeling a wide range of devices containing magnetic cores, it is of a principal significance to provide an adequate description of the transients in the thin sheets or ribbons of conducting ferromagnetic material present in the device. In particular, the important practical problem of magnetization of the wound core, where the lengths of the shortest, l_{min}, and the longest, l_{max}, magnetic paths are close, is immediately reduced to the study of the processes in a separate ribbon of the mean length $l_{\text{a}} = (l_{\text{min}} + l_{\text{max}})/2$. When these paths are substantially different, the problem becomes more complex, however it can also be reduced to a combined solution of several one dimensional problems for each of the separate sheets.

When formulating the boundary value problem (BVP), it should be taken into account that the corresponding numerical scheme is used in a transient simulator, where Maxwell equations are solved together with the ordinary differential equations (ODEs) describing the lumped elements of an electrical circuit. To implement their coupling, the magnetic circuit described by the BVP is considered as a two-pole element, across the terminals of which some arbitrary time-variable voltage is applied. This voltage should be found in the

* The work was supported financially by the GOA-project 99-200/4 and by the research project No 3604209B of FWO-Vlaanderen. Luc Dupre is a postdoctoral researcher of FWO-Vlaanderen

course of solving the BVP with the Neumann boundary conditions (BCs). A comparative analysis of two main approaches to the problem, namely finite-difference (FD) and finite-element (FE), has been carried out. It is shown that the comparison of their possibilities requires the consideration of non-standard FD-numerical schemes with the time derivative distributed over several neighbouring nodes. Two FD-schemes of different accuracy are developed and compared with the corresponding FE-schemes. When solving the linear problem, such schemes are competitive in accuracy with FE-schemes with the same bandwidth of the resulting coefficient matrix. When solving non-linear problems, especially problems characterized by abrupt changes in the differential magnetic permeability, the use of high order numerical schemes (both FD and FE) is not justified. Therefore, the 2nd-order FD-scheme is more suitable.

2 The Model Problem

The quantitative description of the magnetization process in a thin sheet is reduced to the integration of one-dimensional Maxwell equations that link the magnitudes of the magnetic field strength H, the magnetic induction B and the electric field strength E in a material with a conductivity σ and a nonlinear relation $B(H)$. If the x axis is directed into a sheet of thickness d, the z axis is taken along the rolling direction, and the yz plane coincides with the sheet surface, then the penetration of a plane electromagnetic wave into the sheet is described by the simultaneous equations

$$-\frac{\partial H}{\partial x} = \sigma E, \qquad -\frac{\partial E}{\partial x} = \frac{\partial B}{\partial t}, \tag{1}$$

or

$$\frac{\partial^2 H}{\partial x^2} = \sigma \frac{\partial B}{\partial t}, \qquad 0 < x < d,\ t > 0, \tag{2}$$

where $H = H_z(x,t)$, $B = B_z(x,t)$, $E = E_y(x,t)$.

This must be completed by one initial condition (IC) and two BCs. In the absence of magnetic viscosity, the IC can be defined by some function $H_0(x) = H(x,0)$. The type of the BCs is determined by the fact that the partial differential equations (PDEs) above should be coupled with the ODEs describing the lumped elements of an electrical circuit. In order to write Kirchhoff's equations describing the circuit as a whole, each of the circuit elements including the core, should be characterized in terms of the voltage $u(t)$ across the element and its current $i(t)$. It can be shown via Faraday's law that the voltage $u_\mu(t)$ across the magnetizing winding and the electric field strength $E(0,t) = E_\mathrm{b}(t)$ at the surface of the sheet are connected by the relation [1]

$$E_\mathrm{b}(t) = \frac{d}{2w_1 S} u_\mu(t) = \frac{d}{2}\frac{dB_\mathrm{a}}{dt}, \tag{3}$$

where S is the cross-section of the core, w_1 is the number of turns in the magnetizing winding, and B_a is the average induction through the sheet.

Taking into account the first equation of (1) and the known symmetry of the problem [2] we arrive at the Neumann BCs

$$\left.\frac{\partial H}{\partial x}\right|_{x=0} = -\sigma E_\mathrm{b}(t), \qquad \left.\frac{\partial H}{\partial x}\right|_{x=d} = \sigma E_\mathrm{b}(t). \tag{4}$$

Notice that the functions $H(x,t)$ and $B(x,t)$ are symmetric about the plane $x = d/2$, and the function $E(x,t)$ is symmetric about the sheet longitudinal axis. This means that the PDE (2) must only be solved on the interval $[0,d/2]$ and that, correspondingly, the proper 2nd BC is $\partial H/\partial x|_{x=d/2} = 0$.

3 Finite Difference Schemes

To solve the BVP above we first consider time-continuous FD-schemes with respect to the space variable. This is convenient when the BVP has to be solved together with the ODEs in the time variable accounting for the lumped circuit components. We use a uniform grid with the node x_1 at the sheet surface, and the node x_N at point $x = d/2$. Grid functions corresponding to the values of $E(x_i,t)$, $H(x_i,t)$ and $B(x_i,t)$ at ith node, are denoted as $E_i(t)$, $H_i(t)$, and $B_i(t)$, respectively. We also denote $H_i^{(j)} = \partial^j H(x,t)/\partial x^j\big|_{x=x_i}$.

The FD-schemes arise, naturally, from the Taylor expansions

$$H_{i\pm 1} = H_i \pm hH_i^{(1)} + \frac{h^2}{2!}H_i^{(2)} \pm \frac{h^3}{3!}H_i^{(3)} + \frac{h^4}{4!}H_i^{(4)} \pm \frac{h^5}{5!}H_i^{(5)} + O(h^6) \tag{5}$$

which, in particular, imply

$$H_{i-1} - 2H_i + H_{i+1} = h^2 H_i^{(2)} + \frac{h^4}{12}H_i^{(4)} + O(h^6) \tag{6}$$

and hence also

$$\frac{H_{i-1} - 2H_i + H_{i+1}}{h^2} = H_i^{(2)} + O(h^2). \tag{7}$$

Eliminating in (6) the 4th order derivative by means of (7) in terms of 2nd order derivatives, we obtain

$$\frac{H_{i-1} - 2H_i + H_{i+1}}{h^2} = \frac{1}{12}(H_{i-1}^{(2)} + 10H_i^{(2)} + H_{i+1}^{(2)}) + O(h^4). \tag{8}$$

From (2), we are led to a FD-scheme with the 1st order time derivative of the induction B distributed over the considered ith node and its 2 neighbouring nodes. To obtain closed numerical schemes, (7) and (8) must be supplemented by equations for the boundary nodes. It is natural to require the same accuracy for these equations as is for those in the internal nodes.

The first (surface) equation for the numerical scheme based on (7), can be obtained as follows. Consider (5) for i=1 (upper sign) up to the fourth term. Use $H_1^{(3)} = (H_2^{(2)} - H_1^{(2)})/h + O(h)$ and use (7) to approximate $H_2^{(2)}$. We arrive at

$$2h^2 H_1^{(2)} = 8H_2 - 7H_1 - H_3 - 6hH_1^{(1)} + O(h^4). \tag{9}$$

Notice that (9) approximates the first BC of (4) with accuracy $O(h^3)$. At the same time, in view of (2), it approximates dB_1/dt, with the required accuracy $O(h^2)$.

To derive the difference equation in $x = x_1$ for the scheme based on (8), we must retain all six terms in (5). The 3rd, 4rth and 5th order derivatives may be approximated in terms of 2nd order ones leaning upon four-point formulae of numerical differentiation [3], [4]. We obtain

$$h^2(97H_1^{(2)} + 114H_2^{(2)} - 39H_3^{(2)} + 8H_4^{(2)}) = 360(H_2 - H_1 - hH_1^{(1)}) + O(h^6). \tag{10}$$

By (2), this results in an approximation of the 'distributed' derivative of B over $x = x_1$ and its 3 neighbouring grid points which has accuracy $O(h^4)$. In (9) and (10) we insert $H_1^{(1)} = -\sigma E_b$, according to (4).

Finally, the equations for $x = x_N$ are obtained from (7) and (8), respectively, taking into account the mentioned symmetry of H, i.e. $H_{N+1} = H_{N-1}$.

We now can formulate the two FD-schemes considered further. We will not distinguish in the notation between the exact nodal values $H_i(t)$, $B_i(t)$ and their FD-approximations.

The first scheme, referred to as FD-h2 (in accordance with the accuracy $O(h^2)$), is based upon (7) and (9) and reads

$$\begin{aligned} \frac{dB_1}{dt} &= \frac{8H_2 - 7H_1 - H_3}{2\sigma h^2} + \frac{3E_b}{h}, \\ \frac{dB_i}{dt} &= \frac{H_{i-1} - 2H_i + H_{i+1}}{\sigma h^2}, \quad (i = 2, ..., N-1), \\ \frac{dB_N}{dt} &= \frac{2(H_{N-1} - H_N)}{\sigma h^2}. \end{aligned} \tag{11}$$

The scheme FD-h4 with accuracy $O(h^4)$ is built on the basis of (8) and (10):

$$\begin{aligned} \frac{97}{30}\frac{dB_1}{dt} + \frac{19}{5}\frac{dB_2}{dt} - \frac{13}{10}\frac{dB_3}{dt} + \frac{4}{15}\frac{dB_4}{dt} &= 12\frac{H_2 - H_1}{\sigma h^2} + 12\frac{E_b}{h}, \\ \frac{dB_{i-1}}{dt} + 10\frac{dB_i}{dt} + \frac{dB_{i+1}}{dt} &= 12\frac{H_{i-1} - 2H_i + H_{i+1}}{\sigma h^2}, \quad (i = 2, ..., N-1), \\ \frac{dB_{N-1}}{dt} + 5\frac{dB_N}{dt} &= 12\frac{H_{N-1} - H_N}{\sigma h^2}. \end{aligned} \tag{12}$$

Notice that a conventional scheme of accuracy $O(h^4)$,

$$\frac{dB_i}{dt} = \frac{-H_{i-2} + 16H_{i-1} - 30H_i + 16H_{i+1} - H_{i+2}}{12\sigma h^2}, \quad (i = 3, ..., N-1), \tag{13}$$

is built on a five-point stencil instead of the usual three-point stencil sufficient for schemes (11) and (12). For this reason it will be referred to as FD-5pp.

In media without magnetic viscosity, the values of B and H vary synchronously. Therefore,

$$\frac{\partial B}{\partial t} = \frac{\partial B}{\partial H}\frac{\partial H}{\partial t} = \mu_0 \mu_{\mathrm{d}}(H)\frac{\partial H}{\partial t}, \tag{14}$$

where $\mu_0 = 4\pi \cdot 10^{-7}$ H/m, and μ_{d} is the differential magnetic permeability.

Using (14), the FD-schemes above can be written as systems of first order ODEs for $H_i(t)$, $i = 1, ..., N$. The scheme FD-h4, in contrast to (13), is characterized by a distribution of the derivative dB/dt (or dH/dt) over the node considered and over its two neighbouring nodes. Since it is similar in this respect to FE-schemes, it is of interest to compare the results obtained with FD- and FE-techniques.

4 Comparison of FD- and FE-Schemes

The time continuous FD- and FE-schemes lead to a system of 1st order ODEs

$$\boldsymbol{M}\frac{d}{dt}\boldsymbol{H} = \boldsymbol{K}\,\boldsymbol{H} + \boldsymbol{F}, \tag{15}$$

along with an IC. Here $\boldsymbol{H}$ is the column matrix of nodal values H_i; $\boldsymbol{F}$ is a column matrix with only one nonzero element.

The structure of the matrices $\boldsymbol{M}$ and $\boldsymbol{K}$ for the FD-schemes is seen from (11) and (12). When applying the Galerkin FE method, $\boldsymbol{M}$ and $\boldsymbol{K}$ are the usual mass and stiffness matrices, respectively. When using linear interpolating functions, that is for the scheme further referred to as FE-1, the matrices $\boldsymbol{M}$ and $\boldsymbol{K}$ are tridiagonal. When using interpolating functions of the second order, scheme FE-2, in both $\boldsymbol{K}$ and $\boldsymbol{M}$, rows with three and five elements alternate. For the comparison notice that, except for the equations related to the surface node, matrices $\boldsymbol{K}$ of the FD-h2 and FD-h4 are tridiagonal, while the matrices $\boldsymbol{M}$ have one or three diagonals, respectively.

The comparison between FD- and FE-solutions, as well as their comparison with a known analytical solution for the linear problem, requires the representation of the results in normalized (dimensionless) form. In the chosen coordinate system, the analytical solution of the BVP (2), (4) with E_{b} = const, μ_d=const in (14), and zero IC, can be written as [1]

$$H^*\left(\frac{x}{d},\tau\right) = \frac{\tau}{2} + \pi^2\left[\left(\frac{x}{d}-\frac{1}{2}\right)^2 - \frac{1}{12}\right] - \sum_{n=1}^{\infty}\frac{(-1)^n}{n^2}e^{-n^2\tau}\cos\left[2\pi n\left(\frac{x}{d}-\frac{1}{2}\right)\right] \tag{16}$$

where

$$\tau = \frac{t}{\theta},\ \theta = \frac{\mu_0\mu d^2\sigma}{4\pi^2},\ H^* = \frac{H}{H_{\mathrm{norm}}},\ H_{\mathrm{norm}} = \frac{E_{\mathrm{b}} d\sigma}{\pi^2}. \tag{17}$$

We solve (15) with both explicit and implicit methods. When analyzing the linear problem, we compare the solutions obtained with the Crank-Nicolson (implicit) method on a uniform grid $\omega_{h,\Delta\tau}$ with the steps h and $\Delta\tau$.

The problem is written in nondimensional form and solved on the interval [0, 3]. The time step $\Delta\tau$=0.0001 is small so that the integration error for any of the schemes is determined completely by the spatial approximation only. The number of the steps is denoted by K. We eliminate the errors in the discrete norms

$$||\varepsilon||_{\rm rms} = 1000 \left(\sum_{k=1}^{K} \sum_{i=1}^{N} (H^*_{\rm ex} - H^*_{i,k})^2/(NK) \right)^{1/2},$$

$$||\varepsilon||_{\max} = \max \left| H^*_{\rm ex} - H^*_{i,k} \right| .$$

Here, $H^*_{ex} \equiv H^*_{ex}(x_i, \tau_k)$, $\tau_k = k\Delta\tau$, is calculated from (16), while $H_{i,k} \equiv H_i(\tau_k)$ is the approximate nodal value in $x = x_i$ at $\tau = \tau_k$. These errors are represented in Table 1.

Table1. The errors of computational schemes for the linear problem

Error	N	FD-h2	FE-1	FD-h4	FE-2	FD-5pp
$\|\|\varepsilon\|\|_{\rm rms}$	13	2.6629	3.2257	0.4176	0.5034	1.2948
	25	0.7185	0.8076	0.0751	0.0906	0.2334
	49	0.1883	0.2017	0.0130	0.0158	0.0415
$\|\|\varepsilon\|\|_{\max}$	13	0.0471	0.0419	0.0333	0.0336	0.0605
	25	0.0235	0.0210	0.0166	0.0168	0.0302
	49	0.0118	0.0105	0.0083	0.0084	0.0151
N_0		$5N$	$5N$-4	$5N$+4	$8N$-7	$11N$-1

The number N_0 in Table 1 is the number of multiplications and divisions necessary for the solution of the resulting algebraic equations, using the algorithm of the Thomas type. On the basis of the errors, the numerical schemes above can be divided into two groups. The first of them is formed by schemes FD-h2 and FE-1, while the second consists of FD-h4 and FE-2. The conventional scheme FD-5pp takes an intermediate place. The results obtained for the linear magnetic media indicate that the best scheme according to the criterion "accuracy/complexity" is the three-point scheme FD-h4.

A well known feature of the material magnetization in a field of variable direction is the abrupt changes in the differential permeability $\mu_{\rm d}(H)$. They are caused by hysteresis properties of the material and take place at the transition from one hysteresis branch to another, when the value of $\mu_{\rm d}(H)$ can change by several orders of magnitude. To compare the numerical schemes when modeling the abrupt change of $\mu_{\rm d}(H)$, we use a highly nonlinear dependence $B(H)$, shown in Fig. 1 by the curve $1'$-$1''$.

To couple the magnetic and electric circuits of the device, we must use the relation between the current i_μ in the excitation winding (w_1 turns) and

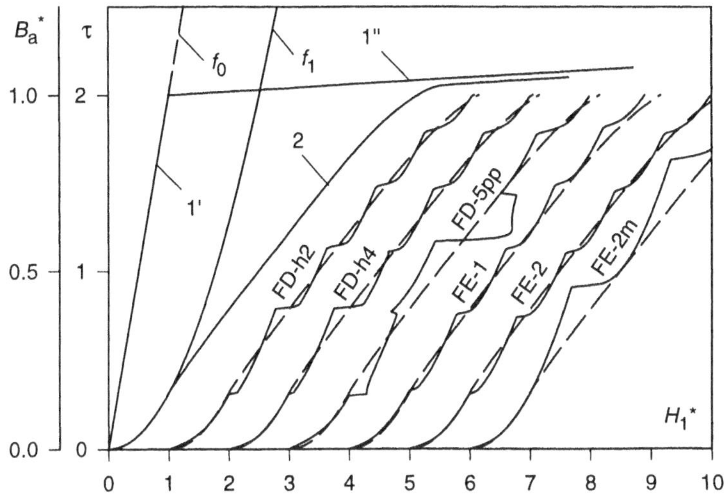

Figure1. Dynamic magnetization curves calculated on a coarse grid

the magnetic field strength H_1 at the surface of the sheet,

$$w_1 i_\mu(t) = l_a H_1(t). \tag{18}$$

Therefore, it is important to analyze the solution of the problem concerning $H_1(t)$. Besides, for any numerical methods and schemes, the solution errors for the first grid node are the largest, regardless of the norm in which these errors are determined.

In the case of the step magnetizing voltage, $u_\mu(t) = U$, $t \geq 0$ and $u_\mu(t) = 0$, $t < 0$, the derivative dB_a/dt is constant. It means that the average induction B_a increases linearly in time, and, in accordance with (3), $B_a = 2E_b t/d$. This allows us to represent the calculated results in the graph (Fig. 1), where the axis of ordinates is used for the two normalized variables, namely τ and B_a^*, where $B_a^* = B_a/(\mu_0 \mu H_{norm}) = \tau/2$.

In order to compare the solutions obtained with the different methods, we numerically solve (15), assuming that at $\tau = 0$ the material was in the demagnetized state, while by $\tau = \tau_s$ the average induction B_a^* reaches saturation. Then, for a chosen number of the grid points (or finite elements), the approximate solution of (15) is completely characterized by the nondimensional saturation time τ_s and by the ratio $R_\mu = \mu_s/\mu$.

The results shown in Fig. 1, correspond to $\tau_s = 2$ and $R_\mu = 0.01$. Curve f_1 corresponds to the analytical solution $H_1^*(\tau)$ for the linear case ($R_\mu = 1$). The straight line $f_0(\tau) = \tau/2$, which coincides with segment $1'$ of the static magnetization curve, defines the field strength at the surface in the absence of eddy currents. Curve 2 represents the solution $H_1^*(\tau)$ obtained with scheme FD-h2, normalized by (17), for N=99. It is further referred to as $H_{1,99}^*(\tau)$. Owing to the large value of N, this solution can be considered as quasi-exact.

Each of the dashed curves, originating from points (1,0), (2,0),..., (6,0), is the curve 2 displaced along the H_1^*-axis by 1, 2,..., 6, respectively. The solid lines, originating from the same points, are the correspondingly displaced curves calculated for N=7 with three FD- and three FE-schemes. The notation FE-2m is used for the scheme FE-2 where the permeability $\mu_{\rm d}$ was assumed to be constant over the finite element and equal to $\mu_{\rm d}$ at its midpoint. The deviations of the calculated results from the solution $H_{1,99}^*(\tau)$ were measured in norm $||\varepsilon_1||_{\rm rms}$, defined by means of 200 equidistant time points τ_i in $[0, \tau_{\rm s}]$:

$$||\varepsilon_1||_{\rm rms} = \left(\sum_{i=1}^{200}[H_{1,99}^*(\tau_i) - H_1^*(\tau_i)]^2/200\right)^{1/2} .$$

These rms errors, for N=7 and N=13, are represented in Table 2.

Table2. Errors $||\varepsilon_1||_{\rm rms}$ of the computational schemes for a nonlinear magnetization curve $1' - 1''$ (Fig. 1)

N	FD-h2	FD-h4	FD-5pp	FE-1	FE-2	FE-2m
7	0.0861	0.0862	0.2475	0.0964	0.0690	0.2746
13	0.0250	0.0245	0.0541	0.0341	0.0225	0.0988

As seen from Table 2, the accuracy of the schemes FD-h2, FD-h4, FE-1 and FE-2 is almost the same. Therefore, the preference can be given to scheme FD-h2 as the simplest one, which has, in addition, the best stability. It can be coupled with the ODEs for the electric circuit by means of relationships (3) and (18). The use of this scheme in a transient simulator is illustrated in [5].

References

1. Karasev, V.V.: Computation of magnetization process of saturable reactors. Electrichestvo. No. 2 (1982) 26-32 (in Russian)
2. Dupre, L.R., Van Keer, R., Melkebeek, J.A.A.: On a magnetodynamic model for the iron losses in non-oriented steel laminations. J.Phys. D: Appl.Phys. **29** (1996) 856-861
3. Chapra, S.C., Canale, R.P.: Numerical Methods for Engineers. McGraw-Hill, Inc. 1988
4. Milne, W.E.: Numerical Calculus. Princeton University Press, Princeton, New Jersey, 1949
5. Dupre, L.R., Van Keer, R., Melkebeek, J.A.A., Moroz, Y.I., Zirka, S.E.: Hysteresis Models for Transient Simulation. These proceedings

Editorial Policy

§1. Volumes in the following four categories will be published in LNCSE:

i) Research monographs
ii) Lecture and seminar notes
iii) Conference proceedings
iv) Textbooks

Those considering a book which might be suitable for the series are strongly advised to contact the publisher or the series editors at an early stage.

§2. Categories i) and ii). These categories will be emphasized by Lecture Notes in Computational Science and Engineering. **Submissions by interdisciplinary teams of authors are encouraged.** The goal is to report new developments – quickly, informally, and in a way that will make them accessible to non-specialists. In the evaluation of submissions timeliness of the work is an important criterion. Texts should be well-rounded, well-written and reasonably self-contained. In most cases the work will contain results of others as well as those of the author(s). In each case the author(s) should provide sufficient motivation, examples, and applications. In this respect, Ph.D. theses will usually be deemed unsuitable for the Lecture Notes series. Proposals for volumes in these categories should be submitted either to one of the series editors or to Springer-Verlag, Heidelberg, and will be refereed. A provisional judgment on the acceptability of a project can be based on partial information about the work: a detailed outline describing the contents of each chapter, the estimated length, a bibliography, and one or two sample chapters – or a first draft. A final decision whether to accept will rest on an evaluation of the completed work which should include

- at least 100 pages of text;
- a table of contents;
- an informative introduction perhaps with some historical remarks which should be accessible to readers unfamiliar with the topic treated;
- a subject index.

§3. Category iii). Conference proceedings will be considered for publication provided that they are both of exceptional interest and devoted to a single topic. One (or more) expert participants will act as the scientific editor(s) of the volume. They select the papers which are suitable for inclusion and have them individually refereed as for a journal. Papers not closely related to the central topic are to be excluded. Organizers should contact Lecture Notes in Computational Science and Engineering at the planning stage.

In exceptional cases some other multi-author-volumes may be considered in this category.

§4. Category iv) Textbooks on topics in the field of computational science and engineering will be considered. They should be written for courses in CSE education. Both graduate and undergraduate level are appropriate. Multidisciplinary topics are especially welcome.

§5. Format. Only works in English are considered. They should be submitted in camera-ready form according to Springer-Verlag's specifications. Electronic material can be included if appropriate. Please contact the publisher. Technical instructions and/or TEX macros are available via http://www.springer.de/author/tex/help-tex.html; the name of the macro package is "LNCSE – LaTEX2e class for Lecture Notes in Computational Science and Engineering". The macros can also be sent on request.

General Remarks

Lecture Notes are printed by photo-offset from the master-copy delivered in camera-ready form by the authors. For this purpose Springer-Verlag provides technical instructions for the preparation of manuscripts. See also *Editorial Policy.*

Careful preparation of manuscripts will help keep production time short and ensure a satisfactory appearance of the finished book. The actual production of a Lecture Notes volume normally takes approximately 12 weeks.

The following terms and conditions hold:

Categories i), ii), and iii):
Authors receive 50 free copies of their book. No royalty is paid. Commitment to publish is made by letter of intent rather than by signing a formal contract. Springer-Verlag secures the copyright for each volume.

For conference proceedings, editors receive a total of 50 free copies of their volume for distribution to the contributing authors.

Category iv):
Regarding free copies and royalties, the standard terms for Springer mathematics monographs and textbooks hold. Please write to Peters@springer.de for details. The standard contracts are used for publishing agreements.

All categories:
Authors are entitled to purchase further copies of their book and other Springer mathematics books for their personal use, at a discount of 33,3 % directly from Springer-Verlag.

Addresses:

Professor M. Griebel
Institut für Angewandte Mathematik
der Universität Bonn
Wegelerstr. 6
D-53115 Bonn, Germany
e-mail: griebel@iam.uni-bonn.de

Professor D. E. Keyes
Computer Science Department
Old Dominion University
Norfolk, VA 23529-0162, USA
e-mail: keyes@cs.odu.edu

Professor R. M. Nieminen
Laboratory of Physics
Helsinki University of Technology
02150 Espoo, Finland
e-mail: rni@fyslab.hut.fi

Professor D. Roose
Department of Computer Science
Katholieke Universiteit Leuven
Celestijnenlaan 200A
3001 Leuven-Heverlee, Belgium
e-mail: dirk.roose@cs.kuleuven.ac.be

Professor T. Schlick
Department of Chemistry and
Courant Institute of Mathematical
Sciences
New York University
and Howard Hughes Medical Institute
251 Mercer Street, Rm 509
New York, NY 10012-1548, USA
e-mail: schlick@nyu.edu

Springer-Verlag, Mathematics Editorial
Tiergartenstrasse 17
D-69121 Heidelberg, Germany
Tel.: *49 (6221) 487-185
e-mail: peters@springer.de
http://www.springer.de/math/
peters.html

Lecture Notes in Computational Science and Engineering

Vol. 1 D. Funaro, *Spectral Elements for Transport-Dominated Equations.* 1997. X, 211 pp. Softcover. ISBN 3-540-62649-2

Vol. 2 H. P. Langtangen, *Computational Partial Differential Equations.* Numerical Methods and Diffpack Programming. 1999. XXIII, 682 pp. Hardcover. ISBN 3-540-65274-4

Vol. 3 W. Hackbusch, G. Wittum (eds.), *Multigrid Methods V.* Proceedings of the Fifth European Multigrid Conference held in Stuttgart, Germany, October 1-4, 1996. 1998. VIII, 334 pp. Softcover. ISBN 3-540-63133-X

Vol. 4 P. Deuflhard, J. Hermans, B. Leimkuhler, A. E. Mark, S. Reich, R. D. Skeel (eds.), *Computational Molecular Dynamics: Challenges, Methods, Ideas.* Proceedings of the 2nd International Symposium on Algorithms for Macromolecular Modelling, Berlin, May 21-24, 1997. 1998. XI, 489 pp. Softcover. ISBN 3-540-63242-5

Vol. 5 D. Kröner, M. Ohlberger, C. Rohde (eds.), *An Introduction to Recent Developments in Theory and Numerics for Conservation Laws.* Proceedings of the International School on Theory and Numerics for Conservation Laws, Freiburg / Littenweiler, October 20-24, 1997. 1998. VII, 285 pp. Softcover. ISBN 3-540-65081-4

Vol. 6 S. Turek, *Efficient Solvers for Incompressible Flow Problems.* An Algorithmic and Computational Approach. 1999. XVII, 352 pp, with CD-ROM. Hardcover. ISBN 3-540-65433-X

Vol. 7 R. von Schwerin, *Multi Body System SIMulation.* Numerical Methods, Algorithms, and Software. 1999. XX, 338 pp. Softcover. ISBN 3-540-65662-6

Vol. 8 H.-J. Bungartz, F. Durst, C. Zenger (eds.), *High Performance Scientific and Engineering Computing.* Proceedings of the International FORTWIHR Conference on HPSEC, Munich, March 16-18, 1998. 1999. X, 471 pp. Softcover. 3-540-65730-4

Vol. 9 T. J. Barth, H. Deconinck (eds.), *High-Order Methods for Computational Physics.* 1999. VII, 582 pp. Hardcover. 3-540-65893-9

Vol. 10 H. P. Langtangen, A. M. Bruaset, E. Quak (eds.), *Advances in Software Tools for Scientific Computing.* 2000. X, 357 pp. Softcover. 3-540-66557-9

Vol. 11 B. Cockburn, G. E. Karniadakis, C.-W. Shu (eds.), *Discontinuous Galerkin Methods.* Theory, Computation and Applications. 2000. XI, 470 pp. Hardcover. 3-540-66787-3

Vol. 12 U. van Rienen, *Numerical Methods in Computational Electrodynamics.* Linear Systems in Practical Applications. 2000. XIII, 375 pp. Softcover. 3-540-67629-5

Vol. 13 B. Engquist, L. Johnsson, M. Hammill, F. Short (eds.), *Simulation and Visualization on the Grid.* Parallelldatorcentrum Seventh Annual Conference, Stockholm, December 1999, Proceedings. 2000. XIII, 301 pp. Softcover. 3-540-67264-8

Vol. 14 E. Dick, K. Riemslagh, J. Vierendeels (eds.), *Multigrid Methods VI.* Proceedings of the Sixth European Multigrid Conference Held in Gent, Belgium, September 27-30, 1999. 2000. IX, 293 pp. Softcover. 3-540-67157-9

Vol. 15 A. Frommer, T. Lippert, B. Medeke, K. Schilling (eds.), *Numerical Challenges in Lattice Quantum Chromodynamics.* Joint Interdisciplinary Workshop of John von Neumann Institute for Computing, Jülich and Institute of Applied Computer Science, Wuppertal University, August 1999. 2000. VIII, 184 pp. Softcover. 3-540-67732-1

Vol. 16 J. Lang, *Adaptive Multilevel Solution of Nonlinear Parabolic PDE Systems.* Theory, Algorithm, and Applications. 2001. XII, 157 pp. Softcover. 3-540-67900-6

Vol. 17 B. I. Wohlmuth, *Discretization Methods and Iterative Solvers Based on Domain Decomposition.* 2001. X, 197 pp. Softcover. 3-540-41083-X

Vol. 18 U. van Rienen, M. Günther, D. Hecht (eds.), *Scientific Computing in Electrical Engineering.* Proceedings of the 3rd International Workshop, August 20-23, 2000, Warnemünde, Germany. 2001. XII, 428 pp. Softcover. 3-540-42173-4

GPSR Compliance

The European Union's (EU) General Product Safety Regulation (GPSR) is a set of rules that requires consumer products to be safe and our obligations to ensure this.

If you have any concerns about our products, you can contact us on ProductSafety@springernature.com

In case Publisher is established outside the EU, the EU authorized representative is:

Springer Nature Customer Service Center GmbH
Europaplatz 3
69115 Heidelberg, Germany

Batch number: 10761599

Printed by Printforce, the Netherlands